W9-AZL-991

Prentice-Hall Industrial Arts Series

ARCHITECTURAL DRAFTING, by William J. Hornung

AUTOMATIC INDUSTRIAL CONTROLS, by Abraham Marcus

BASIC ELECTRICITY, by Abraham Marcus

BASIC ELECTRONICS, by Abraham Marcus and Samuel E. Gendler

ELEMENTS OF RADIO, by Abraham Marcus and William Marcus

ELEMENTS OF TELEVISION SERVICING, by Abraham Marcus and Samuel E. Gendler

GENERAL METAL: PRINCIPLES, PROCEDURES, AND PROJECTS, Roland R. Fraser and Earl L. Bedell

INDUSTRIAL ARTS FOR THE GENERAL SHOP, by Delmar W. Olson

MECHANICAL DRAFTING ESSENTIALS, by Francis T. McCabe, Charles W. Keith, and Walter E. Farnham

RADIO SERVICING: THEORY AND PRACTICE, by Abraham Marcus

WOODS AND WOODWORKING FOR INDUSTRIAL ARTS, by Delmar W. Olson

second edition

BASIC ELECTRONICS

ABRAHAM MARCUS
SAMUEL E. GENDLER

Prentice-Hall, Inc., Englewood Cliffs, N.J.

The cover shows an enlarged photograph of a complex circuit created by the use of large-scale integration (LSI). The circle is a silicon wafer with a diameter of about one and a half inches, on which more than one hundred elements, in the form of transistors and diodes, were deposited. The wafer, which is set in a square frame, had been covered by a mask through which metallic layers were deposited in fine strips. After the mask was removed, the strips remained and interconnected the necessary elements into the circuit shown in the picture. The thin straight lines in the photograph are leads that connect the wafer into the external circuitry. A more detailed description is in the section on integrated circuits, pages 533 and 534.

Cover photograph courtesy of Texas Instruments, Inc.

BASIC ELECTRONICS, Second Edition, **by Abraham Marcus and Samuel E. Gendler**

ISBN 0-13-060384-8

10 9 8 7 6 5

PRENTICE-HALL INTERNATIONAL, INC., *London*
PRENTICE-HALL OF AUSTRALIA, PTY. LTD., *Sydney*
PRENTICE-HALL OF CANADA, LTD., *Toronto*
PRENTICE-HALL OF INDIA PRIVATE LTD., *New Delhi*
PRENTICE-HALL OF JAPAN, INC., *Tokyo*

5

Preface

It is frequently pointed out that the growth of scientific knowledge has been accelerating constantly. If we compressed time so that the five thousand years of man's recorded history were squeezed into one year, then the airplane was invented about five days ago and space travel began late yesterday. Similarly, in the field of electronics the vacuum tube came into being less than five days ago and the transistor is not quite two days old.

The knowledge explosion has been so great that if science had not provided ways of storing, transmitting, and retrieving information, we would have been overwhelmed by its bulk. Electronics has sparked the growth of science and has woven its strands into the fabric of all the sciences—physics, chemistry, medicine, astronomy, and the others.

In the field of electronics, there are many different levels of understanding, ranging from the largely mathematical approach of the physicist to the practical and nonmathematical approach of the beginner. *Basic Electronics* is aimed at the beginner. The approach is simple and direct, with a minimum of mathematics. Where mathematics is used, moreover, examples have been worked out to show how it is applied to the problem in question.

The book explains the basic operation of semiconductor and vacuum-tube devices as well as elements such as resistors, capacitors, and inductors. All of these elements and devices are then combined into functioning circuits, and the circuits are combined into systems.

A thorough understanding of this basic book will prepare the beginner for work in the field, though it is hoped that he will be sufficiently motivated to go on to advanced studies in electronics.

The contribution of illustrative material by the manufacturers of the various devices described in this book is gratefully acknowledged. Thanks are also due to Dr. Harry Lewis, Director of Trade and Technical Education for the Board of Education of the City of New York, for his critical reading of the manuscript and for his helpful suggestions.

S. E. GENDLER

Contents

Section 2

Section 5

Section 6

Prologue

The ancient people of the world had a "magic." Digging in the earth, they sometimes turned up yellow, glasslike stones. When these stones were rubbed with a cloth, they mysteriously acquired the power of attracting small pieces of straw, leaves, feathers, and such.

These stones were amber, a fossilized form of the resins that ooze from certain types of trees. As early as 600 B.C., a Greek philosopher, Thales of Miletus, wrote about experiments he conducted with amber, which the Greeks called "electrum." It is from this word that we get our modern terms "electricity," "electron," and "electronics."

The ancients had an even more wonderful "magic." They occasionally dug up heavy black stones that could attract pieces of iron. These stones were called "magnets," probably because they were particularly plentiful in the province of Magnesia in Asia Minor. It was believed that both amber and magnets had "souls" which gave them their mysterious properties. All sorts of legends arose. For example, these stones were supposed to have magical healing powers. Also, it was believed that if a magnet were rubbed with garlic, it would lose its power of attraction, but if it then were smeared with goat's blood, this power would be restored.

Of course, these legends are false. But man is a restless animal, constantly seeking to learn the reason for things. And it was in this search to find why amber and magnets behaved as they do, that the science of electronics came into existence.

Section 1

Electrical Theory

In order to learn about the electron tube and its associated circuits, it is necessary to understand the electrical nature of these circuits. It would be well if the student had a knowledge of electrical theory before embarking upon a study of electronics. However, since this is a book for beginners, the assumption must be made that the student does not have such knowledge. Accordingly, the electrical theory necessary for an understanding of the electronic phenomena included in the remainder of the book is covered by this first section.

1 The Structure of Matter

For many years it was known that if a piece of amber was rubbed it acquired the ability to attract small bits of straw, paper, and such. About 1600 A.D., William Gilbert, physicist and doctor to the English Queen Elizabeth I, discovered that other substances, such as sulfur, sealing wax, and glass, had properties similar to that of amber. The scientists of his time believed that such substances, when rubbed, or *charged*, exuded a sort of fluid which attracted light objects. They called this "fluid" *electricity*.

Early in the eighteenth century, a French scientist, Charles du Fay, came to the conclusion that there were *two* types of "fluids," or electricity. One type he called "vitreous," the electricity present in charged glasslike substances. The other type he called "resinous," the electricity present when objects such as amber, wax, and rubber are charged.

About 1747, Benjamin Franklin, the American statesman and scientist, came to the conclusion that there was only one type of "fluid," or electricity. The "vitreous" and "resinous" charges, he believed, were only two opposite phases of the same phenomenon. He arbitrarily called the "vitreous" charge *positive* (+) and the "resinous" *negative* (−).

For the next one hundred and fifty years, though dissatisfied with the "fluid" theory of electricity, scientists could find no better explanation. But at the beginning of the twentieth century, while they were investigating the nature of matter in general, the outlines of a better and more rational theory began to take shape.

As man looked about him, he noticed a great number of things. He saw animals, plants, stones, liquids such as water, gases such as air, and lots more. These things we call *matter* and all matter has this in common—it has weight and occupies space.

But what is this matter made of? Well, suppose we start with a drop of water. It has certain properties which we recognize as peculiar to water. Divide this drop and you have two smaller drops. Each droplet still has the properties of water.

Continue dividing the droplets until you come (theoretically, at least) to a particle so small that it can be divided no further. This particle still is water and exhibits all its properties. The smallest particle of matter which retains the properties of that matter, is called a *molecule.* Matter, then, is made up of countless number of molecules, each of which exhibits the properties of that particular type of matter. You can see, accordingly, that there are as many different types of molecules as there are types of matter.

It was found that if an electric current passes through the molecule of water, it breaks the water down further into two gases, *oxygen* and *hydrogen.* Note, however, that these gases do not resemble the original water. The molecule seems to be made up of simpler substances.

A substance whose molecules can be broken down into two or more simpler substances is called a *compound.* If the substance cannot be broken down any further, it is called an *element.* Water is a compound, oxygen and hydrogen are elements. The smallest particle of an element is called an *atom.*

We know of only 92 types of atoms in nature.[1] And just as a few different kinds of bricks may be used to build a great many types of

[1] When scientists succeeded in unlocking the secret of nuclear energy, they also succeeded in creating a number of new types of atoms such as we do not normally find on our earth.

buildings, so these relatively few kinds of atoms may be combined to form the enormous number of different molecules known to man.

The atom was considered to be the smallest particle of matter until in 1897 J. J. Thomson, the famous English scientist, announced he had definite proof that atoms, under certain conditions, shot out still smaller particles of matter, which we now call *electrons.* The amazing thing about these electrons is that they are all alike, regardless of what substance emits them.

Once it was shown that the atom could be broken up, scientists delved deeper into its secrets. As a result, the *electron theory*[2] of the structure of matter was set forth. Scientists today generally believe this theory to be true. But keep in mind that it is only a theory. It may be modified from time to time, and even, if proved to be false, discarded.

According to the electron theory, all matter is composed mainly of three types of particles. These are (1) the *electron,* a particle carrying a negative electrical charge; (2) the *proton,* a particle carrying a positive electrical charge; and (3) the *neutron,* a particle that carries no electrical charge. All atoms are composed of these particles; the atoms differ from one another in the number of particles they contain and in the arrangement of these particles.

In the early part of the twentieth century, Niels Bohr, a Danish scientist, gave us a theoretical picture of atomic structure which, at that time, was widely accepted.[3] According to Bohr, the atom is composed of a central *nucleus,* which is surrounded by revolving *electrons,* somewhat as our sun is surrounded by revolving planets. In fact, the electrons that revolve around the nucleus are called *planetary* electrons. (See Figure 1-1.)

---

[2] When this theory was first stated it was believed that all atoms were made up of electrons and protons. Hence the name "electron theory." Today we know that the atom contains a number of other particles such as neutrons, mesons, positrons, neutrinos, anti-protons, and several more. Thus it is no longer strictly accurate to talk about the "electron theory of atomic structure." However, to avoid confusion between our present-day theory and the old atomistic theories, we will retain the term *electron theory.*

Except for the electrons, protons, and neutrons, the particles that make up the atom do not normally appear before us. They seem to be created and exist for an extremely short period of time as atoms break up. Accordingly, we will ignore these short-lived particles in this book. Also, the theories presented here must, of necessity, be in greatly simplified form.

[3] Although the picture of atomic structure originally presented by Niels Bohr has been considerably modified by the discovery of new facts, for the purpose of this book it may be better if we consider the earlier, and simpler, Bohr atomic structure.

The nucleus contains all the protons and neutrons. An atom of one element differs from an atom of any other in the number of protons contained in the nucleus. The number of protons in the nucleus is called the *atomic number* of the element and varies from 1, for the element *hydrogen,* to 92 for the element *uranium.* (The number is even higher for the new man-made elements.) The atomic number of *helium,* whose structure is shown in Figure 1-1, is 2.

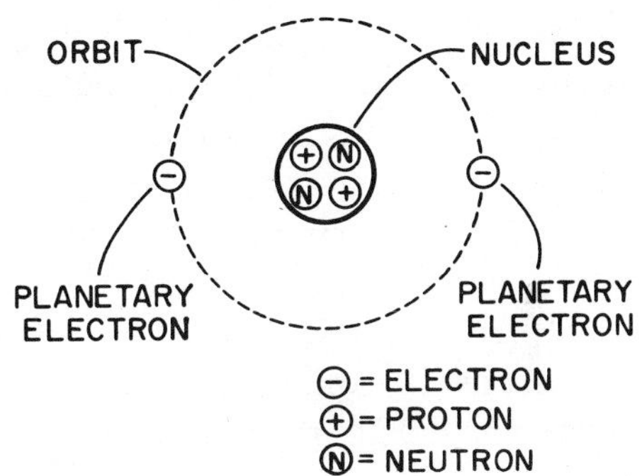

Fig. 1—1.

Theoretical structure of an atom. The nucleus here contains two protons and two neutrons. The two planetary electrons revolve around the nucleus in the orbit indicated. This is the helium atom.

The negative charge on the planetary electron is equal and opposite to the positive charge on the proton. Since the nucleus contains all the protons, it carries a total positive charge that is equivalent to the number of protons present. Inasmuch as a normal atom is *neutral*—that is, it has no external electrical charge—the positive charges on the nucleus are exactly neutralized by the negative charges of the planetary electrons revolving about it.

It follows that the neutral atom has as many planetary electrons as there are protons in its nucleus. Consequently, the number of electrons revolving around the nucleus varies from 1 for hydrogen to 92 for uranium (and higher, for the man-made elements). Note that in the helium atom shown in Figure 1-1 there are two protons and two planetary electrons.

All atoms, except those of ordinary hydrogen, contain one or more neutrons in the nucleus. The helium atom (Figure 1-1) contains two neutrons; the uranium atom may contain 146 neutrons. The neutron carries no electrical charge and in some respects acts as though it were composed of a proton and electron combined, with the positive charge of the proton neutralized by the negative charge of the electron.

Although the opposite electrical charges carried by an electron and a proton are equal in magnitude, the *mass,* or weight, of the proton is about 1,840 times as great as that of the electron. The mass of the neutron is about equal to that of the proton. We can readily see that practically the entire mass, or weight, of the atom is contained in its nucleus.

The number of protons and neutrons in the nucleus of an atom determines its weight, or *mass number.* The mass number of atoms varies from 1 for ordinary hydrogen, which is the lightest of the elements (one proton and no neutrons in its nucleus), to 238 for uranium which, until recently, was the heaviest element (92 protons and 146 neutrons in its nucleus). The new and heavier man-made elements have even greater mass numbers. The mass number for the helium atom shown in Figure 1-1 is 4 (two protons and two neutrons).

All of the atoms of an element contain the same number of protons in their nuclei, and the atoms of one element differ from those of all other elements in the number of protons so contained. For example, each hydrogen atom contains one proton in its nucelus and each uranium atom has 92 protons. But the atoms of the same element may differ in mass number, owing to different numbers of neutrons in their nuclei.

For example, three different types of hydrogen atoms have been found. All of these atoms have the same atomic number of 1 (one proton in the nucleus). However, one of these atoms (the type most commonly found in nature) has a mass number of 1—that is, one proton and no neutron in its nucleus. A second type has a mass number of 2—one proton and one neutron in the nucleus. A third type has a mass number of 3—one proton and two neutrons. Except for the differences in mass, all three types of atoms have identical properties.

We call these different atoms of the same element *isotopes.* Most elements are known to have two or more isotopes. It is interesting to note that scientists have been able to produce artificial isotopes by bombarding the nuclei of atoms with neutrons. In this way, for example, an atom of uranium with a mass number of 238 (92 protons and 146 neutrons in its nucleus) sometimes captures a neutron, raising the mass number to 239 (92 protons, 147 neutrons).

So far, we have concentrated on the nucleus of the atom. Now let us turn our attention to the electrons revolving around the nucleus.

As previously stated, the normal atom has one planetary electron for each proton in the nucleus. Thus, the number of such electrons will vary from 1 for hydrogen to 92 for uranium (and higher for the new elements).

These electrons do not revolve around the nucleus in a disorderly fashion. Their orbits, or paths, form concentric shells, or layers, somewhat like the layers of an onion. There is a certain maximum number of electrons that each shell can contain. If this number is exceeded, the excess electrons arrange themselves in the next outer shell.

The shell nearest the nucleus may contain up to two electrons. If there are more than two electrons, the excess form a second shell around the first. This second shell may hold up to eight electrons. Then a third shell will be formed which may hold up to 18 electrons. The fourth shell may hold up to 32 electrons; the fifth shell up to 50 electrons; and the sixth shell up to 72 electrons. However, in some of the more complicated atoms, electrons may be found in outer shells even before some of the inner ones have been filled up.

Look at Figure 1-2A. This is the theoretical picture of the structure of the helium atom. The nucleus contains two protons and two neutrons. The two planetary electrons revolve in a single shell around the nucleus. (Although the orbits of the electrons generally are considered to be elliptical, they are drawn circular here for convenience.)

In Figure 1-2B the structure of the carbon atom is shown. Around a nucleus containing six protons and six neutrons revolve six electrons. The first two electrons form the shell nearest to the nucleus. Since this shell can contain no more than two electrons, the remaining four electrons form themselves into a second shell outside the first. In Figure 1-2C we see the structure of the lead atom. The 82 electrons are contained in six shells, or layers, around the nucleus.

Except for the electrons in the outermost shell, the particles of the atom are held together tightly, and relatively tremendous forces are required to pry them apart. Most difficult to disrupt is the nucleus. What holds the protons and neutrons together, we are not quite sure, but when we do succeed in exploding the nucleus, as was accomplished in the atomic bomb, an enormous amount of energy is released.

Next in difficulty of removal are the electrons in the shells closest to the nucleus. These electrons are held in place because of the attraction between their negative charges and the positive charges of the protons of the nucleus. When these electrons are disturbed, manifestations, such as X-rays, are obtained.

Least difficult to disturb are the electrons in the outermost layer. Because they are farthest away, they are least attracted to the nucleus. As a matter of fact, under normal conditions electrons are

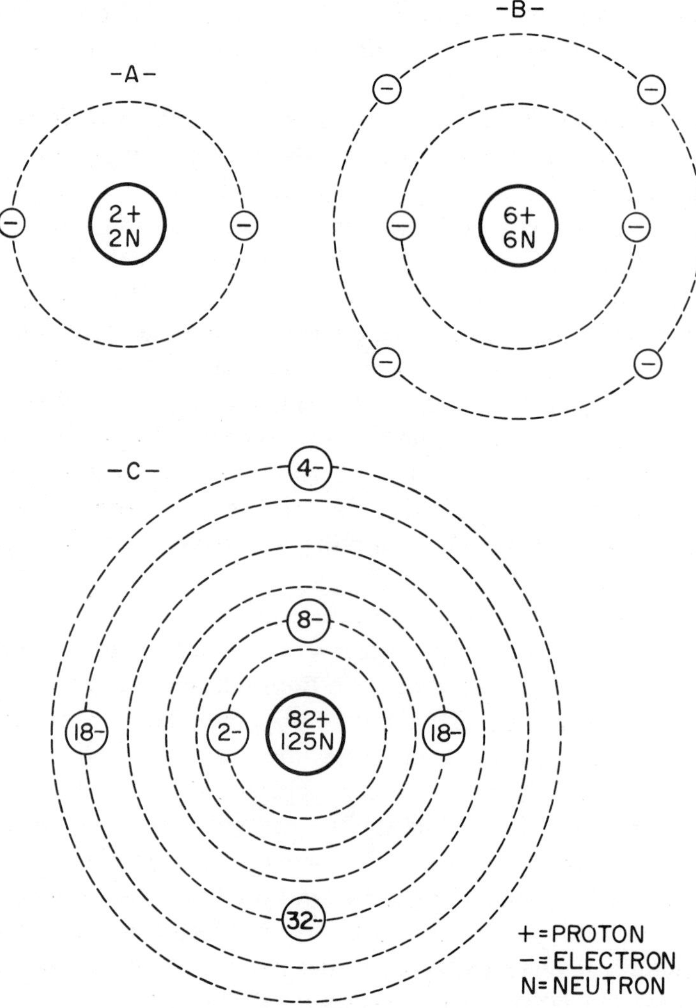

Fig. 1—2. Theoretical structures of atoms.
A. Helium atom.
B. Carbon atom.
C. Lead atom.

constantly leaving and entering this outer shell. Associated with the electrons in the outer shell is the chemical and physical behavior of the atom. And it is with these outer-shell electrons that most of our electrical phenomena are concerned.

Now what about the sizes of the particles we have been discussing? Some molecules are made up of hundreds, and even thousands, of atoms. Using our most powerful microscopes, only a few of the very largest molecules have ever been seen by the human eye. It has been estimated that if 250 million hydrogen atoms were placed side by side, they would extend about one inch. And if 100,000 electrons were placed side by side, they would be as large across as a single hydrogen atom. The electron, proton, and neutron are approximately the same size.

At this point an amazing fact should become apparent. The atom is made up mostly—of empty space! The atom thus becomes analogous to our solar system, which, too, consists of a nucleus (the sun), the planets in their orbits around the sun, and, for the most part, space.

In 1938 scientists succeeded in splitting the atom by bombarding an isotope of uranium (U235 whose nucleus contains 92 protrons and 143 neutrons) with neutrons. The uranium nucleus broke up into several pieces, releasing a relatively large amount of energy in the process, which is known as *fission*. When the weights of these pieces were added up, they were found to total less than the weight of the original uranium nucleus. The energy, therefore, came from the conversion of some of the matter of the uranium nucleus into energy, in accordance with Albert Einstein's famous formula $E = mC^2$, where E is the energy, m is the mass of the missing matter, and C is the velocity of light.

The energy from the fission of a single uranium nucleus is small, but when we consider the millions and billions of nuclei in only a very small piece of uranium we can see that the total amount of energy is tremendous. It has been estimated that, if all the atoms in a kilogram (2.2 pounds) of uranium underwent fission, the energy released would be equivalent to the energy released in the explosion of about 20,000 tons of TNT!

The energy is mainly in the form of heat rays, light rays, and gamma rays. These rays are part of a large group known as *electromagnetic waves* which include, in addition to the above three, radio waves, ultraviolet rays, and X-rays. All the electromagnetic waves resemble each other, differing only in frequency and wavelength.

QUESTIONS

1. **Compare the proton, electron, and neutron in terms of the electrical charge each carries.**
2. **Compare the proton, electron, and neutron in terms of their relative mass.**
3. **Describe the structure of the atom in terms of protons, electrons, and neutrons.**
4. **What is meant by a *neutral* atom?**
5. **What is meant by**
   a) the *mass number* of an atom;
   b) its *atomic number?*
6. **In terms of the electron theory, how do the atoms of an element differ from the atoms of all other elements?**
7. **a) What is an *isotope?***
   b) How do isotopes of the same element differ from one another?
8. **Draw the theoretical picture of the *deuterium* atom whose atomic number is 1 and whose mass number is 2.**
9. **Draw the theoretical picture of the *calcium* atom whose atomic number is 20 and whose mass number is 40.**

The Electric Current

A. *The electric field*

We have seen that in the normal atom the positive charges of the protrons of the nucleus are exactly neutralized by the negative charges of the planetary electrons. Hence the normal atom is neutral, that is, it has no external electrical charge. We have also seen that the electrons in the outermost shell of the atom are loosely held.

What happens when a neutral atom loses one of these electrons? Since there no longer are as many negative charges as there are positive ones, the overall effect is to give the atom a positive charge. We call any particle that has an electrical charge an *ion.* Hence the atom becomes a positive ion.

And what would happen if an extra electron attached itself to a neutral atom? Now there would be more negative charges than positive ones. The overall effect would be to convert the atom into a negative ion.

When two charged particles are brought near each other, an interaction is set up between them. If the particles have similar charges, that is, if they both are charged positive or both negative, the particles tend to repel each other. If, on the other hand, the two particles have opposite charges, that is, one is positive and the other negative, they attract each other.

When two bodies are able to affect each other through space, we say that a *field of force* exists between them. For example, the earth and the sun attract each other even though they are separated by 92,000,000 miles of space. In this case, the field of force is the *gravitational field* which surrounds all matter.

In the case of charged particles the field of force is called the *electrostatic*, or *electric, field*. Under its influence, particles having unlike charges are attracted to each other. Particles with like charges repel each other.

This can be demonstrated by a simple experiment. Rub a small glass rod with a piece of silk cloth. As the rod is rubbed, some of the electrons in the outer shells of its atoms are torn away by the cloth. The cloth, then, has an excess of electrons, or a negative charge. The rod, having lost some of its electrons, has a positive charge. (The positive charge is indicated by + and the negative charge by —.)

Repeat with a second glass rod. We now have two rods each containing a positive charge. Suspend one of the rods so that it may swing freely. As the second positively-charged glass rod is brought near it, the first rod will swing away from it. (See Figure 2-1A.)

Rub a hard-rubber rod with a piece of fur. In this case the atoms of the rubber rod will tear electrons away from the atoms of the fur. Thus the fur will acquire a positive charge and the rubber rod a

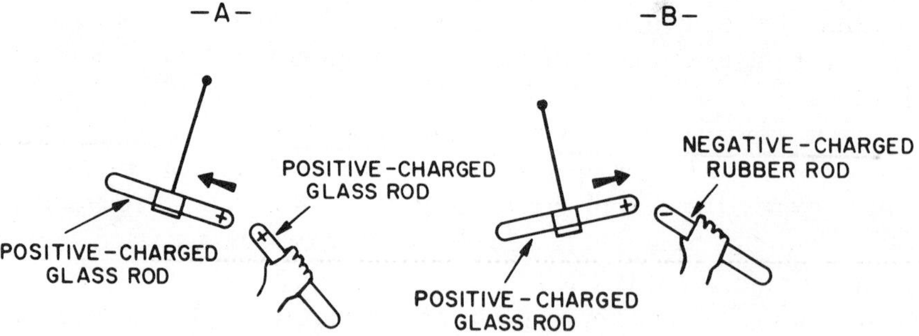

Fig. 2—1. **A.** Bodies having like charges repel each other.
B. Bodies having unlike charges attract each other.

negative charge. Now bring the negatively-charged rubber rod near the suspended positively-charged glass rod. The glass rod will be attracted to the rubber rod. (See Figure 2-1B.)

What determines which substance will seize electrons and which will lose them? The answer lies in the nature of the substance—that is, the number and arrangement of its outermost electrons. Some substances, such as glass and fur, have their outermost electrons so arranged that they can lose them quite easily (and so obtain a positive charge). On the other hand, some substances, such as silk and rubber, have their outermost electrons so arranged that, when they are rubbed, they will seize electrons from the material with which they are stroked, and so will accumulate an excess of electrons (negative charge).

When a glass or rubber rod is being charged by losing or acquiring electrons, the action is local. That is, only the portion of the rod being rubbed is affected. If you take away or add electrons to one end of a glass or rubber rod, the atoms at the other end remain neutral and thus the rod at that end has no charge. We call such materials *insulators.* Examples are glass, rubber, wax, amber, sulfur, and such.

On the other hand, there are certain substances, generally metals, whose outer electrons are held very loosely. As a matter of fact, a certain number of these outer electrons are constantly jumping from atom to atom, even without external influence. We call such moving electrons *free* electrons. Substances which contain many free electrons are called *conductors.*

If an excess of electrons (negative charge) is placed at one end of a conductor, the repulsion between like charges will cause the loosely-held electrons of neighboring atoms to move toward the other end. The movement of these electrons will cause a disturbance among electrons further away and, as a result, the excess of electrons quickly distributes itself throughout the entire conductor.

Similarly, if electrons are removed from one end of a conductor (positive charge), electrons from neighboring atoms are attracted to compensate for the deficiency. Again, all the atoms are affected and soon the deficiency is spread throughout the entire conductor. Thus a charge placed upon any portion of a conductor quickly spreads itself.

Suppose we attach a sort of "pump" to the ends of a conductor, as indicated in Figure 2-2, which removes electrons from one end, thereby creating a deficiency or positive charge (+), and deposits

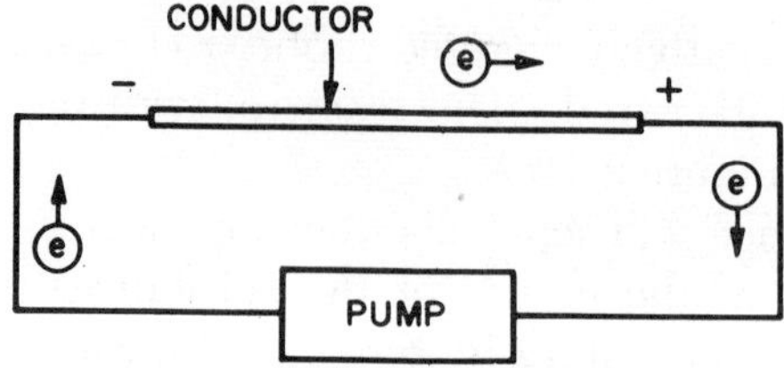

Fig. 2–2.

How an "electron pump" causes electrons to move through a conductor.

them at the other end, creating an excess or negative charge (–). (The electrons are indicated by ⓔ→ and the arrows show their direction of movement.) The free electrons in the conductor are attracted to the positive end and, as they move in that direction, are replaced by the electrons being deposited at the negative end.

These newly deposited electrons, in turn, are attracted to the positive end and electrons will continue to move through the conductor from the negative to positive end as long as the "pump" is in operation. We call this flow of electrons an *electric current.*

The electric current is not confined merely to the movement of electrons. In fact, the flow of any charged particle (ion) under the influence of the electric field constitutes such a current. We shall consider such an ionic flow later in the book.

B. *The electric current*

1. ELECTROMOTIVE FORCE

Whenever an excess of electrons (negative charge) occurs at one end of a conductor and a deficiency (positive charge) at the other end, the electric field between the two ends will set up an electrical pressure that will cause the loosely-held electrons from the outer shells of the atoms of the conductor to stream from the point of excess to the point of deficiency. Thus, an electric current will flow through the conductor from the negative end to the positive one.

To get an idea of *electrical pressure,* let us consider a simple analogy. Assume that we have a U-shaped tube with a valve or stopcock at the center (Figure 2-3). Assume that the valve is closed. We now pour water into arm A to a height represented by X. We pour water into arm B to a height represented by Y. If the valve now is opened, the water will flow from arm A to arm B until X and Y are equal.

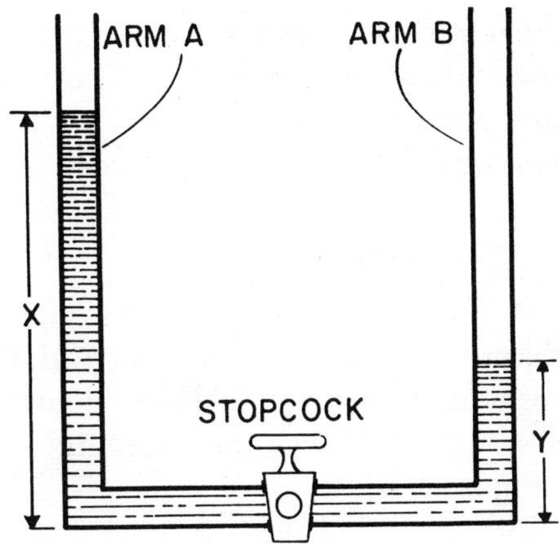

Fig. 2—3.

Diagram illustrating the fact that it is the *difference* in pressure which causes a fluid to flow through a tube.

What caused the water to flow? It was not *pressure* in arm A, because when X and Y are of equal length, no water flows even though the water in arm A still exerts pressure. It was the *difference in pressure* between the two arms that caused the water to flow. The flow continued until the pressures in the two arms were equal and the *difference in pressures* was zero.

So it is with electrons. If, at one end of a conductor, electrons are piled up, and at the other end electrons are fewer in number or are being taken away, the excess electrons will flow toward the point of deficiency.

In Figure 2-3, the water in arm A can do no work until the valve is opened. Nevertheless, it represents a *potential* source of energy —that is, energy due to position. However, the actual work is not done by the potential energy of the water in arm A, but by the *difference in potential energy* between the water in A and the water in B.

Similarly, in Figure 2-4, it is not simply the potential energy of the excess electrons at one end of the conductor that causes the electrons to flow. It is the *difference* between the amounts of potential energy at the two ends of the conductor that does the work. We may say that electrons flow through a conductor (that is to say, an electric

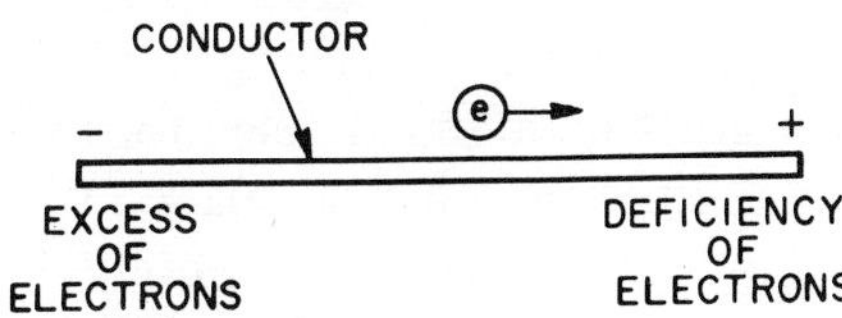

Fig. 2—4.

Diagram illustrating the fact that it is the *difference* in electrical pressure which causes electrons to move through a conductor.

current flows) because of the difference in potential energy between the ends of the conductor. The force that moves the electrons from one point to another is known as the *potential difference,* or *electromotive (electron-moving) force.*

2. RESISTANCE

There is a factor other than electromotive force, or potential difference, that affects electrical flow. Suppose that we were to suspend two metallic balls in air several inches apart, and place a negative charge on one and a positive charge on the other (Figure 2-5).

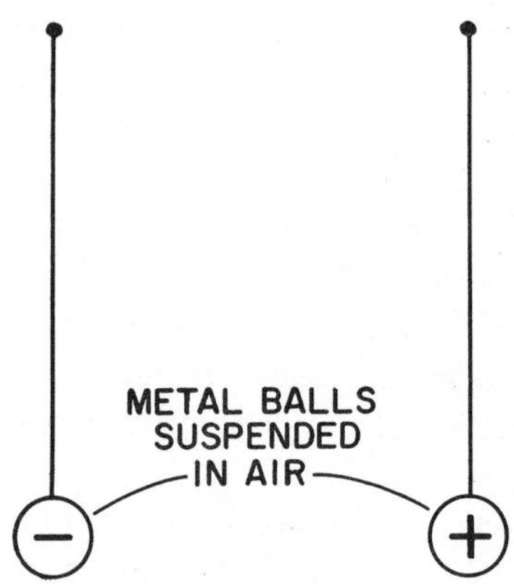

Fig. 2–5.

Two oppositely-charged balls suspended apart in air. Although there is a potential difference between them, no current flows because the resistance of the path between them is too great.

Here we have a potential difference, but very little current flows. The reason is that the air between the two balls offers a great resistance to the flow of current. If the air were very dry, it would take hours for the excess electrons of the negatively-charged ball to completely leak off and neutralize the positive charge in the other. If you connect the two balls by a piece of metal, however, the electric current will be much larger and the negative charge will quickly neutralize the positive charge.

But it is not necessary to connect the two balls with a metal strip to cause the electrons to flow from one to the other. All we need do is increase the charges. When the potential difference becomes great enough, the electrons will jump across through the air in the form of an electric spark. We conclude, therefore, that, for electric current to flow, *the potential difference must be great enough to overcome the resistance of the path.*

Different substances offer different resistances to the flow of electric current. Metals, generally, offer little resistance and are good conductors. Silver is the best conductor known, and copper is almost as good. Insulators, such as glass, rubber, sulfur, and the like, offer a very high resistance. But all substances will permit the passage of some electric current, provided the potential difference is high enough.

3. CURRENT FLOW

For the third factor that affects electrical flow, refer back to Figure 2-3. We measure the flow of water from one arm to the other in terms of quantity per unit of time. We say so many gallons flow past a certain point in a minute. Similarly, we measure the flow of electricity by the number of electrons that flow past a point on a conductor in one second.

What determines the amount of water per unit of time that flows through the valve in Figure 2-3? Obviously, it is the difference between the amounts of potential energy of the water in the two arms of the tube, and it is also the size of the opening in the valve (that is, the resistance the valve offers to the flow of water).

In the case of the electric current, the quantity of electricity (the number of electrons per second) that flows in a conductor depends upon the potential difference and the resistance of the conductor. *The greater the potential difference, the larger the quantity of electricity that will flow; the greater the resistance, the smaller the quantity of electricity.*

C. *Units of measurement*

1. HOW TO WRITE LARGE NUMBERS

When considering the units used to measure the various electrical effects, we meet up with numbers so large that they become awkward to handle. Scientists have worked out several methods to deal with such large numbers.

One such method is to use a series of prefixes which signify certain numerical quantities. Thus, the prefix *meg* or *mega* means a *million* (1,000,000). The prefix *kilo* means a *thousand* (1,000). The prefix *milli* means a *thousandth* (1/1,000). The prefix *micro* means a millionth (1/1,000,000). Using this system a *kilometer* means a thousand meters (a meter is slightly more than a yard), and a *millimeter* means a thousandth of a meter. Similarly, a *megohm* is equal to a million ohms, and a *microfarad* is equal to a millionth of a farad. (The *ohm* and *farad* are electric units which will be explained later in the book.)

Another method used to write large numbers is a sort of mathematical shorthand. For example, multiply 10 by 10. You get 100. Since 100 is formed by *two* 10's multiplied together, scientists express 100 as 10^2. Similarly, 1,000 is formed by *three* 10's multiplied together. It may be expressed, therefore, as 10^3. Using this system, a million (1,000,000) becomes 10^6. If we wish to express the number 5,000,000, we may do so simply by writing 5×10^6 (that is, five times 10^6).

The small figure following the 10 is called the *exponent* and indicates the number of 10's that are multiplied together. Thus, in the case of 10^6 (1,000,000), 6 is the exponent. If we wish to indicate 1/1,000,000, we may write $1/10^6$. Another method is to write 10^{-6}. The figure -6 is called the *negative exponent*. Thus, 1,000 may be expressed as 10^3, and 1/1,000 as $1/10^3$ or 10^{-3}.

2. QUANTITY OF ELECTRIC CHARGE

Adding an electron to a neutral atom gives it a negative charge; taking an electron away from a neutral atom gives it a positive charge. Thus the electrical charge of the electron is our basic unit.

Since the charge of one electron is very small, a *coulomb* is used as a practical unit for measuring the quantity of electric charge. A coulomb is equal to the combined charges of 6,280,000,000,000,000,000 electrons or, using our shorthand system, of 6.28×10^{18} electrons.

Of course, a figure this large is meaningless to the average person. It is stated here merely to emphasize the fact that the electrical charge on the electron is extremely small and that the combined charges of many electrons are employed to make up the coulomb.

3. ELECTRIC CURRENT

When we talk of electric current, we mean electrons in motion. When the electrons flow in one direction only, the current is called a *direct current* (abbreviated *dc*). In the discussion of electricity in this portion of the book we are speaking only of direct currents. (As we shall see later, the electrons may flow, alternately, first in one direction and then in the other. Such a current is known as an *alternating current.*)

It is important to know the number of electrons that flow past a given point on a conductor in a certain length of time. If a coulomb flows past a given point in one second, we call this amount one *ampere* of electric current. Hence, the unit of electric current is the ampere.

Aside from the fact that electrons are too small to be seen, we would find it impossible to count them as they flowed by. Fortunately, we have an electrical instrument, called the *ammeter* (to be described later) that indicates directly the amount of current flowing through it.

Where the ampere is too large a unit to be used, we may employ the *milliampere,* which is a thousandth (1/1,000) of an ampere, or the *microampere,* which is a millionth (1/1,000,000) of an ampere. In an electrical formula, the capital letter I stands for current.

4. RESISTANCE

A number of factors determine the resistance that a substance offers to the flow of electric current. First of all, there is the nature of the substance itself. The greater the number of free electrons present in the substance, the lower is its resistance.

Resistance is also affected by the length of the substance. The longer an object is, the greater its resistance. Another factor is the cross-sectional area of the substance, which is the area of the end exposed if we slice through the substance at right angles to its length. The greater the cross-sectional area, the less the resistance to current flow.

Resistance is also affected by the temperature of the substance. Metals generally offer higher resistance at higher temperatures. Certain nonmetallic substances, such as carbon, on the other hand, offer lower resistance at higher temperatures.

The unit of resistance is the *ohm.* By international agreement, the ohm is the resistance to the flow of electric current offered by a uniform column of mercury, 106.3 centimeters long, having a cross-sectional area of one square millimeter, at 0°C. Where the ohm is too small a unit, we may employ the *kilohm* (1,000 ohms) and the *megohm* (1,000,000 ohms). The symbol for the ohm is the Greek capital letter Ω (*omega*). In an electrical formula, the capital letter R stands for resistance.

5. ELECTROMOTIVE FORCE

The electromotive force creates the electric pressure that causes the current to flow through a conductor. Another name for this force is *voltage.* We can measure the voltage between any two selected points on a conductor by means of an electrical instrument, called the *voltmeter* (to be described later).

The unit of measurement of electromotive force, or voltage, is called the *volt.* The volt is defined as *that electromotive force that is necessary to cause one ampere of current to flow through a resistance of one ohm.* Where the volt is too large a unit, we may use the *millivolt* (1/1,000 of a volt) or the *microvolt* (1/1,000,000 of a volt). Where the volt is too small a unit, we may use the *kilovolt* (1,000 volts). In an electrical formula, the capital letter E stands for voltage.

6. ELECTRIC POWER AND ENERGY

A body at rest tends to remain in this condition or, if it is in motion, it tends to continue this motion in a straight line. This property of the body is known as *inertia.* To overcome the effect of inertia, *force* is required. Thus force may be considered as a push or pull, and, in mechanics, it commonly is measured in units of *pounds.*

Work is the product of the force and the distance through which it acts and is measured in *foot-pounds.* If a 1-pound weight is lifted 1 foot, the work done is 1 foot-pound. If the weight is lifted 5 feet, the work done is 5 foot-pounds.

Power is the rate at which work is done. A child might lift the weight 5 feet in 5 seconds. A more powerful adult might lift it the same distance in 1 second. The formula then may be

$$\text{Power} = \frac{\text{work done}}{\text{time}}.$$

The unit of power (for mechanics) is *foot-pound per second.* Thus, in the case of the child,

$$\text{Power} = \frac{1 \text{ pound} \times 5 \text{ feet}}{5 \text{ seconds}} = 1 \text{ foot-pound per second.}$$

In the case of the adult,

$$\text{Power} = \frac{1 \text{ pound} \times 5 \text{ feet}}{1 \text{ second}} = 5 \text{ foot-pounds per second.}$$

Five hundred and fifty foot-pounds per second, or 33,000 foot-pounds per minute, equals one *horsepower.*

In electricity, work is done as the electromotive force causes electrons to move through a conductor. The *electric power* (the rate of doing work) is the product of the electromotive force and the number of electrons set flowing per unit of time. But the number of electrons flowing per unit of time is the current (measured in units of ampere). Hence, the power, whose symbol is P, is the product of the electromotive force, in volts, and the current, in amperes. The unit of electric power is the *watt.* The formula for power thus becomes

$$P \text{ (watts)} = I \text{ (amperes)} \times E \text{ (volts)}.$$

Example. What is the power required to enable a dry cell whose electromotive force is 1.5 volts to cause a current of 2 amperes to flow through a conductor?

$$P = I \times E = 2 \text{ amperes} \times 1.5 \text{ volts} = 3 \text{ watts.} \quad \textit{Ans.}$$

The dry cell generates the power; the power is consumed as the current flows through the conductor.

Most electrical appliances carry labels stating the voltage at which the appliance is to operate and the power it consumes. Thus an electric lamp may bear a label indicating "60 watts, 120 volts." From this you can calculate that the filament of this lamp has 0.5 ampere of current flowing through it.

As previously stated, the unit of electric power is the *watt,* which is defined as the power used when an electromotive force of one volt causes a current of one ampere to flow through a conductor. Where the watt is too large a unit to be used conveniently, we employ the *milliwatt* (1/1,000 of a watt). Where the watt is too small a unit, we may use the *kilowatt* (1,000 watts).

The horsepower (33,000 foot-pounds per minute) is the unit of mechanical power. The kilowatt is the unit of electric power. Since mechanical energy can be converted to electric energy, and vice versa, the two units of power may be equated. Thus:

1 kilowatt = 1.34 horsepower
1 horsepower = 0.746 kilowatt.

To measure the total electrical energy consumed by an appliance we must know how much power it uses (in watts) and the length of time (in seconds) it continues to consume this power. Thus, the unit of energy is the *wattsecond,* or *joule*—that is, one watt of power applied for one second. Since the wattsecond is a small unit, we frequently use the *watthour* (one watt applied for one hour) or the *kilowatthour* (1,000 watts applied for one hour).

It is on the basis of kilowatthours that we pay our electric bill. For example, how much would it cost to run five 60-watt lamps four hours a day for 30 days if we had to pay at the rate of $.05 per kilowatthour?

Each day we consume

$$5 \times 60 \text{ watts} \times 4 \text{ hours} = 1{,}200 \text{ watthours.}$$

In 30 days we consume

$$1{,}200 \text{ watthours} \times 30 = 36{,}000 \text{ watthours or } 36 \text{ kilowatthours.}$$

At $.05 per kilowatthour, our bill is $1.80.

D. Ohm's law

The relationship between the electromotive force, the current, and the resistance was discovered by a German scientist, George Simon Ohm, at the beginning of the nineteenth century. The unit of resistance was named in his honor.

This relationship, which is called *Ohm's law,* can be expressed mathematically by means of the following formula:

$$\text{Current} = \frac{\text{electromotive force}}{\text{resistance}} \quad \text{or} \quad I = \frac{E}{R},$$

where I is measured in amperes, E in volts, and R in ohms. This means that the greater the electromotive force is, the greater will be the current; and the greater the resistance, the smaller the current.

Let us try a problem. How much current will flow through a conductor whose resistance is 10 ohms, when the electromotive force is 100 volts? Using our formula $I = E/R$, we get

$$I = 100 \text{ volts}/10 \text{ ohms, or } 10 \text{ amperes.} \quad \textit{Ans.}$$

Our Ohm's law formula can be transposed as follows:

$$E = I \times R \quad \text{and} \quad R = \frac{E}{I}.$$

Example. How many volts are required to make a current of 5 amperes to flow through a conductor whose resistance is 2 ohms?

Substituting our known values in the formula $E = I \times R$, we get

$$E = 5 \text{ amperes} \times 2 \text{ ohms, or } 10 \text{ volts.} \quad \textit{Ans.}$$

Example. What is the resistance of a conductor if 100 volts are required to force 2 amperes of current through it?

$$\text{Since } R = \frac{E}{I}, \ R = \frac{100 \text{ volts}}{2 \text{ amperes}}, \text{ or } 50 \text{ ohms.} \quad \textit{Ans.}$$

We also can use these formulas to show the relationship between power (in watts), current (in amperes) and resistance (in ohms). You will recall that

$$\text{Power} = \text{current} \times \text{voltage} \quad \text{or} \quad P = I \times E.$$

If we substitute for E its equivalent $(I \times R)$ we get

$$P = I \times I \times R \quad \text{or} \quad P = I^2 \times R.$$

Example. How much power will be required to force a current of 2 amperes to flow through a conductor whose resistance is 5 ohms?

Since $P = I^2 \times R$, substituting our values we get

$$P = (2)^2 \times 5 = 4 \times 5 = 20 \text{ watts.} \quad \textit{Ans.}$$

In a similar manner, we can show the relationship between power (in watts), electromotive force (in volts), and resistance (in ohms). Since $P = I \times E$, substituting for I its equivalent (E/R) we get

$$P = \frac{E}{R} \times E \quad \text{or} \quad P = \frac{E^2}{R}.$$

Example. If the electrical pressure is 10 volts, how much power will be dissipated by a resistance of 20 ohms?

$$\text{Since } P = \frac{E^2}{R}, \text{ then } P = \frac{(10)^2}{20} = \frac{100}{20} = 5 \text{ watts.} \quad \textit{Ans.}$$

E. *Sources of electromotive force*

1. PRIMARY CELL

In Figure 2-2 we have indicated that an "electric pump" is used to generate an electromotive force which creates a difference of potential between the ends of a conductor. As a result of this difference of potential, the free electrons of the conductor are made to move from the negative to positive end, that is, a current flows through the conductor. There are several types of generators of electromotive force.

In 1789 Alessandro Volta, an Italian physicist, invented such a generator that converts chemical to electrical energy. He noticed that if two dissimilar metal strips are placed in an acid solution, a potential difference (that is, an electromotive force) appears between the two metals. If a conductor connects the two metal strips, the electromotive force causes electrons to flow through the conductor. Such a generator is called a *voltaic cell* in honor of its inventor.

Let us perform an experiment that will duplicate Volta's findings. (The explanation given here is in simplified form.) Pour some hydrochloric acid (a compound of hydrogen and chlorine) into a jar of water. We believe that when hydrochloric acid is placed in water the compound breaks up. The chlorine atom seizes an electron from the hydrogen atom and thus becomes a negative chlorine ion (which appears in Figure 2-6 as Cl^-). The hydrogen atom, having lost an electron, becomes a positive ion (H^+). If, when a substance goes into solution ions are formed, we say that the substance has *ionized* and call the solution an *electrolyte*. Thus the water solution of hydrochloric acid is an electrolyte.

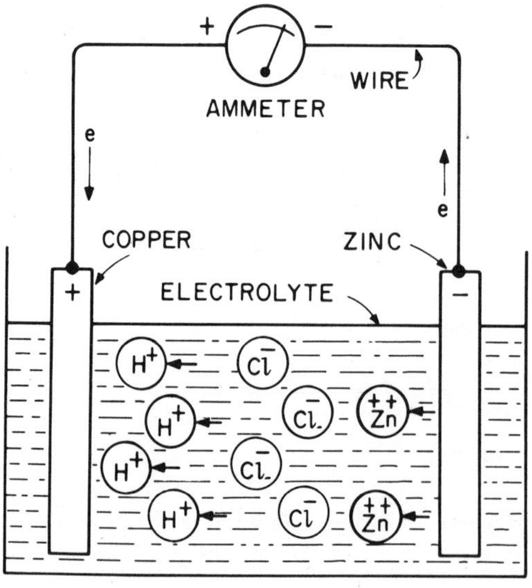

Fig. 2–6.

Voltaic cell.

Into this solution insert a strip of copper and one of zinc. Connect two pieces of wire and an ammeter between the two strips, as indicated in Figure 2-6. The ammeter will show that a current is flowing through the wires from the zinc to the copper strip.

Let us see what happened. When the zinc strip was placed in the acid solution, the zinc started to dissolve—that is, zinc atoms started to leave the strip and enter the solution. As each zinc atom left the strip, however, it left behind two electrons. Thus the zinc atom became a positive zinc ion (Zn^{++}). And the zinc strip, because of the electrons left behind, became negatively charged.

The positive zinc ions repelled the positive hydrogen ions in the solution toward the copper strip. As each positive hydrogen ion reached the copper, it seized an electron from the strip and, becoming in this way a neutral hydrogen atom, bubbled off into the air. The copper strip, having lost electrons, became positively charged.

Thus a difference of potential (electromotive force) was created between the zinc and copper strips. When connected by a conductor, electrons flowed from the zinc to the copper strip, as indicated by the meter. This process will continue until the entire zinc strip is dissolved.

Almost any two dissimilar metals can be used. A carbon rod may be substituted for the copper strip. Almost any acid may be used for the electrolyte and there are a number of other substances, such as sal ammoniac or lye, that may be used as well. An interesting thing about the voltaic cell is the fact that its electromotive force does not depend upon its size. Making the cell larger increases the amount of current that we may draw from it, but it will not increase the electromotive force or voltage. Its voltage depends, mainly, upon the chemical action and this, in turn, depends upon the materials of the strips (*electrodes*) and upon the substance used for the electrolyte. In the cell we have described, the electromotive force is about one volt.

There are a number of disadvantages to the voltaic cell we have described. For example, when the hydrogen ions reach the copper electrode and take away electrons from it, these ions become neutral hydrogen atoms. You can see them as bubbles around the positive electrode when the external circuit is completed. Some of these hydrogen bubbles tend to cling to the positive electrode, forming a sheath completely surrounding it. After a short time, the action of the cell ceases, owing to the insulating action of the hydrogen bubbles which prevent any new hydrogen ions from reaching the positive electrode. We call this effect *polarization.*

In 1866, Georges Leclanché, a French scientist, using a carbon rod as a positive electrode, overcame the effect of polarization by placing this rod in a porous cup containing manganese dioxide. The porous cup and its material did not prevent the positive hydrogen ions from reaching the carbon rod. But after these hydrogen ions became hydrogen atoms, they combined chemically with the manganese dioxide and thus could not form the insulating sheath around the rod. The manganese dioxide is called, appropriately enough, a *depolarizer.*

The voltaic cell we have been discussing presents another disadvantage. It is a *wet* cell, that is, the electrolyte is a liquid and can be spilled easily. To overcome this defect, the *dry cell* was created. In this cell the electrolyte is a paste, instead of a liquid, and thus cannot be spilled so readily.

Look at Figure 2-7. A zinc can is used as the negative electrode and as a container for the cell. A carbon rod in the center of the cell forms the positive electrode. The space between the zinc shell and the carbon rod is filled with a paste containing *sal ammoniac* (a compound of ammonia and chlorine) which is used as an electrolyte. In addition to the electrolyte, this paste contains manganese dioxide, which is used as a depolarizer, and some material, such as sawdust, which is used as a filler. The top is sealed with a cap made of metal, pitch, or sealing wax to prevent the paste from coming out. Immediately below this cap is an air space in which the gases formed by the cell may collect. The entire cell is enclosed in a cardboard case.

Binding posts are attached to the zinc can and to the carbon rod to facilitate connecting the dry cell to an external circuit. The action of this cell is similar to that of the ordinary voltaic cell and it is called a "dry cell" only because the electrolyte is a paste, rather than a liquid. As a matter of fact, should the water of this paste finally evaporate, the cell would cease to function.

The electromotive force that this type of dry cell can produce is approximately 1.5 volts. This voltage remains fairly constant as long as the cell is in good operating condition, regardless of the cell's size. How much current can be drawn from such a cell? The amount of current will depend upon two factors—the resistance of the external circuit and the internal resistance of the cell itself. The greater the external resistance, the less will be the current that is drawn from the cell.

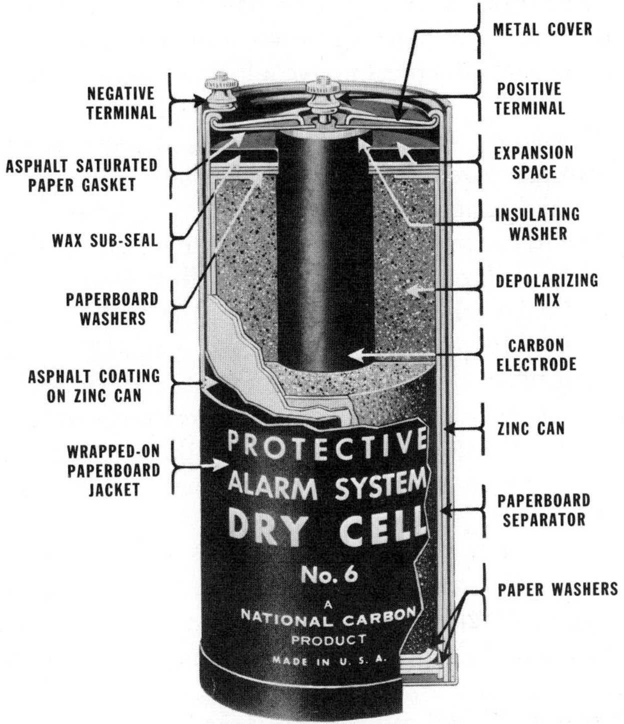

Fig. 2–7. Cross-sectional view of dry cell.

If the resistance of the external circuit is reduced to practically zero (we call this a *short circuit*), the amount of current drawn then would be limited by the internal resistance of the cell. (You realize, of course, that the movement of ions within the cell encounters a certain amount of resistance.) This internal resistance depends upon the material and structure of the cell. In general, the larger the size of the cell, the lower the internal resistance, and hence the greater the amount of current that can be drawn. Also, the larger cell has more active material and thus is able to function for a longer period of time than a smaller cell.

Dry cells generally are used for fairly light, intermittent work such as operating flashlights. Under such conditions a dry cell may last about a year. If, however, a heavy drain is placed on the cell for an appreciable length of time, its life may be shortened or it may even be ruined.

In recent years the dry cell has been improved considerably. One such improvement is the so-called *mercury cell.* Here the negative electrode is made of an amalgam of powdered zinc and mercury pressed into shape. The depolarizer is composed of mercuric oxide and graphite powder. Such cells have a longer life than the ordinary dry cell.

Frequently, two or more cells are connected together to form a *battery.* There are several ways in which these cells may be connected. We may connect them in *series,* that is, with the positive terminal of one to the negative terminal of the other. In Figure 2-8, a battery of three cells connected in series is illustrated. The electromotive force of each cell is added to that of the others and thus the total electromotive force of this battery is 4.5 volts (3×1.5).

But note that when the three cells are connected in series, so are their internal resistances. Thus, the internal resistance of the battery is three times that of a single cell. Since both the electromotive force and internal resistance are tripled, the current that can be drawn from this battery is the same as that which a single cell is capable of producing. In certain applications, such as for use in transistor radios, we often use a number of single cells by snapping the individual cells into springlike mounts that connect them in series or a single compact battery of 9 volts which contains 6 cells in series (6×1.5 volts $= 9$ volts).

The cells may be joined to form a battery by connecting them in *parallel* (see Figure 2-10). In this method of connection, all the positive terminals are joined together, and in like manner all the negative terminals. The electromotive force of this battery is equal to that of a single cell—1.5 volts.

Now the internal resistances of all three cells are connected in parallel. Consequently, the total internal resistance of the battery is equal to one-third the internal resistance of a single cell. Since the electromotive force of the battery is equal to that of a single cell and since its internal resistance is only one-third that of a single cell, the current that may be drawn from the battery is three times the current that a single cell may produce.

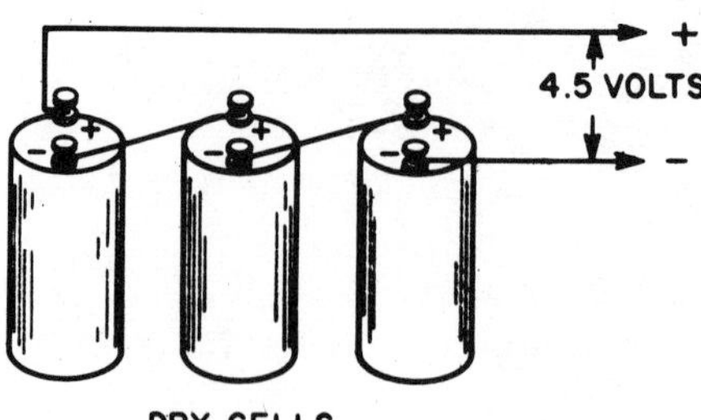

Fig. 2–8.

Three dry cells in series

Fig. 2–9.

Dry cell battery of 22½ volts.

National Carbon Co., Inc.

Where a large current is desired, the cells are connected in parallel. Where a large voltage is wanted, they are connected in series. Sometimes, a number of cells are connected in series to produce a greater voltage, and a number of blocks of such series-connected cells are joined in parallel to produce a higher current. We call such a connection a *series-parallel battery.*

Cells such as the voltaic cell and the dry cell are called *primary cells* and are a source of direct current. When used in an electrical diagram, the symbol for a cell is —ı|— , the small vertical line indicating the negative terminal and the long vertical line the positive terminal. The symbol for a battery of cells connected in series is —ı|ı|ı|— .

2. STORAGE CELL

In the primary cell chemical energy is converted to electrical energy. In 1859, Gaston Planté, a French scientist, discovered he could convert electrical energy into chemical energy which would be

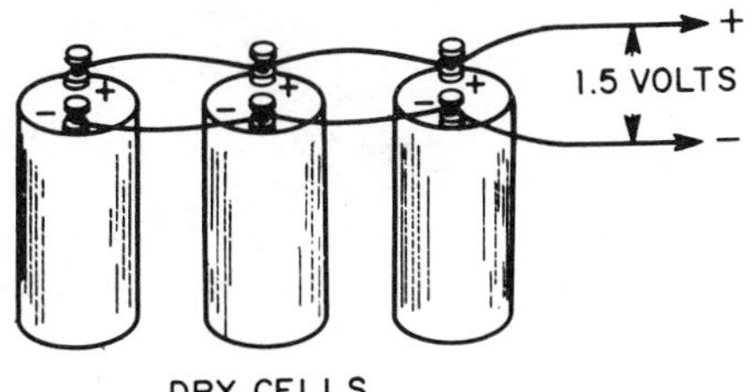

Fig. 2–10.

Three dry cells in parallel.

stored in a cell. Then, by connecting this cell to an external circuit, he was able to reconvert the chemical energy to electrical energy as he needed it. In a sense, he was storing electrical energy. Naturally enough, such a cell is called a *secondary*, or *storage, cell.*

There are several types of storage cells, but the most common type is the *lead-acid cell.* This cell consists of two plates placed in an electrolyte of sulfuric acid and water. The negative plate consists of a lead grid whose openings are filled with a paste of spongy lead (see Figure 2-11A). The positive plate consists of a similar grid filled with lead peroxide (Figure 2-11B).

When we draw current from the cell (this process is called *discharging*), a chemical action takes place that terminates with both plates being converted to lead sulfate. The cell then is *discharged* and, as long as it remains in this condition, no more current may be drawn from it.

Electrical energy is "pumped" back into the cell by connecting the cell, in reverse (+ to +, − to −), to a source of direct current whose voltage is higher than that of the cell. A reverse current flows through the cell which creates a chemical action which reconverts the negative plate back to spongy lead and the positive plate to lead peroxide. The cell then is *charged* and ready to deliver electric current on demand.

A cell of this type produces an electromotive force of about two volts when fully charged. As is true of primary cells, the voltage of

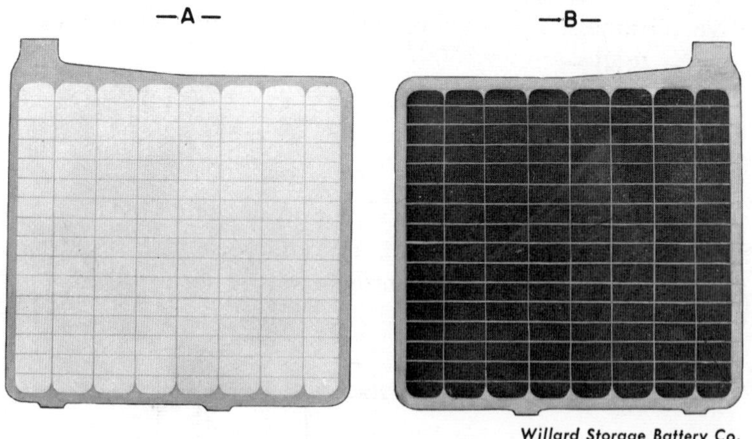

Willard Storage Battery Co.

Fig. 2–11. **A.** Negative plate of lead-acid storage cell. **B.** Positive plate.

General Motors Corp.

Fig. 2—12. How plates and separators are meshed.

the storage cell does not depend upon the size of its plates. However, the larger the plates, the more chemical energy that can be stored in the cell and, hence, the greater the current that can be drawn from it. To increase the effective size of the plates and yet not make the cell too bulky, alternate positive and negative plates are sandwiched together with insulators (called *separators*) of wood, rubber, or other materials between plates (Figure 2-12). The negative plates are connected together, as are the positive plates, producing the effect of a single cell with very large plates. Commercial cells may contain 13, 15, 17, or more, plates.

Storage cells generally, are connected in batteries, the chief use for which is in the automobile. Such batteries usually are of the 6-volt and 12-volt types. Thus, three similar cells are connected in series to form the 6-volt type and six cells are series-connected for the 12-volt type.

The battery container usually is made of hard rubber or some other material that is able to withstand mechanical shock, extremes of heat and cold, and is resistant to the action of the acid electrolyte. A separate compartment is provided for each cell and each compartment has space at its bottom for any sediment that may drop from the plates. Each set of plates of the cells has a heavy lead terminal post and these posts are connected in series (that is, the positive terminal of one cell to the negative terminal of its adjacent cell) by means of heavy lead connectors. (See Figure 2-13.)

Each cell compartment has a cover, usually of molded hard rubber. Openings are provided in these covers for the two terminal posts and for a vent. Each vent has a plug so constructed that the electrolyte cannot splash out, although gases may escape from the cell. The joints between covers and containers are sealed with an acid-resistant compound.

As indicated, there are other types of storage cells. One is the Edison cell, invented by Thomas A. Edison around 1908. This cell consists of nickel peroxide and iron plates in an electrolyte of potassium hydroxide. A more recent type consists of nickel and cadmium plates in a similar electrolyte. The storage cell, as the primary cell, is a source of direct current.

3. OTHER SOURCES OF ELECTROMOTIVE FORCE

Not only chemical energy may be converted to electrical energy. Other types of energies may be converted as well. The chief source of electrical energy, especially where large currents and high

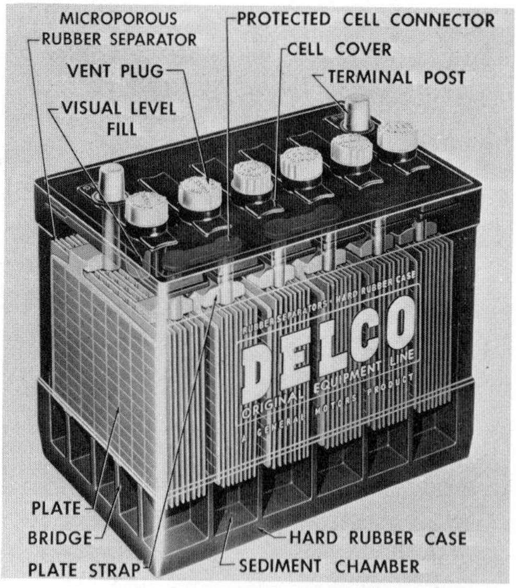

General Motors Corp.

Fig. 2—13. "Phantom" view of the structure of lead-acid storage battery with six cells.

voltages are required, is the *mechanical generator* where mechanical energy is transformed to electrical energy as a conductor is moved through a magnetic field. We will consider this generator in a later chapter.

There are other types of generators that are coming into use. For example, the solar battery uses banks of cells that create electrical energy from light. While these cells are relatively inefficient and depend on clear days for full power, they are excellent sources of power for satellites and instruments which operate in outer space or on the surface of the moon. In outer space, there are no clouds to obstruct the sunlight.

Many companies are busy developing fuel cells which convert fuels directly into electrical energy. In one type, hydrogen and oxygen are fed to the cell, which then combines them to form water as well as electrical energy. In other types, more complex fuels, such as petroleum products, are fed to the cell. Much research work is necessary to overcome such problems as overheating while operating, build-up of waste products, and high cost. Fuel cells are now used for the larger power demands of space-craft operation. Experimental automobiles and tractors have also been built and tested.

QUESTIONS

1. Explain the relationship between the electrical charges carried by the proton and the electron.
2. What is the effect on a neutral atom
   a) when an electron is added;
   b) when an electron is removed?
3. What is the effect between particles that carry
   a) like electrical charges;
   b) unlike electrical charges?
4. a) What is a *field of force?*
   b) What is an *electrostatic field of force?*
5. a) In terms of electrons, what is a *conductor?*
   b) Name three good conductors.
6. a) In terms of electrons, what is an *insulator?*
   b) Name three good insulators.
7. a) What is an *electric current?*
   b) Define the unit in which it is measured.
8. a) What is electrical *resistance?*
   b) Define the unit in which it is measured.
9. List four factors determining the resistance of a substance.

10. a) What is *electromotive force?*
    b) Define the unit in which it is measured.
11. What is the unit
    a) of electrical *power;*
    b) of electrical *energy?*
12. a) State the Ohm's law formula for the relationship between electromotive force, current, and resistance.
    b) In what unit is each quantity measured?
13. State the formula for *power*
    a) in terms of electromotive force and current;
    b) in terms of current and resistance;
    c) in terms of electromotive force and resistance.
14. a) How much current will flow through a circuit where the electromotive force is 100 volts and the resistance is 10 ohms?
    b) How much power will be consumed by the resistor?
15. How much would it cost to operate a 24-ohm heater from a 120-volt line for 10 hours if the cost of electricity is $.05 per kilowatt hour?
16. a) What energy transformation takes place in a primary cell?
    b) What is the advantage of connecting cells in series?
    c) What is the advantage of connecting cells in parallel?
17. Explain the energy transformations that take place in a storage cell.
18. What is the prefix used to denote
    a) one thousand;
    b) one thousandth;
    c) one million;
    d) one millionth?

3 How Current Flows Through a Circuit

A. *The electric current in solids*

If electrons are added to one end of a solid conductor, such as a piece of copper wire, and some are taken away from the other end, a difference of potential is set up between the ends of the wire. This potential difference (voltage) tends to cause free electrons in the wire to move from the negative end to the positive one. As previously stated, this movement of electrons is an electric current.

The free electron moves comparatively slowly through the wire and travels but a short distance before it collides with an atom. This collision generally knocks an electron free from the atom, and this new free electron travels a short distance toward the positive end of the wire before it collides with another atom. Thus, there is a slow drift of electrons from the negative to the positive end of the wire.

But although the drift of electrons is comparatively slow, the disturbance that causes this drift travels through the wire at a very

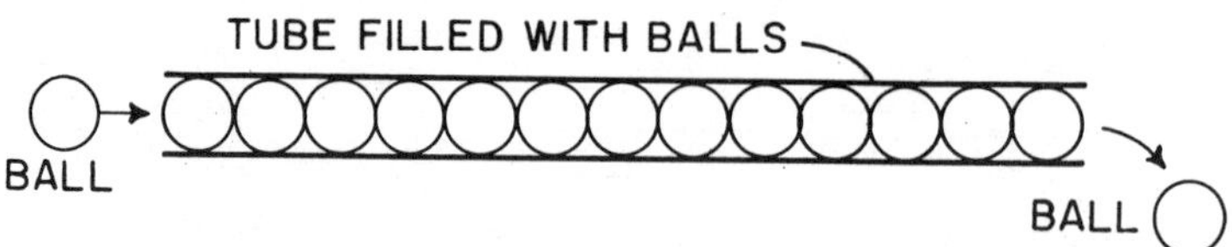

Fig. 3–1. Diagram illustrating the motion of electrons through a solid conductor.

high speed (thousands of miles per second). This action may be understood by visualizing a long, hollow tube completely filled with balls. (See Figure 3-1.) If a ball is added to one end of the tube, a ball at the other end is forced out immediately. Thus, although each ball moves slowly and for only a short distance, the disturbance is transmitted almost instantaneously through the entire tube.

Note that the electric current consists, essentially, of the movement of the electrons of the conductor itself. The generator that furnishes the electromotive force may be considered as a sort of "pump" that removes electrons from one end of the conductor (thus creating a deficiency or positive charge at that end) and piles them up at the other end (creating an excess or negative charge).

B. *The electric current in liquids*

A molecule of ordinary table salt is made up of an atom of sodium and an atom of chlorine. When these two atoms combine to form a molecule of salt (whose chemical name is *sodium chloride*) they are joined by a bond formed by one electron from the outer shell of each atom.

When the molecule of salt is dissolved in water, the two atoms separate, or *disassociate.* The chlorine atom seizes both electrons of the bond and, since it now has an extra electron, becomes a negative ion (Cl^-). The sodium atom, having lost an electron, becomes a positive ion (Na^+). The salt solution thus is an electrolyte containing positive and negative ions (see Figure 3-2).

If two metal plates (called *electrodes*) are set at opposite ends of the solution and a source of electromotive force is connected to these plates so that one becomes a positive (electron-deficient) electrode and the other a negative (electron-excess) electrode, an electric field is created between these two electrodes.

Since opposite charges attract, the negative chlorine ion is attracted to the positive electrode and the positive sodium ion is attracted to the negative electrode. Upon reaching the positive electrode, the chlorine ion surrenders its extra electron to the electrode and becomes a neutral chlorine atom. As the sodium ion reaches the negative electrode, it obtains an electron from the electrode and becomes a neutral sodium atom.

The effect of the electromotive force, then, is to cause a movement of ions through the solution. This movement of charged particles constitutes an electric current and in this way the electric current flows through a liquid.

An electrolyte is a liquid containing substances that are able to disassociate into free ions. The resistance of an electrolyte depends upon the degree to which these ions are released (ionization). The greater the amount of ionization, the less is the resistance of the liquid. Thus, for a given electromotive force, the greater the ionization the greater is the current transmitted by ions through the liquid.

C. *The electric current in gases*

The molecules of a gas are in a perpetual state of motion, constantly colliding with one another. These collisions knock off electrons, producing free electrons and converting the atoms that have lost electrons into positive ions. Since, by definition, a charged particle is called an ion, we may consider the free electrons as negative ions. Thus the gas contains positive and negative ions, just as an electrolyte does. If positive and negative electrodes are placed in

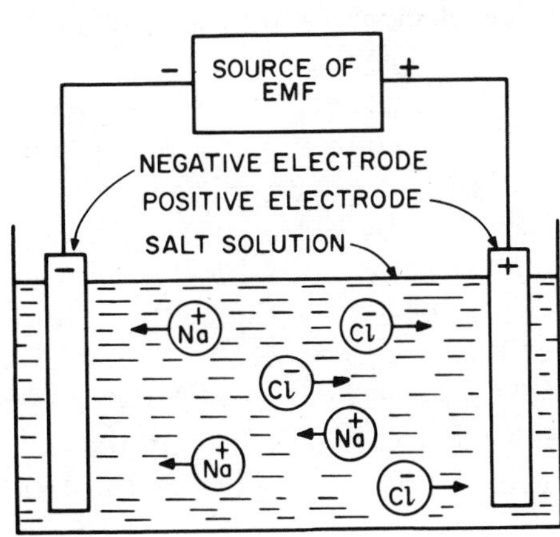

Fig. 3–2. Movement of ions through a liquid. The symbol Na^+ represents the positive sodium ion, Cl^- the negative chlorine ion.

the gas, the free electrons tend to travel to the positive electrode, and the positive ions to the negative electrode, thereby producing an electric current.

Normally, a positive or negative ion cannot travel very far in a gas before meeting an ion of opposite charge. This meeting would tend to produce neutralization and would result in neutral molecules. Since neutral molecules are not affected by the electric field between the two electrodes, the current would tend to cease flowing.

But if the gas is placed in a sealed container (such as a glass tube or bulb, with the two electrodes sealed in), and if most of the gas is pumped out, then the ions can travel considerable distances without being obstructed. The effect of the electric field is to accelerate, or speed up, the motion of the ions, so the farther they travel, the more velocity they attain. If a fast-moving ion collides with a neutral molecule, the ion tends to knock electrons off the neutral molecule, thereby creating more ions. This process is cumulative and tends to keep a constant stream of ions moving toward the electrodes. In this manner, an electric current flows through a gas.

When the positive ions reach the negative electrode, they acquire electrons to become neutral molecules once more. The negative ions (electrons) are attracted to the positive electrode. Then the entire process is repeated.

D. *The electric current in a vacuum*

If a free electron were in a vacuum within the electric field set up between positive and negative electrodes, the negatively-charged electron would be attracted to the positive electrode. The movement of the electron would constitute a flow of electric current. It is upon this principle that the electron tubes used in our radio and television receivers operate.

We can construct an electron tube by sealing a pair of metal electrodes into opposite ends of a glass bulb and evacuating the air from within the bulb, leaving a vacuum. Connecting the electrodes to a source of electromotive force makes them positive and negative, respectively. (See Figure 3-3.) A question now arises. How can we get the free electron into the tube?

As previously described, there always is a disorderly movement of free electrons within all substances, especially metals. If the difference of potential between the two sealed-in electrodes be made great enough, some of the free electrons of the negative electrode will be attracted so strongly to the positive electrode, that they will leave the former and fly through the vacuum to the latter.

Or else, if a substance is heated, the movement of free electrons within that substance is increased. If the temperature is raised high enough, the movement of free electrons is increased to the point where some of the electrons actually fly off from the substance. We call this process *thermionic electron emission.*

In most electron tubes, the negative electrode is heated to the point where it emits electrons. These electrons are attracted to the positive electrode and constitute a one-way flow of electric current through a vacuum from the negative to the positive electrode.

Further, certain substances, such as sodium, potassium, and cesium, will emit electrons if they are exposed to light. This phenomenon is known as *photoelectric emission.* If the negative electrode of the tube is made of such a material and light is permitted to fall on it, it will emit electrons that will be attracted to the positive electrode. Such a tube is called a *photoelectric cell.* We will delve deeper into the flow of current through gases and through a vacuum in the section on electron tubes.

E. *Types of electric circuits*

Just as water flows downhill from a point of high potential energy to a point of low potential energy, so the electric current flows from a high-potential point (excess of electrons) to a low-potential point (deficiency of electrons). The path or paths followed by the current flow is called the *electric circuit.*

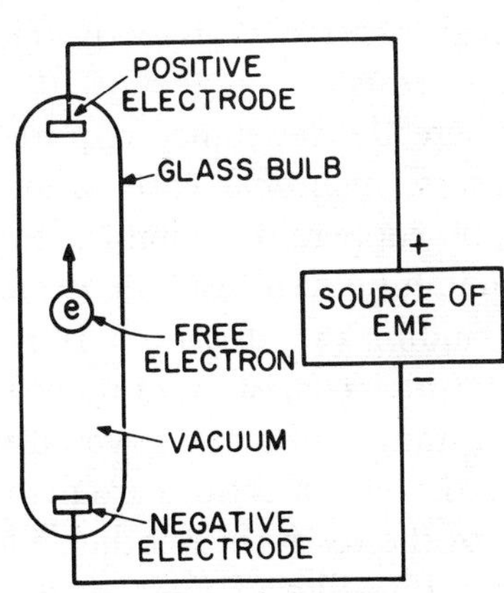

Fig. 3–3.

Diagram illustrating the basic principle of the electron tube. The symbol (e) represents a free electron.

Benjamin Franklin arbitrarily assumed that electricity flows from positive to negative. A number of books have been written that perpetuate this notion. Today we know this to be false; electricity flows from negative (high potential) to positive (low potential). Franklin may be forgiven; he did not know about electrons. But there is no excuse for us to repeat this error. Accordingly, in this book we shall consider the electric current as flowing from *negative to positive*.

All circuits must contain a source of electromotive force to establish the difference of potential that makes possible the current flow. This source may be a dry cell, a mechanical generator, or any of the other devices that will be discussed later in the book. All paths of the circuit lead in closed loops from the high-potential (negative) end of the electromotive-force source to its low-potential (positive) end.

The current in a circuit may flow through solid conductors, liquids, gases, vacuums, or any combination of these. Its path may include lamps, toasters, motors, or any of the thousand-and-one electrical devices available in this electrical age. But, regardless of the type of circuit and the devices through which the current flows, all circuits offer some resistance to the current.

This resistance may be high or low, depending upon the type of circuit and the devices employed. Sometimes this resistance is undesirable, as, for example, the resistance of the wires connecting the various devices in a circuit. Accordingly, we keep this resistance at a low level by using wires made of copper having a large cross-sectional area, and by keeping their lengths as short as possible. Sometimes, however, it is desirable to introduce concentrated, or lumped, resistances into the circuit. Such a lumped resistance is called a *resistor*.

Where the resistance required is not too great, the resistor may consist of *nichrome* wire, whose resistance is more than 50 times that of copper, of suitable length and thickness, wound upon a ceramic tube. An insulating ceramic coating usually is applied over the winding (see Figure 3-4A). If the resistance is to be large, the resistor may consist of a thin coat of a carbon composition deposited on a ceramic tube and covered with some insulating material (Figure 3-4B). Connections are made by means of wires attached to the ends of the resistor. Another popular type is the metal-film resistor. A very thin film of metal is evaporated or "sputtered" under high

– A –

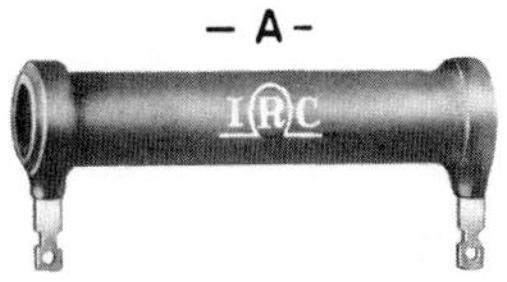

– B –

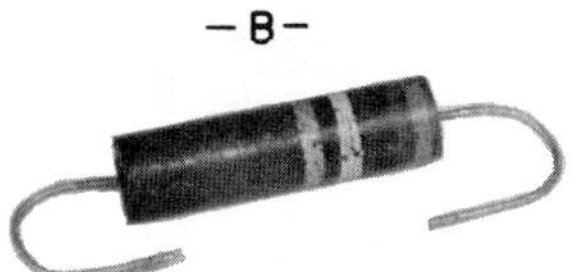

International Resistance Co.

Fig. 3—4. Fixed resistors.

A. Wire-wound type.
B. Composition type.

vacuum on to a ceramic or glass tube. Careful control of the film thickness regulates the resistance. Connections are made by means of wires attached to the ends of the resistors. In electrical diagrams the symbol for the resistor is .

In electrical circuits it is frequently very important that the resistance value when selected remain constant under varying external temperatures. The wire-wound and metal-film types of resistors are the most stable.

Where the resistor is to be variable, the wire is wound on a fiber strip which may be bent into a circular form. A metal arm or slider, manipulated by a knob, is made to move over the wire, thus making contact with any desired point. In this way the length of the wire, and hence the resistance, between one end of the wire and the slider may be varied. Such variable resistors are called *rheostats* or *potentiometers* (see Figure 3-5A). Where the resistance is to be large, a carbon composition deposited on the fiber strip may replace the coiled wire (Figure 3-5B). Another type uses a ceramic material called Cermet as the resistance element. The electrical symbol for the variable resistor is or .

Fig. 3—5. Variable resistors.

A. Wire-wound type.
B. Composition type.

– A –

Ohmite Mfg. Co.

– B –

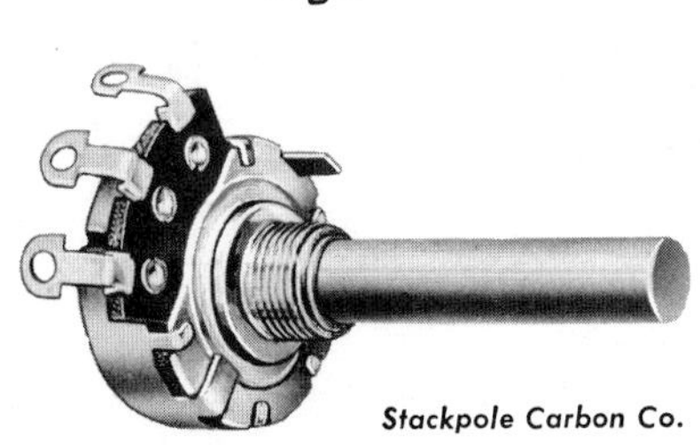

Stackpole Carbon Co.

In this chapter we will consider three general types of circuits. One is the *series* circuit which offers a single, continuous, external path for current flow from the negative side of the electromotive-force source to the positive side. Another is the *parallel* circuit which offers two or more parallel paths for current flow from negative to positive. The third type is the *series-parallel* circuit, a combination of the other two.

1. THE SERIES CIRCUIT

The series circuit, we have stated, offers a single, continuous, external path for current flow from the negative side of the electromotive-force source to the positive side. Such a circuit is illustrated in Figure 3-6. Electrons flow (as indicated by the symbol e →) from the negative side of the source, through resistor R, and back to the positive side of the source.

As the electrons leave the battery and flow through the resistor, a voltage is developed across the resistor which opposes the supply or source voltage. There is a law (Kirchoff's law) which states that the sum of all the voltages around any closed circuit path must add up to the supply voltage. In this simple circuit of one battery and one resistor, therefore, sufficient current must flow so that the voltage across the resistor must equal the battery voltage. This is computed from Ohm's law which states $E = I \times R$ or since we know the voltage of the source (the supply voltage) we can determine the current by another version $I = E/R$ where the resistance R is in ohms, the voltage E is in volts, and the current I is in amperes.

Since there is one path over which the current may flow, the current is the same in all parts of the series circuit. This is a particular example of another part of Kirchoff's law which states that the sum of all the currents entering any junction of a circuit must equal the sum of all the currents leaving that junction. In this simple case, the only junctions are those two where the battery terminal joins the

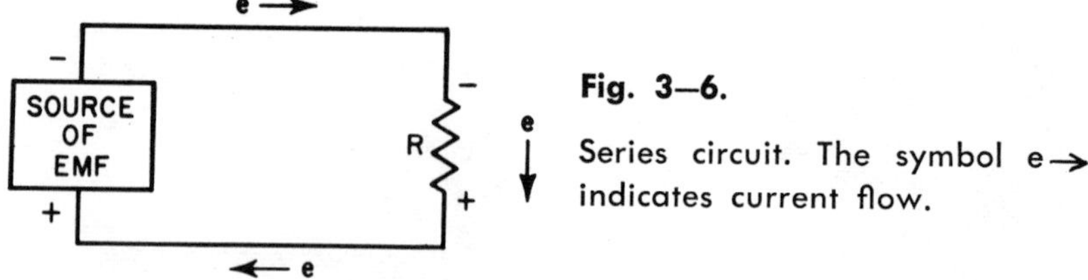

Fig. 3–6.

Series circuit. The symbol e→ indicates current flow.

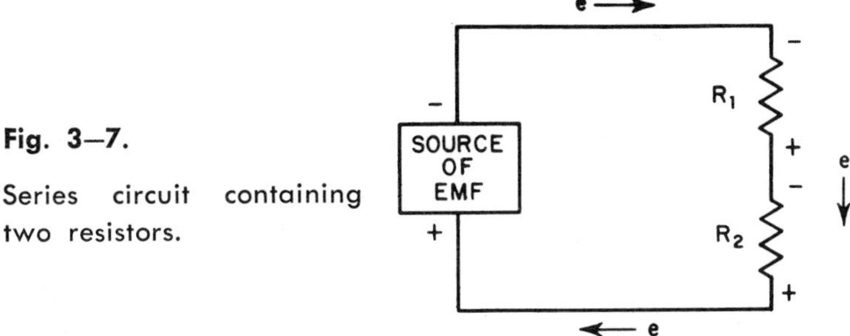

Fig. 3–7. Series circuit containing two resistors.

resistor. If we wanted to prove this case, we could insert a current-measuring device called an ammeter in different portions of the circuit and note that the reading remains the same.

Example. In Figure 3-6, if the electromotive force is 100 volts and the resistance of R is 20 ohms, what current will flow through the circuit (neglecting, for the sake of simplicity, the resistance of the source and the connecting wires)?

$$I = \frac{E}{R} = \frac{100 \text{ volts}}{20 \text{ ohms}} = 5 \text{ amperes.} \quad \textit{Ans.}$$

The series circuit, though offering a single, continuous path for current flow, may contain more than one resistor. For example, examine the circuit illustrated in Figure 3-7. As before, the current flow is the same in all parts of the circuit.

The total resistance of the series circuit is equal to the sum of all the individual resistances. Thus the total resistance of the series circuit shown in Figure 3-7 is equal to the sum of the resistance of R_1, the resistance of R_2, the resistance of the connecting wires, and the resistance of the electromotive-force source (all expressed in ohms). The total resistance of a series circuit may be determined by the following formula:

$$R_{\text{total}} = R_1 + R_2 + R_3 + R_4, \text{ and so forth.}$$

Example. In Figure 3-7, if the resistance of R_1 is 30 ohms and R_2 is 20 ohms, what is the total resistance of the circuit (neglecting the resistance of the source and the connecting wires)?

$$R_{\text{total}} = R_1 + R_2 = 30 \text{ ohms} + 20 \text{ ohms} = 50 \text{ ohms.} \quad \textit{Ans.}$$

If the voltage of the source is 100 volts, what current will flow through the circuit?

By Ohm's law,

$$I = \frac{E}{R} = \frac{100 \text{ volts}}{50 \text{ ohms}} = 2 \text{ amperes.} \quad \textit{Ans.}$$

As we have learned, the current is the same in all portions of the series circuit. If there is more than one resistor in this series circuit, the sum of all of the voltages across the resistors must, according to Kirchoff's law, add up to the supply voltage.

Let us see how this applies to the circuit illustrated in Figure 3-7. We again assume that R_1 is 30 ohms, R_2 is 20 ohms, and the voltage of the source is 100 volts. We also know that the current is equal to 2 amperes. For the sake of simplicity we again neglect the resistances of the connecting wires and of the source.

As current flows through R_1 (Figure 3-7), the top of the resistor becomes negative with respect to its bottom. Similarly, the top of R_2 becomes negative with respect to its bottom. This is indicated by the $+$ and $-$ signs at each resistor and shows the polarity of the voltage drop across each.

Note that the voltage across R_1 is 60 volts and the current flowing through it is 2 amperes. From the formula P (watts) $= E$ (volts) $\times I$ (amperes) we may determine the electric power dissipated by that resistor. Thus:

$$P = E \times I = 60 \text{ volts} \times 2 \text{ amperes} = 120 \text{ watts.}$$

The 2 amperes passing through the 30-ohm resistor creates a voltage

$$E = I \times R = 2 \text{ amperes} \times 30 \text{ ohms} = 60 \text{ volts.}$$

We repeat for the 20-ohm resistor

$$E = I \times R = 2 \text{ amperes} \times 20 \text{ ohms} = 40 \text{ volts.}$$

If we add the 2 voltages across the resistors

$$E\ \textit{Total} = 60 \text{ volts} + 40 \text{ volts} = \text{the supply voltage.}$$

Note, once again, that the current flowing through a resistor causes a voltage or difference of potential to appear between the ends of the resistor. This voltage is sometimes called the IR drop. This means that R_1 must be able to dissipate safely at least 120 watts. The voltage drop across R_2 is 40 volts and the current flowing through it is 2 amperes. Thus:

$$P = E \times I = 40 \text{ volts} \times 2 \text{ amperes} = 80 \text{ watts}$$

and R_2 need dissipate only 80 watts.

The total power consumed by the entire circuit may be found by multiplying the voltage of the source by the total current. Thus:

$$P = E \times I = 100 \text{ volts} \times 2 \text{ amperes} = 200 \text{ watts.}$$

Note that this is equal to the sum of the power dissipated by R_1 and R_2.

(There is an interesting sidelight we may consider. Suppose that R_2 becomes defective, opening the circuit. Since the circuit is broken, no current flows. Hence there are no *IR* drops. However, since the ends of R_2 are still connected to the source, the full 100 volts appears across those ends.)

Assume we have three resistors R_1, R_2, and R_3 connected in series across a voltage source, as illustrated in Figure 3-8. Further assume that the voltage of the source is 100 volts and that the resistance of R_1, R_2, and R_3 are 20 ohms, 30 ohms, and 50 ohms, respectively. The total resistance of the circuit (neglecting the resistances of the source and the connecting wires) is $20 + 30 + 50 = 100$ ohms. Accordingly, since by Ohm's law $I = E/R$, the current flowing through the circuit is 100 volts/100 ohms = 1 ampere.

The *IR* drop across R_1 is equal to 1 ampere $\times$ 20 ohms = 20 volts. Similarly, the *IR* drops across R_2 and R_3 are 30 volts and 50 volts, respectively. Thus the voltage of the source is divided across the network of R_1, R_2, and R_3 according to their resistances, that is, in a 2, 3, 5 ratio. Note the polarities of the voltage drops across these resistors.

Such a network is called a *voltage divider* and is used where we wish to obtain a portion of the source voltage. If, for example, we wish 20 volts, we may obtain it from across R_1. If we wish 30 volts, we may obtain it from across R_2, and we may obtain 50 volts from

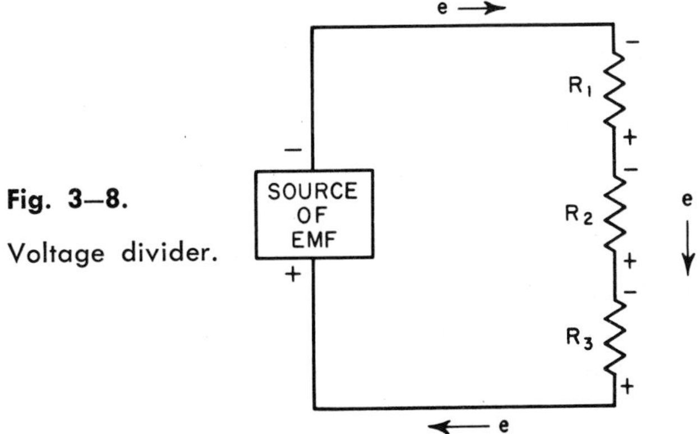

Fig. 3–8. Voltage divider.

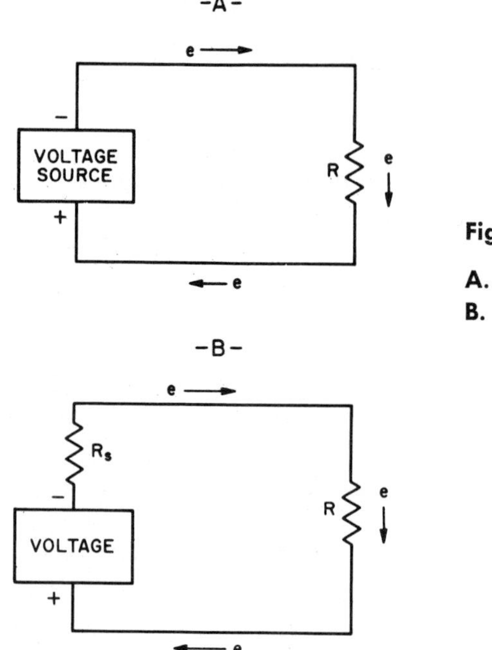

Fig. 3–9.

A. Series circuit.

B. Its equivalent circuit.

across R_3. If we desire 80 volts, we may obtain it from across R_2 and R_3 in series.

Of course, any combination of resistors may be used to obtain other voltages. In practice, the voltage divider may be a single resistor whose resistance is equal to the sum of all, with taps at suitable points.

You may have noticed we have been saying "neglecting the resistance of the source." Actually, the resistance of the source is quite important. Here is why.

In Figure 3-9A, we represent the voltage source as a box. In practice, we find that if the current flowing from this box is large, the voltage of the source is reduced. For very small currents or for an open circuit (no current at all), the voltage of the source is essentially constant. It was shown by Thevenin many years ago that any voltage source could be represented by an equivalent circuit—a constant voltage source in series with a fixed resistor. This is shown in Figure 3-9B. If the voltage and the resistance could be determined, we could calculate the behavior of the source under varying external conditions.

As current flows through the circuit a voltage drop occurs across R_s which reduces the voltage left for resistor R. If the current drawn by R (which is called the *load*) is low, the voltage drop across R_s, too, is low. If, on the other hand, the current drawn by the load is high, the voltage drop across R_s is high. Thus it can be seen that the resistance of R_s (the internal resistance of the source) must be very low if extreme voltage fluctuations with variations in the load are to be avoided.

2. THE PARALLEL CIRCUIT

The parallel circuit is ilustrated in Figure 3-10. Note that the electromotive force across all components of such a circuit is the same. Current flows (as indicated by the symbol **e** →) from the negative side of the source to the upper junction of resistor R_1 and R_2. Here it divides, part flowing through R_1 and part through R_2. At the lower junction of R_1 and R_2 both currents reunite and flow back to the positive side of the source.

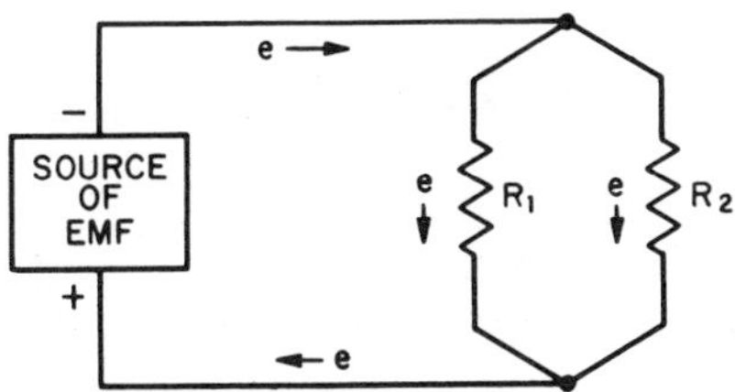

Fig. 3—10. Parallel circuit.

[At this point, it might be well to mention a few facts concerning electrical circuit diagrams. Often, we must show two wires crossing each other. There are several methods of designation, but in this book, if two wires cross and connect with each other, this connection will be indicated by a dot at the point of crossover (┿). If there is no connection, a loop (⌒) will be used to indicate this fact.]

The total resistance of resistors connected in parallel may be expressed by the following formula:

$$\frac{1}{R_{total}} = \frac{1}{R_1} + \frac{1}{R_2} + \frac{1}{R_3} + \frac{1}{R_4} +, \text{ and so forth.}$$

Example. Assume, as in Figure 3-10, that R_1 is 90 ohms, and R_2 is 45 ohms. What is the total resistance of R_1 and R_2 in parallel?

$$\frac{1}{R_{total}} = \frac{1}{R_1} + \frac{1}{R_2} = \frac{1}{90 \text{ ohms}} + \frac{1}{45 \text{ ohms}} = \frac{3}{90}$$

$$R_{total} = \frac{90}{3} = 30 \text{ ohms.} \quad \textit{Ans.}$$

Now let us carry our problem a step further. Assuming a voltage source of 90 volts, what will be the total current flowing in the circuit (neglecting the resistance of the source and connecting wires)?

Since the total resistance is 30 ohms and the voltage is 90 volts, by Ohm's law ($I = E/R$) we get

$$I = \frac{E}{R} = \frac{90 \text{ volts}}{30 \text{ ohms}} = 3 \text{ amperes.} \quad \textit{Ans.}$$

Knowing that a total of 3 amperes are flowing through the circuit, how much current flows through R_1 and through R_2? Note that the same voltage is applied across R_1 as R_2, that is, the IR drop across R_1 and R_2 is 90 volts. Again, by Ohm's law, we can determine the current flowing through each resistor. Thus, for R_1:

$$I = \frac{E}{R} = \frac{90 \text{ volts}}{90 \text{ ohms}} = 1 \text{ ampere.} \quad \textit{Ans.}$$

And for R_2:

$$I = \frac{E}{R} = \frac{90 \text{ volts}}{45 \text{ ohms}} = 2 \text{ amperes.} \quad \textit{Ans.}$$

Note that the greater current flows through the resistor with the lower resistance. And because R_2 has half the resistance of R_1, twice as much current flows through R_2 as through R_1.

Power dissipated by each resistor may be determined from the formula $P = E \times I$. For resistor R_1, E equals 90 volts and I equals 1 ampere. Hence the power is 90 watts. For resistor R_2, E equals 90 volts and I equals 2 amperes. Hence the power is 180 watts. The power consumed by the entire circuit may be found by multiplying the voltage of the source by the total current. Thus:

$$P = E \times I = 90 \text{ volts} \times 3 \text{ amperes} = 270 \text{ watts.}$$

Note that this is equal to the sum of the power dissipated by R_1 and R_2. Note, too, that in the series circuit the total power dissipated is also equal to the sum of the power dissipated by each of the individual resistors.

3. THE SERIES-PARALLEL CIRCUIT

As its name implies, this circuit is a combination of the other two types. A simple example of such a circuit is illustrated in Figure 3-11. Note that R_2 and R_3 are connected in parallel with each other and that both together are connected in series with R_1. To find the total resistance of this circuit, we first find the joint resistance of R_2 and R_3 in parallel and then add this joint resistance to that of R_1, as in any other series circuit.

Let us assume that the voltage of the source is 150 volts. Further assume R_1 is 20 ohms, R_2 is 90 ohms, and R_3 is 45 ohms. What is the total resistance of the circuit (neglecting the resistance of the connecting wires and of the source)?

First we find the joint resistance of R_2 and R_3 in parallel.

$$\frac{1}{R_{\text{joint}}} = \frac{1}{R_2} + \frac{1}{R_3} = \frac{1}{90 \text{ ohms}} + \frac{1}{45 \text{ ohms}} = \frac{3}{90}$$

$$R_{\text{joint}} = \frac{90}{3} = 30 \text{ ohms.}$$

Next, we find the total resistance of R_{joint} and R_1 in series.

$$R_{\text{total}} = R_{\text{joint}} + R_1 = 30 \text{ ohms} + 20 \text{ ohms} = 50 \text{ ohms.} \quad \textit{Ans.}$$

To find the current flowing through R_1, R_2, and R_3, first find the total current of the circuit. Knowing the voltage of the source and the total resistance, we may find the total current by Ohm's law. Thus:

$$I = \frac{E}{R} = \frac{150 \text{ volts}}{50 \text{ ohms}} = 3 \text{ amperes.}$$

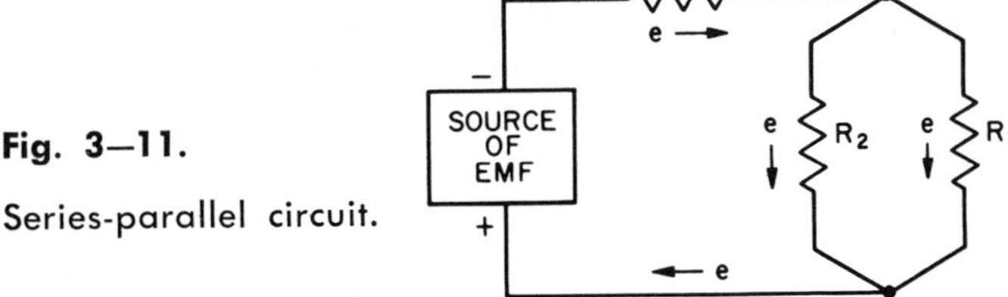

Fig. 3—11.

Series-parallel circuit.

Since R_1 is in series with the rest of the circuit, the total current flows through it. Thus the current flowing through R_1 is 3 amperes.

We now can calculate the voltage drop across R_1.

$$E = I \times R = 3 \text{ amperes} \times 20 \text{ ohms} = 60 \text{ volts.}$$

Since 60 volts are expended forcing the current through R_1, 90 volts are left for the rest of the circuit. Because R_2 and R_3 are connected in parallel, the same voltage is applied to each. Hence the voltage drops across R_2 and R_3 are 90 volts each.

Since the resistance of R_2 is 90 ohms and its voltage drop is 90 volts, then:

$$I = \frac{E}{R} = \frac{90 \text{ volts}}{90 \text{ ohms}} = 1 \text{ ampere flowing through } R_2.$$

Similarly, since R_3 is 45 ohms and its voltage drop, too, is 90 volts, then:

$$I = \frac{E}{R} = \frac{90 \text{ volts}}{45 \text{ ohms}} = 2 \text{ amperes flowing through } R_3.$$

To find the total power consumed by the circuit, multiply the total voltage (150 volts) by the total current (3 amperes). This comes to 450 watts. The power dissipated by R_1 may be found by multiplying its voltage drop (60 volts) by its current (3 amperes). This comes to 180 watts. To find the power dissipated by R_2, multiply its voltage drop (90 volts) by its current (1 ampere). This comes to 90 watts. To find the power dissipated by R_3, multiply its voltage drop (90 volts) by its current (2 amperes). This comes to 180 watts. If we add up the power dissipated by R_1, R_2, and R_3, it equals the total of 450 watts consumed by the entire circuit. Thus, regardless of the method of connection, the total power consumed by the circuit is equal to the sum of the power consumed by all of its parts.

QUESTIONS

1. **Explain how current flows through a solid.**
2. **Explain how current flows through a liquid.**
3. **Explain how current flows through a gas.**
4. **Explain how current flows through a vacuum.**
5. **Explain what is meant**
   a) **by *thermionic electron emission;***
   b) **by *photoelectric electron emission.***

6. a) What is meant by an *electric circuit?*
   b) What causes the current to flow through a circuit?
7. a) What is meant by *voltage drop?*
   b) Upon what factors does it depend?
8. What is meant by
   a) a *series circuit;*
   b) a *parallel circuit;*
   c) a *series-parallel circuit?*
9. What is the formula for determining the total resistance of several resistors connected in series?
10. Two resistors, one 20 ohms and the other 5 ohms, are connected in series across a 100-volt source.
    a) What is their total resistance?
    b) What is the total current flowing through the entire circuit?
    c) How much current will flow through each resistor?
    d) What will be the voltage drop across each resistor?
    e) What is the total power dissipated by the circuit?
    f) How much power is dissipated by each resistor? Check this answer against (e) above.
11. What is the formula for determining the total resistance of several resistors connected in parallel?
12. Two resistors, one 20 ohms and the other 5 ohms, are connected in parallel across a 100-volt source.
    a) What is their total resistance?
    b) What is the total current flowing through the entire circuit?
    c) How much current will flow through each resistor?
    d) What will be the voltage drop across each resistor?
    e) What is the total power dissipated by the circuit?
    f) How much power is dissipated by each resistor? Check this answer against (e) above.
13. Two resistors, R_1 (3 ohms) and R_2 (2 ohms), are connected in series and then both together are connected in parallel with R_3 (20 ohms). The whole circuit is connected to a 20-volt source.
    a) What is the total resistance of the circuit?
    b) What is the total current flowing through the entire circuit?
    c) How much current will flow through each resistor?
    d) What will be the voltage drop across each resistor?
    e) What is the total power dissipated by the circuit?
    f) How much power is dissipated by each resistor? Check this answer against (e) above.

4 Effects of Electric Current

A. *Thermal effect*

When mechanical power is applied to a machine, it meets a kind of resistance called friction. Mechanical power is "lost" overcoming this friction. However, it is not really lost since it shows up as heat at the point or points of friction. Mechanical power has been changed to heat power.

Similarly, when an electric current is passed through a resistor, electrical power is converted into heat power and the resistor may get very hot.

You will recall that the electrical power consumed by a circuit is equal to the product of the current and the electromotive force. Thus:

$$P \text{ (in watts)} = E \text{ (in volts)} \times I \text{ (in amperes)}.$$

Since, by Ohm's law, $E = I \times R$, by substituting for E in the first equation its equivalent $(I \times R)$, we may indicate the power equation in terms of current (I) and resistance (R). Thus:

$$P = E \times I = (I \times R) \times I = I^2 \times R.$$

Similarly, since $I = E/R$, we may indicate the power equation in terms of electromotive force (E) and resistance (R). Thus:

$$P = E \times I = E \times \frac{E}{R} = \frac{E^2}{R}.$$

Thus, if we wish to determine the power consumed or "lost" as current flows through a resistor, we may multiply the square of that current by the resistance of the resistor. Still another method is to divide the square of the voltage drop by the resistance. Both these methods produce the same result.

Since the voltage of a circuit usually is kept at a constant value, the two variables generally are the resistance of the resistor and the current flowing through it. Hence the power loss most frequently is expressed in terms of $I^2 \times R$. Since this power produces heat, the heating effect of an electric current often is called the I^2R loss.

In many instances the heat so produced is undesirable and steps are taken to keep it at a minimum. For example, where large currents are to flow, conductors may be made of heavy copper bars to keep the resistance low. Where these steps are not sufficient to keep the heat at safe levels, the heat itself may be conducted away. Thus many motors are constructed with a built-in fan to blow cool air over the wires heated by the current flowing through them.

However, there are instances where the heat is desirable. Certain devices, such as toasters, heaters, electron-tube filaments, and such, are constructed with special conductors made of alloys that offer a fairly high resistance to current flow. This resistance, in turn, produces heat.

One such device is the *fuse* (see Figure 4-1) which is a protective device used to prevent too much current from flowing through an electrical circuit. Since, by Ohm's law, $I = E/R$, should the resistance of the circuit be reduced for any reason, the current flowing through it will increase. If the resistance is reduced sufficiently, the current may rise to a value that may damage by overheating the various components of the circuit or even the wires themselves.

It is here that the fuse comes to the rescue. It consists of a strip of metal that will melt at a comparatively low temperature, placed in

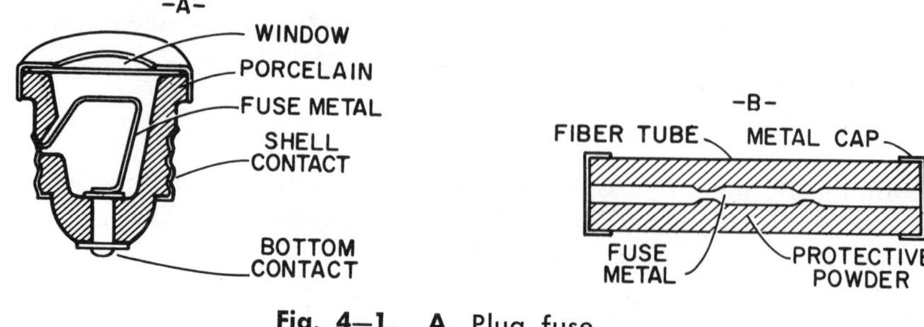

Fig. 4—1. **A.** Plug fuse.
B. Cartridge fuse.

series with the circuit. As the current rises, the heating effect causes the fuse to melt, thus opening the circuit and cutting off the flow of current, well before any of the other components of the circuit are damaged.

B. *Luminous effect*

If we heat a substance, such as a metal wire, for example, the molecules of the substance are made to move faster. As we continue to add heat, the molecules move faster until a point is reached where light is emitted. (We believe the light is produced as a result of the rearrangement of the electrons around the nucleus. However, this matter is not within the scope of this book.)

You now know that as a current flows through a conductor, heat is produced as a result of the I^2R loss. If the current and resistance be large enough, the heat so produced may be great enough to make the conductor emit light. This is the principle of the *incandescent lamp,* invented by Thomas A. Edison in 1879.

To provide for a high enough resistance, Edison used a wire, or filament, made of carbon. However, if this filament be heated until it emits light—that is, to *incandescence*—it burns up in the air, which supports combustion. Accordingly, Edison sealed the carbon filament in a glass bulb from which he pumped out the air. (See Figure 4-2.)

The passage of an electric current may heat a gas, as well as a solid, to incandescence. This principle underlies the *carbon-arc light* which was used extensively for street lighting at the beginning of the

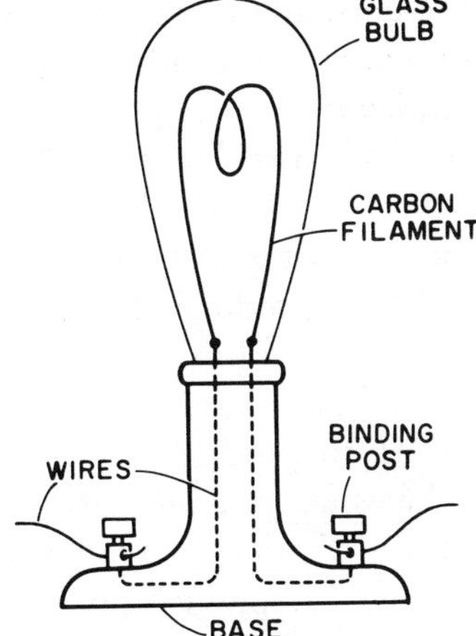

Fig. 4–2.

Edison incandescent lamp.

twentieth century. In this type of street light, current is led to two carbon rods. (See Figure 4-3). The tips of these rods are touched together and then slightly separated. As a result, a hot electric spark, or *arc,* jumps from tip to tip. The heat of this arc vaporizes some of the carbon, and the passage of current through this carbon vapor heats it to burning and incandescence. (The tips of the rods, too, are heated to incandescence, thus furnishing an additional source of light.)

So far, we have discussed light produced when a solid or a gas is heated to incandescence. There are other electrical methods for producing light without the necessity for heating a substance. You will recall that electric current flows through a gas by the movement of ions to oppositely-charged electrodes sealed into opposite ends of a tube from which most of the gas has been evacuated (Chapter 3, subdivision C). When the positively-charged gas ion reaches the negative electrode, it receives an electron and becomes a neutral atom once more. As it does so, light is emitted (as a result of the rearrangement of the electrons of the gas atom).

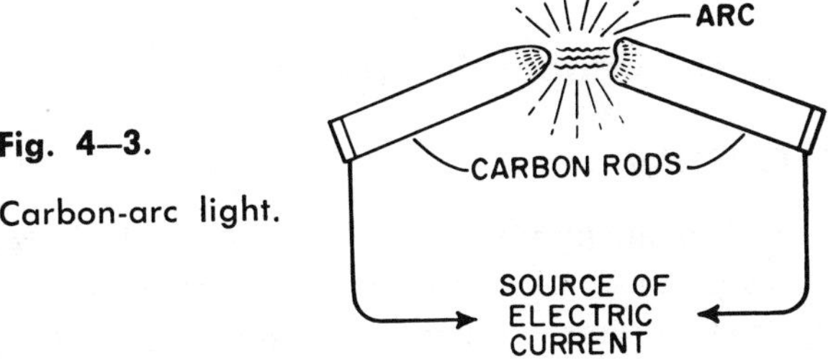

Fig. 4–3.

Carbon-arc light.

It is not only at the negative electrode that the gas atom emits light. Anywhere in the tube, when a positive gas ion meets an electron, neutralization takes place and light is emitted. Hence the tube glows over its entire length. This process is continuous since the neutral atom soon is struck by a speeding ion and becomes an ion once more. Each type of gas produces light of characteristic color. Neon produces a reddish light. Helium produces a pinkish light; argon produces a bluish-white light; mercury vapor produces a greenish-blue light.

In addition to producing a visible greenish-blue light, mercury vapor produces an invisible ultraviolet light. When ultraviolet light strikes certain chemicals, called *phosphors,* these chemicals glow with a color that depends upon their composition. This is the principle upon which the fluorescent lamps so widely used today operate.

A long glass tube is coated on its inner surface with one of these phosphors. Within the tube, mercury is heated until it vaporizes and forms a mercury vapor. Current is passed through this mercury vapor, producing ultraviolet light. As the ultraviolet light strikes the phosphor, visible light is produced by the phosphor glow.

C. *Chemical effect*

We touched upon the transformation of electrical energy into chemical energy when we discussed the storage cell (Chapter 2, subdivision E,2) and the breaking up by the electric current of the salt molecule into its component sodium and chlorine atoms (Chapter 3, subdivision B). This type of energy transformation is used in electroplating where the atoms of the metal to be used for plating are removed from a solution and deposited upon a negative electrode (which is the object to be plated). Similarly, aluminum is removed from its ore by an electric current. These are but a few examples of the chemical effects produced by the electric current.

D. *Magnetic effect*

1. MAGNETISM

Ever so often, ancient man dug up a heavy black stone that had the mysterious property of attracting small pieces of iron. Also, if a bar of iron or steel were stroked with one of these stones, the bar

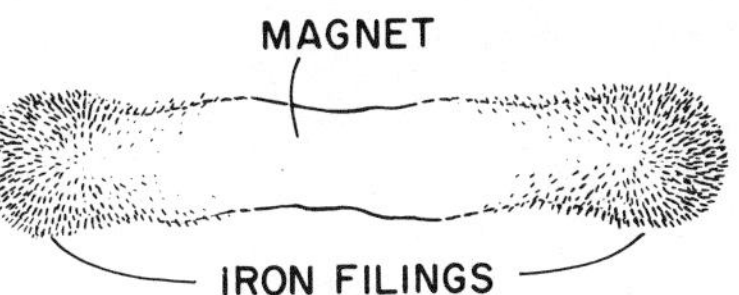

Fig. 4–4.

How iron filings cling to a magnet. Note concentration at both ends.

would acquire the same mysterious power of attracting other pieces of iron. Further, if the stone, or the iron bar that was stroked by it, were suspended so that it could swing freely, the stone or bar would turn until one end faced north and the other south. In addition, if the north-facing end of one such stone or bar were to be brought near the north-facing end of another, the two stones or bars would be repelled. Similarly, if two south-facing ends were brought together, they also would repel each other. On the other hand, if a north-facing end were brought near a south-facing end, they would attract each other.

No wonder the ancients thought that these stones, which we now know to be a type of iron ore, had magical power. A great many legends grew up about these stones, which were called *lodestones* or *magnets*. Of course, these legends, such as the ability of a magnet to cure disease, were false. Nevertheless, the ancient Chinese did invent an extremely useful navigation device, the *compass*, that made use of the north-facing property of the magnet.

Today we know that a magnet can attract, not only pieces of iron, but also certain other metals, such as nickel and cobalt, although with less force. We call substances that can be attracted by a magnet *magnetic*, and the ability of a magnet to attract magnetic substances we call *magnetism*.

If a magnet is sprinkled with iron filings, we notice that these filings are not attracted uniformly to the whole surface, but rather tend to cluster at either end of the magnet. (See Figure 4-4). It would seem that the magnetism is concentrated at these two ends of the magnet. We call these two ends of concentration the *poles* of the magnet. It has been found that the earth itself is a huge magnet with its two magnetic poles located in the Arctic region (near the north geographic pole) and in the Antarctic region (near the south geographic pole).

We have confirmed that if a magnet is suspended so that it can swing freely, one pole will face toward the earth's north magnetic pole and the other toward its south magnetic pole. We call the pole of the magnet that turns to the earth's northern magnetic pole the *north-seeking pole,* or, more simply, the *north pole.* We generally designate it by *N.* The pole that turns to the earth's southern magnetic pole is called the *south-seeking,* or *south pole.* It is designated by S.

Thus, every magnet has a north pole and a south pole. We have confirmed the fact that like poles (that is, two north poles or two south poles) tend to repel each other. On the other hand, unlike poles tend to attract. It is for this reason that the north pole of the compass always turns to the earth's north magnetic pole. (Thus the compass' north pole is, in reality, a south pole, though misnamed.)

a. The magnetic field

This matter of attraction and repulsion between poles of magnets warrants close attention. It was found that the poles need not touch each other. Even if they are a distance apart, like poles will repel each other and unlike poles will attract each other. If a nonmagnetic substance is placed between the poles, their attraction or repulsion is unchanged. Thus, if a sheet of glass or copper is placed between two unlike magnetic poles, they continue to attract each other as thought the glass or copper were not there.

We may understand this phenomenon a little more clearly if we consider the electrostatic, or electric, field existing between charged bodies (see Chapter 2, subdivision A). Here, under the influence of the field of force existing between them, bodies containing similar charges repel each other. If the bodies have unlike charges they attract each other.

It would appear, then, that a *magnetic force,* or *field,* exists between the two opposite poles of a magnet. We may find out more about this magnetic field by means of a simple experiment. Place a magnet on a wooden table. Over it, place a sheet of glass. Sprinkle iron filings on the glass and tap the glass lightly. The iron filings will assume a definite pattern on the glass sheet (see Figure 4-5).

The iron filings are attracted to the magnet through the glass sheet. Although the glass prevents these iron filings from touching the magnet, nevertheless the filings will form a pattern which will show the form of the magnetic field. Note that the iron filings arrange themselves outside the magnet in a series of closed loops that extend from pole to pole.

It is in this way that we visualize the magnetic field around a magnet. We must be careful to understand that the magnetic field does not exist in lines but is smooth and continuous throughout the space surrounding the magnet. Upon observing the clumping of filings as shown in Figure 4-5, an erroneous conclusion may be drawn that the magnetic field exists in the form of magnetic lines of force. Although this is not true, it is convenient to depict the presence of a magnetic field by sketching these lines and making them continuous from pole to pole and even continuing within the magnet. The crowding of the lines indicates the region of strongest magnetic field while the fact that they are continuous lines is in agreement with the mathematics that describes the magnetic field.

Note that the lines of force do not cross, but actually appear to repel each other. It would seem that these lines try to follow the shortest distance from pole to pole, at the same time repelling each other. It might help if we think of them as a bundle of stretched rubber bands. Thus they tend to shorten and, at the same time, push the others away sidewise.

A force must have a direction in which it acts. In the case of the electrostatic field we consider the force as acting from the point of negative charge to the point of positive charge. In the case of the magnetic field we arbitrarily assume the force acts from the north pole to the south pole. It is as if lines of force "flowed" from the north pole into the south pole. If, theoretically, we were to place

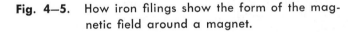

Fig. 4—5. How iron filings show the form of the magnetic field around a magnet.

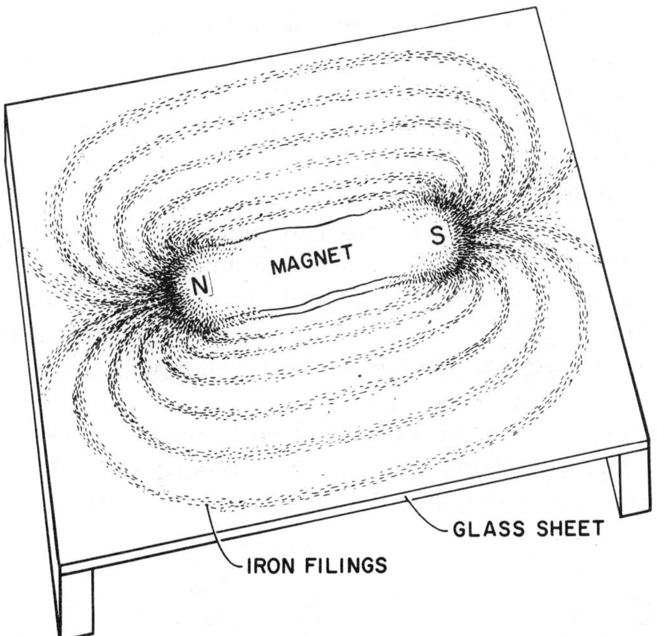

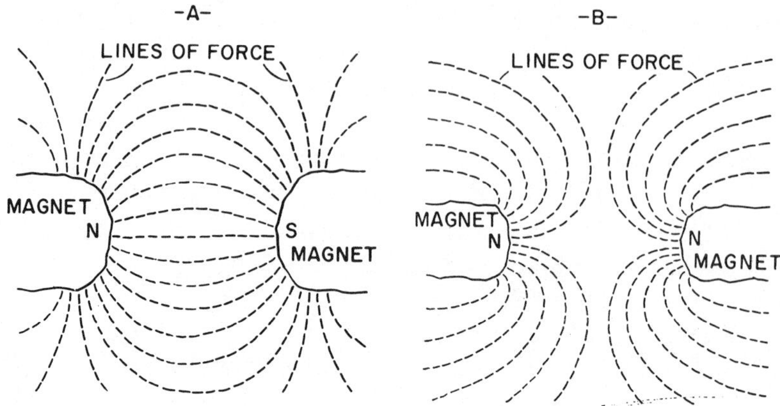

Fig. 4—6. **A.** Pattern of the resulting magnetic field when two unlike poles are placed near each other.
B. Pattern of the resulting magnetic field when two like poles are placed near each other.

a small north pole in the magnetic field, it would be repelled from the north pole of the magnet and move along the path of a line of force until it reached the south pole.

We can see now why like poles repel and unlike poles attract. If we place two unlike poles near each other, as in Figure 4-6A, the lines of force flow from the north pole to the south pole. Since these lines of force tend to shorten, the two magnets are pulled to each other. If, on the other hand, we place two like poles near each other, as in Figure 4-6B, the lines of force tend to repel each other and the two magnets are pushed apart.

The magnetic lines of force are known as the *magnetic flux* and are considered as flowing in a magnetic circuit, somewhat as current flows in an electrical circuit. Like the current, the flux flows in closed loops. This is illustrated in Figure 4-7 which shows the flux around a magnet. The lines of force flow out of the north pole, through the air, and back into the south pole of the magnet. Within the magnet, they flow back to the north pole, thus completing the loop. (In the previous figures the flow of flux within the magnets was omitted for the sake of simplicity.)

Continuing our analogy with the electric current, we find that the magnetic flux encounters less opposition flowing through magnetic substances, such as iron, than through nonmagnetic substances, such as air, glass, or copper. We call the opposition to the flow of magnetic flux, *reluctance*. The reluctances of all nonmagnetic substances are the same. Thus the flux will flow with the same ease (or difficulty) through air, glass, copper, etc.

On the other hand, it will flow much more readily through a piece of iron placed in its path. Thus, if a piece of iron is placed within the magnetic field, the lines of force will distort their pattern to take the easier path through the iron, as shown in Figure 4-8A. If, as in Figure 4-8B, an iron washer is placed within the magnetic field, the flux will flow through the iron of the washer and no lines of force will be found in the air space within the washer.

b. Theories of magnetism

It is not enough to smile at the fanciful legends of the ancients. We must arrive at an acceptable explanation of the mysterious powers of the magnet. Let us start by considering a pendulum, which consists of a weight suspended by a string from a fixed point. Normally, the weight hangs straight down, attracted toward the center of the earth by gravity. (See Figure 4-9.)

Force must be applied to raise the weight to position A. At this position the weight has acquired the ability to do work since, if released, it would fall back toward its original position. We say the weight has acquired *mechanical energy* and, since this is due to its position, we say that it has *position*, or *potential, energy*.

If the weight is released, it falls back toward its original position. As it does so, it gradually loses the advantage of its position and, accordingly, its potential energy. At the original position, the weight has lost all its potential energy. Nevertheless, it continues to swing upwards toward position B. Where does it get the energy for this rise?

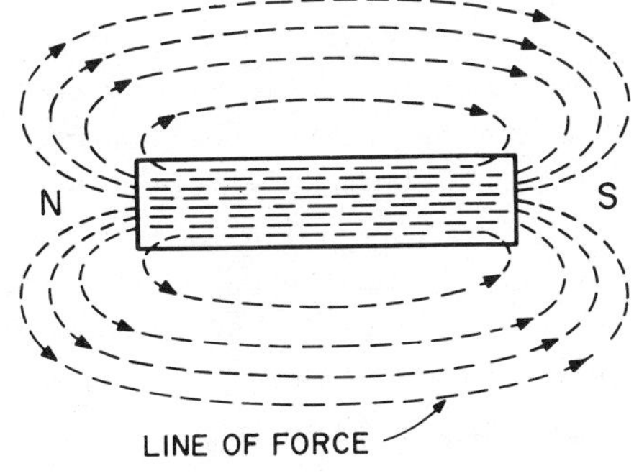

Fig. 4–7.

Magnetic flux around a magnet.

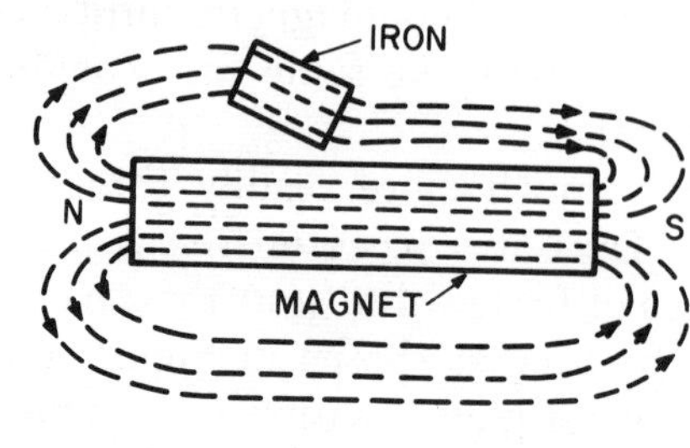

-B-

IRON WASHER

N

S

MAGNET

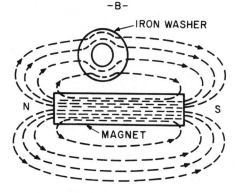

Fig. 4–8.

A. Distortion of the magnetic field around a magnet by a piece of iron.

B. Distortion of the magnetic field around a magnet by an iron washer. Note that there are no lines of force in the air space within the washer.

The answer is that, as the weight starts to fall from position A, its position, or potential, energy starts to transform into a *motion,* or *kinetic, energy.* The faster the weight moves, the more its potential energy is changed to kinetic energy. When the weight reaches its original position at the bottom of its swing it is traveling at its fastest and all its potential energy has been changed to kinetic energy. It is this latter energy that keeps the weight moving and raises it to position B.

As the weight passes through its original position and rises toward position B, it travels slower and slower, and finally comes to rest at position B. With the slowing up of the weight its kinetic energy gradually changes back to potential energy. At position B the weight is at rest and all the kinetic energy has been transformed to potential energy. Then the entire cycle is repeated.

We have seen that motion changes potential energy to kinetic energy. The greater the motion, the greater is the change. Further, this action is reversible. Reducing the motion changes the kinetic energy to potential energy. At rest, all the kinetic energy is transformed to potential energy.

A somewhat analogous situation exists between the electrostatic and magnetic fields. An electrostatic field, you will recall, exists around a charged body. If this charged body is made to move, some of the energy of this field is transformed into *magnetic energy*. The faster the charged body moves, the more the electrostatic energy is changed to magnetic energy.

A magnetic field always accompanies the motion of a charged body. Another way of saying this is that the motion of the electrostatic field produces a magnetic field. This action is reversible—the motion of a magnetic field produces an electrostatic field. (This matter will be discussed later when we consider *induced voltage*.)

Let us turn, now, to the structure of the atom. It consists, you will recall, of electrons revolving in concentric shells around a central nucleus. Since each moving electron is a charged particle, each electron, therefore, should be surrounded by a magnetic field. And each atom, accordingly, should have a magnetic field that is the resultant of all the magnetic fields around its electrons.

This theory may explain why certain substances exhibit magnetic properties, but since all substances are made up of atoms with moving electrons, why are not all substances magnetic? The answer lies in the fact that the direction of the magnetic field around a moving electron depends upon whether the electron is rotating clockwise or counterclockwise around the nucleus. All atoms contain electrons moving in both directions. The magnetic field around an electron moving in a clockwise direction will cancel out the field around an electron moving in a counterclockwise direction.

If half the electrons of an atom rotate in one direction and half in the other, the magnetic fields will cancel each other and there will be no resultant field around the atom. Hence it is nonmagnetic.

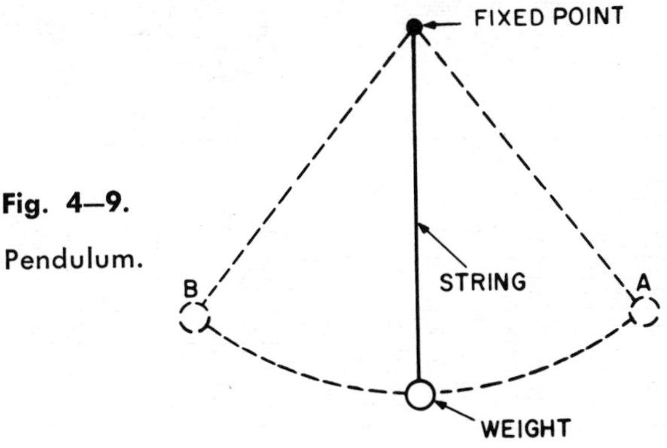

Fig. 4–9. Pendulum.

If, on the other hand, more electrons rotate in one direction than in the other, the atom will have a resultant field and will exhibit magnetic properties. It is for this reason that iron is magnetic.

The manner in which atoms are arranged in molecules also has a bearing upon the magnetic properties of these molecules. Because of this, certain alloys—such as *alnico,* an alloy of aluminum, nickel, cobalt, and iron—exhibit excellent magnetic properties. (However, we will not go further into the subject of molecular structure at this point.)

Another question now arises. How is it that a magnetic substance, such as a piece of iron, does not always act as a magnet? For example, you may find that an iron bar will not attract another piece of iron until the bar is converted into a magnet by stroking it with another magnet. What happens to the bar when it is so stroked?

We now know that the molecules of a magnetic substance are tiny magnets, each with a north and south pole and with a surrounding magnetic field. These molecules may be arranged in a disorderly fashion within the substance (see Figure 4-10A). The magnetic fields around the molecules then cancel each other out and, as a result, there is no external magnetic field. We say that, although the substance is magnetic, it is *unmagnetized.*

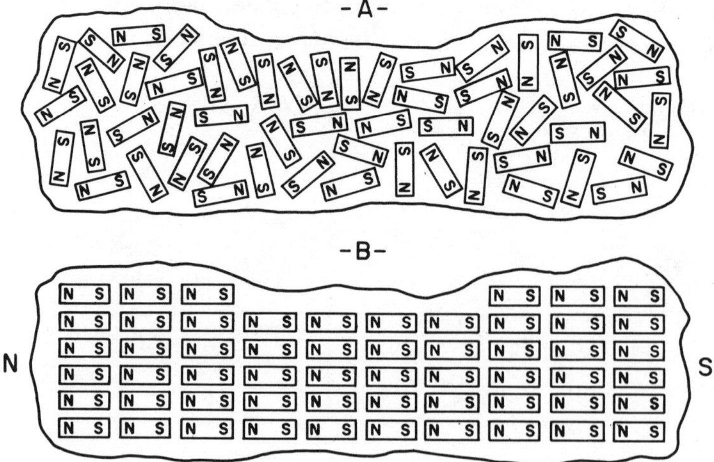

Fig. 4—10. Arrangements of molecules in a piece of magnetic material.
A. Unmagnetized.
B. Magnetized.

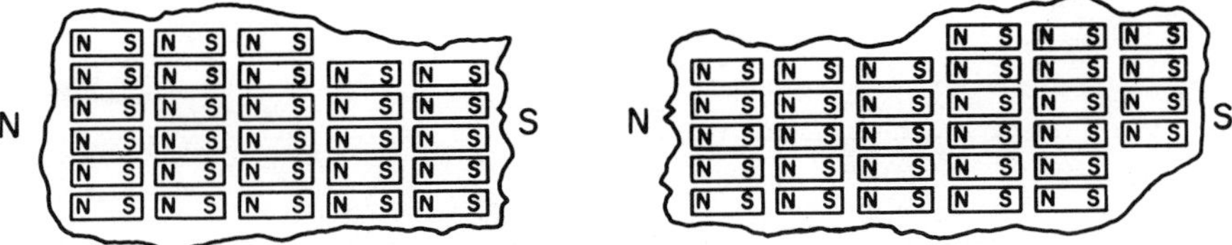

Fig. 4–11. Arrangement of molecules showing why two magnets are formed when a magnet is broken in two.

If the substance is magnetic, it is possible to line up these molecules in an orderly array, with the north pole of one molecule facing the south pole of another (see Figure 4-10B). You will notice that all the north poles are facing one way, and all the south poles are facing the opposite way. The result, then, is that we have a magnet with an external magnetic field and whose magnetism is concentrated at the two opposite poles. We say that the substance now is *magnetized.*

There is considerable evidence in favor of this theory. If you break a magnet in two, you obtain two magnets, each with a set of poles, as illustrated in Figure 4-11. Furthermore, you can destroy the magnetism of a magnet by any means that will disarrange the orderly array of the molecules, such as by heating or jarring the magnet.

On the other hand, you can make a magnet of any magnetic substance by placing it within the magnetic field of another magnet. For example, you may stroke the magnetic substance with a magnet. The attraction between the external magnet and the molecules of the magnetic substance causes these molecules to line up in the necessary orderly array (Figure 4-12). This also explains why a magnet will attract a piece of magnetic material in its field.

A nonmagnetic substance, such as glass or copper, resists all attempts to align its molecules in orderly fashion. Nor do all magnetic substances submit to this lining-up process to the same degree. In some substances, such as soft iron, the molecules are easily moved and will line up readily under the influence of the magnetic field of another magnet. However, once the external magnetic field is removed, the molecules of the soft iron revert to their original, disorderly condition. The soft iron forms a *temporary magnet* which is magnetized only so long as it is acted on by an external magnetic field.

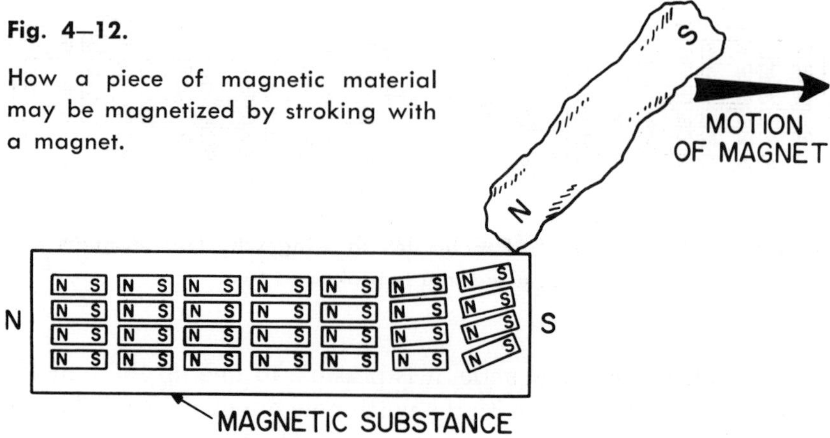

Fig. 4–12.

How a piece of magnetic material may be magnetized by stroking with a magnet.

On the other hand, the molecules of some substances, such as steel, require a much greater magnetic force to produce an orderly arrangement. However, when the external magnetic field is removed, these molecules will retain their positions, and consequently these substances form *permanent magnets.* As we have seen, however, heating or jarring the magnet will disarrange its molecules and thus destroy its magnetic properties.

Magnets may be made in a great variety of shapes. The earliest type, of course, was the natural one dug from the earth. Because the ore could be used as a compass to guide the mariner, it was called *lodestone* (lead stone). Common shapes of manufactured magnets are the *bar magnet,* which has a pole at each end, and the *horseshoe magnet,* which really is a bar magnet that has been bent into a U or horseshoe shape so that both poles are close together. Thus, practically all its magnetism is concentrated into a smaller space.

Magnets are also made in a great many other shapes, such as disks, rings, cylinders, and balls. The earth is a ball-shaped magnet. A peculiarity of the earth is that its magnetic poles are not fixed but vary somewhat from year to year. Furthermore, the earth's lines of force do not describe regular curves from pole to pole and, consequently, a compass does not point directly to the earth's magnetic poles at all places. In most locations, a slight deviation is noted. This deviation is believed to be due to masses of magnetic material within the earth, which attract the compass and deflect it from a true bearing on the magnetic pole. Mariners must take such deflections into consideration when plotting their courses.

An important use for the permanent magnet is the compass. In addition, the magnet is used in many toys and novelties. Industry uses the permanent magnet for a great many devices such as the magnetic tool holder, certain types of electrical measuring instruments, and the radio loudspeaker. Recently-developed alloys, such as *cobalt-steel* and *alnico,* make possible permanent magnets many times more powerful than those made from steel alone.

2. ELECTROMAGNETISM

In 1819, Hans Christian Oersted, a Danish physicist, brought a small compass near a wire that was carrying an electric current. He noticed that the compass was deflected. When he turned the current off, the compass assumed its original position. This discovery started a chain of events that has helped shape our industrial civilization.

Let us examine the significance of Oersted's discovery. The deflection of the compass while current was flowing through the wire indicated that it was being acted upon by an external magnetic field. Where did this magnetic field come from?

Not from the copper wire, which we know is nonmagnetic. Obviously, it could come only from the electric current flowing through the wire. The compass was deflected only when the current flowed through the conductor and continued to be deflected only so long as the current continued to flow. When the flow of current ceased, so did the deflection.

Fig. 4—13. Permanent magnets can be constructed in many different shapes.

General Electric Company

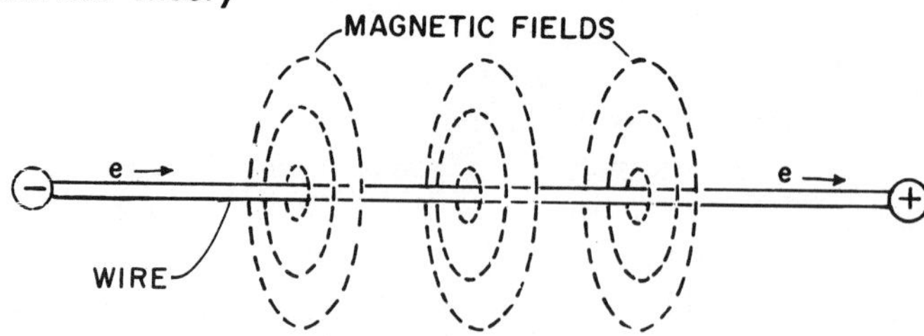

Fig. 4—14. Magnetic field formed around a conductor carrying a current. (For the sake of simplicity, the field is shown as a series of concentric circles instead of cylinders.)

We believe that a magnetic field always accompanies the motion of a charged particle. It is in this way that we explain the magnetic fields around the electrons revolving about the nucleus of the atom. There is no reason why a magnetic field should not surround an electron moving in a conductor. Since the flow of current consists of the movement of electrons, we should expect a magnetic field around a conductor through which a current is flowing. This Oersted found to be true.

As might be expected, the greater the current flow—that is, the greater the number of electrons flowing per second—the greater is the magnetic field. The magnetic field surrounds the conductor and is depicted by continuous lines surrounding the conductor. The lines are more crowded close to the conductor but are spaced farther apart away from the conductor. These indicate a magnetic field whose strength (flux density) is greatest close to the surface of the conductor and weakens as we move away. (See Figure 4-14.)

Further experimentation produced a simple method for determining the direction of this magnetic field. If the conductor is grasped in the left hand with the extended thumb pointing in the direction of the current flow, the fingers then circle the conductor in the direction of the magnetic lines of force. This is illustrated in Figure 4-15.

Suppose we bend the current-carrying conductor into a loop. The magnetic field then would appear as illustrated in Figure 4-16. Note

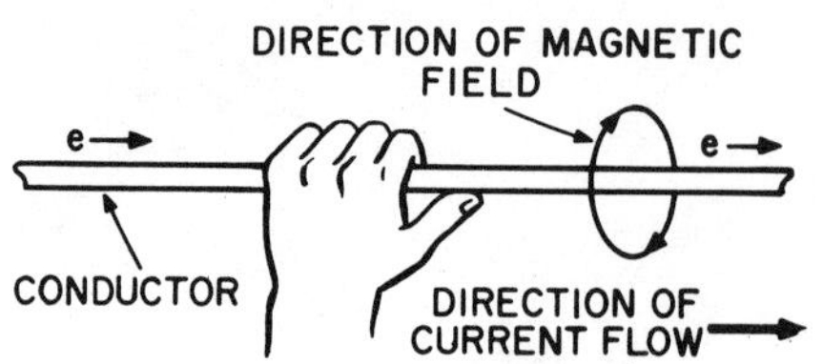

Fig. 4—15. Left-hand rule for finding the direction of the magnetic field around a conductor carrying a current.

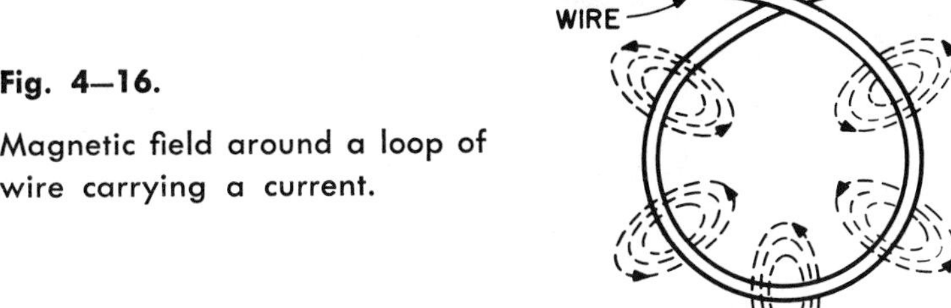

Fig. 4—16.

Magnetic field around a loop of wire carrying a current.

that the lines of force all pass through the enclosed portion of the loop. If we add more loops, each loop adds its magnetic field, thus producing a greater overall magnetic effect. The resulting magnetic field would appear as shown in Figure 4-17.

Note that the coil becomes a temporary magnet (that is, it is a magnet only while current flows through it) with a set of north and south poles. The greater the number of turns, the stronger the magnetic field will be. Also, the greater the current flowing through it, the stronger the field will be. Accordingly, we say that the strength of the magnetic field depends upon the *ampere-turns* of the coil. (The ampere-turns are designated by the symbol NI, where N stands for the number of turns and I for the current flowing through these turns.)

The polarity of the magnet formed by the coil may be determined by grasping it in the left hand so that the fingers follow around the coil in the direction in which the electrons are flowing. The extended thumb then will point toward the north pole. (See Figure 4-18.)

We may increase the strength of the magnetic field by winding the coil on a core of magnetic material. Then the magnetism of the core is added to that of the coil. Since we generally desire the coil to

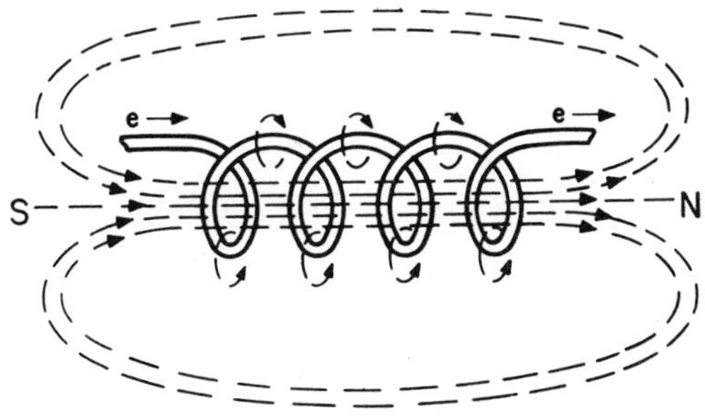

Fig. 4—17.

Magnetic field around a coil of wire through which a current is flowing.

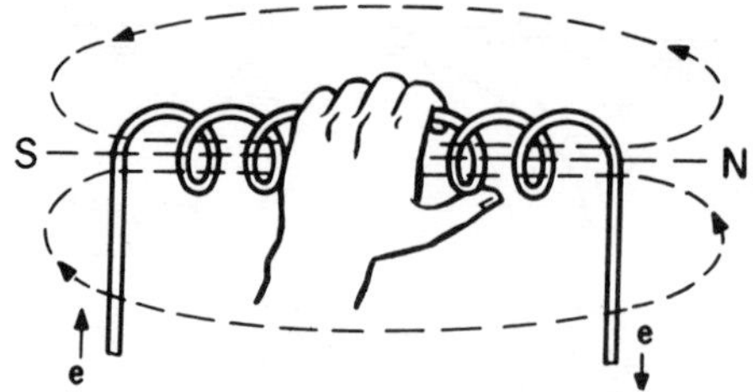

Fig. 4–18.

Left-hand rule for finding the polarity of a coil through which a current is flowing.

act as a temporary magnet, this core usually is made of some material, such as soft iron, which becomes a magnet only when under the influence of the magnetic field of the coil. The coil with its core is called an *electromagnet.* Without its core the coil usually is called a *solenoid,* or *helix.*

The main factors that determine the magnetic strength of an electromagnet are the number of turns of the coil, the current flowing through it, and the material and size of its core. Other determining factors are the shape and size of the windings and the mechanical arrangement of the core. There are a great many practical applications of the electromagnet. Some of these will be discussed later in this book.

There is another magnetic effect produced as current flows through a conductor. Since the conductor is surrounded by a magnetic field, if two current-carrying conductors are placed near each other the two magnetic fields will react with each other. Thus the conductors will be brought closer together or pushed apart, depending upon the directions of the two fields. This principle is employed in a number of applications, such as the electric motor.

3. MAGNETIC UNITS

We may best understand the magnetic circuit, perhaps, by comparing it to the electrical circuit. In the latter, current is made to flow through a closed loop, or circuit, because of an electrical pressure (electromotive force, or voltage). In flowing through this circuit, the current must overcome opposition or resistance.

The magnetic circuit, too, is a closed one. Corresponding to current is the *magnetic flux* which is designated by the Greek letter ϕ (phi). There are several systems in use. In the *English* system the unit of flux is the *line of force.* For most scientific work the *metric,* or *cgs* (centimeter, gram, second) system is employed. In this system the unit of flux (line of force) is called the *maxwell.*

Of greater importance to the engineer who designs magnetic circuits is the *flux density* (designated by B), which is a measure of the flux units per unit area taken at right angles to the direction of flux. In the English system its unit is the *number of lines of force per square inch.* In the metric, or cgs, system the unit of flux density is the *gauss,* which means *maxwells per square centimeter.*

Corresponding to the electromotive force of the electrical system, the *magnetomotive force* (designated by F) is the force that tends to produce the magnetic field. This force may result from a magnetized body (magnet), or it may be produced by a current flowing through a coil of wire. In the latter case, you know that this force is dependent upon the ampere-turns.

The unit of magnetomotive force in the English system is the *ampere-turn.* In the cgs system its unit is the *gilbert* and it may be derived from the following formula:

$$F = \mathrm{K} \times NI$$

where F is the magnetomotive force (in gilberts), N is the number of turns of the coil, and I is the current (in amperes) flowing through it. The letter K stands for a constant which is equal to 0.4π. (The Greek letter π, pronounced *pi,* stands for 3.14.)

Corresponding to resistance in an electrical circuit is the *reluctance* (designated by $\mathcal{R}$) of the magnetic circuit. You will recall that in an electrical circuit the resistance depends upon the length, cross-sectional area, and material of the path. The same holds true for the reluctance in a magnetic circuit. The longer the path, the greater is the reluctance; and the greater the cross-sectional area, the less is the reluctance.

We may set up a sort of Ohm's law for magnetic circuits. In electrical circuits

$$I = \frac{E}{R}.$$

In the magnetic circuit

$$\phi = \frac{F}{\mathcal{R}}$$

or, if we are using cgs units,

$$\phi = \frac{\mathrm{K} \times NI}{\mathcal{R}}.$$

The reluctance of a magnetic circuit is seldom considered. For practical purposes it is more useful to employ the *permeability,* designated by the Greek letter μ (pronounced *mew*). If reluctance is a measure of the opposition to the flux in a magnetic circuit, permeability is the ease with which the flux will flow. Thus, permeability is the reciprocal of reluctance, or

$$\mu = \frac{1}{\mathcal{R}} \quad \text{and} \quad \phi = \mu \times F.$$

Actually, the permeability of any material is a measure of the ease with which the molecules of which it is composed may be lined up under the action of a magnetic field. Since the molecules of all nonmagnetic materials cannot be lined up, their permeabilities are practically the same. The permeability of all such materials is taken to be unity, that is, 1. When considering a magnetic substance, its permeability is a measure of the number of times easier it is to line up its molecules as those of some nonmagnetic substance, such as air. The permeability of certain magnetic substances may run over 25,000.

As a magnetizing force is applied to a magnetic material, the greater the magnetizing force, the greater will be the degree of magnetization. But only up to a point. In Figure 4-19 you see the magnetization curve for a sheet of silicon steel. As the magnetizing force (in units of ampere-turns per inch) is increased, the magnetization (in units of lines of force per square inch) increases rapidly. When a certain point (the *saturation point*) is reached, however, the steel acquires practically all the magnetization of which it is capable. Increasing the magnetizing force beyond this point results in very little more magnetization.

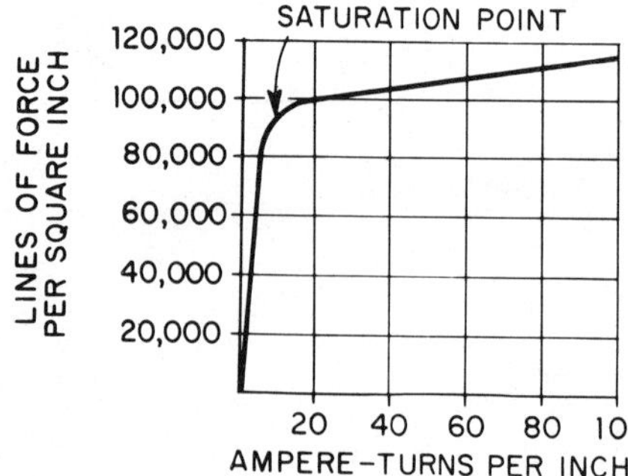

Fig. 4—19.

Magnetization curve for sheet of silicon steel.

E. *Measuring Instruments*

It is extremely important to have instruments by means of which we may measure directly the quantity of certain factors, such as current, voltage, and resistance, that may be present in various portions of the electrical circuit. These instruments are called *meters,* and they operate by measuring the various effects produced by the electric current.

Generally, the meter consists of a *movement* which translates current flowing through it into a displacement of a *pointer,* which indicates this displacement on a *scale.* The whole is enclosed in a *case* for protection. Since the displacement is proportional to the current causing it, the excursion of the pointer over the scale indicates this amount of current. Depending upon the circuit in which the meter is connected, the scale may be calibrated in amperes, volts, ohms, and so forth.

1. THE d'ARSONVAL MOVEMENT

As indicated, there are meters based upon all the various effects of the electric current. Most common, however, is a movement that depends upon the magnetic effect of the current and was invented by a French physicist, Arsene d'Arsonval.

In essence, this instrument consists of an electromagnet pivoted between the two opposite poles of a permanent horseshoe magnet (see Figure 4-20). As current flows through the turns of the electromagnet, the latter becomes a magnet. If the current flows as indicated in Figure 4-20A, the left-hand side of the electromagnet becomes a north pole, and the other side a south pole. Since like poles are facing each other, they repel. However, the permanent magnet is fixed and cannot move. Accordingly, the electromagnet rotates around its pivot, as indicated in Figure 4-20B.

The amount of repulsion between the like poles depends upon the relative strengths of the magnetic field around the permanent magnet and that around the electromagnet. Since the magnetic field around the permanent magnet is a fixed value, the stronger the magnetic field around the electromagnet, the greater will be the repulsion and rotating effect. But the strength of the magnetic field around the electromagnet will depend upon the current flowing through it. Thus, the amount of repulsion and the rotating effect

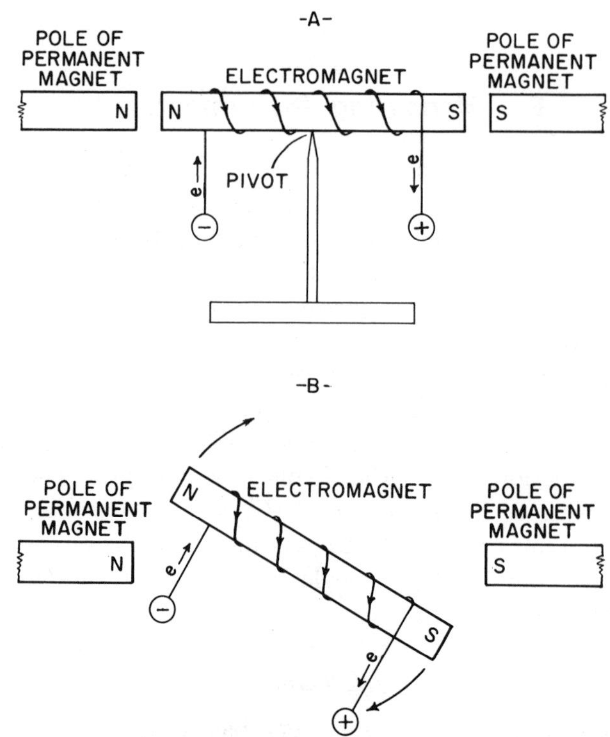

Fig. 4—20. Diagram illustrating the d'Arsonval principle.
A. The electromagnet is pivoted to rotate between two fixed poles of a magnet.
B. Looking down on the top of the electromagnet.

will depend upon the strength of the electric current flowing through the coil of the electromagnet.

Here, then, is a convenient way to measure the current flowing through the coil of the electromagnet. All we need do is fix a pointer to this electromagnet and measure the amount of rotation on a suitable scale. Edward Weston, an American scientist, developed a practical instrument using this d'Arsonval principle. (See Figure 4-21.)

The electromagnet, called the *armature,* consists of a coil of very fine wire wound on a soft-iron cylinder as a core. The complete armature is delicately pivoted upon jewel bearings and is mounted between the poles of a permanent horseshoe magnet. Attached to these poles are two soft-iron *pole pieces* which concentrate the magnetic field.

As current flows through the armature coil, a magnetic field is set up around it in such a way that it opposes the field of the permanent magnet. As a result, the armature is rotated in a clockwise direction, carrying the pointer which is attached to it. The *spring* opposes the rotation and brings the pointer back to the no-current position when the current ceases. The greater the current, the greater will be the deflection of the pointer. This deflection is indicated on the scale over which the pointer passes.

Note that rotation will take place only when the poles of the permanent magnet face like poles on the armature. Thus the current must flow through the coil of the armature in such a way that this condition prevails. If the current flows in the opposite direction, unlike poles will face and attract each other, and there will be no rotation. Accordingly, the terminals of the instrument are suitably marked *plus* (+) or *minus* (−). If the minus terminal is connected to the high-potential (negative) side of the circuit under test and the plus terminal to the low-potential (positive) side, current will flow through the instrument in the proper direction.

This instrument is called, appropriately, a *moving-coil movement.* Such instruments can be constructed to be highly accurate and extremely sensitive. Its efficiency, too, is high as it requires very little power for its operation. Another advantage is its uniform scale since the deflection of the pointer is directly proportional to the amount of current flowing through the coil of the armature.

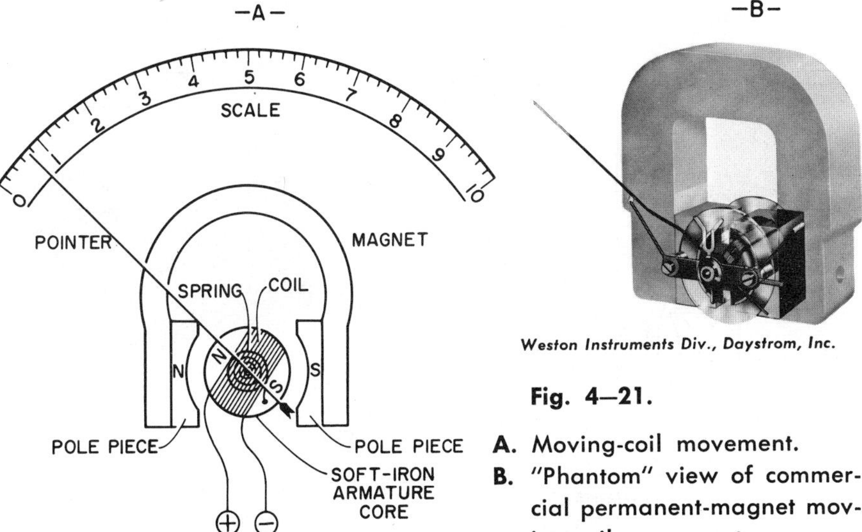

Weston Instruments Div., Daystrom, Inc.

Fig. 4—21.

A. Moving-coil movement.

B. "Phantom" view of commercial permanent-magnet moving-coil movement.

2. TYPES OF METERS

There are a great many types of electrical meters. Most of them employ movements that are mere variations of the one we have discussed. Since it is impossible to take up all types of electrical instruments in this book, we shall cover only those that are most common.

a. The galvanometer

The *galvanometer* is an instrument used to indicate the presence, strength, and direction of very small currents in a circuit. For this purpose the Weston-type movement is well suited. It has the required sensitivity and, you will recall, its pointer will be deflected only when current flows through its moving coil from its minus terminal to its plus terminal. Thus we can tell, by the manner in which the pointer is deflected, the direction of the current.

Actually, should the instrument be connected so that current flows through it from the plus terminal to the minus one, the pointer will be deflected backwards (to the left), as an examination of Figure 4-21 will show. This may bend the pointer and otherwise damage the instrument. Accordingly, this practice should be avoided.

However, it is possible to modify the instrument so that this danger is avoided. The coil and pointer are so mounted that, with no current flowing through the instrument, the pointer comes to rest (zero)

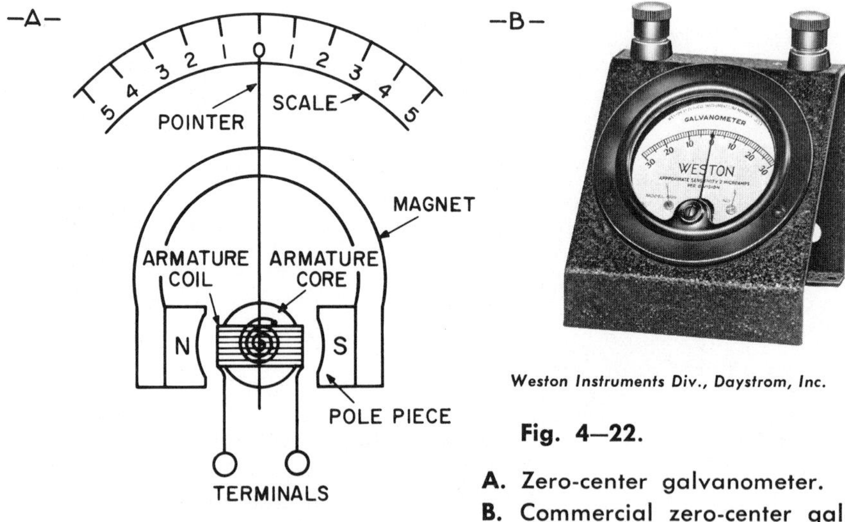

Weston Instruments Div., Daystrom, Inc.

Fig. 4—22.

A. Zero-center galvanometer.

B. Commercial zero-center galvanometer.

at the center of the scale (see Figure 4-22). Then a deflection of the pointer to the right indicates a current flow from the right-hand terminal to the left-hand one. A deflection of the pointer to the left indicates a flow in the opposite direction. Such an instrument is known as a *zero-center* galvanometer.

The numbers on the scale generally are arbitrary ones and merely indicate the *relative* strength of the current. Thus, if the pointer is deflected two divisions, the current is twice as great as one that deflects the pointer one division. Some galvanometers are further calibrated so that the deflection of the pointer may be translated into an indication of the *absolute* strength of the current. For example, a deflection from zero to the end of the scale of a certain galvanometer may indicate a current strength of 0.0005 ampere (0.5 milliampere or 500 microamperes). Generally, however, galvanometers are used where only relative values of current are desired.

The moving coil must be light and, hence, is wound with very fine wire. Accordingly, care must be taken that only very small currents flow through it lest the wires burn up. The coil generally is wound with wire that can safely carry a maximum current of about 0.03 ampere (30 milliamperes).

The symbol for the galvanometer, when used in an electrical diagram, is -(G)-.

b. The ammeter

The *ammeter* is used to measure the flow of current through a conductor somewhat as a flowmeter is used to measure the flow of water through a pipe. In both cases the circuit is broken and the meter is inserted in the break so that all the water (for the water circuit) or all the current (for the electrical circuit) flows through the meter. (See Figure 4-23.) In both cases the meter is inserted in *series* with the circuit under test.

If the current is small enough so that it does not burn up the coil, a meter similar to the moving-coil galvanometer previously described may be used. The scale is calibrated to read milliamperes or microamperes, depending on the magnitude of the current. Generally, zero-center scales are not employed. Instead, a scale with the zero position at the extreme left is used. Values increase the further the pointer is deflected to the right. Such an ammeter can be used up to about a 30-milliampere full-scale deflection.

As in the case of the flowmeter, which must be connected so that the water flows through it only in one direction, so the moving-coil

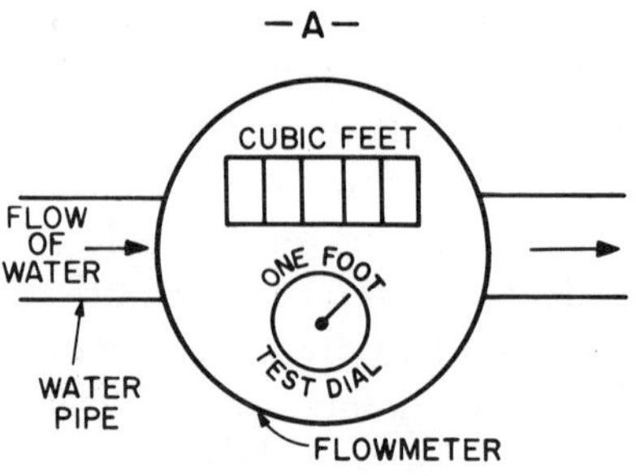

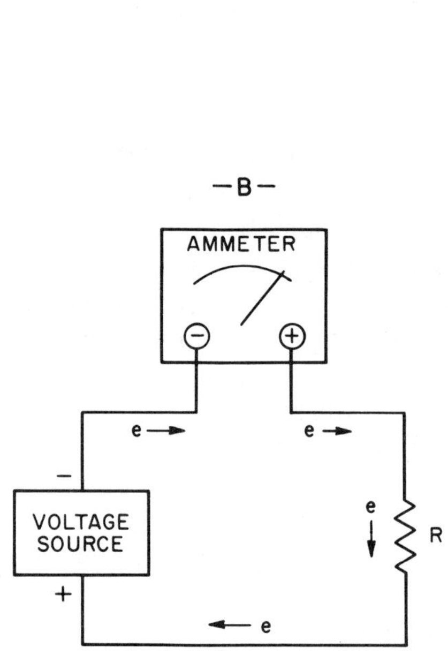

Fig. 4—23.

A. How the flowmeter is connected in the water circuit.
B. How the ammeter is connected in the electrical circuit.

ammeter must be connected so that the current flows through it only in one direction. Thus the minus (—) terminal must be connected to the high-potential (negative) side of the circuit and the plus (+) terminal must be connected to the low-potentiaı (positive) side.

Where the current to be measured is greater than that which the moving coil can safely pass, a device called a *shunt* is used (see Figure 4-24). This shunt consists of a metal wire or ribbon, usually of some alloy such as *manganin,* which is placed in parallel with the moving coil. Thus the current in the circuit divides (see Figure 4-24A), the greater portion flowing through the coil or shunt, depending upon which has the lower resistance. If the shunt has the same resistance as the coil, half the current will flow through the shunt and half through the coil. If the shunt has half the resistance of the coil, two-thirds of the current will flow through the shunt and one-third through the coil. The scale of the meter, of course, must be calibrated accordingly.

By choosing the proper resistance ratio between the coil and the shunt, a meter can be used to measure any quantity of current. Assume, for example, that we have a meter whose maximum range is 0.001 ampere and that the resistance of the moving coil is 50 ohms. Suppose we wish to use this meter to measure currents up to 0.1 ampere. We must arrange for a shunt that will be able to carry 0.099 ampere and permit 0.001 ampere to flow through the coil. Hence the resistance of the shunt must be $\frac{1}{99}$ that of the coil. The resistance of this shunt then will be $\frac{50}{99}$, or 0.5 ohm (approximately). The scale of the meter must be recalibrated to indicate 0.1 ampere for full-scale deflection.

Where ammeters are used to measure currents in the order of milliamperes, they are usually called *milliammeters.* Where the full-scale deflection is less than one milliampere, the meter usually is called a *microammeter.* Moving-coil ammeters may be obtained in ranges of from about 10 microamperes (0.00001 ampere) to thousands of amperes. Where shunts are used, they generally are enclosed in the meter case for ammeters up to about 30 amperes. Beyond that, the shunts usually are external to permit adequate heat dissipation. Alloys such as manganin are used for shunts because their resistances vary little with the temperature changes caused by the passage of current.

The symbol for the ammeter, when used in an electrical diagram is -(A)-. If the instrument be a milliammeter, its symbol is -(MA)-. If it be a microammeter, its symbol is -(μA)-.

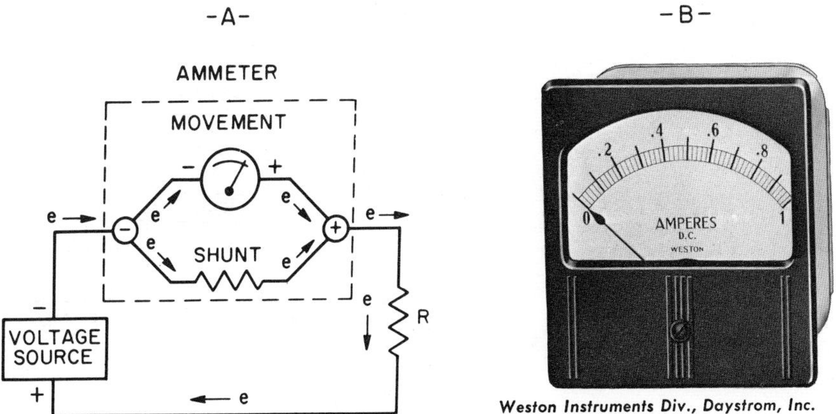

Weston Instruments Div., Daystrom, Inc.

Fig. 4–24. **A.** How the shunt is connected in the ammeter circuit. **B.** Commercial ammeter.

c. The voltmeter

The *voltmeter* is used to measure the difference of potential (electrical pressure or voltage drop) between two points in a circuit somewhat as the pressure gage is used to measure the water pressure in a pipe (see Figure 4-25). In Figure 4-25B the voltmeter is used to measure the voltage drop across resistor R. In both cases the measuring instrument is connected in *parallel* with the circuit under test.

Note that in the electrical circuit the meter actually measures the *current* flowing through it. But since this current depends upon the voltage drop across R, we may calibrate the scale in volts rather than amperes. Since the moving-coil movement will operate only if the current flows through it in the proper direction, care must be taken to connect its minus terminal to the negative side of the resistor and the plus terminal to the positive side.

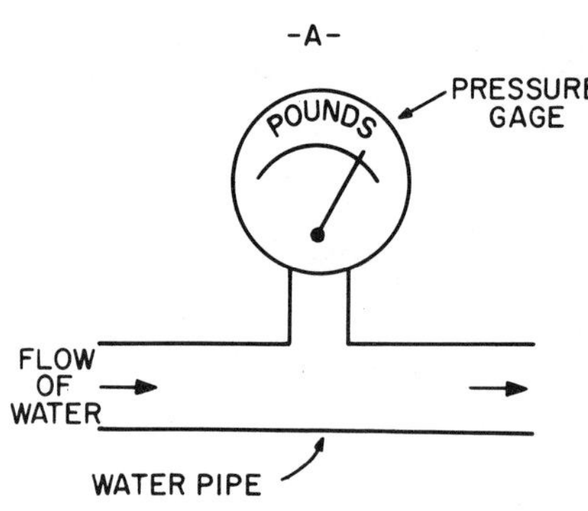

Fig. 4–25.

A. How the pressure gage is connected in the water circuit.

B. How the voltmeter is connected in the electrical circuit.

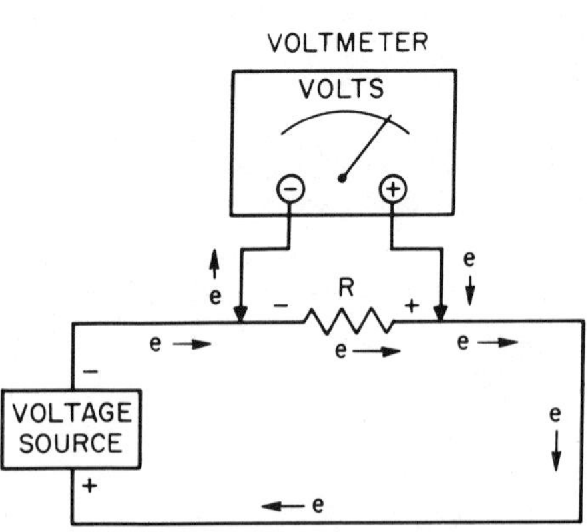

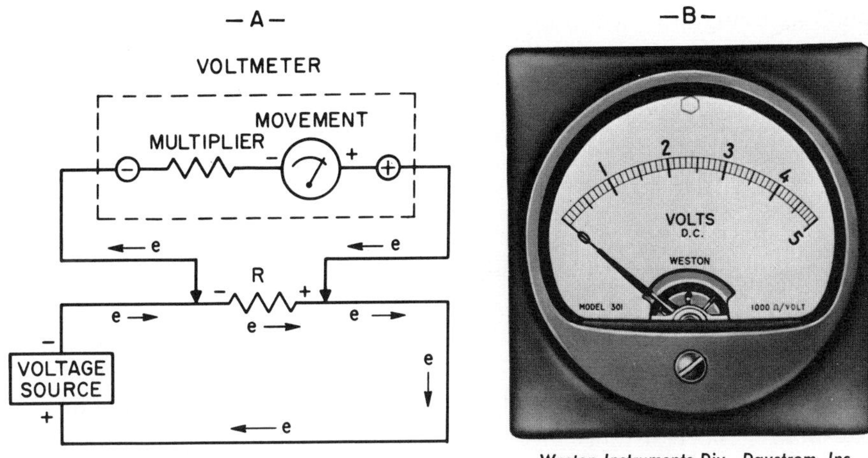

Weston Instruments Div., Daystrom, Inc.

Fig. 4—26. **A.** How the multiplier is connected in the voltmeter circuit. **B.** Commercial voltmeter.

The resistance of the meter must be large for two reasons. First, since the current flowing through the parallel circuit divides inversely according to the ratio between the resistance of the meter and R, the meter resistance must be large so that very little current will flow through the instrument. Otherwise it will take too much current from the circuit and thus give a false reading of the voltage drop across R. Second, the meter resistance must be large to prevent too much current from flowing through its movement and burning up its movable coil. Since the resistance of the moving coil is low, large resistors, called *multipliers,* are connected in series with it. Thus, the voltmeter consists of the movement and the series multiplier (generally enclosed in the same case.) (See Figure 4-26.)

Assume, for example, that we wish to convert the 1-milliampere meter whose coil has a resistance of 50 ohms to a voltmeter whose full-scale deflection would indicate 100 volts. This means that with 100 volts across the voltmeter, one milliampere (0.001 ampere) of current must flow through it. From Ohm's law we get $R = E/I$, or $R = 100/0.001$. Thus the total resistance of the voltmeter must be 100/0.001 or 100,000 ohms. Since the resistance of the coil is 50 ohms, the resistance of the series multiplier must be 99,950 ohms. (Generally, the resistance of the coil is neglected and a series multiplier of 100,000 ohms is employed.)

The multipliers usually are constructed of high-resistance wire, such as manganin, wound on wooden spools. Where the resistance is so high as to make the windings prohibitively large and costly, composition resistors are employed. Some of the less expensive types of voltmeters use composition resistors throughout.

Frequently the total resistance of the meter and its multiplier is divided by the full-scale voltage and the answer is listed as ohms per volt. This is constant regardless of the size of the multiplier. For instance, in the example just given, 100,000 ohms was the total resistance and the full-scale voltage was 100 volts. The result is 1,000 ohms per volt. This number is useful as an indicator of the total resistance of the voltmeter circuit when various multipliers are used. When measurements of circuits with high internal resistance are made, the resistance of the voltmeter circuit may substantially alter the circuit and therefore erroneous readings may result.

The symbol for the voltmeter when used in an electrical diagram is -(V)-. If the instrument be a millivoltmeter its symbol is -(MV)-.

d. The ohmmeter

The measurement of resistance is based upon current flowing through the circuit under test and on the voltage drop across the circuit produced by that current. If the current and voltage drop are known, the resistance can be calculated from Ohm's law ($R = E/I$). The basic circuit for this resistance measurement is shown in Figure 4-27.

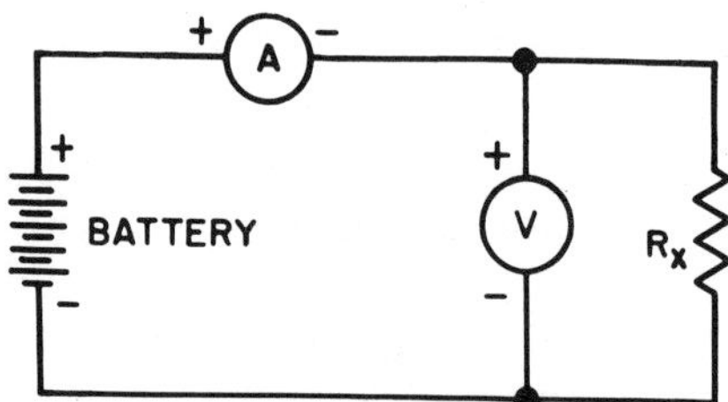

Fig. 4–27. Basic circuit for the resistance test.

Assume that a voltage source, such as a battery of 100 volts, is connected in series with an ammeter (A) and an unknown resistor (R_x) whose resistance we wish to measure. The voltmeter (V) will indicate the voltage drop across R_x. Since we will neglect here the slight voltage drop across the ammeter and connecting wires and the slight amount of current flowing through the voltmeter, the reading

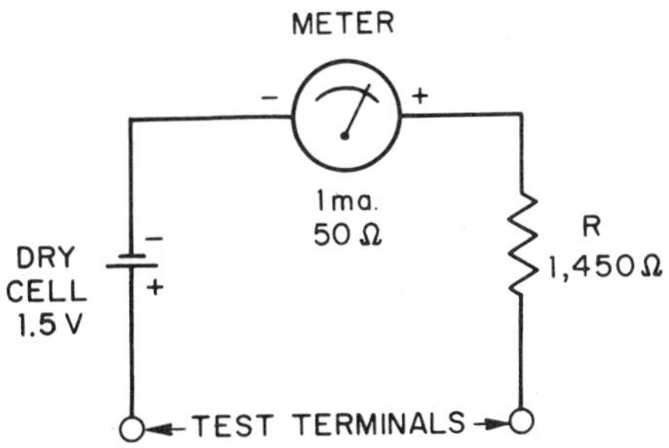

Fig. 4—28.

Typical ohmmeter circuit.

on the voltmeter will be 100 volts, the voltage of the battery. If the ammeter reads, say, 2 amperes, then, since $R = E/I$, $R_x = 100/2 =$ 50 ohms.

You can readily see that if the voltage remains constant, we may remove the voltmeter and can calibrate the ammeter to read resistance directly. The greater the resistance, the less the current flowing in the circuit and the less will be the reading on the ammeter. We call the recalibrated ammeter with its voltage supply an *ohmmeter*.

A typical ohmmeter circuit is shown in Figure 4-28. A 1-milliampere, 50-ohm milliammeter is connected in series with a fixed resistor (R) of 1,450 ohms and a 1.5-volt dry cell. When the test terminals are connected together the circuit is completed and current flows. Since the voltage is 1.5 volts and the total resistance (R plus the resistance of the meter) is 1,500 ohms, then, by Ohm's law ($I = E/R$), one milliampere of current flows through the circuit. Thus the meter will show full-scale deflection.

If, now, any additional resistance is introduced into the circuit, less current will flow and the deflection no longer will be full-scale. The greater this new resistance, the less will be the reading on the meter. Thus, the meter can be calibrated directly in terms of the amount of resistance introduced.

With no resistance introduced (the test terminals connected together) one milliampere of current flows and the deflection is full-scale. If a resistance equal to that of the circuit (1,500 ohms) is connected between the test terminals, the total resistance of the circuit is 3,000 ohms. Hence only one-half milliampere of current flows and the deflection is half-scale. If 3,000 ohms are introduced, only one-third milliampere of current flows and the deflection is only one-third of full-scale. And so forth.

Note that the scale runs backward; zero resistance produces full-scale deflection and the greater the resistance under test (between the test terminals), the smaller the deflection. Note, too, that the scale is not uniform. A resistance equal to that of the ohmmeter circuit produces half-scale deflection. A resistance equal to twice that of the ohmmeter circuit produces one-third of full-scale deflection; and so on. Hence the scale crowds together at the left-hand side.

QUESTIONS

1. **Name four effects produced by the electric current. Give one example of each.**
2. **Draw a diagram showing the arrangement of the magnetic lines of force around a bar magnet.**
3. **State and explain the law of magnetic attraction and repulsion.**
4. **In terms of the electron theory, explain why some substances are magnetic and others nonmagnetic.**
5. **Explain the difference between a magnetized and an unmagnetized piece of iron.**
6. **a) Explain how a magnet can attract an unmagnetized piece of iron.**
   b) Why will it not attract a piece of wood?
7. **What is the difference between a temporary and permanent magnet?**
8. **What is *electromagnetism?* Explain.**
9. **Explain the left-hand rule for determining the direction of the magnetic field around a current-carrying conductor.**
10. **Explain the left-hand rule for determining the magnetic poles of an electromagnet.**
11. **Name three factors that determine the strength of an electromagnet.**
12. **Name the magnetic factors that correspond to electromotive force, current, and resistance of an electric current.**
13. **Explain what is meant by the *permeability* of a magnetic substance.**
14. **By means of a simple diagram, explain the d'Arsonval moving-coil meter movement.**
15. **Explain how a d'Arsonval-type galvanometer may be converted into**
    a) an ammeter;
    b) a voltmeter.
    In each case, tell how you would connect the instrument in the circuit.
16. **Explain how you would find the resistance of a resistor using an ammeter and voltmeter.**

Alternating Current

A. *Induced current*

Oersted discovered that an electric current flowing through a conductor can create a magnetic field. Can a magnetic field create an electric current in a conductor? In 1831, Michael Faraday, the famous English scientist, discovered that this could be done. Strangely enough, at about the same time, Joseph Henry, an American scientist and teacher, working independently, discovered the same thing.

Let us try an experiment to illustrate this point. Connect the terminals of a zero-center galvanometer to a coil of about 50 turns of wire wound in the shape of a cylinder of about two inches in diameter. Now plunge the north end of a permanent magnet into the center of the coil (Figure 5-1A). You will observe that the pointer is deflected to the right, showing that an electric current was set flowing for a moment in the coil and galvanometer.

When the magnet comes to rest inside the coil (Figure 5-1B), the pointer swings back to zero, showing that the current has ceased flowing. Now remove the magnet from the coil (Figure 5-1C). As you do so, the pointer swings to the left, showing that once more an electric current is set flowing, but this time in the opposite direction. The same effect may be obtained if the magnet is held stationary and the coil moved.

How can we explain this? The magnetic field created by the permanent magnet has to be pictured again by the handy concepts of lines of force. These lines emerge from one pole packed closely together (picturing a strong field); they spread out through space and then re-enter the other pole packed closely together. Remembering that the lines are continuous, it is necessary that the concentrated magnetic field extend through the body of the permanent magnet. This is in line with the theory we discussed earlier, of the nature of the magnetic field within a magnetic material.

As the permanent magnet enters the coil, each turn successively surrounds the full magnetic field within the permanent magnet. A voltage is developed (induced) in series with each turn. This voltage depends upon the rate of change of the magnetic field, or in other words, on how quickly the magnetic field is changing. If we had all the turns tightly packed together so that the magnetic field changed in exactly the same manner through each turn at the same time, then the induced voltage in series with each turn would be equal. By the rules for series circuits, all of these voltages would add

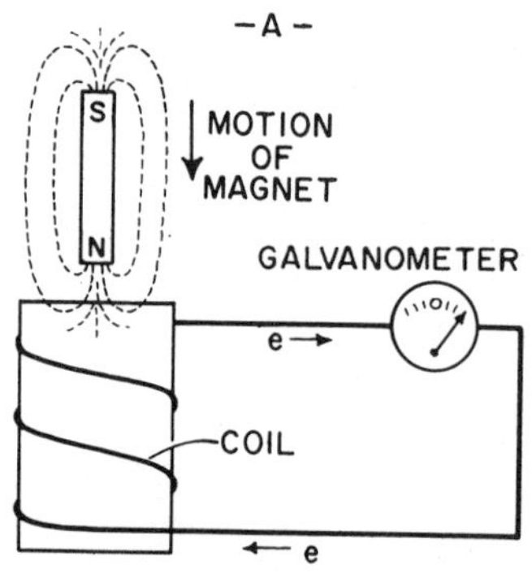

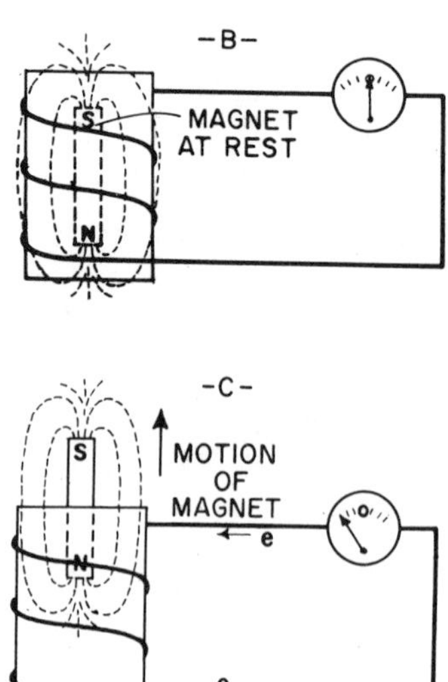

Fig. 5–1.

Demonstration of how a current is induced in a conductor as it cuts across a magnetic field.

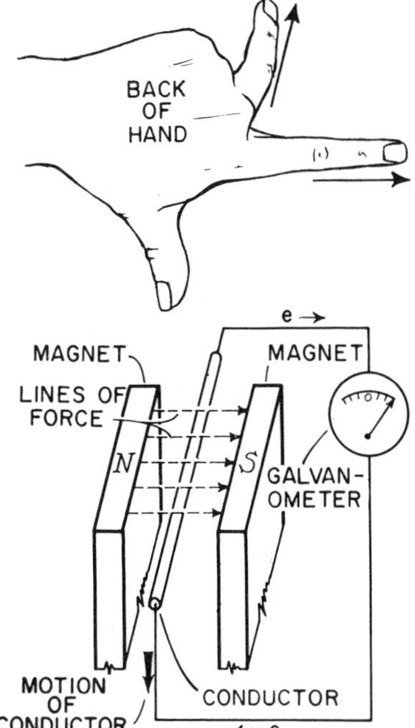

Fig. 5—2.

Left-hand rule for determining the direction of induced current flowing in a conductor.

together into one total induced voltage so that if the total induced voltage is called E and the number of turns is N, then $E = N \times$ the rate of change of the magnetic field.

Please note that no matter how strong the field, if there is no relative motion there is no rate of change and therefore no induced voltage. The idea of rate of change not only needs an amount of magnetic field change per second which determines the amount of induced voltage per turn but when the field is increasing, the rate of change is positive creating a plus voltage, and when the field is decreasing, it is a negative quantity which creates a minus voltage. If the coil is connected to an external circuit, say a resistor, the induced voltage will cause a current to flow. We say this is the induced current.

Experimentation has evolved a rule to determine in which way an induced current will flow. Examine Figure 5-2, where a conductor is moving across a magnetic field set up between two poles of a horseshoe magnet.

Assume that the conductor is moving down between the poles of the magnet. Extend the thumb, the forefinger, and the middle finger of the left hand so that they are at right angles to one another. Let the thumb point in the direction in which the conductor is moving (down). Now let the forefinger point in the direction of the magnetic lines of force (you will recall that we assume they go from the north to the south pole). The middle finger then will indicate the direction in which electrons will be set flowing by the induced electromotive force (away from the observer).

There is another important principle in connection with induced currents. Look at Figure 5-1. As the north pole of the magnet enters the coil, a current is induced in this coil. You already know that when a current flows through the conductor, it sets up a magnetic field around this conductor. Thus, the coil becomes an electromagnet. The induced current in this coil is set flowing in such a direction that the end of the coil facing the north pole of the magnet becomes a north pole, too. Since like poles repel, this arrangement of magnets tends to prevent the insertion of the north pole of the magnet into the coil. Work must be done to overcome the force of repulsion.

When you try to remove the magnet from the coil, the induced current is reversed. The top of the coil becomes a south pole and, by attraction to the north pole of the magnet, tends to prevent you from removing it. Thus, once again, work must be done—this time to overcome the force of attraction. You see, you must perform work to create the induced electric current. Of course, the same holds true if the magnet is stationary and the coil is moved.

These results may be summarized in a law formulated by Heinrich Lenz, a German scientist who investigated this phenomenon. According to Lenz's law:

> *An induced current set up by the relative motion of a conductor and a magnetic field always flows in such a direction that it forms a magnetic field that opposes the motion.*

This is to be expected from the more universal law of the conservation of energy which states that energy can neither be created nor destroyed. Therefore, if the motion of the magnet through a coil causes a current to flow through a resistor thereby supplying electrical energy which heats the resistor, this energy must be supplied externally through the mechanical energy necessary to push the magnet into the coil or withdraw it.

B. *The simple generator*

Here, then, is the beginning of our electrical age. We could convert the mechanical energy of a steam engine or water turbine to electric current for light, heat, and power to operate the marvelous machines that soon were invented.

From a mechanical point of view, it is not practical to move our magnet in or out of a stationary coil of wire, or to move the coil over a stationary magnet. The same thing can be accomplished more simply by rotating a loop of wire between the poles of a magnet, thereby inducing a current in the wire as the magnetic field is cut. Let us examine such an arrangement, as illustrated in Figure 5-3.

A simple loop of wire (called an *armature coil*) is mounted so that it may be rotated mechanically on a shaft between the north and south poles of a magnet. The two ends of the loop are connected to two brass or copper rings, A and B respectively, called *collector rings,* which are insulated from each other and from the shaft on which they are fastened. These collector rings rotate with the loop. Two stationary *brushes* (A_1 and B_1) make a wiping contact with these rotating collector rings and lead the current that has been induced in the loop to the external circuit. These brushes usually are made of copper or carbon. This arrangement of loop, magnetic field, collector rings, and brushes constitutes a simple *generator.* Since the two poles are relatively close to each other, the magnetic field is very strong and fairly uniform. This is because most of the magnetic path is high-permeability iron or steel with a very short high-reluctance path (the air gap).

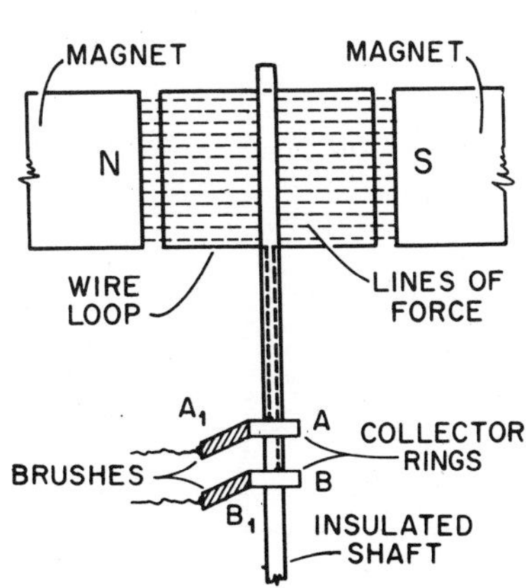

Fig. 5—3.

Simple generator consisting of a loop of wire revolving between the poles of a magnet.

Let us assume that the loop starts from the position shown in Figure 5-4A, and rotates at a uniform speed in a counterclockwise direction. In its initial position the loop surrounds the field, but the magnetic field is not changing because conductors 1-2 and 3-4 (the arms of the loop) are moving parallel to the field, not across it.

As the loop revolves, however, the conductors begin to cut across the lines of force at an increasing rate creating a rapidly changing magnetic field, and therefore the induced electromotive force becomes larger and larger. At the position shown in Figure 5-4B, the loop has the maximum electromotive force induced in it because conductors 1-2 and 3-4 cut across the maximum number of lines of force per second. This is so since the conductors are moving at right angles to the magnetic field.

As the loop rotates to the position of Figure 5-4C, the electromotive force is still in the same direction, but is diminishing in value, until it is zero again. The loop now has made one-half turn during which the induced electromotive force increased to a maximum and then gradually fell off to zero. Since conductors 1-2 and 3-4 are now

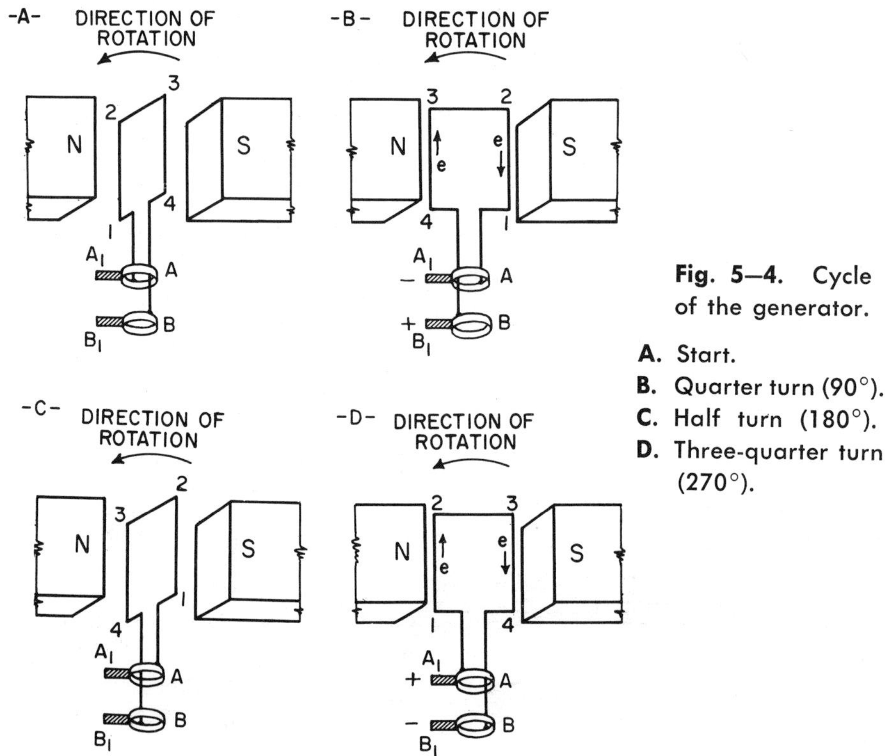

Fig. 5–4. Cycle of the generator.

A. Start.
B. Quarter turn (90°).
C. Half turn (180°).
D. Three-quarter turn (270°).

in reversed positions, the induced electromotive force changes direction in both conductors. The electromotive force, however, again increases in strength and becomes maximum when the loop is again cutting the lines of force at right angles (Figure 5-4D).

Finally, the last quarter of rotation brings the loop back to its original position (Figure 5-4A), at which point the electromotive force is zero again. As the rotation is continued, the cycle is repeated.

This, then, is how the generator operates. Of course, practical generators are not constructed as simply as the one illustrated here. However, it will suffice to illustrate the principle involved and to help us understand the alternating-current phenomenon.

C. *The alternating-current cycle*

The term *cycle* really means "circle"—a circle or series of events which recur in the same order. A complete turn of the loop of the generator is a cycle. So, also, is the series of changes in the induced electromotive force and the current set flowing by it. As the loop of the generator makes one complete revolution, every point in the conductors describe a circle. Since the circle has 360 degrees (360°), a quarter turn is equal to 90°; a half turn to 180°; a three-quarter turn, 270°; and a full turn, 360°. The number of degrees, measured from the starting point, is called the *angle* of *rotation.*

Thus, Figure 5-4A represents the starting point, or zero-degree (0°) position; Figure 5-4B, the 90° position; Figure 5-4C, the 180° position; Figure 5-4D, the 270° position; and Figure 5-4A again (after a complete revolution), the 360° position. Of course, positions in between these points may be designated by the corresponding degrees. However, it is customary to use the *quadrants* (that is, the four quarters of a circle) as the angle of rotation for reference.

We now are ready to examine more closely the induced electromotive force in the loop of the generator as it goes through a complete cycle or revolution. Note that during half the cycle, the direction of the induced electromotive force is such as to cause electrons to move onto Brush A_1. During the next half-cycle, the direction of the induced electromotive force is reversed so that the electrons move onto Brush B_1. To avoid confusion, let us designate the induced electromotive force in one direction by a plus (+) and in the other direction by a minus (−).

Let us assume that the armature loop makes a complete revolution (360°) in one second. Then, at ¼ of a second the loop will be at the 90° position, at ½ of a second the loop will be at the 180° position, and so on. Assume, too, that the maximum electromotive force generated by this machine is 10 volts. Now we are able to make a table showing the electromotive force being generated during each angle of rotation.

Time in seconds	0	¼	½	¾	1
Angle of rotation	0°	90°	180°	270°	360°
Induced EMF (volts)	0	+10	0	—10	0

You will note that in one complete revolution of the loop there are two position (Figures 5-4A and C) at which there is no induced voltage and, therefore, no current flowing to the brushes (or to the external circuit that is connected to them). There are also two positions (Figures 5-4B and D) at which the induced voltage is at its maximum value, although in opposite directions. At intermediate positions the voltage has intermediate values. Note that as the loop rotates, there are two factors that are continuously changing—the position of the loop and the value of the induced electromotive force, or voltage.

1. THE SINE CURVE

Now let us return to our table above, which shows the relationship between the induced electromotive force in the loop of the generator and the degrees of rotation of the loop (or, what amounts to the same thing, the time in seconds which the loop rotated). Let us try to show this relationship by means of a graph.

First draw the horizontal line of zero voltage (Figure 5-5) and divide it into four equal sections of a quarter-second each. Since we assume that the loop makes one revolution (360°) in one second, each quarter-second will correspond to 90°. Accordingly, these sections may be marked in degrees as well.

Next, draw the vertical line showing the induced electromotive force in volts. You will recall that this voltage is in one direction for half the cycle and then reverses and is in the other direction during the next half-cycle. You will recall, too, that we decided to indicate the voltage in one direction by a plus (+) sign and that in the other direction by a minus (—) sign. All the plus values of voltage will appear above the line of zero voltage, and all the minus values below it. Accordingly, all the numbers indicating induced voltage that appear above the line of zero voltage bear plus signs, and all those below bear minus signs.

We now are ready to transpose the values of our table to the graph. At 0° rotation, the table shows the time to be zero and the induced electromotive force to be zero as well. This point is located on the graph where the horizontal line meets the vertical (#1). At the 90° or ¼-second mark, the induced electromotive force has risen to a value of +10 volts. We find this point on the graph by drawing a vertical line up from the ¼-second (90°) mark and a horizontal line to the right from the +10-volts mark. Where these two lines intersect is the required point (#2).

At the ½-second (180°) mark, the electromotive force has fallen to zero again and the point (#3) on the graph lies on the line of zero voltage. At the ¾-second (270°) mark, the electromotive force has risen once more to 10 volts, but this time it is in the opposite (−) direction. We find the corresponding point (#4) on the graph by dropping a vertical line down from the ¾-second mark and drawing a horizontal line to the right from the −10-volt mark. The point

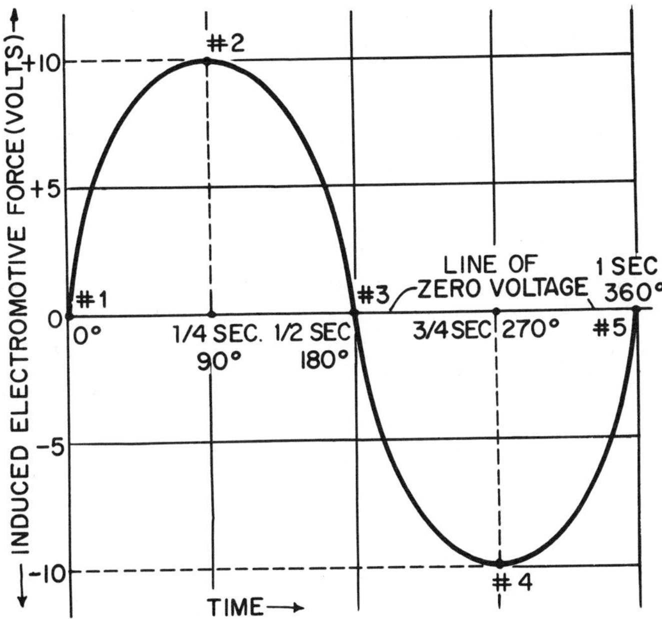

Fig. 5—5. Graph showing the sinusoidal waveform of the generator's alternating-voltage output.

lies on the intersection of these two lines. At the 1-second (360°) mark, the electromotive force again has become zero. Accordingly, this point (#5) lies on the line of zero voltage.

Having thus established the points on the graph corresponding to the quadrants of the circle, how do we go about filling in the rest of the curve? Well, you will recall that as the loop rotated from its original position, as illustrated in Figure 5-4A, a quarter-revolution (90°), as illustrated in Figure 5-4B, the induced electromotive force rose gradually from zero to its maximum value (+10 volts). Accordingly, we indicate this rise by the curve appearing between points #1 and #2 on the graph illustrated in Figure 5-5.

During the next quarter-revolution (Figure 5-4B to Figure 5-4C), the electromotive force dropped gradually from its maximum value to zero. This is indicated by the curve connecting points #2 and #3 on the graph. During the next quarter-revolution (Figure 5-4C to Figure 5-4D), the electromotive force reversed its direction and again rose to its maximum value (−10 volts), as indicated by the curve connecting points #3 and #4. During the next quarter-revolution (Figure 5-4D back to Figure 5-4A), the electromotive force dropped gradually to zero again (as indicated by the curve connecting points #4 and #5).

Thus we obtain a graph that illustrates the electromotive force generated by the armature coil as it makes one complete revolution (and the flow of current in the external circuit connected to the brushes resulting from this electromotive force). Now you must not get the impression that the current is flowing in this roller-coaster type of path. Actually, the current is flowing back and forth through the external circuit. What this curve does show, however, is the strength of the induced electromotive force (and the resulting current flow) and its direction (+ or −) at any instant during one revolution.

So at the ¼-second mark, the electromotive force is 10 volts and is acting in the direction indicated by plus (+). At the ¾-second mark, the electromotive force again is 10 volts, but this time it is acting in the opposite direction, as indicated by minus (−). In the interval between the ¼-second mark and the ¾-second mark, the electromotive force changes from +10 volts to −10 volts, dropping to zero at the ½-second mark, at which instant the electromotive force changes its direction.

We call an electromotive force whose strength and direction as indicated by the curve of Figure 5-5, an *alternating electromotive force.* The current that is set flowing by such an electromotive force will show a similar curve and is called an *alternating current* (abbreviated *ac*).

Note well the curve of Figure 5-5. This curve is typical for alternating currents and is known as the *sine curve.* This curve is said to be the *waveform* of the alternating current produced by the generator. Usually, we desire alternating current to be of this *sinusoidal* (that is, in the form of a sine curve) waveform.

Each *cycle* of alternating current represents one complete revolution of the armature coil of the generator. The number of cycles per second, which is known as the *frequency,* depends upon the number of revolutions per second. In Figure 5-5, the alternating current has a frequency of one cycle per second. Alternating current supplied to house mains in this country generally has a frequency of 60 cycles per second. In radio, we encounter frequencies that run into millions and billions per second. Of course, no generator can be rotated at such a tremendous number of revolutions per second, but we have other means of producing high-frequency currents. These methods will be discussed later in the section on oscillators.

To facilitate discussing such high frequencies, we use the term *kilocycle* (abbreviated *kc*) which means 1,000 cycles per second; and *megacycle* (abbreviated *mc*) which means 1,000,000 cycles per second. The symbol for the cycle is the sine curve (∿). Thus 60-cycle alternating current may appear as 60∿ ac. It is from this symbol of a sine curve that we obtain the symbol for the alternating-current generator which is -(∿)-.

Alternating electromotive force and current are changing constantly in magnitude—that is, the *instantaneous* values are changing. From the sine curve, you can see that there are two *maximum,* or *instantaneous peak,* values for each cycle; a positive maximum and a negative maximum. We call the magnitude of the peak values—that is, the values represented by the distance of these peaks from the zero line in the graph shown in Figure 5-5—the *amplitude.* Thus, in our illustration, the amplitude of the generated voltage is 10 volts.

If you observe the sine curve of alternating electromotive force or current, you will see that the true average value for a full cycle is zero, because there is just as much of the curve above the zero line (+) as there is below it (−). But when we use the *average values* in connection with alternating electromotive force or current, we do not refer to the average of a full cycle, but to the average of a half-cycle (or *alternation,* as it is also called).

It can be proven mathematically that the average value of a half-cycle of a sine curve is equal to 0.636 times the maximum or instantaneous peak value. Thus:

$$\text{Average current} = 0.636 \times \text{maximum current}$$

and

$$\text{Average emf} = 0.636 \times \text{maximum emf}.$$

In practice, we generally use neither the instantaneous nor average values of the electromotive force or current. To make alternating current compare as nearly as possible to direct current, it is necessary to use an *effective value.* In other words, we must find the value for the sine curve of alternating electromotive force or current that would have the same effect in producing *power* as a corresponding direct-current value. You will recall that the direct-current formulas for power are

$$P = I^2R \quad \text{and} \quad P = \frac{E^2}{R}.$$

From this relationship, you can see that the power is proportional to the square of the current (I^2) or to the square of the electromotive force (E^2). Thus, we must get the average (or *mean*) of the instantaneous values squared (instantaneous value × instantaneous value), and then calculate the square root of this average.

Because of the method used to determine the effective value, it is known as the *root-mean-square* (abbreviated *rms*) value. By means of mathematics, it can be proven that the effective value is equal to 0.707 times the maximum, or peak, value and that the peak value is equal to 1.41 times the rms, or effective, value.

2. PHASE RELATIONSHIPS

Flow of current through a circuit is caused by the electrical pressure, or electromotive force. If this electromotive force is direct, the current, too, is direct. If the electromotive force is alternating, the resulting current is alternating as well.

If an alternating electromotive force, or voltage, of sinusoidal waveform is applied to a circuit, the natural tendency will be to cause to flow a current with a similar sinusoidal waveform. Current and voltage will reach zero together, rise and fall together, and attain peak values together. We say that the voltage and current are in step, or *in phase.* This relationship is illustrated in the graph shown in Figure 5-6.

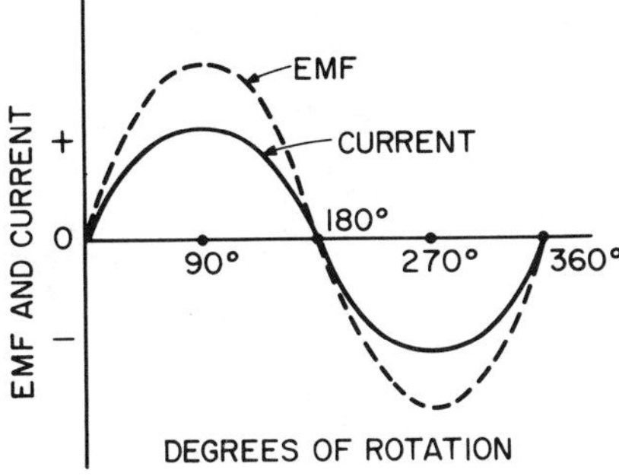

Fig. 5–6.

Graph showing voltage and current in phase.

(Note that the relative sizes of the curves have no significance here. They are plotted for different units—the EMF curve in volts and the CURRENT curve in amperes. They are shown of different sizes primarily for the sake of clarity in presentation. What is important is the fact that they are in phase.)

In practical circuits, however, for reasons we will discuss later, the electromotive force and current may not be in step with each other. The current may either lag behind or lead the electromotive force. We then say that the electromotive force and the current are *out of phase* with each other.

In Figure 5-7 we see such a condition. Note that the EMF curve reaches its peak 90 degrees before the CURRENT curve and that it crosses the zero line 90 degrees ahead of the current. We say the electromotive force is *leading* the current by 90°, or that the current is lagging 90° behind the electromotive force. Another way of looking at this is that the phase angle represents a time difference between the two sine waves. In Figure 5-7, let us say that the frequency is 1,000 cycles per second. The length of time, T, for one whole cycle, called the period, is $\frac{1}{1000}$ second (in general, T in seconds $= \frac{1}{f}$). The figure shows the current lagging by 90° out of

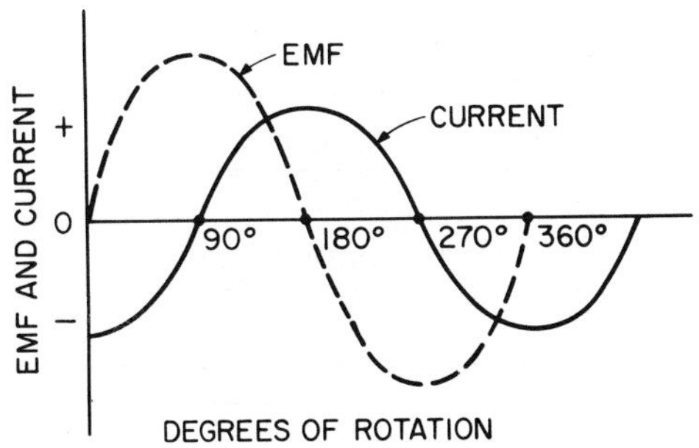

Fig. 5–7.

Graph showing current lagging 90° behind voltage.

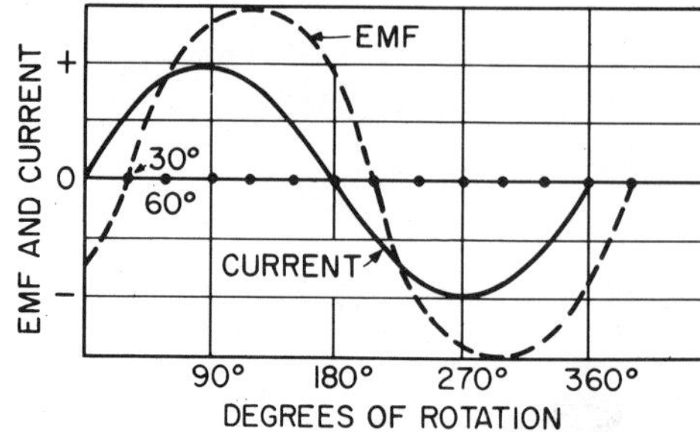

Fig. 5–8.

Graph showing current leading voltage by 30°.

a total of 360°; therefore, the time delay, T_d, is $\frac{90°}{360°}$ or ¼ of the total time of the cycle.

$$
\begin{aligned}
T_d &= \frac{1}{4} \times \frac{1}{1000} \text{ second} \\
&= \frac{1}{4} \text{ millisecond} \\
&= 250 \text{ microseconds}
\end{aligned}
$$

In Figure 5-8 the current is leading the electromotive force by 30°. Another way of describing this difference in phase between electromotive force and current is to say that the *phase angle* is 30°. The electrical symbol for phase angle is the Greek letter θ (*theta*).

So far, we have been considering phase relationships between voltage and current coming from the same source. We may have phase difference between two or more currents (or voltages) coming from different sources.

In direct-current circuits (where there are no phase differences), if current is supplied from two or more sources, the resulting current (or voltage) is obtained merely by algebraic addition. (Look at Figure 5-9.)

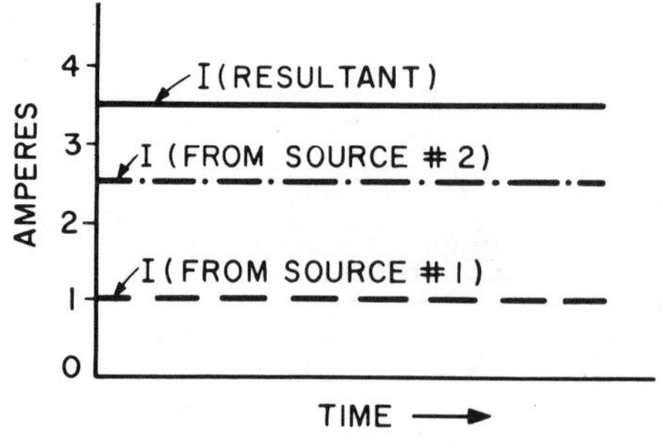

Fig. 5–9.

Graph showing resultant of two direct currents.

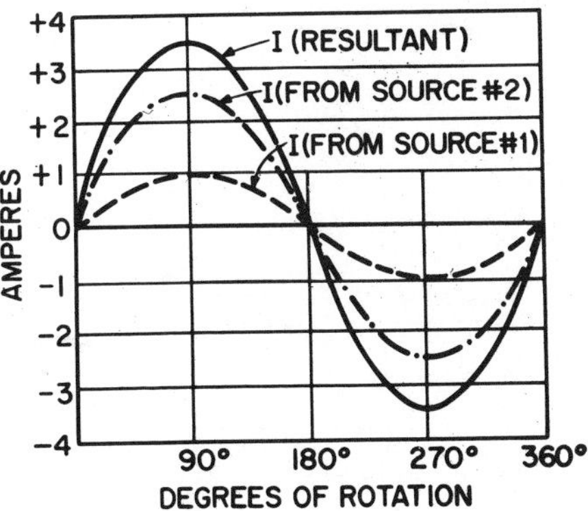

Fig. 5—10.

Graph showing the resultant of adding two alternating currents in phase.

Here, we have assumed that source #1 supplies a steady, direct current of 1 ampere to the circuit. Source #2 supplies a similar current of 2.5 amperes. As a result, a steady direct current of 3.5 amperes flows through the circuit. (In Figure 5-9 we have assumed that the direction of current flow is the same from each source. Hence the two currents are added. If the direction of current flow is different for each source, we subtract the smaller current from the larger, and the direction of flow of the resultant current is that of the larger.)

A similar effect is produced with alternating currents only when the currents from sources #1 and #2 are in step, or in phase, with each other. (See Figure 5-10.) The current from source #1 rises and falls in step with the current from source #2. The resultant current (which is found by the algebraic addition of the two) also is in phase.

However, if the two currents are out of phase with each other, the result is different. (Look at Figure 5-11.) Here the two currents

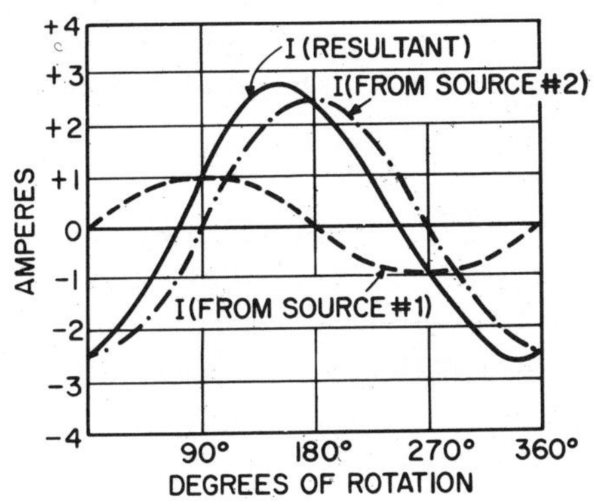

Fig. 5—11.

Graph showing the resultant of adding two alternating currents 90° out of phase.

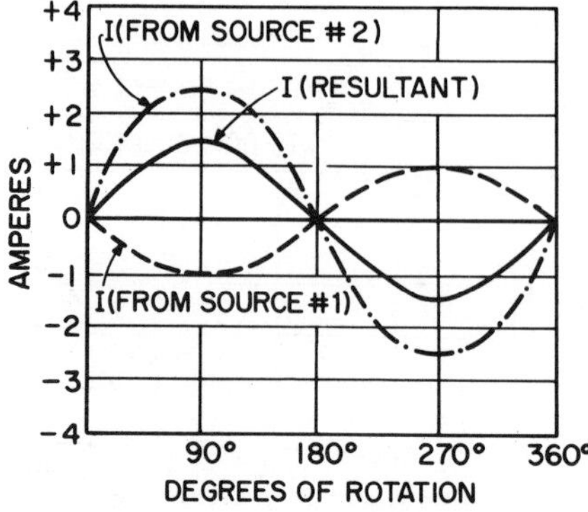

Fig. 5—12.

Graph showing the resultant of adding two alternating currents 180° out of phase.

are out of phase, with the current from source #1 leading the current from source #2 by 90°. Note that the resultant current is out of phase with both of the others.

In Figure 5-12 you see the result when the currents are 180° out of phase with each other. The resultant current is obtained by subtracting one from the other (since one is always positive when the other is negative, and vice versa) and it is in phase with the larger current. But note that if the two currents were of equal value, the resultant would be zero since the positive and negative loops would cancel out. Of course, similar results are obtained when we add voltages from different sources.

D. *Ohm's law for alternating-current circuits*

You will recall that when we were studying Ohm's law for direct-current circuits we found the relationship between current (I), electromotive force (E), and resistance (R) expressed in the equation $I = E/R$ where I is measured in amperes, E in volts, and R in ohms. In d-c circuits the resistance expresses the total opposition to current flow.

In alternating-current circuits, however, except where the voltage and current are in phase, there are other factors besides resistance that oppose the flow of current. (We shall learn about these factors a little later.) Accordingly, our Ohm's law does not apply to a-c circuits unless we substitute for the resistance some value that will take into consideration the increased opposition to current flow.

This new value of total opposition to current flow is called *impedance* and is represented in electrical equations by the capital letter Z. Since impedance, like resistance, measures the opposition to current flow, it has the same unit of measurement, the *ohm.*

Where the current and voltage are in phase, the impedance of the circuit is equal to the resistance. But where a phase difference exists, the impedance is different from the resistance.

If, now, we substitute impedance (Z) for resistance (R), our Ohm's-law equations apply equally well to alternating-current circuits. Thus:

$$I = \frac{E}{Z}, \quad E = I \times Z, \quad \text{and} \quad Z = \frac{E}{I}$$

where I is the current (in amperes), E is the electromotive force (in volts), and Z is the impedance (in ohms).

E. *Power in a-c circuits*

In any circuit, the electrical power consumed at any instant equals the product of the voltage and current at that instant. The equation may be written as

$$p = e \times i$$

where p is the instantaneous power (in watts), e is the instantaneous voltage (in volts), and i is the instantaneous current (in amperes). (It is common practice to use the small-letter equivalent of the capital letter to indicate an *instantaneous* value. Thus, whereas I stands for current, i stands for instantaneous current.)

The instantaneous power equation applies regardless whether the current is direct or alternating. If the current be a steady direct current, the power, too, will be steady. But if the current be alternating, the power will vary from instant to instant with the changing current.

Let us examine the graph in Figures 5-13, which shows the relationship between voltage, current, and power in an alternating-current circuit. Note that the power curve is the result of the product of instantaneous values of voltage and current. Where both voltage and current are positive, the power, too, is positive since the product of two positive values is another positive value. Where both voltage and current are negative, the power, again, is positive since the

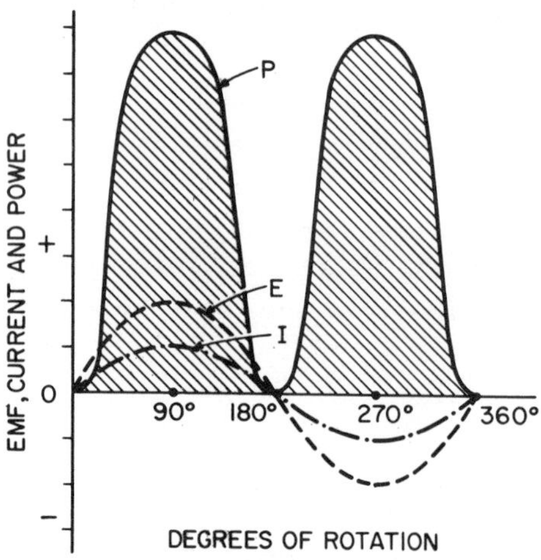

Fig. 5—13.

Graph showing the relationship between voltage, current, and power when voltage and current are in phase. Shading is used to emphasize the power curve.

product of two negative values is a positive value. Thus, except when it drops momentarily to zero, the power is always positive. That means that the source (which may be a generator) is constantly delivering electrical energy to the circuit.

Note, however, this situation exists if the current and voltage are in phase. Let us see what happens to the power in an out-of-phase circuit.

A graph depicting the relationship between current and voltage in an out-of-phase circuit was shown in Figure 5-7. By multiplying the instantaneous values of current and voltage we may obtain the power curve, as shown in Figure 5-14.

For the first 90° of the cycle, the voltage is positive and current is negative. The product of a positive value and a negative value is a negative value. Accordingly, the resulting power is negative. That means that electrical energy is flowing from the circuit back to the source.

During the next 90° of the cycle, both current and voltage are positive. Accordingly, the power, which is the product of the two, also is positive, and energy is flowing from the source into the circuit.

For the next 90° of the cycle, the current is positive and the voltage is negative. The power is negative and energy again flows from the circuit back to the source. During the last 90° of the cycle, both circuit and voltage are negative. The power, thus, is positive and energy flows once more from the source into the circuit.

Examination of the graph shows that the positive power is equal and opposite to the negative power. As a result, they cancel out, and

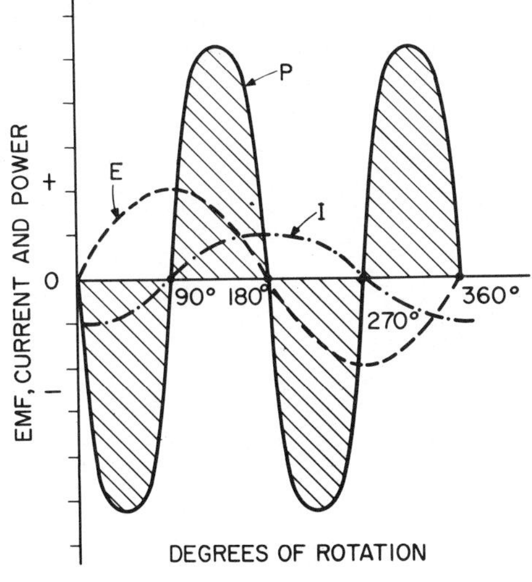

Fig. 5—14.

Graph showing the relationship between voltage, current, and power when voltage and current are 90° out of phase.

the net result is zero. That is, the circuit consumes, or dissipates, no power. The electrical energy merely flows from the source into the circuit and back again. Note, however, that this condition holds only when the phase angle is 90°.

Now, what if the phase angle is some other value such as, for example, 30°, as shown by the graph of Figure 5-8? Examine Figure 5-15.

Fig. 5—15.

Graph showing the relationship between voltage, current, and power when voltage and current are 30° out of phase.

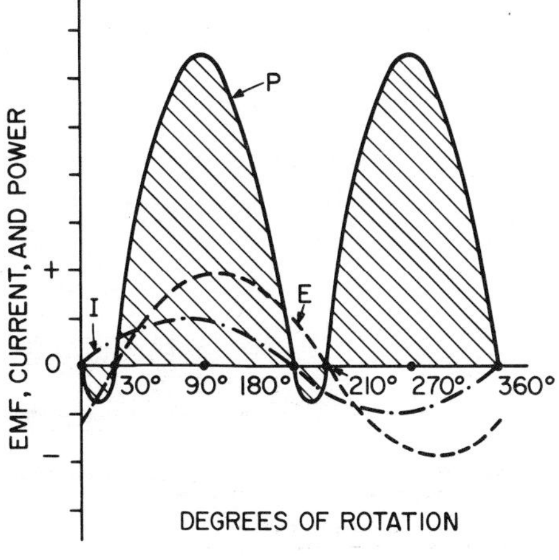

Note that the power is negative only during the intervals between 0° and 30° and between 180° and 210°. During the rest of the cycle the power is positive. Thus, the average power is not zero, as in the case of a circuit where the phase angle is 90°, but some positive value. However, it is less than the average power dissipated by a comparable circuit where the current and voltage are in phase.

F. *Vectors*

The magnitude and direction of factors such as force, pressure, etc., may be shown graphically by means of an arrow called a *vector*. The direction of the force is shown by the arrowhead. The magnitude of the force is indicated by the length of the arrow, choosing any convenient scale. Thus, a vector one inch long may be taken to represent, say, a force of 10 pounds. Then, using the same scale, a vector two inches long would represent 20 pounds.

When we wish to add the electromotive forces (or currents) from two direct-current sources, such as batteries, the process is a simple problem of addition or subtraction, as illustrated vectorially in Figure 5-16. In Figure 5-16A we have assumed that the two batteries have been connected so that their electromotive forces reinforce each other (that is, the direction of electromotive force is the same for each). The electromotive force of the first battery is represented by the vector E_1, using some suitable scale. The electro-

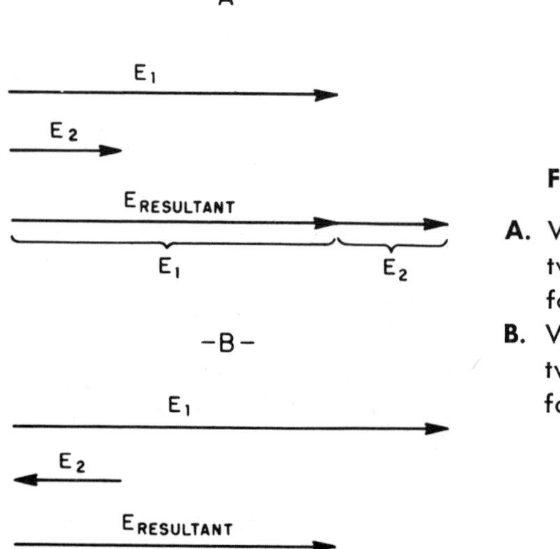

Fig. 5—16.

A. Vectorial diagram showing two direct electromotive forces reinforcing each other.

B. Vectorial diagram showing two direct electromotive forces opposing each other.

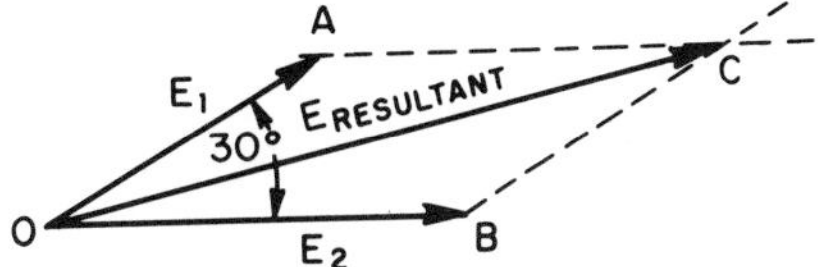

Fig. 5—17.

Vectorial diagram showing the addition of two alternating voltages 30° apart in phase.

motive force of the second battery is represented by the vector E_2, using the same scale. The resultant vector ($E_{resultant}$), then, is the sum of the two (drawn to the same scale).

In Figure 5-16B we have assumed that the two electromotive forces oppose each other, as indicated by the opposing directions of E_1 and E_2. The resultant, then, is the difference between the two.

But when we wish to add together two alternating electromotive forces (or currents), we have a different problem. The electromotive forces (or currents) may not be in phase. For example, suppose we wish to add together two alternating electromotive forces with one (E_1) leading the other (E_2) by 30°.

The vectorial diagram for this problem is illustrated in Figure 5-17. Using any convenient scale, draw a horizontal line (OB) to represent the vector for E_2. From point O and at an angular distance of 30° (vectorial diagrams are read in a counterclockwise direction) draw vector OA to represent E_1, using the same scale. Thus you will note that we have represented E_1 as leading E_2 by 30°.

From point A draw a line parallel to OB, and from point B draw a line parallel to OA. These lines intersect in point C. Line OC represents the vector for the resulting electromotive force (measured on the same scale as the other two electromotive forces). $E_{resultant}$ leads E_2 by the angular distance between OC and OB, but lags behind E_1, by the angular distance between OC and OA. We follow the same procedure in adding together two alternating currents.

We can also show the phase relationship between the electromotive force and current from the same source by means of a vectorial diagram. Assume we wish to show the voltage leading the current by 30°. Draw a horizontal vector for the current, using any convenient scale. From the tail end of the current vector, and at the proper angular distance (30°) draw the voltage vector, as shown in Figure 5-18A. Since we are dealing here with two different factors, we need not use the same scale for the voltage vector.

Note that the diagram shows the voltage leading the current by 30°. Should we wish to show the voltage lagging, say, 90° behind the current, the vector diagram would appear as in Figure 5-18B.

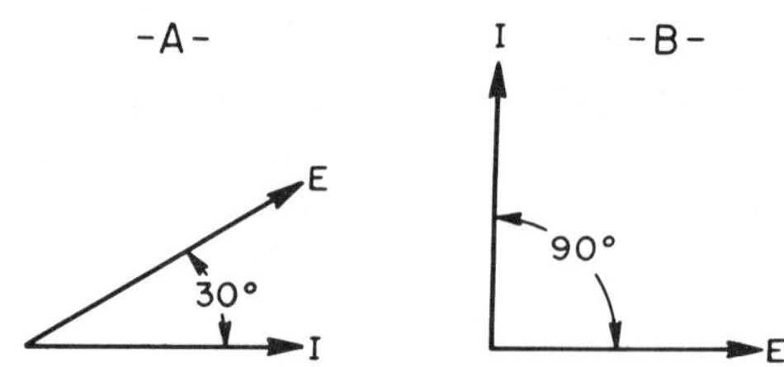

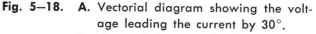

Fig. 5–18. **A.** Vectorial diagram showing the voltage leading the current by 30°.
B. Vectorial diagram showing the voltage lagging behind the current by 90°.

G. *Waveforms*

So far we have considered currents and voltages with either of two types of waveforms. One is the steady direct current whose waveform is illustrated in Figure 5-19A. Since it is a direct current, it flows through its circuit only in one direction, as indicated by the fact that its curve lies only in the positive (+) half of the graph. That it is steady is indicated by the fact that its amplitude (the distance of its curve from the zero line) is constant.

The other is the regular alternating current whose waveform is illustrated in Figure 5-19B. Since an alternating current flows through its circuit first in one direction and then in the other, its curve lies in both the positive (+) and negative (−) halves of the graph. Its frequency (that is, the number of cycles per second) and its amplitude (the distance of its peaks from the zero line) are constant.

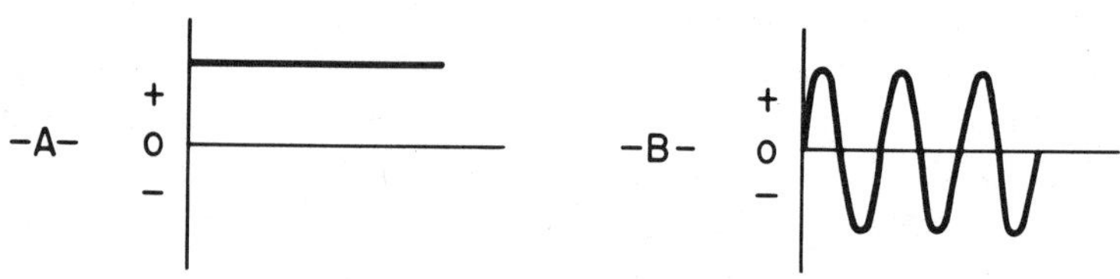

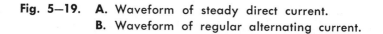

Fig. 5–19. **A.** Waveform of steady direct current.
B. Waveform of regular alternating current.

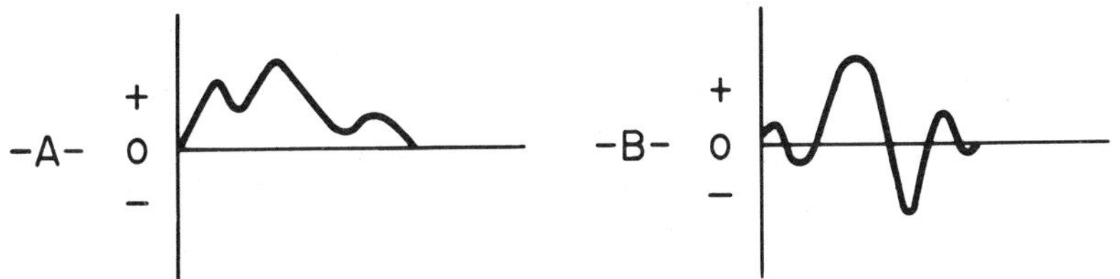

Fig. 5—20. A. Waveform of fluctuating direct current.
B. Waveform of varying alternating current.

The direct current need not be steady. Look at the graph of Figure 5-20A. Here we have a direct current (as indicated by the fact that its curve appears in only the positive half of the graph). But note that its amplitude varies from time to time. Hence a current having this type of waveform is known as a *fluctuating direct current.*

Similarly, the waveform of a *varying alternating current* is illustrated in Figure 5-20B. Here both the frequency and amplitude fluctuate.

Nor need the waveforms be sinusoidal. In Figure 5-21A you see a *triangular,* or *sawtooth,* waveform. In Figure 5-21B is illustrated the *rectangular,* or *square,* waveform. Because both curves lie in both halves of the graph, we are dealing here with alternating current. If the curves were to lie only in one half of the graph, we would have direct current.

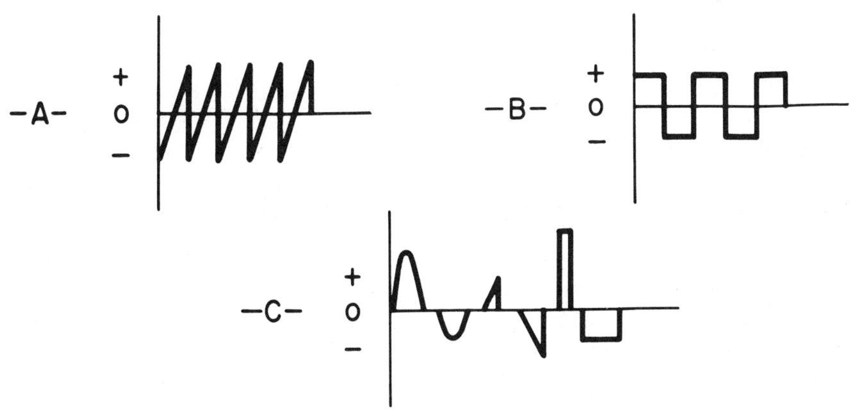

Fig. 5—21. A. Triangular, or sawtooth, waveform.
B. Rectangular, or square, waveform.
C. Waveforms of various types of pulses.

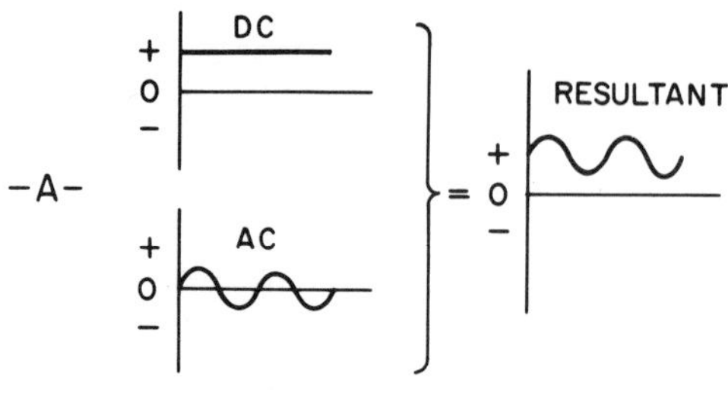

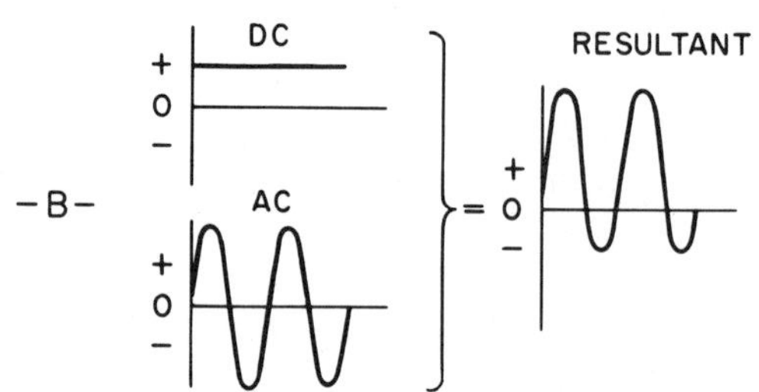

Fig. 5—22.

A. How a steady direct current and a regular alternating current are added to produce a fluctuating direct current.

B. How a steady direct current and a regular alternating current are added to produce a varying alternating current.

Also, in Figures 5-21A and B the waveforms are continuous, that is, a cycle starts where the previous one has left off and is repeated. However, the currents need not be continuous. In Figure 5-21C you see the waveforms of *pulses* of current or voltages. These waveforms may be sinusoidal, triangular, or rectangular, as illustrated, or may be irregular. These pulses may be *positive-going* (where they extend into the positive half of the graph) or *negative-going* (extending into the negative half). They may be *periodic*, recurring at regular intervals of time, or they may be *transient*, appearing at irregular intervals.

In Figure 5-22A we see the resultant of the addition of a steady direct current and a regular sinusoidal alternating current where the amplitude of the direct current is larger than that of the alternating current. The waveform of this resultant current is that of a fluctuating direct current.

In Figure 5-22B, two similar currents are added, but in this case the amplitude of the alternating current is larger than that of the direct current. The resultant is an alternating current, but note that the two halves of the cycle are no longer equal in amplitude.

Over two centuries ago, a brilliant French mathematician named Fourier showed that any complex waveform can be constructed

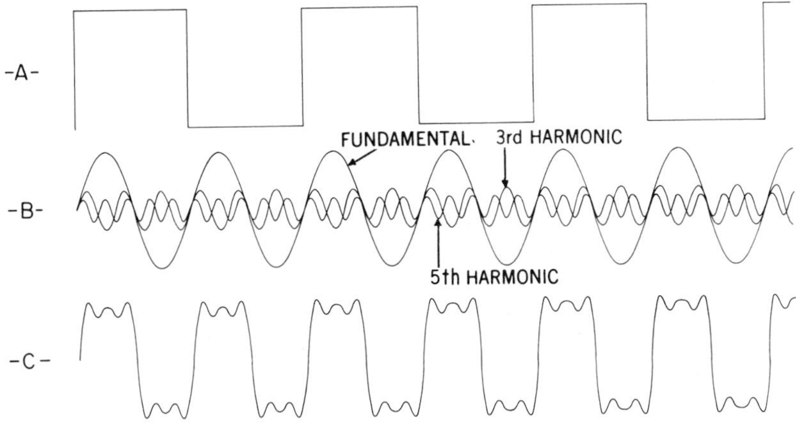

Fig. 5—23. **A.** Square wave.
B. Three components.
C. Sum of three components.

from a group of sine waves. Cosine waves are sine waves with a phase shift of 90°. The group of sine waves must be in harmonic order, which means that if the first one has a frequency called f, the next one has a frequency of 2 × f; then 3 × f; then 4 × f and so forth. If the complex waveform has very steep sides like a series of pulses, then the harmonics have appreciable amplitude out to very high frequencies. As an example, see Figure 5-23A, which shows a square wave. In Figure 5-23B we see the first three components consisting of a fundamental sine wave, the third harmonic and the fifth harmonic (a square wave has no even harmonic components). In Figure 5-23C, we see the sum of the three components. The more components there are, the closer the approximation. The advantage of breaking down a complex waveform into a Fourier series of individual sine waves is that we can study the behavior of an electronic circuit to one frequency at a time and then add up the results.

Thus you can see that any current or voltage having a varying or fluctuating waveform may be considered as being made up of steady direct-current and alternating-current components. Under certain circumstances, which will be considered later in the book, such currents may be separated into their various components.

QUESTIONS

1. What is the difference between direct current and alternating current?

2. Explain what is meant by
   a) an *induced electromotive force;*
   b) an *induced current.*
3. Explain the left-hand rule for determining the direction of induced current flowing in a conductor.
4. State and explain Lenz's law.
5. What are the three factors that determine the strength of an induced current?
6. Draw a labeled diagram of a simple a-c generator and explain the function of each of its parts.
7. Explain the reversal in direction of current flow as the armature coil of the generator rotates through a complete revolution.
8. Draw the waveform of the alternating voltage produced as the armature coil rotates through a complete revolution.
9. What is meant by
   a) a *cycle* of alternating current;
   b) its *frequency;*
   c) its *amplitude?*
10. Explain the relationship between the *peak value,* the *average value,* and the *effective value* of a sinusoidal alternating current.
11. a) What is meant when we say that an alternating current and voltage are *in phase?*
    b) Draw a graph illustrating this condition.
12. a) What is meant when we say that an alternating current and voltage are *out of phase?*
    b) What is meant by *phase angle?*
    c) Draw a graph showing the current *leading* the voltage by 90°.
    d) Draw a graph showing the current *lagging behind* the voltage by 90°.
13. a) What is a *vector?*
    b) What two conditions does it describe?
14. a) Draw a graph showing the resultant of the addition of two steady direct voltages, one +3 volts and the other −5 volts.
    b) Draw the vector diagram of the same.
15. a) Draw a graph showing the resultant of the addition of an alternating current whose maximum value is 3 amperes leading another alternating current whose maximum value is 5 amperes by 45°.
    b) Draw the vector diagram of the same.
16. In any a-c circuit, what must be the phase angle between the voltage and current for the power to be:
    a) at its maximum;
    b) at its minimum?

17. a) What is meant by *impedance?*
    b) In what units is it measured?
    c) State Ohm's law for a-c circuits.
18. a) Draw the waveform of a steady direct current;
    b) of a fluctuating direct current.
19. Draw the waveform
    a) of a regular, sinusoidal alternating current;
    b) of a varying, sinusoidal alternating current.
20. Draw a graph showing the resultant of the addition of a steady direct voltage of +5 volts and a regular, sinusoidal alternating voltage whose amplitude is 3 volts.
21. Draw the resultant of the addition of a sine wave of 1 volt and a sine wave at twice the frequency with an amplitude of 0.5 volt.

Factors Affecting Alternating Current

A. *Resistance*

Resistance, we have learned, opposes the flow of current through a circuit. This is true, regardless whether the current be direct or alternating. If the circuit contains nothing but resistance, Ohm's law for direct current ($I = E/R$) applies equally well for alternating current, and the impedance (Z) is equal to the resistance (R). However, when dealing with alternating-current circuits we encounter instantaneous, peak, average, and rms values. Accordingly, care must be taken to employ similar values for both current and voltage in each case. Furthermore, resistance does not affect the phase relationship between current and voltage in an a-c circuit. That is, the current and voltage remain in phase.

B. *Inductance*

In discussing induced currents (Chapter 5, subdivision A), we have seen that if the magnetic field changes through a coil, an induced voltage is developed in the coil. It makes no difference how the magnetic field is produced. It may be produced by a permanent magnet, as illustrated in Figure 5-1. Or else, it may be produced by current sent flowing through the turns of the coil by some voltage source. As the current starts flowing through it, a magnetic field is built up within and around the coil. As this field is built up, it acts just like an externally-caused field to the coil, inducing therein an electromotive force. When the current reaches a steady value, the magnetic field, too, becomes steady, and the induced electromotive force drops to zero. When the current drops, the magnetic field collapses, but in the reverse direction, and the induced electromotive force comes into being once more.

We have also learned, from Lenz's law, that the induced electromotive force and the induced current it sets flowing are always in such a direction as to oppose any change in the existing magnetic field. Thus, if the original (source) current is increasing and is causing the magnetic field around the coil to expand, the induced electromotive force sets the induced current flowing in such a direction as to build up a magnetic field in opposition to the one set up by the source current. If the source current is decreasing and causing the magnetic field to collapse, the induced electromotive force sets the induced current flowing in such a direction as to build up a magnetic field that aids the original magnetic field and thus tends to prevent its collapse. (See Figure 6-1.) Because it acts in opposition to any change in the source current and the source voltage that causes this current to flow, the induced electromotive force is called a *counter voltage* or *counter electromotive force (cemf).*

The property of a circuit to oppose any change in the current flowing through it is called *inductance.* This property is due to the voltages induced in the circuit itself by the changing magnetic field. Components of the circuit that produce inductance, such as the coil we have been discussing, are called *inductors* or *reactors.* As we have learned, the induced voltage depends upon the strength of the magnetic field set up around the coil, the number of its turns, and the speed with which the changing magnetic field cuts across these turns. Hence the inductance of the inductor depends upon these factors.

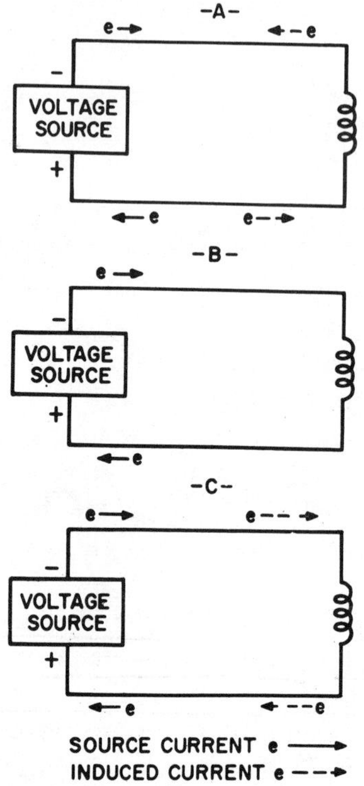

Fig. 6–1. Flow of current in a circuit containing an inductor.

A. Source current is increasing. Induced voltage opposes source voltage.
B. Source current is steady. There is no induced voltage.
C. Source current is decreasing. Induced voltage aids source voltage.

Anything that affects the magnetic field also affects the inductance of the inductor. For example, increasing the number of turns of the inductor increases its inductance. Similarly, substituting an iron core for an air core also increases its inductance.

The speed with which the magnetic field around an inductor is changing also affects its inductance. Thus, if a steady direct current flows through a circuit, there is no inductance except for those instants when the circuit is completed and broken. However, when dealing with alternating currents, the current strength is constantly changing and inductance becomes a factor in such a circuit.

The faster the current changes (that is, the greater its *frequency*), the larger the inductance will be. Since even a straight wire has a surrounding magnetic field when current is flowing through it, the wire has some inductance. However, unless the frequency of the current is very great or the wire extremely long, the inductance of a straight wire is small enough to be neglected.

The symbol for inductance, when used in electrical formulas, is L. Its unit is the *henry* (abbreviated h) which is the inductance of a circuit or component that will produce an induced electromotive force of one volt with a change of one ampere per second in the current flowing through it. (Since current variations seldom are at a uniform rate, a circuit or component has an inductance of one henry if it develops an *average* induced electromotive force of one volt as the current changes at an *average* rate of one ampere per second.) Where the henry is too large a unit, we may use the *millihenry* (mh) which is 1/1,000 of a henry, or the *microhenry* (μh) which is 1/1,000,000 of a henry.

Inductors employed for high-frequency applications generally have no cores (that is, are air-cored) or else have one made of powdered iron or ferrite material. Those employed for low-frequency operation have cores made of stacks of sheet iron. The symbol for the air-core inductor is . If the inductor has a core of powdered iron, its symbol becomes . If its core is made of stacks of sheet iron, its symbol becomes . (See Figure 6-2.)

Inductors, like resistors, may be connected in series, in parallel, or in series-parallel circuits. The total inductance of several inductors connected in series (provided the magnetic field of one inductor cannot act upon the turns of another) is equal to the sum of the inductances of the individual inductors. Thus:

$$L_{total} = L_1 + L_2 + L_3 +, \text{ and so forth.}$$

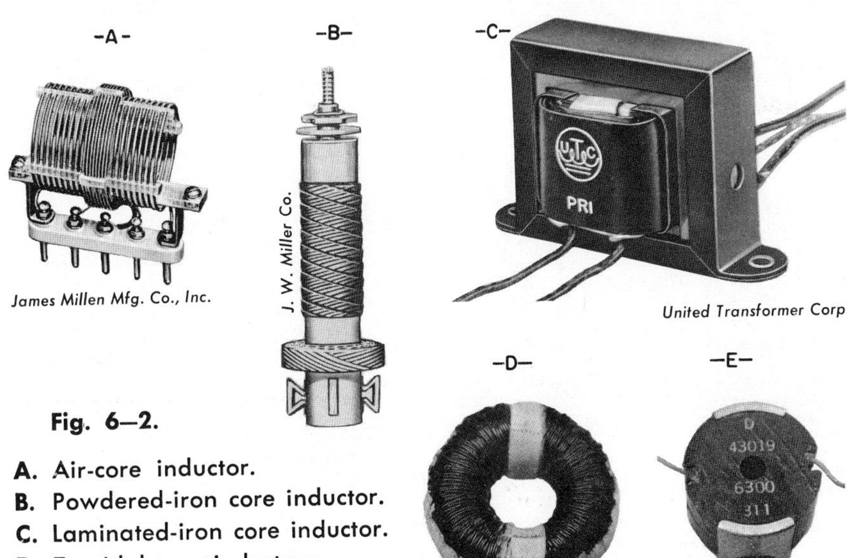

Fig. 6–2.

A. Air-core inductor.
B. Powdered-iron core inductor.
C. Laminated-iron core inductor.
D. Toroidal core inductor.
E. Ferrite cup-core inductor.

If two or more inductors are connected in parallel (again, providing there is no interaction, or coupling, of their magnetic fields) we can find the total inductance from the following formula:

$$\frac{1}{L_{\text{total}}} = \frac{1}{L_1} + \frac{1}{L_2} + \frac{1}{L_3} +, \text{ and so forth.}$$

Note the similarity to the formulas for resistors connected in series and parallel.

As in the case of resistors, the total inductance of inductors connected in a series-parallel circuit (if there is no coupling of magnetic fields) may be obtained by first finding the joint inductance of the inductors in parallel and then adding this inductance to the inductances in series with it as though it were a straight series-inductor circuit.

1. MUTUAL INDUCTANCE

We have seen that when a changing current flows through a circuit that possesses the property of inductance, an induced voltage is generated in that circuit which opposes the changes in current. The ability of the circuit to act in this manner is called *self inductance.* Since the counter electromotive force developed across an inductor depends, in part, upon the speed with which the magnetic field around it changes (that is, the speed of changes in the current), if the rate of change is great, the counter electromotive force, too, may be large. In fact, if the rate of change is great enough, the counter electromotive force so developed may be many times as large as the voltage of the source driving the current through the circuit. This is the principle upon which the ignition coil of the automobile, the device that produces the high-voltage spark used to ignite the gasoline mixture in the cylinders, operates.

We have seen, too, that when two or more inductors are connected in series in such a way that their magnetic fields do not interact, the total inductance is the sum of the inductances of all the inductors. That is to say, the total inductance is the sum of the self inductances of all the inductors.

If, however, two inductors are placed close to each other, their magnetic fields will interact and the magnetic field of one may induce a voltage in the other. We say, then, that the inductors are *coupled.* Each inductor has its own self inductance, but, in addition, there is a further inductance due to the induced voltage produced by coupling between the inductors. We call this further inductance, *mutual inductance*. We say the two coils are coupled together by mutual inductance. The terms *magnetic*, or *inductive*, *coupling* are sometimes used.

Mutual inductance, whose electrical symbol is M, is measured in the same unit as self inductance, the *henry*. When a change of one ampere per second in one inductor induces one volt in the other, the two inductors have a mutual inductance of one henry.

Inductors can be series-connected in two ways. Figure 6-3A shows the inductors connected so that the two magnetic fields aid each other. The effect of mutual inductance is to increase the total inductance, and our formula becomes

$$L_{\text{total}} = L_1 + L_2 + 2M.$$

The two inductors may also be connected in series in such a way that the magnetic fields oppose each other. Figure 6-3B shows this circuit. The effect of the mutual inductance is to decrease the total inductance, and our formula becomes

$$L_{\text{total}} = L_1 + L_2 - 2M.$$

A similar relationship holds true for two inductors connected in parallel. If the two magnetic fields aid each other, the formula for the total inductance becomes

$$\frac{1}{L_{\text{total}}} = \frac{1}{L_1 + M} + \frac{1}{L_2 + M}.$$

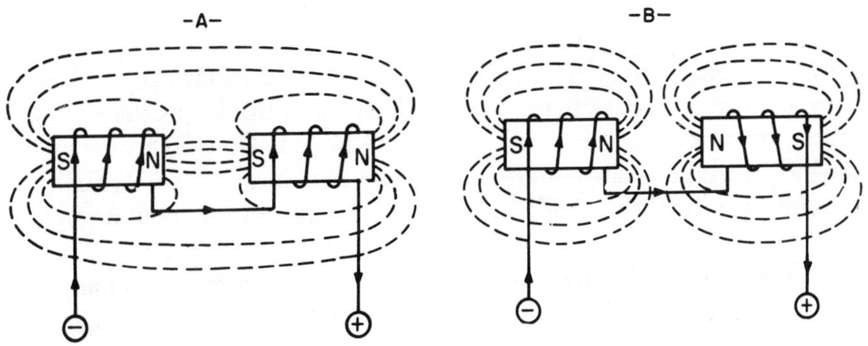

Fig. 6–3. Two inductors connected in series.
A. Magnetic fields aid each other.
B. Magnetic fields oppose each other.

Where the magnetic fields oppose each other, the formula becomes

$$\frac{1}{L_{\text{total}}} = \frac{1}{L_1 - M} + \frac{1}{L_2 - M}.$$

Inductors are said to be *closely coupled* when a large portion of the magnetic field set up by one of the inductors passes within the turns of the other. When only a small portion of the field of one cuts the turns of the other, they are said to be *loosely coupled.* If all the magnetic field of one passes within the turns of the other, we say we have *maximum coupling.*

In practice, such a condition never exists since some part of the field of one inductor escapes. Accordingly, an expression is used to give the degree of coupling. Maximum coupling is considered *100 per cent coupling* or, as it is often called, *unity coupling.* If only half of the field passes within the other coil, the degree of coupling is said to be 50 per cent. Only when the two coils are wound on the same iron core that tends to concentrate the lines of force does the coupling approach 100 per cent.

The coupled inductors need not both be parts of the same circuit. One may be part of one circuit and the other part of another. In this way electrical energy may be transferred from one circuit to another as the current flowing through one inductor induces a voltage in the other. A device that operates in this manner is the *transformer,* which will be discussed further later in this book.

2. EFFECT OF INDUCTANCE ON PHASE RELATIONSHIP

In an a-c circuit containing only resistance, the voltage and current constantly remain in phase. This relationship is shown graphically in Figure 6-4A. (The relative amplitudes of the voltage and current curves have no significance here.) Now let us consider an a-c circuit that contains nothing but inductance. (Such a circuit is only theoretically possible since all circuits contain some resistance.)

The electromotive force is greatest when the current is changing at its fastest rate. If you examine Figure 6-4B, you will see that this occurs as the current passes through zero at the 90° position. Since the source voltage in a circuit containing only inductance is equal and opposite to the induced electromotive force, the source voltage is at its positive peak at the same 90° position.

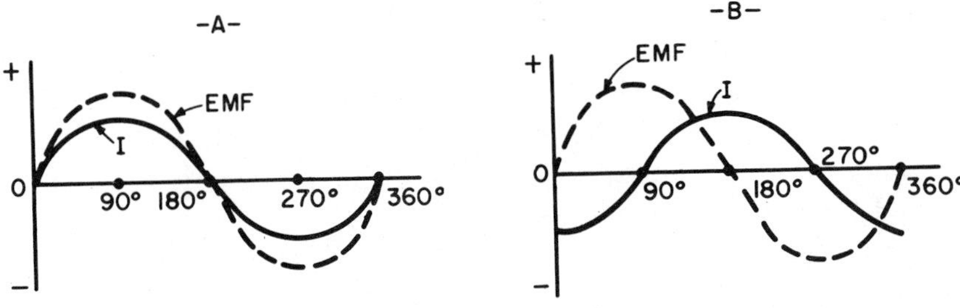

Fig. 6–4. **A.** Phase relationship between current and voltage in a circuit containing only resistance.
B. Phase relationship in a circuit containing only inductance.

The induced electromotive force drops to zero as the current reaches its lowest rate of change. This occurs at the 180° position when the current reaches its positive peak. Since the source voltage is equal and opposite to the induced electromotive force, it, too, is zero at that point. Thus, you can see that, in an a-c circuit containing only inductance, the current lags behind the source voltage by 90° or, what is the same thing, the source voltage leads the current by 90°. In practical circuits, because of the presence of resistance, the current lag is some value less than 90° behind the voltage.

3. INDUCTIVE REACTANCE

In Chapter 5, subdivision D, we learned that the impedance of an a-c circuit is the total opposition that circuit offers to the flow of current. Where only pure resistance is present in the circuit, the impedance is equal to the resistance. But we have seen that the presence of an inductor in the circuit causes a counter electromotive force to be built up which further opposes the flow of current. Under such conditions, the impedance of the circuit is greater than the resistance.

The factor which, in an a-c circuit, causes the impedance (Z) to be larger than the resistance (R) is called the reactance (X). Since this reactance is due to the presence of inductance, we call it the *inductive reactance.* To show that it is inductive reactance, we add the subscript L to the symbol for reactance (X) and we now get X_L as the symbol for inductive reactance.

[This method of adding a subscript to identify an electrical value is commonly used. Thus, the current (I) flowing through the inductor is shown as I_L. The voltage (E) across the inductor becomes E_L. This type of notation is not restricted to inductors. Thus, for example, the voltage drop across a resistance may be designated as E_R and the current flowing through it as I_R.]

Since impedance represents an opposition to the flow of current and has the ohm as its unit, the inductive reactance, which increases the impedance, also has the ohm for its unit.

The inductive reactance depends upon two factors: the inductance of the circuit (L) and the rate or frequency (f) at which the current (and, therefore, the magnetic field) is changing.

The formula for inductive reactance is

$$X_L = 2\pi fL$$

where X_L is the inductive reactance in ohms, f is the frequency in cycles per second, and L is the inductance in henrys. The factor 2π is necessary to make the result come out in ohms. Since π is equal, approximately, to 3.14, 2π therefore equals 6.28.

Example. What is the inductive reactance of a coil of 2 henrys as a 60-cycle alternating current flows through it?

$$X_L = 2\pi fL = 6.28 \times 60 \text{ cycles per second} \times 2 \text{ henrys}$$

$$= 753.6 \text{ ohms.} \quad \textit{Ans.}$$

If we assume a theoretical circuit that has only inductance, we may substitute the inductive reactance (X_L) for the impedance (Z) in the formulas that state Ohm's law for a-c circuits. These formulas are expressed as follows:

$$I = \frac{E}{Z}, \quad E = I \times Z, \quad Z = \frac{E}{I}.$$

Substituting inductive reactance for impedance, we get for a theoretical circuit with inductance only,

$$I = \frac{E}{X_L}, \quad E = I \times X_L, \quad X_L = \frac{E}{I}.$$

C. Capacitance

The outermost electrons of a conductor, we have learned, are loosely held and easily removed. Insulators, or *dielectrics*, on the other hand, have their electrons more firmly fixed. If a dielectric is placed between two conductors (see Figure 6-5A) a *capacitor* is formed. (Formerly, a capacitor was called a *condenser*. However, since the latter term is misleading, the term *capacitor* is preferred.)

Normally, the electrons of the atoms of the dielectric revolve around their nuclei in more or less circular orbits, as indicated in Figure 6-5A. If, however, the capacitor is connected to a voltage source, as in Figure 6-5B, electrons will flow onto plate B and away from plate A. Thus plate B will receive a negative charge and plate A a positive charge. This is called *charging the capacitor*.

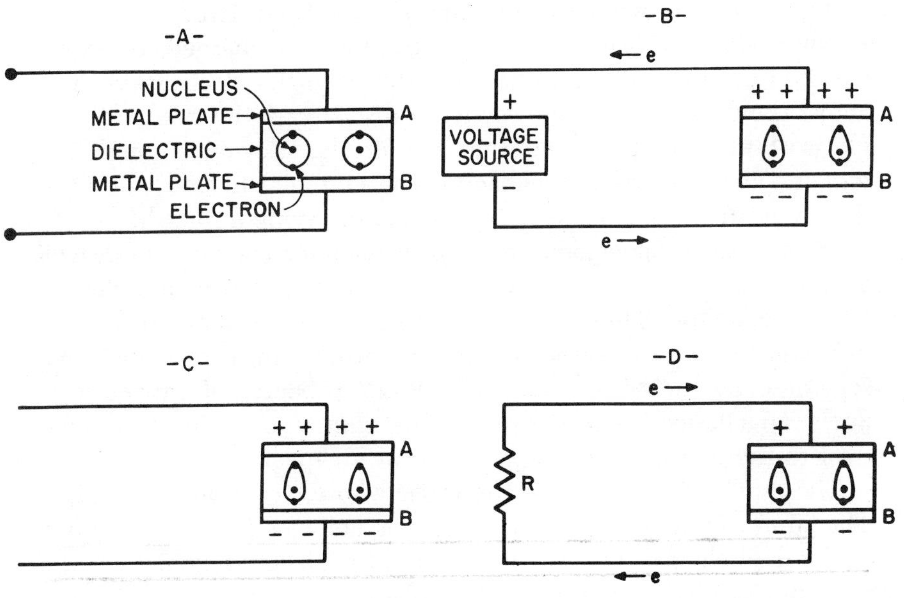

Fig. 6—5.

A. Uncharged capacitor.
B. Charging the capacitor.
C. Voltage source removed, capacitor retains its charge.
D. Discharging the capacitor.

In spite of these charges, electrons cannot flow through the dielectric. However, its electrons will be attracted toward the positive plate A and their orbits will be distorted, as indicated in Figure 6-5B. If the voltage source now is removed (Figure 6-5C), the charges will remain on the plates. Accordingly, the electrons of the dielectric will retain their distorted orbits. The capacitor remains in its charged condition.

If, now, a resistor (R) is placed across the capacitor, it will start to *discharge* through this resistor (Figure 6-5D). Electrons will move from plate B around the circuit toward plate A until they are distributed equally over the entire circuit. As the charges are removed from the plates of the capacitor, the electrons of the dielectric will gradually resume their normal unstrained positions.

Note what happened. As the capacitor was charged, electrical energy from the voltage source was stored in the electrostatic field across the dielectric. As the capacitor discharged, the energy of this electrostatic field furnished the current that flowed through the circuit. The property of a capacitor (or a circuit) to store electrical energy in this way is called *capacitance,* the symbol for which is *C.*

(Note that we have mentioned the capacitance of a circuit as well as that of a capacitor. Any two conductors separated by a dielectric form a capacitor. Thus two of the connecting wires separated by air may constitute a capacitor. Even two adjacent turns of a wire coil, separated by their insulation, exhibit the property of capacitance. Such capacitance is called *stray,* or *distributed,* capacitance in contrast to the *concentrated* capacitance of a capacitor.)

The unit of capacitance is the *farad* whose electrical symbol is f. The farad can be defined as being the capacitance present when one coulomb of electrical energy is stored in the electrostatic field of the capacitor or circuit as one volt is applied.

So far, we have been discussing capacitance in direct-current terms. When considering a-c circuits we must take a somewhat different point of view. In Figure 6-5B, as the capacitor was charged, electrons flowed from the voltage source toward plate B, placing a negative charge on that plate, and from plate A toward the voltage source, placing a positive charge on that plate. Because of these charges, a counter electromotive force was set up that was opposed to the voltage from the source. Electrons, then, would continue to flow until the counter electromotive force of the capacitor equalled the voltage of the source.

If, for any reason, the voltage of the source should now rise, more electrons would flow from the source to the capacitor. If the source voltage should fall below that of the counter electromotive force, electrons would flow back from the capacitor to the source. Thus, if the source voltage were changing, electrons would flow back or forth, depending upon which voltage were higher. (Remember that in a-c circuits the voltage is changing constantly.)

If the capacitor had a capacitance of one farad, a change of one volt in the source would cause the charge on the plates to increase or decrease by one coulomb. That is, it would cause one coulomb to flow through the circuit. Should the voltage change by one volt per second, it would cause one coulomb per second to flow through the circuit. Since one coulomb per second is equal to one ampere, the change of one volt per second would cause one ampere to flow through the circuit. We now can say that a capacitor (or circuit) has a capacitance of one farad if one ampere of current flows through the circuit when the applied voltage changes at the rate of one volt per second.

The farad is too large a unit for ordinary purposes. Accordingly, we have *microfarad* (whose symbol is μf) which is 1/1,000,000 of a farad. Where even the microfarad is too large a unit, we may use the *micromicrofarad* ($\mu\mu$f) which is 1/1,000,000 of a microfarad.

If a capacitor is placed in series with a source of steady direct current, current flows for an instant until the capacitor is sufficiently charged to develop a counter electromotive force equal to the voltage of the source. Then there would be no more current flow in the circuit.

If the d-c source is replaced by an a-c one, a different situation arises. As the voltage rises from zero to its maximum positive value, current flows from the source to the capacitor, building up a counter electromotive force across its plates. As the source voltage starts decreasing, this counter electromotive force sends current flowing back from the capacitor to the source.

The source voltage next reverses its direction and rises to its maximum negative value. Now the current set flowing by the counter electromotive force and the current from the source are flowing in the same direction. As a result, the currents flow through the circuit to the capacitor, building up a new counter electromotive force across its plates (though opposite in direction to the original counter electromotive force). As the source voltage decreases from its negative peak value, the counter electromotive force sends current from the capacitor back to the source. Then the entire cycle is repeated.

Note that during the entire cycle current is flowing through the circuit (excepting the dielectric of the capacitor). If a lamp, or any other device, were placed in series in this circuit it would indicate a continuous current flow through it. It is for this reason that we say that current can flow through an a-c series circuit containing a capacitor.

1. CAPACITORS

A capacitor is a device that is purposely constructed and inserted in the circuit to introduce the desired capacitance. Any two conductors separated by a dielectric will have the property of capacitance. Thus, to make a capacitor, all that is necessary is to have two or more metallic plates separated by air or some other insulating material. The dieletrics in general use are air, mica, waxed paper, polyester, polystyrene, polycarbonate, polypropylene, glass, ceramics, oil, and, in certain types, gas films.

Assume we have a capacitor made of two metal plates separated by a dielectric sheet. The ability of this capacitor to store electrical energy—that is, its capacitance—depends upon the electrostatic field between the plates and the degree of distortion of the orbits of the electrons of the dielectric. Thus, increasing the area of the plates increases the capacitance. Also, if the plates are brought closer together, thus intensifying the electrostatic field, the capacitance increases.

The degree of distortion of the orbits of the electrons of the dielectric depends upon the nature of the substance and is known as the *dielectric constant.* The dielectric constant of a substance is a measurement of its effectiveness when used as the dielectric of a capacitor. Air is taken to have a dielectric constant of unity, or 1. If, in a certain capacitor using air as a dielectric, as the air is replaced by mica and all other things remain equal, the capacitance becomes six times as great, the mica is said to have a dielectric constant of 6. Paper has a dielectric constant of from 2 to 3; paraffin about 2; titanium oxide from 90 to 170. The electrical symbol for dielectric constant is K.

The formula for calculating the capacitance of a capacitor may be stated as follows:

$$C = \frac{0.0885 \times K \times A}{T}$$

where C is the capacitance in micromicrofarads ($\mu\mu$f), K is the dielectric constant of the dielectric, A is the area (in square centimeters) of one side of one of the plates that is in actual contact with the dielectric, and T is the thickness of the dielectric (in centimeters).

Example. Calculate the capacitance of a capacitor having two tin-foil plates each 2.5 centimeters wide and 250 centimeters long. The waxed paper that separates these plates has a thickness of 0.025 centimeter and has a dielectric constant of 2.

Substituting these values in the formula we get

$$C = \frac{0.0885 \times 2 \times 625}{0.025} = 4{,}425\ \mu\mu\text{f}.$$

Thus the capacitance of this capacitor is 4,425 micromicrofarads or 0.004,425 microfarad. *Ans.*

If the voltage across the plates of the capacitor becomes too large, the dielectric may be ruptured and the capacitor ruined. The ability of the dielectric to withstand such rupture is called its *dielectric strength* and is measured in the maximum number of volts a one-centimeter thickness of the dielectric can withstand. Air has a dielectric strength of about 30,000 volts and mica a dielectric strength of about 500,000 volts.

Capacitors may be fixed or variable. In its simplest form the fixed capacitor consists of two metal plates separated by a dielectric of mica or some ceramic material and enclosed in a plastic case for protection. Wire leads through the case make contact with the plates. Sometimes the metallic plates are deposited electrolytically directly upon the opposite sides of the dielectric sheet.

To increase the plate area, and, hence, the capacitance, a number of plates may be sandwiched between a number of sheets of dielectric. Alternate plates are connected together, thus forming two sets with a larger effective area. The capacitances of the capacitors we have described generally are quite small.

A fixed capacitor of greater capacitance can be made by placing a strip of dielectric between two strips of metal foil about an inch wide and several feet long. The large area of the metal-foil plates will permit this capacitor to have a large capacitance. To save space, the whole is rolled up and encased in cardboard. This is called a tubular capacitor. For still greater capacitance, a conducting material consisting of aluminum or a lead alloy is deposited directly onto the dielectric material.

Another type of fixed capacitor, commonly used where larger capacitances are required, is the *electrolytic capacitor*. In such a capacitor, a sheet of aluminum is kept immersed in a borax solution (called the *electrolyte*). An extremely thin coating of aluminum oxide and oxygen gas forms on the surface of the aluminum. If we consider the aluminum as one plate of the capacitor and the borax solution as the other, the coating of aluminum oxide and oxygen gas, which will not conduct electricity, becomes the dielectric. The aluminum need not be a straight sheet, but may be folded over many times or loosely rolled to save space. Because the "plates" are separated by an extremely thin dielectric, the capacitance of such a capacitor is very high.

A variation of this type of electrolytic capacitor (called a *wet* type because of the solution) is the *dry* type. Although this capacitor is not strictly dry, it is so called because, instead of the liquid electrolyte, a gauze saturated with borax solution is used. This "dry" electrolytic capacitor has a definite advantage in that the solution cannot spill.

Care must be taken to always connect the aluminum plate of the electrolytic capacitor to the positive (+) side of the line; otherwise,

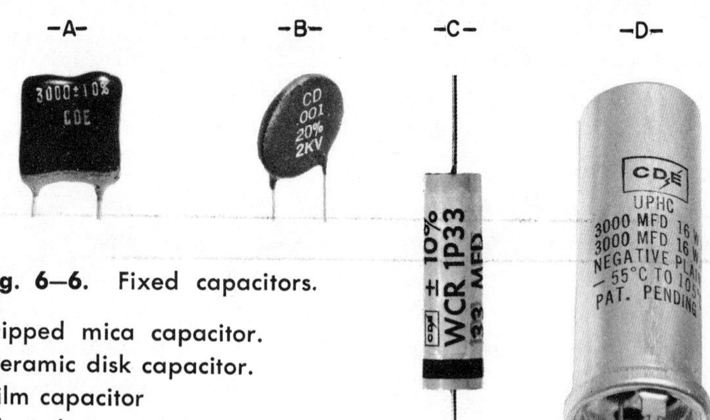

Fig. 6–6. Fixed capacitors.

A. Dipped mica capacitor.
B. Ceramic disk capacitor.
C. Film capacitor
D. Electrolytic capacitor.

Cornell-Dubilier Elect. Corp.

the dielectric will be punctured and the capacitance destroyed. Fortunately, the "wet" type of capacitor is self-healing and a new coating of oxide and gas will form once the proper connections are made. In the "dry" type, however, puncturing the dielectric may permanently damage the capacitor.

If the electrolytic capacitor is connected in an a-c circuit, the dielectric will be punctured constantly since the voltage continuously reverses its direction. There is a variation of the electrolytic capacitor, however, that may be used safely in a-c circuits. This is the *a-c capacitor* formed by mounting two self-healing electrolytic capacitors back to back—that is, with the solution joining one to the other. Connections are made to the two outside plates. Now, whatever the direction of the voltage, one capacitor will always be connected properly (and the other punctured).

For very high quality electrolytic capacitors, tantalum is used as the active material.

When rating a capacitor, we must take into consideration its capacitance and the dielectric strength of its dielectric. The *breakdown voltage* is the maximum voltage that the dielectric of a particular capacitor can stand without being punctured or "breaking down." The *working voltage* of the capacitor is the maximum safe d-c voltage that the manufacturer recommends be placed across its plates. Thus, for example, a capacitor may be rated as "0.1μf, 600 volts, d-c working voltage. [Remember that, unless otherwise stated, a-c voltages generally are stated in rms values. Thus an a-c voltage of 600 volts (rms) has peak values of 846 volts. If this voltage is applied to a capacitor rated at 600 volts, d-c working voltage, the capacitor may be destroyed.]

The capacitance of a capacitor may be varied by changing the effective areas of the plates or the distance between them. One type of variable capacitor is the *trimmer* or *padder* capacitor illustrated in Figure 6-7A. Two metal plates are arranged like the pages of a book. The springiness of the metal keeps the "book" open. Between the "pages" is a sheet of mica or other dielectric material. The metal plates may be brought closer together or further apart by the adjustment of a screw. Thus the capacitance may be varied.

Another type of variable capacitor that frequently is employed to "tune in" radio stations in the radio receiver is illustrated in Figure 6-7C. Here, two sets of meshing metal plates use air as their dielectric. The effective areas of the opposing sets of plates (and, hence, the capacitance of the capacitor) are varied by sliding one

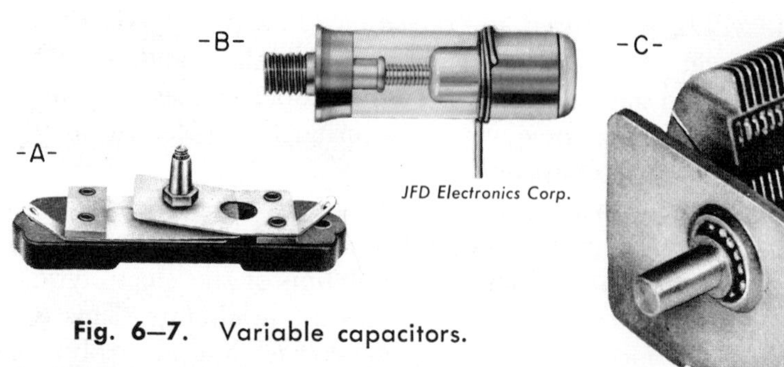

Fig. 6–7. Variable capacitors.

A. Trimmer capacitor.
B. Glass piston capacitor.
C. Variable air capacitor.

set of plates (called the *rotor*) between the other set (called the *stator*).

When used in electrical diagrams, the symbol for the fixed capacitor is . The curved element of this symbol represents the plate that is connected to the negative portion of the circuit. Sometimes, where electrolytic capacitors are involved, the polarity is further identified by plus and minus signs. The symbol for the variable capacitor is . The curved element generally indicates the movable portion of the capacitor.

Capacitors may be connected in series, in parallel, or in series-parallel circuits. Where capacitors are connected in series, they act as though we were adding to the thickness of the dielectric. Accordingly, the total capacitance decreases. Thus, for capacitors connected in series, the following formula applies:

$$\frac{1}{C_{\text{total}}} = \frac{1}{C_1} + \frac{1}{C_2} + \frac{1}{C_3} +, \text{ and so forth.}$$

If we connect capacitors in parallel, they act as though we were adding to the areas of their plates. Accordingly the total capacitance increases. Thus, for capacitors connected in parallel, the following formula applies:

$$C_{\text{total}} = C_1 + C_2 + C_3 +, \text{ and so forth.}$$

Note that this is the reverse action of resistors and inductors connected in series and parallel. Where capacitors are connected in series-parallel circuits, we first find the joint capacitance for the series-connected capacitors and add it in series to the joint capacitance of the parallel-connected capacitors.

2. EFFECT OF CAPACITANCE ON PHASE RELATIONSHIP

We have seen that the effect of inductance on an a-c circuit is to make the current lag 90° behind the voltage. Now let us see the effect of capacitance on such a circuit. Assume we have a circuit containing nothing but capacitance (again, this is only theoretical). We know that current will flow from the source to the capacitor only while the source of voltage is rising, and that the greatest current will be flowing when the voltage is rising most rapidly. If we examine Figure 6-8 we will see that the source voltage is rising most rapidly at the 0° position (and again at the 180° position, though in a negative direction). Thus the current flow at the 0° position is at its maximum positive value.

From the 0° to the 90° position the source voltage continues to rise at an ever-decreasing rate. At 90° this rise has come to a halt. Hence, the current gradually falls from its maximum positive value to zero at the 90° point. Meanwhile the capacitor has become fully charged and the counter electromotive force (cemf) has reached its maximum value.

Now the source voltage begins to fall. Since the counter electromotive force finds itself greater than the source voltage, current starts flowing from the capacitor to the source, as indicated by the negative-current loop of Figure 6-8. At the 180° point the source voltage falls to zero again and the current reaches its maximum negative value. The capacitor, however, has become discharged and the counter electromotive force falls to zero.

The source voltage now changes its direction and continues to rise, at an ever-decreasing rate, and the currents starts to decrease from its maximum negative value toward zero. Meanwhile, the capacitor becomes charged again (this time, in the opposite direction to the original charge) and the counter electromotive force rises once more.

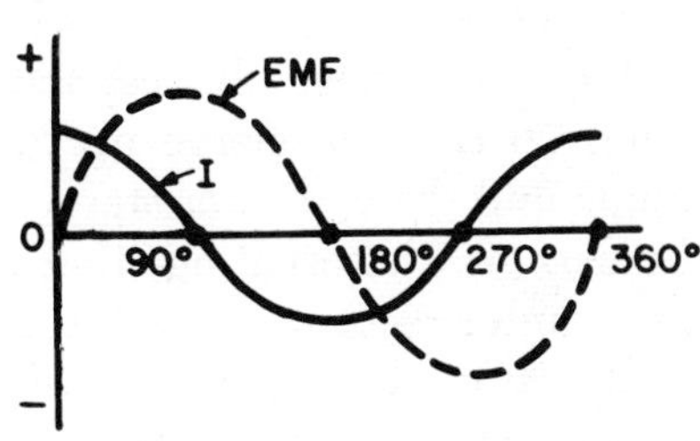

Fig. 6–8.

Phase relationship between the current and voltage in a circuit containing only capacitance.

At the 270° mark the rise of the source voltage ceases and the current reaches zero. As the source voltage starts to fall, the counter electromotive force starts the current flowing in the opposite (positive) direction until this current reaches its positive peak at the 360° mark. Then the entire cycle commences again.

If you examine Figure 6-8, you will see, then, that the effect of capacitance is to make the current *lead* the source voltage by 90° or, what is the same thing, to make the source voltage lag behind the current by 90°. In practical circuits, because of the presence of resistance, the current lead is some value less than 90° ahead of the voltage.

3. CAPACITIVE REACTANCE

Just as inductance, because of its counter electromotive force, increases the opposition to current flow in a circuit by a factor known as inductive reactance (X_L), so capacitance, for the same reason, increases the opposition by a factor known as *capacitive reactance* (X_C). As is true for inductive reactance, the unit for capacitive reactance is the *ohm*.

As a capacitor is charged, it builds up a counter electromotive force across its plates. The larger the plates of the capacitor, the more thinly the charge (that is, the number of electrons) is spread, that is, the smaller the counter electromotive force becomes. But the smaller the counter electromotive force, the smaller is the capacitive reactance. Hence, the larger the plates of the capacitor (or, what is the same thing, the larger its capacitance), the smaller is its capacitive reactance.

If a capacitor is placed in series in a d-c circuit, current flows shortly until the capacitor is fully charged and then the counter electromotive force becomes equal and opposite to the applied electromotive force. Current ceases to flow in the circuit. We might say that a capacitor in a d-c circuit has infinite capacitive reactance.

But if we place the capacitor in a rapidly-alternating circuit, it has no time to charge up fully before the source voltage starts falling and the capacitor starts discharging. Thus, the more quickly the current is alternating—that is, the higher its frequency—the less the counter electromotive force and, hence, the less the capacitive reactance of the capacitor will be.

From the above, we may say that the capacitive reactance is inversely proportional to the capacitance of the capacitor and the frequency of the current. This relationship may be stated in the following formula:

$$X_C = \frac{1}{2\pi f C}$$

where X_C is the capacitive reactance in ohms, f is the frequency of the current in cycles per second, and C is the capacitance in farads. The constant 2π (6.28) is necessary to make the result come out in ohms.

Example. What is the capacitive reactance of a 10-microfarad capacitor in a 60-cycle alternating-current circuit?

$$X_C = \frac{1}{2\pi f C} = \frac{1}{6.28 \times 60 \times 0.00001} = 265.3 \text{ ohms.} \quad \textit{Ans.}$$

As is true of a purely inductive circuit, in a purely capacitive circuit we may substitute the capacitive reactance (X_C) for the impedance (Z) in the formulas that state Ohm's law for a-c circuits. Thus, instead of

$$I = \frac{E}{Z}, \quad E = I \times Z, \quad \text{and } Z = \frac{E}{I},$$

we may state

$$I = \frac{E}{X_C}, \quad E = I \times X_C, \quad \text{and } X_C = \frac{E}{I}.$$

QUESTIONS

1. a) **What is *inductance?***
   b) **What is its unit of measurement?**
   c) **Explain the meaning of this unit.**
2. **Assuming there is no interaction between their magnetic fields, what would be the total inductance of an inductor of 4 henrys and one of 2 henrys if**
   a) **they are connected in series;**
   b) **they are connected in parallel?**
3. **Explain what is meant by**
   a) ***self inductance;***
   b) ***mutual inductance.***
4. **Explain what is meant by**
   a) ***close coupling;***
   b) ***loose coupling;***
   c) ***unity coupling.***

5. Assuming there is interaction between their magnetic fields, what is the formula for finding the total inductance of two inductors if
   a) they are connected in series;
   b) they are connected in parallel?
6. a) What is *inductive reactance?*
   b) In what unit is it measured?
7. What is the inductive reactance of a coil of 10 henrys as a 400-cycle alternating current flows through it?
8. In an a-c circuit containing only inductance, what will be the current if the voltage is 100 volts and the inductive reactance is 25 ohms?
9. a) What is *capacitance?*
   b) What is its unit of measurement?
   c) Explain the meaning of this unit.
10. a) What is a *capacitor?*
    b) What factors determine its capacitance?
11. Explain how alternating current can flow in a series circuit containing a capacitor.
12. a) Describe the construction of the *electrolytic capacitor.*
    b) What two precautions must be taken when employing an electrolytic capacitor in a circuit?
13. What would be the total capacitance of a capacitor of 6 microfarads and one of 3 microfarads if
    a) they are connected in series;
    b) they are connected in parallel?
14. a) What is *capacitive reactance?*
    b) In what units is it measured?
15. What is the capacitive reactance of a 5-microfarad capacitor placed in a 400-cycle a-c circuit?
16. In an a-c circuit containing only capacitance, what will be the current if the voltage is 100 volts and the capacitive reactance is 200 ohms?
17. What is the phase relationship between the voltage and current in an a-c circuit containing only
    a) resistance;
    b) inductance;
    c) capacitance?

    Draw a vector diagram illustrating each case.

7 Circuits Containing Resistance, Inductance, and Capacitance

When we consider practical electrical circuits, expecially those operating on alternating currents, we are confronted with resistance, inductance, and capacitance. All three phenomena are present at the same time. Thus, in the case of an inductor we have, in addition to its inductance, the resistance of the wire constituting its turns and the capacitance between turns. Similarly, the plates and dielectric of the capacitor offer some resistance and, as current flows through them, are surrounded by a magnetic field and, therefore, exhibit some inductance. Even the connecting wires exhibit resistance, inductance, and capacitance.

Ordinarily, we may neglect these stray inductances and capacitances and the incidental resistances may be minimized by using heavy conductors of short lengths. Under certain conditions, however, such stray inductances and capacitances play an important part. In addition, there are times when lumped resistance, inductance, and capacitance are deliberately inserted into the circuit.

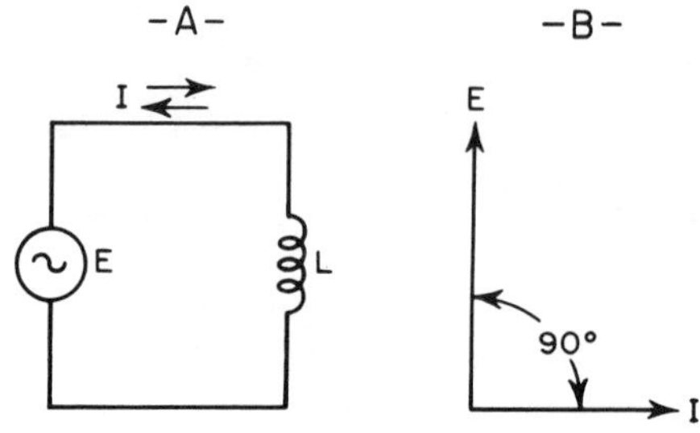

Fig. 7—1.

A. A-c circuit containing only inductance.

B. Vector diagram showing the phase relationship between current (*I*) and voltage (*E*) in this circuit.

A. *Circuits containing resistance and inductance*

We have already learned that the use of vectors furnishes us with a convenient means for picturing the relationships between currents and voltages. Thus, if an alternating-current circuit has, theoretically, nothing but inductance in it, the vector diagram appears as in Figure 7-1B.

In this diagram, the length of the voltage vector (E) is independent of the length of the current vector (I), and the length of each vector depends upon the scale selected for each. Further, because of the inductance, the voltage leads the current (we always read vector diagrams in a counterclockwise direction). The phase angle is 90°.

Pure resistance has no effect on the phase relationship between current and voltage. Thus, if our circuit had nothing but resistance in it, the vector diagram shows us that the voltage and current are in phase (Figure 7-2B).

Another use of the vector diagram is to enable us to add voltages and currents. If we have two resistors in series, it is a simple arithmetical problem to add the voltage drops across each of the resistors in order to calculate the total voltage supplied by the

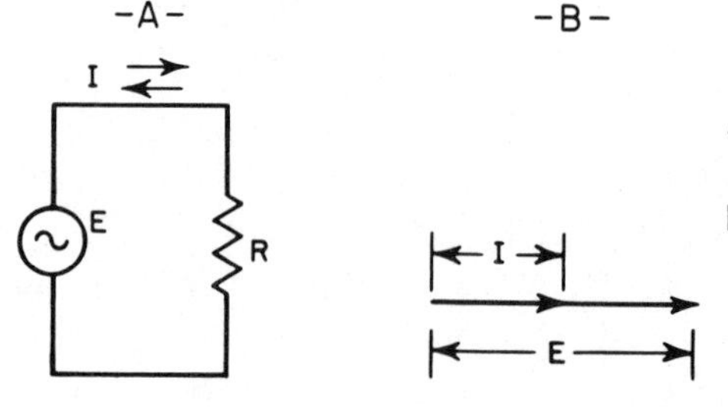

Fig. 7—2.

A. A-c circuit containing only resistance.

B. Vector diagram showing the phase relationship between current and voltage in this circuit.

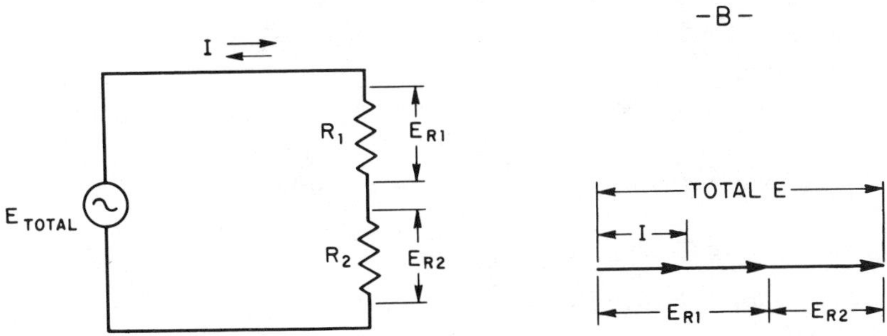

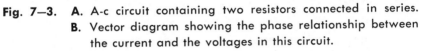

Fig. 7—3. **A.** A-c circuit containing two resistors connected in series. **B.** Vector diagram showing the phase relationship between the current and the voltages in this circuit.

source. See Figure 7-3B. (Although the current and voltage vectors employ different scales, the vectors of all voltages here must use the same scale.)

If, however, we have an inductor and a resistor in series, we cannot simply add the voltage across each to give us the total voltage. We must take into consideration the fact that inductance affects the phase relationships. This situation appears in the vector diagram of Figure 7-4B. Note that I and E_R are in phase, but that I and E_L are 90° out of phase. To obtain the total voltage supplied by the source, we must make use of the parallelogram method described in Chapter 5, subdivision F.

Note, too, that the voltage of the source (E_{total}) is out of phase with the current (I). But the phase difference between the two no longer is 90°. It is some lesser value depending upon the relative values of E_L and E_R. By varying these relative values, the phase

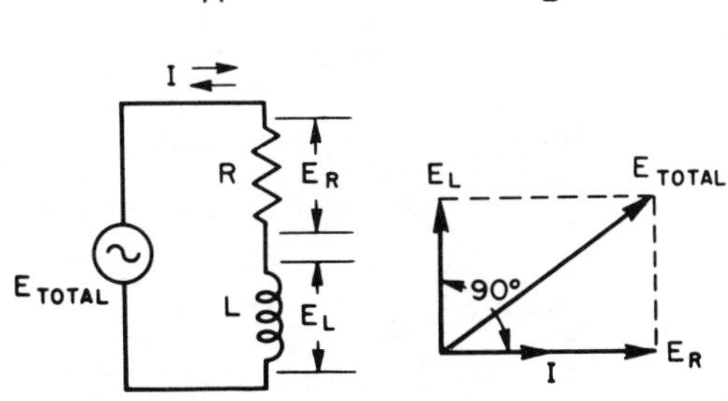

Fig. 7—4.

A. A-c circuit containing a resistor and inductor connected in series.

B. Vector diagram showing the phase relationship between the current and the voltages in this circuit.

difference between the current and the voltage of the source may be varied. Further, note that whereas E_L leads the source voltage, E_R lags behind it.

In a series circuit containing only resistance and inductance, both the resistance (R) and the inductive reactance (X_L), both expressed in ohms, impede the flow of current. However, since the effect of inductance is to cause the current to lag behind the voltage, we cannot use simple addition to find the impedance (Z). Instead, we apply the following formula:

$$Z = \sqrt{R^2 + (X_L)^2}.$$

Example. What is the total impedance of a series circuit containing a resistor of 3 ohms and an inductor whose inductive reactance is 4 ohms?

$$Z = \sqrt{R^2 + (X_L)^2} = \sqrt{9 + 16} = 5 \text{ ohms. } \textit{Ans.}$$

B. *Circuits containing resistance and capacitance*

The vector diagram for an a-c circuit containing nothing but capacitance (theoretically) appears in Figure 7-5B. As you will note, the current now leads the voltage and the phase angle is 90°. If, however, we have a capacitor and resistor in series, the vector diagram appears as in Figure 7-6B. Note that I and E_R are in phase, but that I and E_C are 90° out of phase. Once again, we employ the parallelogram method to obtain the total source voltage (E_{total}).

Note, too, that E_{total} is out of phase with I and that, again, this phase difference is less than 90°, depending upon the relative values of E_C and E_R. Further note that, whereas E_R leads the source voltage, E_C lags behind it.

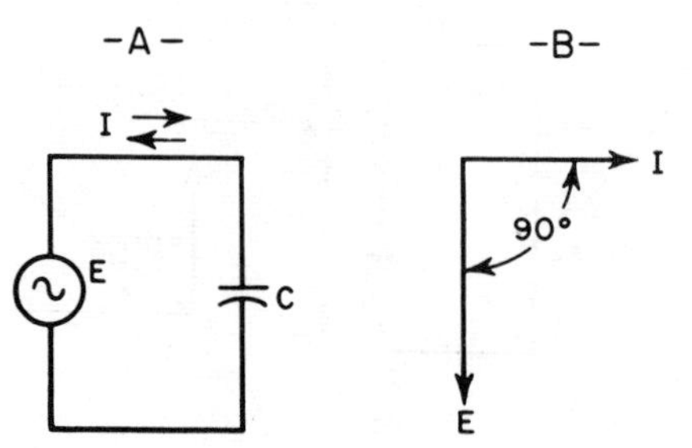

Fig. 7–5.

A. A-c circuit containing only capacitance.

B. Vector diagram showing the phase relationship between current and voltage in this circuit.

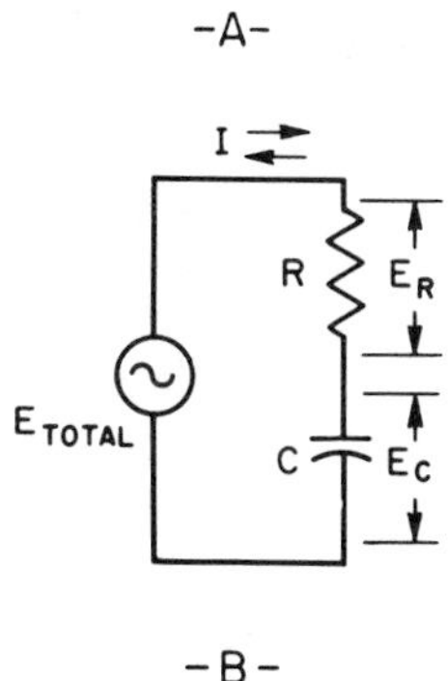

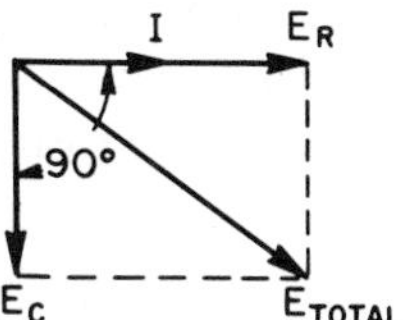

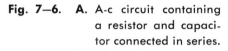

Fig. 7—6. **A.** A-c circuit containing a resistor and capacitor connected in series. **B.** Vector diagram showing the phase relationship between the current and the voltages in this circuit.

As is true for the inductance-resistance circuit, the total impedance offered to alternating current by a series circuit containing resistance and capacitance may be found by means of the following formula:

$$Z = \sqrt{R^2 + (X_C)^2}.$$

There is an aspect of the resistance-capacitance (*R-C*) circuit that is of particular interest in electronics. If a capacitor is connected to a d-c source, such as battery, the moment the circuit is completed, a heavy charging current will flow. This current will fall off quickly as the capacitor becomes charged and a counter electromotive force is built up. When the capacitor is fully charged, this counter electromotive force will equal the electromotive force of the battery and no more current will flow.

If a resistor is placed in series with the battery and the capacitor (Figure 7-7A), the time it would take the capacitor to reach full charge would depend upon the values of the capacitor and the resistor. Actually, the capacitor never succeeds in becoming fully charged. Accordingly, we calculate the time it takes the capacitor to reach 63 per cent of its full-charge value and call that the *R-C time constant* of the circuit.

The time constant of a circuit may be calculated from the following formula:

$$t \text{ (in seconds)} = C \text{ (in farads)} \times R \text{ (in ohms)}.$$

Example. What would be the *R-C* time constant of a series circuit containing a 0.001-microfarad capacitor and a resistor of 50,000 ohms?

$$t \text{ (seconds)} = 0.000{,}000{,}001 \text{ farad} \times 50{,}000 \text{ ohms.}$$
$$t = 0.000{,}05 \text{ second, or } 50 \text{ microseconds.} \quad \textit{Ans.}$$

Actually, we are more concerned usually with the counter electromotive force built up across the capacitor than the current flowing into it. How this counter electromotive force is built up is shown graphically in Figure 7-7B. The time constant, as before, is taken at the point where the counter electromotive force reaches a value of 63 per cent of its full-charge value.

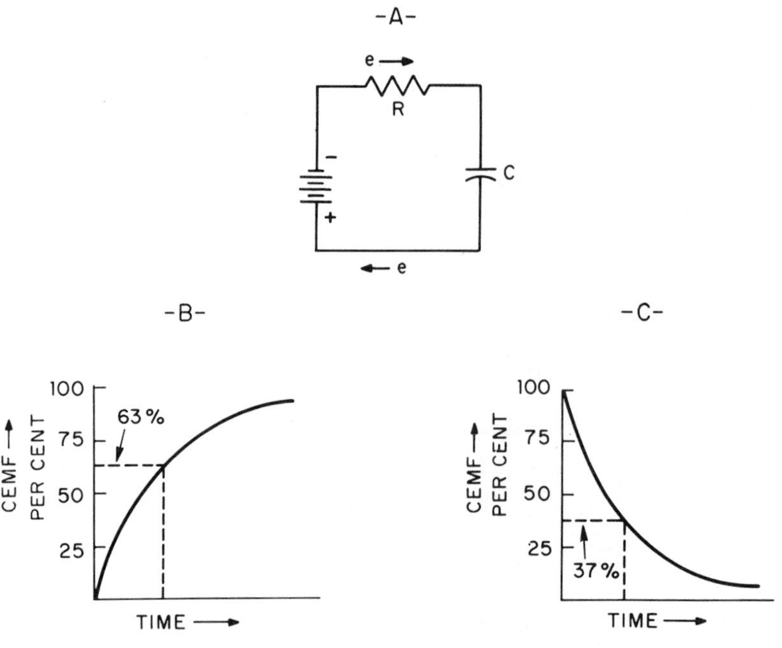

Fig. 7—7. **A.** *R-C* circuit.
B. Graph showing counter electromotive force as capacitor is charged.
C. Graph showing counter electromotive force as capacitor is discharged.

When a fully-charged capacitor discharges through a resistor, the counter electromotive force drops rapidly at first and then more slowly. (See Figure 7-7C). Thus the time constant would be the time required to discharge the capacitor to 37 per cent of its full-charge value. The procedure for calculating this time constant is the same as before.

Later in the book we shall see how the *R-C* circuit is employed.

C. *Circuits containing inductance and capacitance*

If an a-c circuit contains only inductance and capacitance, the only opposition to current flow will be the inductive reactance (X_L) and capacitive reactance (X_C). (We assume here the theoretical position that no resistance is present.) Note, however, that whereas the inductive reactance tends to cause the current to lag 90° behind the voltage, the capacitive reactance tends to cause the current to lead by a similar amount. Hence the two reactances tend to cancel each other out. Thus, the total reactance (X) is equal to the difference between the inductive and capacitive reactances. This may be expressed in the following formula:

$$X = X_L - X_C.$$

Subtraction is performed algebraically. If the answer is a positive value, the resulting reactance (X) has the characteristics of an inductive reactance. That is, the current tends to lag behind the voltage. If the answer is a negative value, the resulting reactance has the characteristics of a capacitive reactance.

Example. What is the reactance of a 60-cycle, a-c circuit containing an inductor of 2.5 henrys and a capacitor of 10 microfarads connected in series?

$$X_L = 2\pi f L = 6.28 \times 60 \times 2.5 = 942 \text{ ohms}$$

$$X_C = \frac{1}{2\pi f C} = \frac{1}{6.28 \times 60 \times 0.00001} = 265.3 \text{ ohms}$$

$$X = X_L - X_C = 942 - 265.3$$
$$= 676.7 \text{ ohms (inductive reactance).} \quad \textit{Ans.}$$

We may illustrate the above vectorially as shown in Figure 7-8. Since the voltage across the inductor (E_L) leads the current by 90°

and the voltage across the capacitor (E_C) lags 90° behind the current, the two voltages are 180° out of phase and their vectors are shown on the same straight line. Since these vectors are in opposite directions, we may subtract one from the other to get the vector for the resulting voltage (E_{total}).

Assume (as illustrated in Figure 7-8B) X_L is larger than X_C and hence the voltage drop (E_L) across the inductor is larger than the voltage drop (E_C) across the capacitor. The result, then, is as if we had nothing in the circuit but an inductor whose voltage-drop vector is the difference between vector E_L and vector E_C. If, as in Figure 7-8C, E_C is larger than E_L, the result is as if we had nothing in the circuit but a capacitor whose voltage-drop vector is the difference between vectors E_C and E_L.

Since inductive and capacitive reactances tend to cancel out, if a circuit contains inductive reactance, we may reduce the total

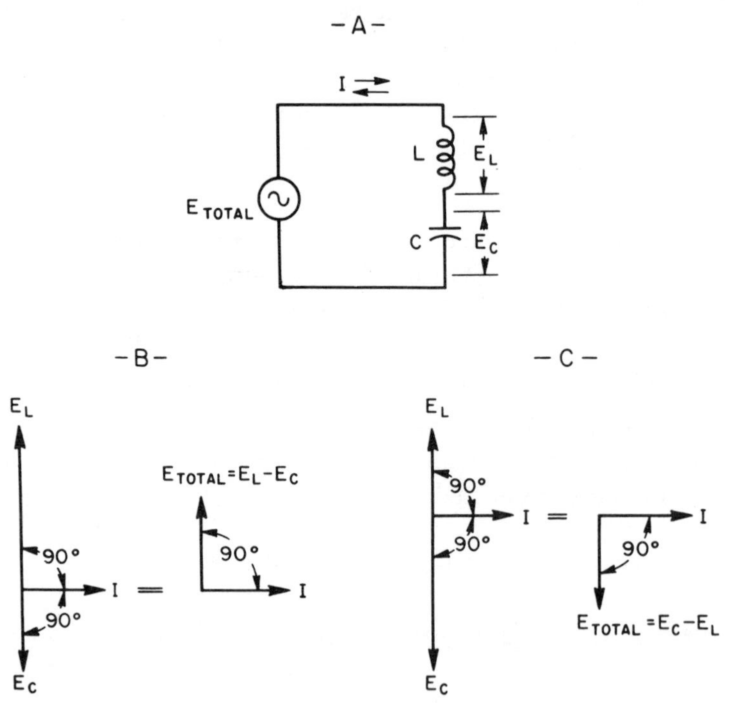

Fig. 7—8. **A.** A-c circuit containing inductance and capacitance.
B. Vectorial diagram when E_L is greater than E_C.
C. Vectorial diagram when E_C is greater than E_L.

reactance by adding some capacitive reactance. This may be done by placing a capacitor in series in the circuit. On the other hand, if the circuit contains capacitive reactance, we may reduce the total reactance by adding some inductive reactance (by placing an inductor in series).

Practical circuits always contain some resistance. To find the total impedance (Z) of a circuit containing resistance as well as inductance and capacitance, we use the following formula:

$$Z = \sqrt{R^2 + (X_L - X_C)^2}.$$

Example. What is the total impedance of an a-c circuit containing a resistor, inductor, and capacitor in series if the resistance is 30 ohms, the inductive reactance is 210 ohms, and the capacitive reactance is 250 ohms?

$$\begin{aligned} Z &= \sqrt{R^2 + (X_L - X_C)^2} \\ &= \sqrt{(30)^2 + (210 - 250)^2} = 50 \text{ ohms.} \quad \textit{Ans.} \end{aligned}$$

1. RESONANCE

Assume we connect an inductor (L) and a capacitor (C) in series with a generator, as in Figure 7-9. Note that some resistance (R) is always present. This is mainly due to the wire of the inductor as well as the connecting leads.

The opposition to the flow of current in this circuit comes from the inductive reactance (X_L), the capacitive reactance (X_C), and the resistance (R). Since the inductive reactance causes the current to *lag* behind the voltage and the capacitive reactance causes the current to *lead* the voltage, these two reactances tend to cancel out each other.

Since $X_L = 2\pi fL$ and $X_C = 1/(2\pi fC)$, the higher the frequency, the greater the inductance reactance and the smaller the capacitive reactance. The lower the frequency, the smaller the inductive

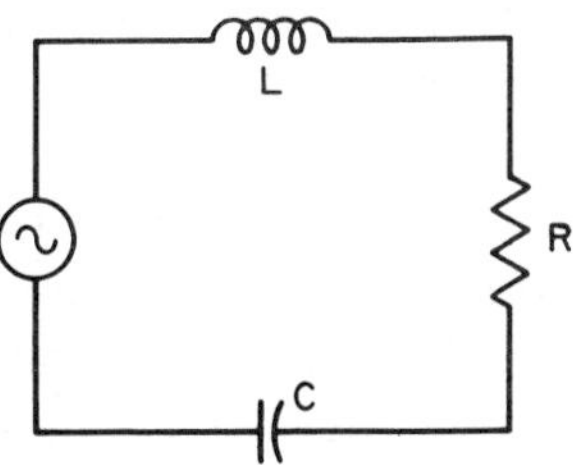

Fig. 7–9.

Inductor L, resistor R, and capacitor C connected in series across a source of alternating current.

reactance and the greater the capacitive reactance. Thus, we can see that at a certain frequency (depending upon the values of L and C), the two reactances will cancel out each other, leaving only the resistance (R) to oppose the flow of current. Since the resistance may be very small, the flow of current may become quite large. This condition is known as *resonance* and the critical frequency, as the *resonant frequency.*

Note that at resonance the inductive and capacitive reactances are not each equal to zero. Rather, X_L and X_C are equal and opposite to each other and thus cancel out. As a matter of fact, these reactances may be quite high. Since the current flowing in the circuit is quite large, the voltage drops across the inductor and capacitor may be very large, too. Under certain conditions, these voltage drops at resonance may even be greater than the applied voltage of the generator.

The resonant frequency (f_r) of the circuit depends upon the values of inductance (L) and capacitance (C). Since, at resonance, $X_L = X_C$, then substituting, we get $2\pi f_r L = 1/(2\pi f_r C)$. Solving this equation for the resonant frequency, we get

$$f_r = \frac{1}{2\pi\sqrt{L \times C}}$$

where the frequency is in cycles per second, inductance is in henrys, and capacitance is in farads.

Example. What will be the resonant frequency of a series circuit containing an inductor of 200 microhenrys (0.0002 henry) and a capacitor of 200$\mu\mu$f (0.000,000,0002 farad)?

$$f_r = \frac{1}{2\pi\sqrt{L \times C}} = \frac{1}{6.28 \times 0.000{,}0002} = \frac{1}{0.000{,}001{,}256}$$

f_r = 796,178 cycles per second, or 796.178 kilocycles per second. *Ans.*

As you know, the inductive and capacitive reactances cancel out each other at the resonant frequency, leaving only the resistance to oppose the flow of current. Since this resistance may be very small, the flow of current may be very large. If the frequency is reduced below resonance, the capacitive reactance becomes larger than the inductive reactance, and the net reactance no longer is zero. If the frequency is increased above resonance, the inductive reactance becomes larger than the capacitive reactance. Again the net reactance no longer is zero. In both cases, the current flowing in the circuit decreases.

A graph showing the relationship between the current and the frequency may be drawn as in Figure 7-10. Note that the current reaches its maximum amplitude at the resonant frequency (f_r), falling off rapidly on either side of that point.

Note also the effect of resistance in the circuit. The smaller the resistance, the greater the current at resonance and also the sharper the curve—that is, the greater the discrimination against nonresonant frequencies. On the other hand, the greater the resistance, the lower the resonant peak and the flatter the curve. The sharply peaked tuned circuit is called a high Q circuit. Q is defined as X_L divided by R.

We call a circuit such as illustrated in Figure 7-9, a *tuned circuit*, tuned to a resonant frequency determined by the values of inductance and capacitance. Thus, if alternating currents of various frequencies were being supplied by the generator, only those close to the resonant frequency would get through the tuned circuit without much opposition. If the frequencies were higher than the resonant frequency, the currents would be opposed by the inductive reactance of L. If the frequencies were lower, they would be opposed by the capacitive reactance of C. The resistance (R) is the same for currents of all frequencies.

The ability of the tuned circuit to discriminate against currents of nonresonant frequency (as indicated by the sharpness of the resonant curves illustrated in Figure 7-10) is called *selectivity*. The sharper the curve, the greater is the selectivity of the circuit, and the greater is the discrimination against currents of the nonresonant frequency.

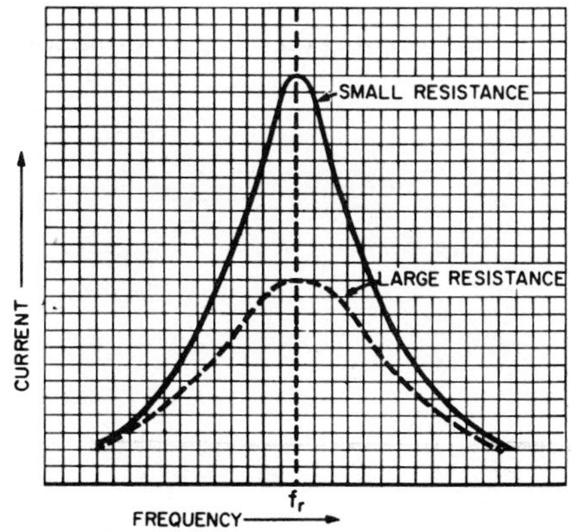

Fig. 7—10.

Curves showing the current-frequency relationship around the resonant frequency f_r of a circuit containing inductance and capacitance in series. Note the effect caused by resistance.

We employ this principle when we tune the antenna circuit of a radio receiver to receive signals of one frequency and to reject all others. Usually, the antenna circuit consists of a tuned circuit, such as described, with the voltage of the generator being replaced by an electromotive force induced in the antenna circuit as the radio wave crosses the antenna. The values of C and L are adjusted to produce resonance at the frequency of the desired signal.

The circuit described above is a series circuit; hence, at the resonant frequency, it is called a *series-resonant circuit.* At this frequency, there is very little opposition to the flow of current from the source, and so very large currents may flow.

Another type of resonant circuit, illustrated in Figure 7-11, is called a *parallel-resonant circuit.* Current flowing from the generator divides into two paths; one through the capacitor and the other through the inductor and resistor. (This resistor represents the resistance of the wire of the inductor and will be disregarded for the time being.) The greater current will flow through the path of lower impedance.

At low frequencies, the capacitive reactance will be higher, and thus more current will flow through the inductor. At high frequencies, the inductive reactance will be higher, and now more current will flow through the capacitor. At a certain frequency (the resonant frequency), both reactances will be the same, and thus equal current will flow through both paths.

But the inductive reactance would tend to cause the current to lag behind the voltage by 90°, and the capacitive reactance would tend to cause the current to lead the voltage by 90°. Thus, the currents flowing in each path would be 180° out of phase—that is, would be flowing in opposite directions. The net result would be that they would cancel out each other, and no current would flow from the generator. This is the same as saying that a *parallel-resonant circuit offers an infinite impedance to the source of current.*

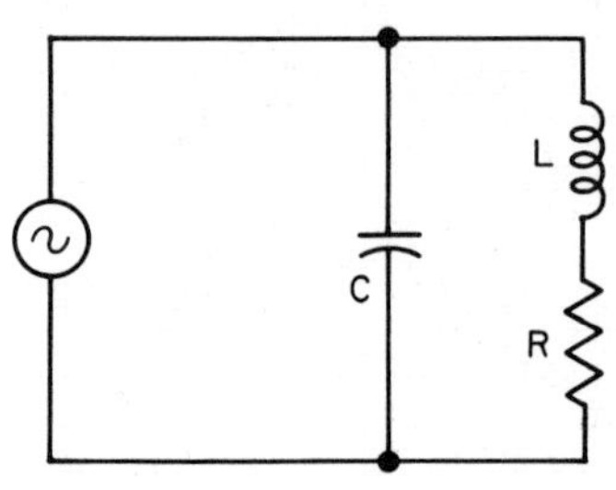

Fig. 7–11.

The parallel-resonant circuit.

Keep in mind, however, that there may be large currents flowing around the loop formed by the capacitor and the inductor. It is not that there are no currents present, but rather that the currents are equal and opposite, to explain why the net result is zero.

We may obtain a better picture, perhaps, if we consider what is happening in the closed loop at resonant frequency. As the capacitor becomes charged, a counter electromotive force is generated which causes current to flow through the inductor. This, in turn, produces a counter electromotive force across the inductor that charges the capacitor. Thus, electric energy is stored up in the electrostatic field of the capacitor and then in the magnetic field of the inductor. This energy continually changes from one field to the other. As a result, current flows back and forth. We call this back-and-forth flow of current, *oscillation* and, theoretically, it should continue indefinitely, once started.

The presence of resistance upsets this theoretical picture. As the current flows in the loop, some electric power is dissipated by this resistance, and this loss must be replaced from the outside source. Hence, some current flows from the generator, and the parallel-resonant circuit acts as a very high, rather than an infinite, impedance.

We may plot an impedance-frequency graph for this type of circuit just as we plotted a current-frequency graph for the series-resonant circuit. This is shown in Figure 7-12. As in the case of the series-resonant graph, the resistance affects the amplitude, sharpness, and the selectivity of the curve. Here again, the sharply peaked curve is called the high Q curve.

In radio receivers, the signal generally is made to pass through a number of circuits in succession, all tuned to the same resonant

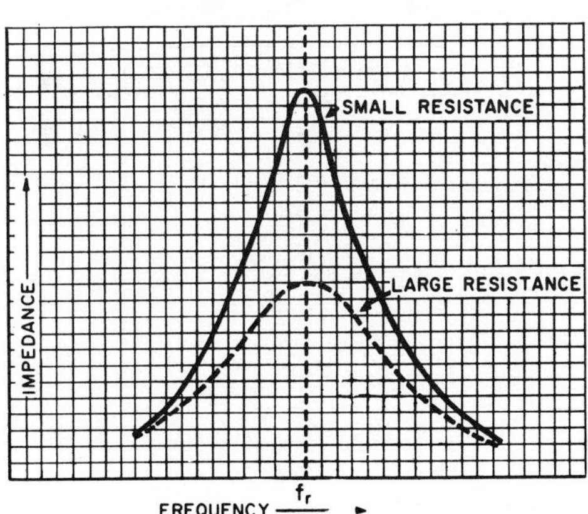

Fig. 7—12.

Curves showing the impedance-frequency relationship around the resonant frequency f_r of a parallel-resonant circuit. Note the effect caused by resistance.

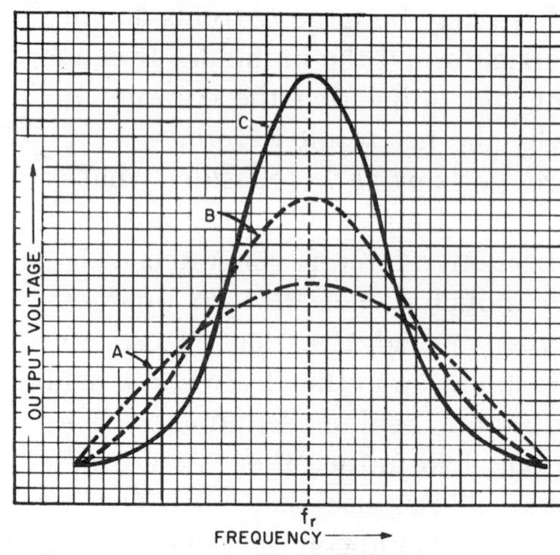

Fig. 7—13.

Curves showing how selectivity is increased as tuned circuits are added.
Curve A—one tuned circuit.
Curve B—two tuned circuits.
Curve C—three tuned circuits.

frequency. The effect of this is to increase the discrimination of the receiver against unwanted signals whose frequencies are other than the frequency of the desired signal to which the circuits are tuned. The effect is to increase the selectivity of the receiver. In Figure 7-13, curve A shows the resonant curve for a single tuned circuit; curve B, the resonant curve of two tuned circuits; and curve C, that of three tuned circuits, all tuned to the same resonant frequency. Note that the curves become sharper (more selective) as the number of tuned circuits is increased. The increased height of the curves is caused by amplification resulting from the amplifiers generally associated with the tuned circuits.

D. *Coupled circuits*

1. THE TRANSFORMER

In electrical circuits, we generally pass power from a source, such as a generator, to a load. It can be shown that the greatest transfer of power takes place when the impedance of the load is equal to that of the source.

Examine Figure 7-14. The source is a generator producing a voltage which we may call E_{gen} and causing a current (I) to flow. For the sake of simplicity, we will consider all impedances to be simple resistors. The impedance of the generator is represented by R_1, which is in series with the load (R_2).

The voltage drop across the load ($I \times R_2$) is represented by E_{R2}. Let us assume E_{gen} to be equal to 100 volts, R_1 to be equal to 10 ohms, and R_2 to be one ohm. The total resistance of the circuit ($R_1 + R_2$) is equal to 11 ohms. From Ohm's law, we can determine the current (I) flowing through the circuit.

$$I = \frac{E_{gen}}{R_{total}} = \frac{100}{11} = 9.09 \text{ amperes.}$$

The voltage drop (E_{R2}) across the load resistor can be determined as follows:

$$E_{R2} = I \times R_2 = 9.09 \times 1 = 9.09 \text{ volts.}$$

If we assume different values for the load resistance, we can draw up the following table:

E_{gen} (Volts)	R_1 (Ohms)	R_2 (Ohms)	I (*Amperes*)	E_{R2} (*Volts*)	P_{R2} (*Watts*)
100	10	1	9.09	9.09	82.62
100	10	5	6.66	33.30	221.77
100	10	10	5.00	50.00	250.00
100	10	20	3.33	66.60	221.77
100	10	100	0.90	90.00	81.00

Note that the voltage of the generator was kept at 100 volts under all conditions of load. From this table, we can see that when the resistance (or impedance) of the load (R_2) is equal to the resistance (or impedance) of the source (R_1), the maximum transfer of power from the source to the load occurs. This rule applies to any circuit where electric energy is transferred from one circuit to another.

At times, a maximum transfer of power is desired even though the impedance of the load does not match that of the source. Under such circumstances, we may employ an impedance-matching device, called a *transformer*.

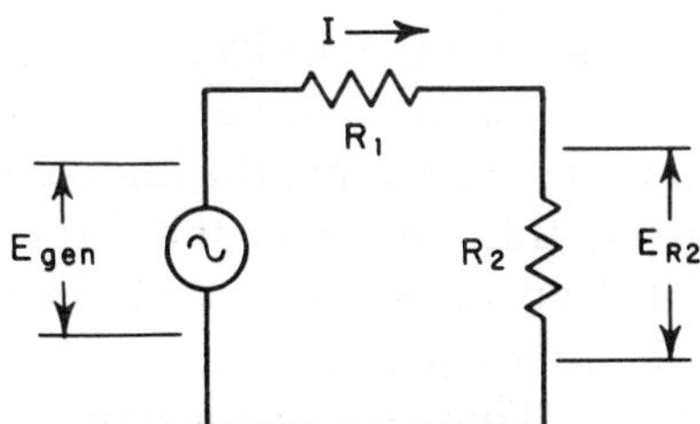

Fig. 7–14.

Diagram to illustrate the transfer of electric power.

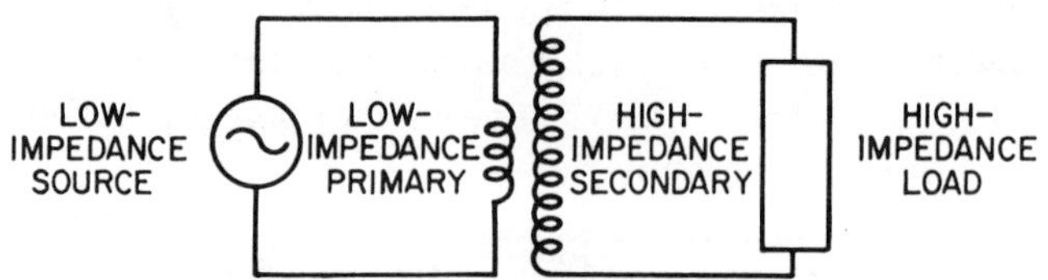

Fig. 7—15. How a transformer is used to match a high-impedance load to a low-impedance source.

The transformer consists of two coils of wire that are magnetically coupled to each other. If an alternating current is set flowing in one of these coils (called the *primary winding*), the varying magnetic field set up within that coil will pass within the turns of the second coil (called the *secondary winding*). As a result, an induced voltage is produced in this secondary winding. In this way, electric energy is transferred from the primary to the secondary circuit.

In a high quality transformer, the impedances are very high with the secondary and primary disconnected. However, when the secondary is connected to a load and the number of turns of the primary and secondary is proper, the primary will act like a resistance equal to the impedance of the source. In this condition the maximum power is delivered to the primary, which merely transfers it to the load. An impedance match is said to occur (Fig. 7-15).

The transfer of energy from the primary to the secondary winding of the transformer depends upon the magnetic field developed by the current through the primary winding linking with the turns of the secondary winding. In an ideal transformer, all these lines of force would link up with all the turns of the secondary winding, achieving 100 per cent (or unity) coupling. Since this is impossible to achieve in practice, a certain amount of the magnetic lines of force will leak off into the air. We call this *leakage flux,* and its effect is termed *leakage reactance.*

To reduce this leakage flux, the primary and secondary windings may be wound on a core of iron or other magnetic substance. This tends to concentrate the lines of force and to keep them from leaking off. If this core is in the form of a straight bar, we have what is called an *open-core* transformer.

More frequently, this core is in the form of a closed ring or square. Thus, a closed magnetic circuit is furnished, and the leakage flux is reduced further. We call this type a *closed-core* transformer (Figure 7-16). A variation that is greatly used is the *shell-core* type illustrated in Figure 7-16B. The primary and secondary windings are placed one over the other on the center arm of the core. The shell-core transformer can be designed to produce a coupling that closely approaches unity, or 100 per cent.

In addition to the losses due to imperfect coupling, we have *copper* and *iron* losses in a transformer. The copper loss is due to the resistance of the wire making up the turns of the windings. The iron loss may be divided into two parts.

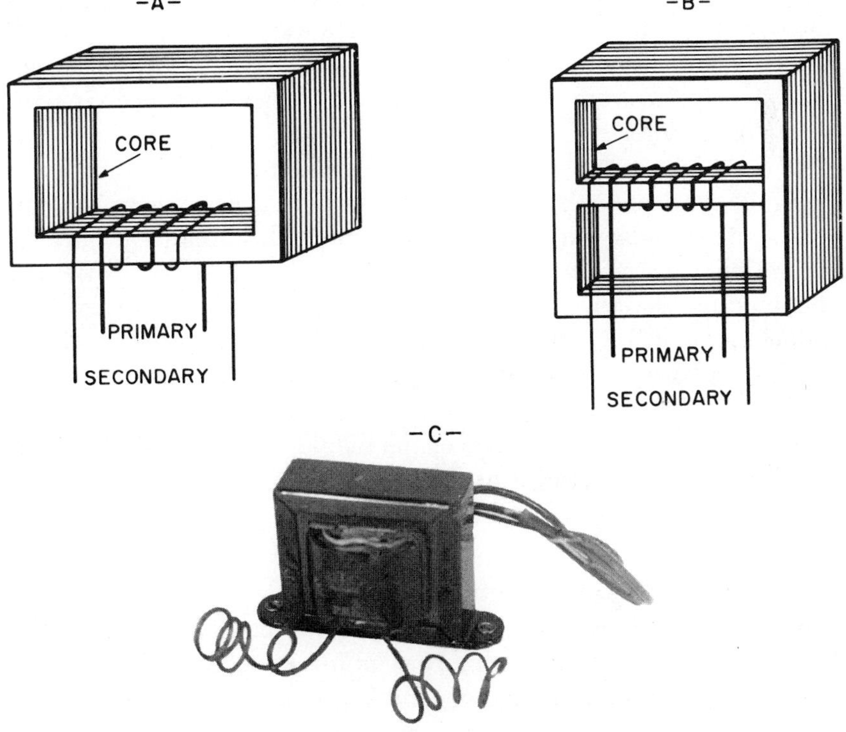

Fig. 7—16. Closed-core transformers.

A. Simple type.
B. Shell-core type.
C. Commercial shell-core transformer.

Since the core is in the magnetic field, it is magnetized. But the alternating current causes the iron core to change the polarity of its poles in step with the frequency of the current. A certain amount of energy is required to make this change. This energy comes from the electrical source and, therefore, is a loss. We call this loss the *hysteresis loss*. This loss may be partially reduced by using cores of silicon steel or of certain other alloys that are much more permeable than iron—that is, are easier to magnetize and demagnetize.

The other iron loss is due to the electric current that is induced in the iron core by the changing magnetic fields. This induced current is called the *eddy current*. Since the eddy current must come from the electrical source, it, too, is a loss. To reduce eddy-current losses, these cores are not made of solid metal, but are built up of very thin strips called *laminations*. Each lamination is coated with an insulation of oxide or varnish so that the eddy current cannot circulate through the core.

If the current flowing through the windings of the transformer becomes great enough, a condition may be reached where the core reaches its utmost limit of magnetization. This condition is called *saturation*. Where this condition occurs, any increase in current in the windings cannot magnetize the core any more. As a result, the inductance falls off. To prevent saturation, the cross-sectional area of the core must be made large enough to handle the currents present in the windings. Sometimes, a very narrow air gap is left in the core to increase the reluctance of the magnetic path, and thus more current may flow before saturation takes place.

In addition to its impedance-matching function, the transformer may perform another function. Assume that we have an ideal transformer with no losses and with coupling of 100 per cent. Further assume that the primary winding contains 100 turns and that it is connected to an a-c generator producing 100 volts.

If we assume that the secondary circuit is open, no power will be consumed by this secondary circuit. Since, in the ideal transformer, the power consumed in both the primary and secondary circuits are equal, no power is consumed in the primary circuit either. Electrical energy is changed to magnetic energy and back again.

The counter electromotive force produced by the primary winding is equal to the voltage of the source, 100 volts. Hence, as the magnetic field passes through this 100-turn winding, each turn has one volt induced into it. But the same magnetic field is linking the turns of the secondary winding as well. Hence, each turn of this winding, too, has one volt induced into it. If the secondary winding also has 100 turns, the induced voltage across the secondary is 100 volts.

Suppose, however, we construct our transformer with only 10 turns in the secondary. Since one volt is induced in each turn, the secondary voltage will be 10 volts. This is called a *step-down* transformer. If we construct the transformer with 1,000 turns in the secondary, the secondary voltage will be 1,000 volts. This is a *step-up* transformer. Thus the transformer can be used to step up or step down alternating voltage.

From the above, we can see that the ratio between the voltage across the primary winding and that across the secondary winding is equal to the ratio between the number of turns of the primary winding and the number of turns of the secondary winding. This relationship may be expressed in a formula as follows:

$$\frac{E_p}{E_s} = \frac{N_p}{N_s}$$

where E_p is the voltage across the primary winding, E_s is the voltage across the secondary winding, N_p is the number of turns in the primary winding, and N_s is the number of turns of the secondary winding.

Example. A transformer is required to deliver a 330-volt alternating current across the secondary winding. Assume a primary winding of 1,000 turns connected across the 110-volt a-c line. How many turns must we have in the secondary winding?

$$\frac{E_p}{E_s} = \frac{N_p}{N_s}, \text{ or } \frac{110}{330} = \frac{1{,}000}{N_s}$$

$$N_s = \frac{1{,}000 \times 330}{110} = 3{,}000 \text{ turns.} \quad \textit{Ans.}$$

In our ideal transformer, we assume no losses. Thus, the power $(E \times I)$ of the secondary circuit is equal to the power of the primary circuit. If, as in the above example, the voltage across the secondary winding has been stepped up three times, the current set flowing in the secondary winding will be reduced to one-third that of the primary. From this, we may obtain the following formula:

$$\frac{I_p}{I_s} = \frac{N_s}{N_p}$$

where I_p is the current flowing in the primary winding, I_s is the current flowing in the secondary winding, N_s is the number of turns in the secondary winding, and N_p is the number of turns of the primary winding. Thus, the transformer can be used as well to step up or step down alternating current. Of course, if the voltage is stepped up, the current is stepped down, and vice versa.

A variation of the two-winding transformer is the *autotransformer* illustrated in Figure 7-17. This transformer consists of a single tapped coil. The turns between the tap and one end constitute one winding of the transformer, and the entire coil constitutes the other winding. In Figure 7-17A, a step-up transformer is illustrated; in Figure 7-17B, we have the step-down version. The turn ratios hold for the autotransformer just as they do for the two-winding type.

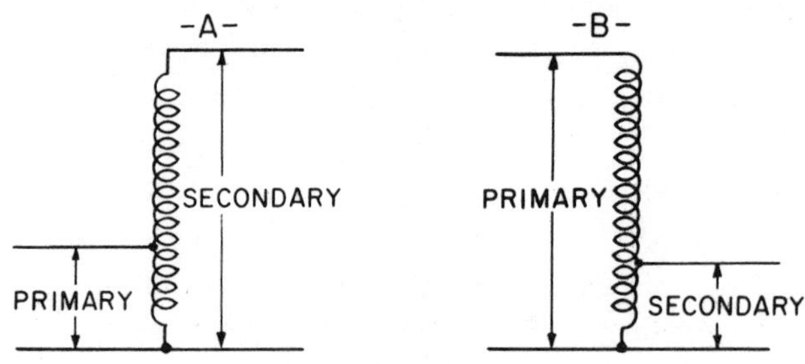

Fig. 7—17. The autotransformer.
A. Voltage step-up connection.
B. Voltage step-down connection.

Since the iron losses increase with the frequency of the current flowing through the windings, the laminated, iron-core transformer can be used only with currents of relatively low frequencies (15,000 cycles per second and lower). At higher frequencies, air-core or powdered iron-core transformers must be used.

In air-core transformers, only a small portion of the magnetic field links the primary and secondary windings. Thus, the voltage and current ratios described for the ideal transformer do not hold true. The powdered iron-core transformer is another high-frequency type employing a special core of powdered iron. Although this type does not achieve the coupling that is possible with laminated cores of silicon steel, it provides much better coupling than does the air-core type.

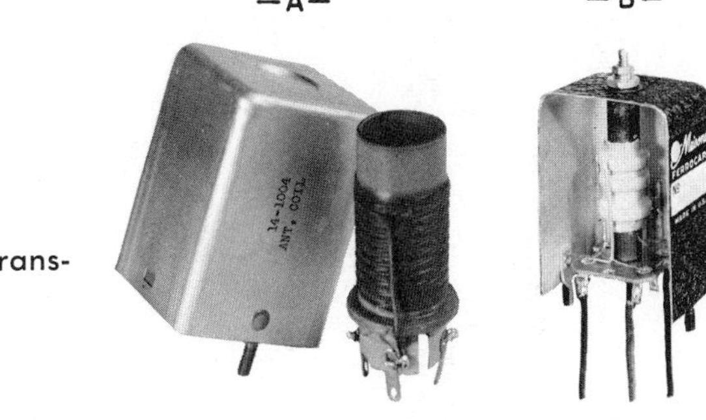

Fig. 7—18.

A. Air-core transformer.
B. Powdered iron-core transformer.

Meissner Mfg. Co.

The symbol for the air-core transformer is ; for the powdered iron-core transformer ; and for the laminated iron-core transformer .

2. COUPLING FOR HIGH-FREQUENCY CIRCUITS

In high-frequency circuits (such as used in radio receivers) it often becomes necessary to transfer energy from one resonant circuit to another. A number of methods exist by which this may be accomplished.

In the *direct-coupled* method, a component, such as an inductor, capacitor, or resistor, is made common to both resonant circuits (Figure 7-19). The input resonant circuit (composed of L_1 and C_1) is called the *primary* circuit. The output resonant circuit (composed of L_2 and C_2) is called the *secondary* circuit. The inductors are so placed that there is no inductive coupling between them.

As current flows in the primary circuit, a voltage drop occurs across the common component (L, C, or R). This voltage drop, in turn, causes current to flow in the secondary circuit.

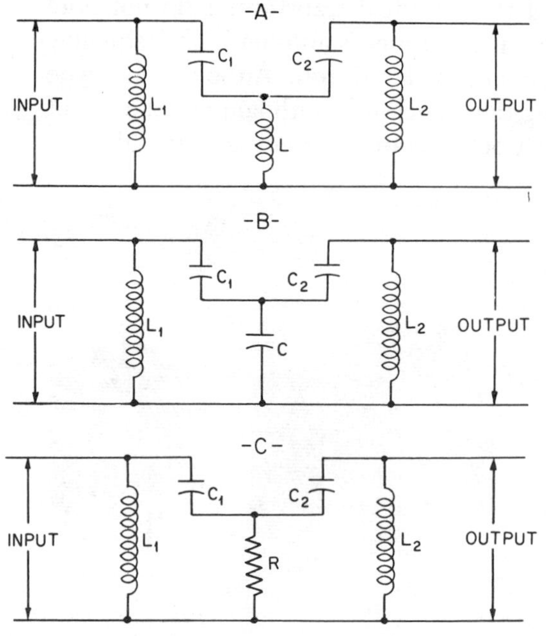

Fig. 7—19. Direct-coupled circuits.

A. Inductor L is the common component.
B. Capacitor C is the common component.
C. Resistor R is the common component.

Another method of coupling is the *capacitive-coupling* method illustrated in Figure 7-20. Energy from the primary circuit is passed on to secondary circuit through the coupling capacitor C. As before, L_1 and L_2 are not inductively coupled.

Still another method of coupling is the *inductive-coupling* method illustrated in Figure 7-21. In this case, coils L_1 and L_2 are inductively coupled, forming the primary and secondary windings, respectively, of a transformer.

Since such a transformer operates at high frequencies, it is of the air-core or powdered iron-core type. As such, the coupling between the primary and secondary windings is considerably less than unity, and the step-up or step-down effects arising out of turn ratios do not hold. Coupling between the inductors can be varied by placing the windings closer or further apart.

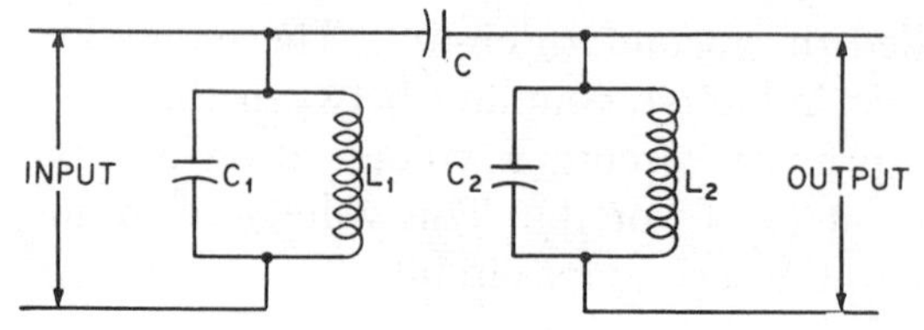

Fig. 7—20. Capacitive-coupled circuits.

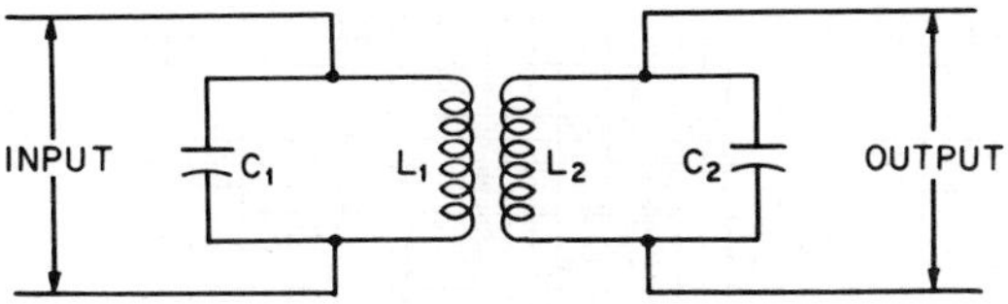

Fig. 7–21.

Inductive-coupled circuits.

The reflected impedances or resistances from one circuit to the other become quite important. Assume that both the primary and secondary circuits are tuned to the same resonant frequency and that the coils are placed a considerable distance apart. As current flows in the primary circuit, a certain amount of voltage will be induced in the secondary circuit. Because the coupling is very small, the energy transferred will be small too, and so will be the voltage output of the secondary circuit.

Since, at resonance, the inductive reactance cancels out the capacitive reactance, both the primary and secondary circuits are resistive in nature. But at all frequencies above and below the resonant frequency (f_r), X_L does not cancel out X_C, and these circuits, therefore, become reactive in nature.

At the resonant frequency the reflected resistance from each circuit reduces the *selectivity* of the other. At frequencies other than the resonant, the reflected impedance from one circuit to the other upsets the *tuning* and makes each circuit resonant for some other frequency than the original resonant frequency. Thus, the voltage output becomes maximum for these new frequencies, whereas it drops off at the original resonant frequency because of the effect of the reflected resistance.

This is illustrated in the series of curves shown in Figure 7-22. Curve A shows the effect of having the coils far apart. Both the reflected resistance and reflected impedance are quite small. The selectivity, therefore, is fairly high, as witnessed by the sharp peak of the curve. But since the energy transfer is low, the output voltage is not very great.

As the coils are brought closer and closer together, the energy transfer becomes greater and the output voltage increases (curves B and C). Note that the curves tend to flatten out, indicating the loss of selectivity resulting from the increased reflected resistance.

Curve C represents the maximum coupling possible before the effect of reflected impedance becomes apparent. This point is called the *critical coupling*. Note that the output voltage has reached its maximum.

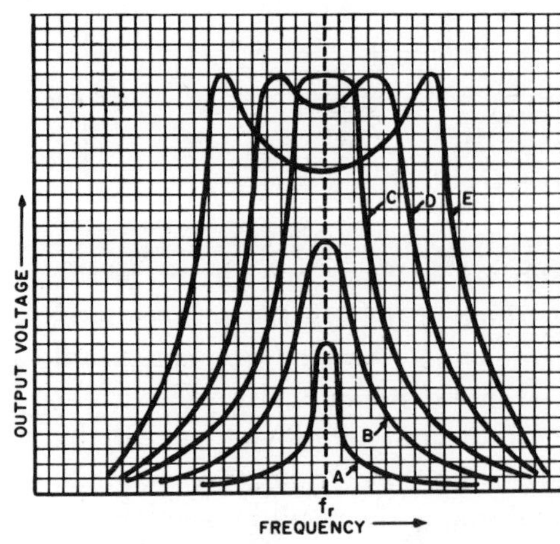

Fig. 7—22.

Resonance curves showing the effects of changes in coupling.

As the coupling is increased, the reflected impedance becomes greater. The selectivity is reduced, flattening the top of the curve, and a double peak appears, one on each side of the original resonant frequency. This indicates the new resonant frequencies. (See curve D.)

As the coupling is increased still further, the effect becomes more pronounced. (See curve E.) The curve becomes flatter and broader, and the two new resonant peaks are further apart. Because of the greater reflected resistance from the closer coupling, the dip at the original resonant frequency becomes greater.

The flow of current in the secondary circuit depends, not only on the output voltage, but on the resistance of the load as well. If the resistance of the load is low, a relatively large current will flow. This larger flow of current will create a more powerful magnetic field around the secondary inductor which, in turn, will result in a larger reflected resistance into the primary circuit. As a result, the selectivity of the circuit will be reduced.

E. *Filters*

Electrically, *filters* are used to separate currents of certain frequencies from those of other frequencies. (For our discussion here, we may consider direct current as having a zero frequency.) Assume that you have as an electrical source a battery (whose symbol is −|+) supplying direct current, a generator of low-frequency

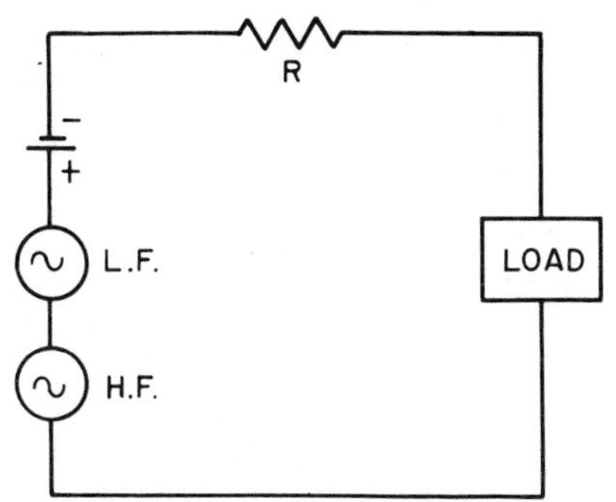

Fig. 7—23.

Circuit showing direct, low-frequency, and high-frequency currents feeding a load through resistor R in series.

alternating current (∼ L.F.), and a generator of high-frequency alternating current (∼ H.F.). Assume that they are all connected in series and supply current to a load through a resistor (R) in series, as shown in Figure 7-23. The resistor will not have any filtering action, since it impedes equally all currents that pass through it, regardless of frequency.

Now assume that you replace the resistor with a capacitor (Figure 7-24). The direct current will be completely filtered out, since the capacitor offers infinite impedance to its passage. Since $X_C = 1/(2\pi f C)$, the higher the frequency, the less will be the impedance. Hence, the low-frequency current will be more strongly impeded (or *attenuated*) than the high-frequency current.

Now assume that the capacitor is replaced by an inductor (Figure 7-25). The direct current will be only slightly impeded, owing to the resistance of the inductor. Since $X_L = 2\pi f L$, the higher the frequency, the greater the impedance will be. Hence, the high-frequency current will be more strongly attenuated than the low-frequency current.

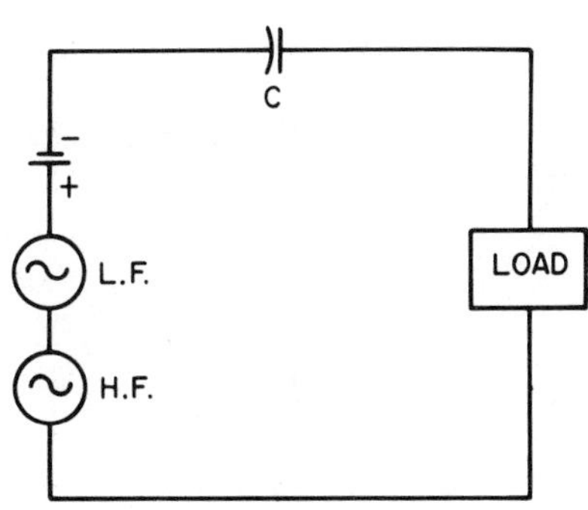

Fig. 7—24.

Circuit showing direct, low-frequency, and high-frequency currents feeding a load through capacitor C in series.

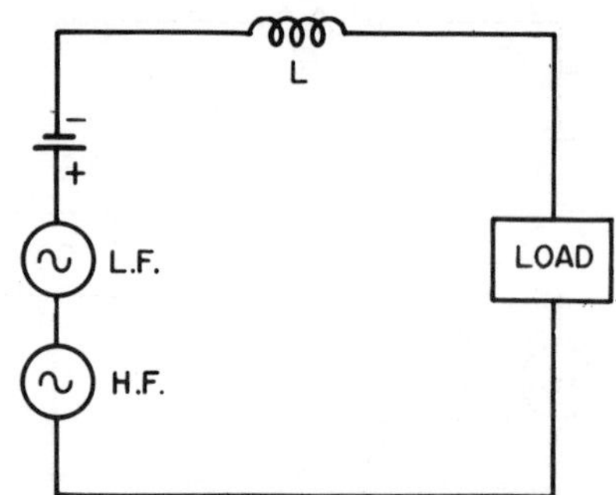

Fig. 7—25.

Circuit showing direct, low-frequency, and high-frequency currents feeding a load through inductor L in series.

Suppose we connect the capacitor across the load (Figure 7-26). None of the direct current will flow through the capacitor, all of it going through the load. Since the capacitor offers a fairly high impedance to low-frequency current, most of this, too, will flow through the load. But since the capacitor offers a low impedance to high-frequency current, most of this current will flow through the capacitor rather than through the load. This type of circuit is called a *low-pass filter,* since it passes the low-frequency currents on to the load, bypassing the high-frequency currents (that is, high-frequency currents bypass the load).

In Figure 7-27, the capacitor is replaced by an inductor. This inductor offers a low-impedance path to the direct and low-frequency currents. Hence, most of these currents are bypassed and do not reach the load. But the high-frequency current finds the inductor a high-impedance path, and thus most of this current flows through the load. This type of circuit is called a *high-pass filter,* since it passes the high-frequency currents on to the load. The resistor R limits the current that can be drawn from the generators by the capacitor at high frequencies or the inductor at low frequencies.

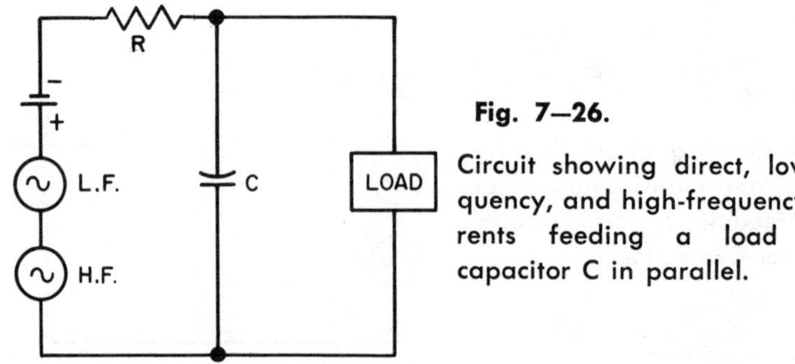

Fig. 7—26.

Circuit showing direct, low-frequency, and high-frequency currents feeding a load with capacitor C in parallel.

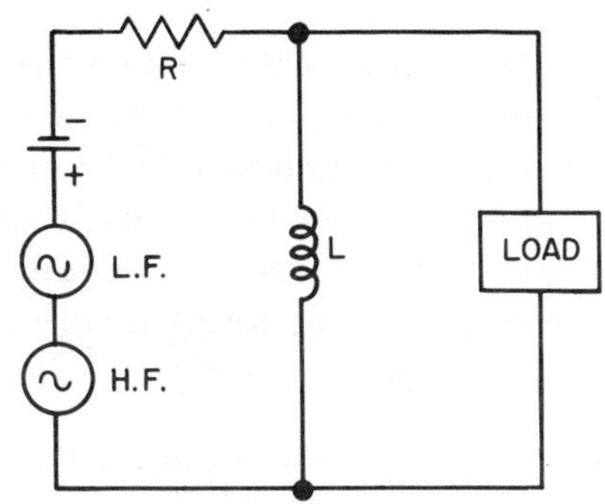

Fig. 7–27.

Circuit showing direct, low-frequency, and high-frequency currents feeding a load with inductor L in parallel.

1. RESONANT CIRCUITS AS FILTERS

Where the filtering action is to be limited to a single frequency or, at most, to a narrow band of frequencies, resonant circuits are used. Filters of this type are used extensively in electronic circuits.

Figure 7-28 shows such a circuit, which is an example of a *band-pass filter* used to pass only current of a single frequency (or a narrow band of frequencies) to a load. The values of the inductors (L) and capacitors (C) are such as to form resonant circuits at the frequency we wish to pass through the filter. The series-resonant circuit offers a very low impedance to currents of the resonant frequency and a relatively high impedance to currents of other frequencies.

The parallel-resonant circuit, on the other hand, offers a very high impedance to currents of the resonant frequency and a relatively low impedance to currents of other frequencies. Thus, currents of the resonant frequency pass easily through the filter while currents of all other frequencies are bypassed through the parallel-resonant circuit.

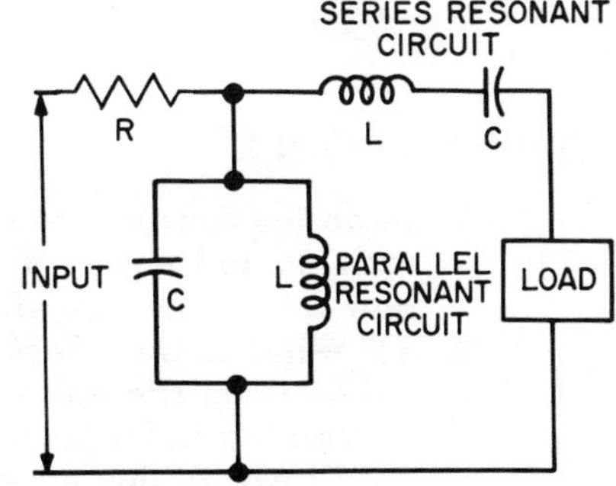

Fig. 7–28.

Band-pass filter.

In Figure 7-29 is shown the circuit used to filter out current of a particular frequency (or a narrow band of frequencies). Here, too, L and C are shown to form resonant circuits at the desired frequency. Currents of this frequency will find the parallel-resonant circuit an extremely high-impedance path, whereas the series-resonant circuit forms a very low-impedance path.

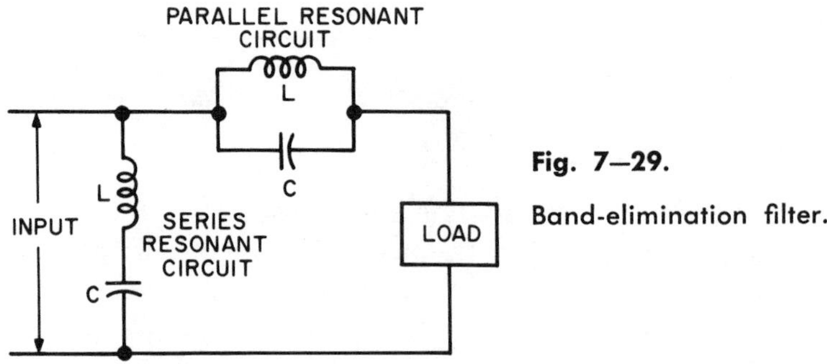

Fig. 7—29.

Band-elimination filter.

Currents of other frequencies, however, will find the parallel-resonant circuit a fairly low-impedance path, but the series-resonant circuit furnishes a fairly high-impedance path. Thus, currents of the resonant frequency will be stopped and bypassed, while currents of all other frequencies will pass through. This type of filter is called a *band-stop*, or *band-elimination, filter*.

QUESTIONS

1. In an a-c circuit containing a resistor of 12 ohms in series with an inductor whose inductive reactance is 16 ohms, how much current will flow if the voltage is 100 volts?

2. a) What is the effect of resistance upon the phase angle between the voltage and current in an a-c circuit containing inductance?

b) Illustrate this effect by means of a vector diagram.

3. A capacitor and a 4,000-ohm resistor are connected in series in an a-c circuit. If, with a voltage of 100 volts, 20 milliamperes of current flow through this circuit, what is the capacitive reactance of the capacitor?
4. a) What is the effect of resistance upon the phase angle between the voltage and current in an a-c circuit containing capacitance?
   b) Illustrate this effect by means of a vector diagram.
5. What is the *time constant* of an *R-C* circuit containing a 0.01-microfarad capacitor and a resistor of 100,000 ohms?
6. a) What is the total reactance in an a-c circuit containing an inductor whose inductive reactance is 2,500 ohms and a capacitor whose capacitive reactance is 3,000 ohms, connected in series?
   b) What is the nature of this reactance?
7. What is the total impedance of an a-c circuit containing a resistor, inductor, and capacitor in series, if the resistance is 150 ohms, the inductive reactance is 300 ohms, and the capacitive reactance is 500 ohms?
8. a) Explain what is meant by the *resonant frequency* of an a-c circuit.
   b) State the formula for resonance in terms of inductance and capacitance.
9. At the resonant frequency, what is the theoretical impedance of
   a) a *series-resonant circuit;*
   b) a *parallel-resonant circuit?*
10. What is the resonant frequency of an a-c circuit containing a 5-henry inductor and a 5-microfarad capacitor?
11. Explain how, in the radio receiver, resonant circuits are used to accept signals from the desired station and tune out those of all others.
12. a) In the radio receiver, what is the effect upon the *selectivity* of the receiver produced by increasing the number of tuned circuits?
    b) What effect does resistance have upon the selectivity?
13. Explain what is happening in an *oscillatory circuit* at resonance.
14. a) Describe the construction of an *iron-core transformer.*
    b) Explain the function of each of its parts.
15. a) What are the *copper* and *iron losses* of a transformer due to?
    b) Explain what can be done to reduce these losses?
16. a) What is meant by *saturation* of the transformer core?
    b) What can be done to minimize its effect?
17. Explain what is meant by
    a) a *step-up transformer;*
    b) a *step-down transformer.*

18. **A transformer has a primary winding of 500 turns and a secondary winding of 5,000 turns. If the primary winding is connected across a 120-volt a-c line, what will be the voltage across the secondary winding?**

19. **Explain three functions of the transformer.**

20. **a) Explain the operation of a *step-up autotransformer*, using a simple diagram.**

b) How may this autotransformer be used as a *step-down type?*

21. **a) What is the function of an *electrical filter?***

b) Draw the circuit of a *low-pass filter* and explain its action.

c) Draw the the circuit of a *high-pass filter* and explain its action.

22. **Draw and explain a *band-pass filter* using resonant circuits.**

23. **Draw and explain a *band-stop filter* using resonant circuits.**

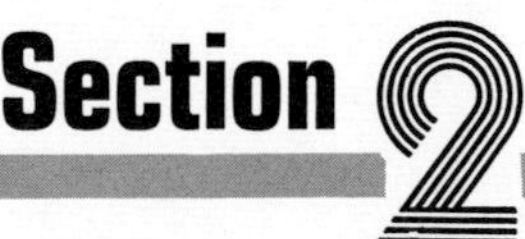

Section 2

The Electron Tube

The heart of the science of electronics is the electron tube—the twentieth century Aladdin's Lamp. In this section we shall consider this tube in some of its most common forms.

Introduction

In 1883, the great American genius, Thomas A. Edison, made a glorious fumble. About four years earlier he had invented the electric lamp by placing a filament in a glass bulb from which the air had been removed. As an electric current flowed through the filament, it was heated due to its resistance until it became incandescent, that is, it gave off light.

Edison's lamp suffered from a serious drawback. As the filament was heated, some of its particles boiled off and were deposited as an opaque coating on the inside of the bulb. This gradually reduced the amount of light that could come through. Edison sealed a metal plate into the bulb, hoping that the particles would be attracted to this plate, rather than distribute themselves over the glass. This attempt failed.

However, in the course of his experiments, he connected a milliammeter between the plate and the positive side of the filament. To his surprise, the meter showed that a current was flowing. (See Figure 8-1.)

Edison could not explain why the current should be there. He merely entered in his notebook a description of this phenomenon which now is known as the *Edison effect*.

Today we can explain it. When the plate is connected through the meter to the positive side of the filament, electrons are pulled away from the plate to the positive side of the electrical source. Hence the plate acquires a positive charge. As the filament is heated, it emits a cloud of electrons. These electrons are attracted to the positive plate and on to the positive side of the source. As these electrons flow through the meter, a current is indicated. Should we connect the plate to the negative side of the filament, the plate would be charged negatively, the electrons would be repelled, and no current would flow through the meter.

We can forgive Edison his fumble. At that time the electron was unknown. In 1904, some seven years after J. J. Thomson discovered the electron, another English scientist, J. Ambrose Fleming, came up with the correct explanation of the Edison effect and thus the electron tube was born.

Essentially, the electron tube consists of an *emitter* of electrons and a positively-charged *plate*, sealed in a tube or bulb. All the air is evacuated from the tube to minimize the interference between the air molecules and the electrons in their flight from emitter to plate. (In some tubes a very small amount of gas, such as mercury vapor or argon, is introduced after all the air has been removed. More about this later.)

When the electrons flow through a conductor, such as a copper, wire, they are confined within the boundaries of the conductor and,

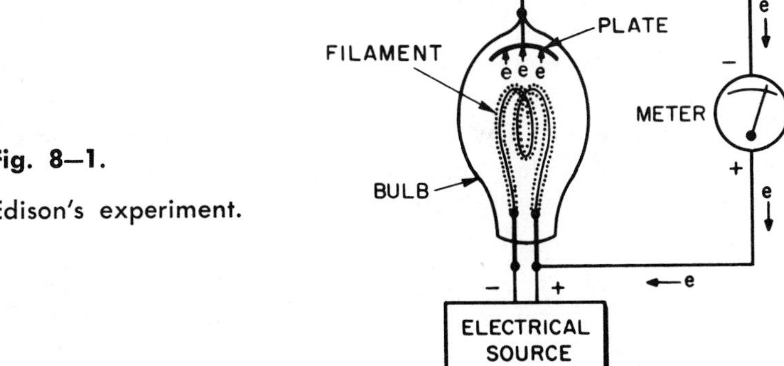

Fig. 8–1.

Edison's experiment.

therefore, control over them is limited. But when electrons move through a vacuum or gas, they are much freer and can be more easily controlled by external means.

For example, since electrons are charged particles, their flow may be controlled by the electric fields set up by charged grids placed in the path of the moving electrons. Further, since the flow of electrons constitutes an electric current and, as such, is surrounded by a magnetic field, the interaction between its own magnetic field and another external magnetic field may be used to control the flow of electrons.

The two-element electron tube (emitter and plate) is the simplest type. In more complicated tubes, one or more grids, or other types of electrodes, may be inserted between the emitter and the plate to control the electrons in their flight.

A. *The envelope*

Since we are dealing with the movement of electrons through a vacuum or gas, this flow must take place in a sealed tube or bulb, called the *envelope.* These envelopes are made of glass, metal, ceramic, or a combination of these. They are constructed in many different sizes and shapes. (See Figure 8-2.)

For use in circuit diagrams, the symbol for the envelope is ○. If, instead of containing a vacuum, the envelope contains a gas, a dot is added, thus ⊙.

B. *The emitter*

Since the envelope is sealed, all the electrons that flow through the electron tube must come from the sealed-in emitter, or *cathode.* There are a number of ways in which the emitter may be made to give off electrons. Most common is the use of heat.

As previously described, there always is a disorderly movement of free electrons within all substances, especially metals. If heated, the movement of free electrons within a substance is increased. If the temperature is raised high enough, the movement is increased to the point where some of the electrons actually fly off from the substance. We call this process *thermionic electron emission.*

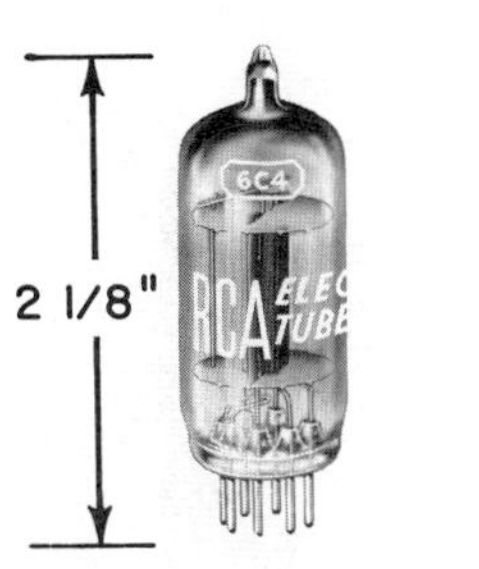

Fig. 8—2. Various types of electron tubes.

A. Miniature, glass-envelope type.
B. Octal-base, metal-envelope type.
C. Metal-ceramic power tube.

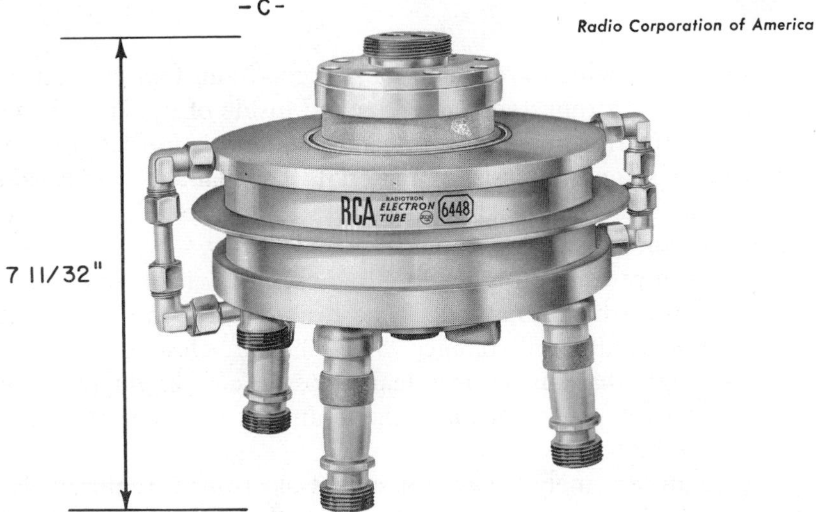

Radio Corporation of America

The most convenient way for obtaining this emission is to construct the emitter in the form of a filament and heat it by passing a current through it. Then, as the electrons are shot out, they are attracted to the positively-charged plate.

Where a relatively small electron emission is required, the filament is made of thin tungsten wire, somewhat similar to the filament of an electric lamp. (See Figure 8-3A). Where a greater electron flow is desired, the filament may be made of heavy tungsten ribbon (Figure 8-3B). In some tubes a small quantity of thorium oxide and carbon are dissolved into the tungsten filament to make it a better emitter of electrons.

A great many tubes employ the *indirect-heater cathode* as a source of electrons. (See Figure 8-4.) The cathode is a metal tube, generally coated with certain chemicals, such as the oxides of barium

-A- -B-

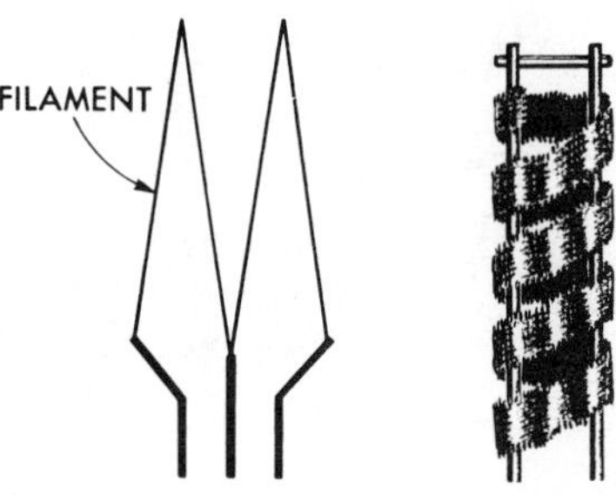

Fig. 8—3. Filament emitters.

A. Thin-wire type.
B. Heavy-ribbon type.

and strontium, which are exceptionally good emitters of electrons. It is heated by a tungsten filament coiled inside of it. This filament has no electrical connection to the cathode and is insulated from it by a tube of some heat-resisting material. The filament here is called a *heater* and its only function is to heat the cathode until it is hot enough to emit a sufficient quantity of electrons.

There is a precaution that must be observed with tubes employing the indirect-heater cathode. The heater and cathode generally are operated at different voltages. Since they are close to each other, there is danger that current may leak across. Accordingly, the difference in potential between the cathode and its heater should never become too high.

Where an extremely heavy emission of electrons is required, as in the case of certain industrial applications, the cathode consists of a pool of mercury. The mercury is heated by an electric arc formed between a special electrode (the *ignitor*) and the mercury. As a result, a heavy cloud of electrons is emitted.

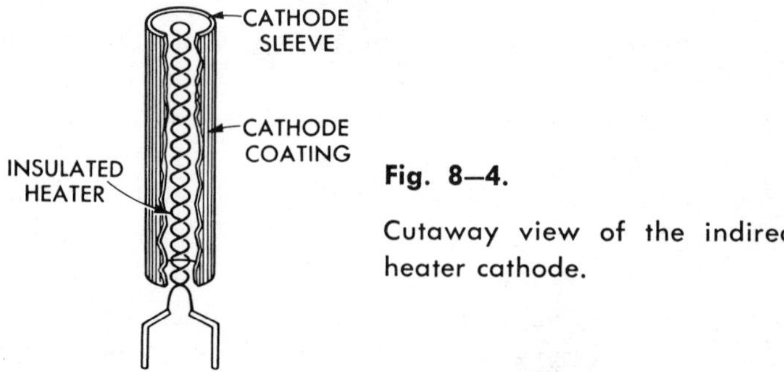

Fig. 8—4.

Cutaway view of the indirect-heater cathode.

For use in circuit diagrams, the symbol for a filament or heater is . The symbol for the cathode is , and where the indirect-heater cathode is employed, its symbol becomes . The symbol for the mercury-pool cathode, including its ignitor, is  .

Voltages used for the filaments or heaters of electron tubes generally range from about 1.5 to 117 volts. The indirect-heater cathode frequently is employed in low-power tubes as, for example, those used in radio and television receivers. For these, heater voltages of 6.3 or 12.6 volts are most common. The heater currents most commonly used are 0.30 or 0.45 ampere for the 6.3-volt tubes and 0.15 ampere for the 12.6-volt tubes.

High-power tubes generally employ heavy, ribbon-type filaments. Filament voltages for these tubes rarely exceed 30 volts. The filament currents, however, may range up to several hundred amperes.

There are other methods besides the use of heat to obtain electronic emission. One makes use of a high positive charge on the plate to pull the negative electrons free from the cathode. Because there is no need for heat, this method is called *cold-cathode emission.* Certain types of X-ray tubes, for example, utilize this method. The symbol for the cold cathode is .

Another method makes use of the fact that when light (visible, ultraviolet, or infrared) falls upon certain substances, such as potassium, sodium, or cesium, the light energy causes electrons to be emitted from these substances. This is called *photoelectric emission.*

The cathode is coated with the light-sensitive material and, as light penetrates the tube envelope and strikes the cathode, electrons are emitted. Tubes using this method of electron emission are called *phototubes.* The symbol for the light-sensitive cathode is  .

C. *The plate*

The plate, or *anode,* is a positively-charged electrode that attracts the electrons emitted by the cathode. In some tubes, the plate is merely a metallic button near the emitter, as shown in Figure 8-5A. In other tubes it may be constructed as a cylindrical or rectangular can surrounding the emitter (Figure 8-5B). The symbol for the plate is .

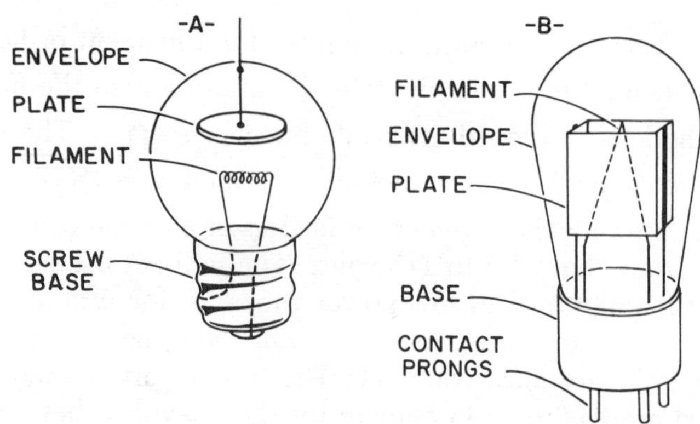

Fig. 8—5. Electron tubes, showing two types of plates.

The flow of electrons through the tube and its associated circuit is called the *plate current.* The greater the positive charge on the plate, the greater will be the flow of electrons to it. The plate current is limited by the ability of the cathode to emit electrons.

As the electrons strike the plate, the force of impact creates heat. Therefore, the plate must be made of some material capable of withstanding heat. The materials generally employed are nickel, tungsten, graphite, molybdenum, or tantalum. The size and shape of the plate are determined by the amount of heat that must be dissipated.

The heat represents a power loss, called *plate dissipation,* which comes from the electric power supplied to the plate. The plate dissipation is only a few watts for the low-power tubes, but may be as great as 100 kilowatts for the high-power ones.

Where small tubes, such as those used in the radio receiver, for example, are employed, sufficient heat can be removed by radiation into the surrounding air. For some of the larger tubes that pass relatively large quantities of current, however, cooling by forced air or circulating water is employed. (See Figure 8-6.) In the case of the air-cooled tube, a finned radiator helps remove the heat from the plate. In the water-cooled tube, the plate is surrounded by a water jacket.

If the plate current becomes too high, there is danger that, even with the cooling precautions, the plate may be overheated because of excessive electronic bombardment. Or else the cathode may be damaged because of excessive emission. The plate current, generally, is continuously variable. Hence we have two values of plate current flow in the tube: the *average current* and the *peak current* (see Chapter 5, subdivision C,1). The maximum average plate current is the safe limit of the tube before the plate is made too hot by electronic bombardment. The maximum peak plate current is the safe limit of the tube before excessive emission damages the cathode.

Under certain conditions (which will be discussed later) the plate may become alternately positive and negative with respect to the cathode. Current flows in the tube only when the plate is positive; when the plate is negative, no current will flow. But if, in this latter condition, the potential between the plate and cathode becomes high enough, current may arc across from the negative plate to the positive cathode. The maximum voltage which may be applied without danger of breakdown is called the *maximum peak inverse voltage.*

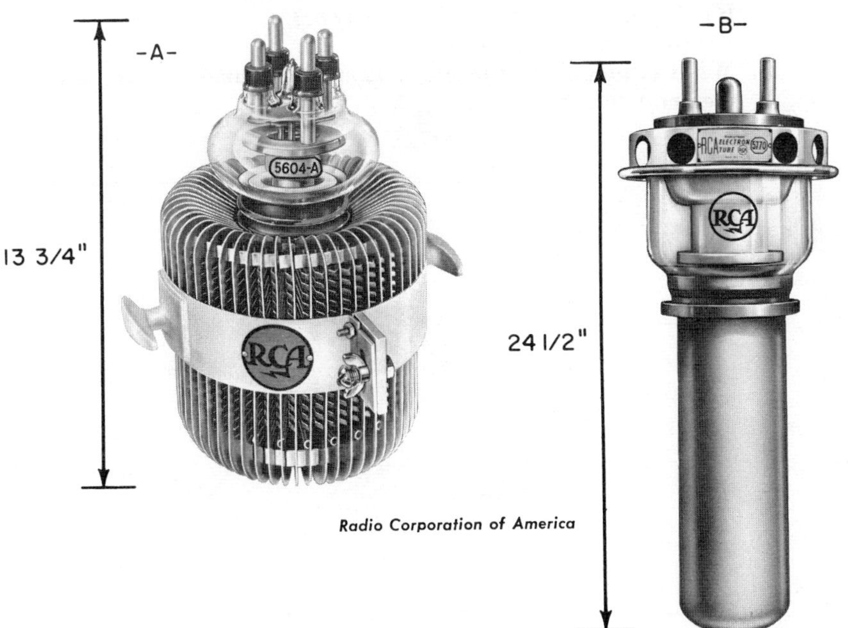

Fig. 8–6. **A.** Forced-air-cooled power tube.
B. Water-cooled power tube.

Connections are made to the various electrodes of the tube by means of wires sealed into the envelope. These wires may terminate in any number of different types of contacts, such as lugs or prongs. Special sockets frequently are provided into which these contacts fit. In this way, the tube may be removed from its socket without the necessity for unsoldering or unscrewing connections.

QUESTIONS

1. **Explain the *Edison effect* and show how it illustrates the basic principle of the electron tube.**
2. **a) Explain the function of the *cathode* of the electron tube.**
   b) What types of cathodes are used for low-power tubes?
   c) What types of cathodes are used for high-power tubes?
3. **Explain the following types of electron emission:**
   a) thermionic;
   b) cold-cathode;
   c) photoelectric.
4. **a) Explain the function of the *anode* of the electron tube.**
   b) What is meant by *plate current;*
   c) by *plate dissipation?*
5. **What is meant by the *maximum peak inverse voltage* of an electron tube?**

The Diode

Simplest of the electron tubes is the one containing only a cathode that emits electrons and an anode, or plate, that attracts them, enclosed in an envelope. Because such a tube contains only two electrodes, it is called a *diode* (*di* meaning two, and *ode* from electrode). The symbol for a diode using filament cathode is . Where an indirect-heater cathode is employed, the symbol becomes . Frequently, the heater symbol is omitted for the sake of simplicity. Where the tube is gas-filled, a dot is added, thus .

Assume we have a diode consisting of a filament cathode and a plate sealed into a tube from which the air had been removed. As current is passed through the filament, it becomes incandescent and starts to emit electrons. As more current is sent through, the filament becomes hotter and emits more electrons. This process continues until a certain limit is reached. Beyond this point the filament will disintegrate.

As the electrons are shot off, they form a cloud (known as the *space charge*) about the filament. Since like charges repel, this cloud of electrons tends to force back to the filament other electrons that are being emitted. The limit of emission, therefore, is reached well before the filament disintegrates.

If a positive charge is placed upon the plate, the electrons of the cloud nearest this electrode are attracted to it. This reduces the electron-repelling action of the space charge, and more electrons are emitted.

The electrons attracted to the positively-charged plate would soon neutralize this charge, but if a battery is connected, as shown in Figure 9-1, the positive terminal of the battery would drain electrons off from the plate and maintain it positive, the electrons being returned to the cathode through the external circuit.

The external circuit from plate to cathode is called the *plate circuit,* and the flow of electrons through this circuit is called the *plate current.* The meter has been inserted in the plate circuit to measure the flow of plate current.

The internal flow of electrons from the cathode to the plate is called the *space current.* In this type of tube, the plate current is equal to the space current.

Note that the space current flows in a one-way path from the cathode to plate. The plate normally does not emit electrons. Note, too, that the space current flows from cathode to plate under the influence of an electric field resulting from a difference of potential between the two electrodes. In other words, space current flows only when the plate is positive relatively to the cathode.

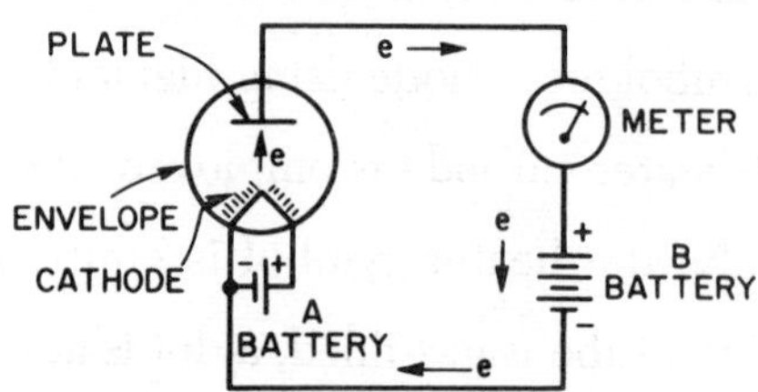

Fig. 9—1. Circuit showing how current flows through the tube and its plate circuit.

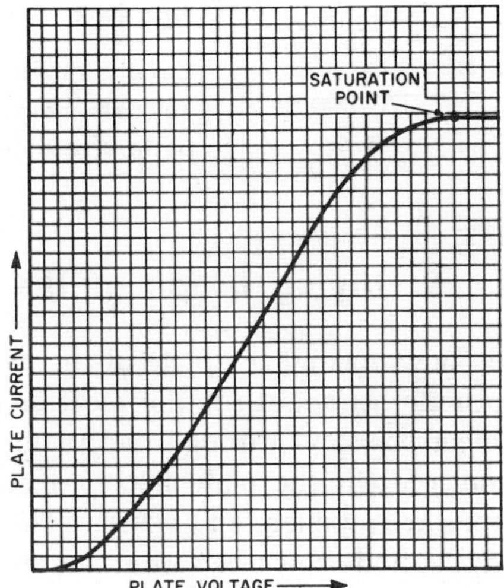

Fig. 9–2. Characteristic curve of tube showing relationship between plate voltage and plate current.

The battery that heats the filament is called the *filament,* or *A, battery.* The battery that maintains the plate positive, relative to the cathode, is called the *plate,* or *B, battery.* Note that the filament and plate circuits are independent of each other, though they touch at one point.

As the positive voltage on the plate is made larger and larger, more and more of the electrons of the space charge are removed. Thus more and more plate current flows. At a certain point, the entire space charge is removed, and electrons emitted by the cathode may travel directly and without impediment to the plate. Increasing the plate voltage beyond this point will not result in any increase in plate current, since the entire output of the cathode is continuously attracted to the plate. This maximum plate current is called the *saturation current.* No greater plate current will flow unless the temperature of the emitter is raised, increasing the emission.

The graph of Figure 9-2 shows the relationship between the plate current and the voltage applied to the plate. Note how the plate current rises as the plate voltage is increased until the saturation point is reached. (To obtain the data for this graph the filament temperature was kept constant.) The curve is called the *plate voltage–plate current characteristic curve* of the tube.

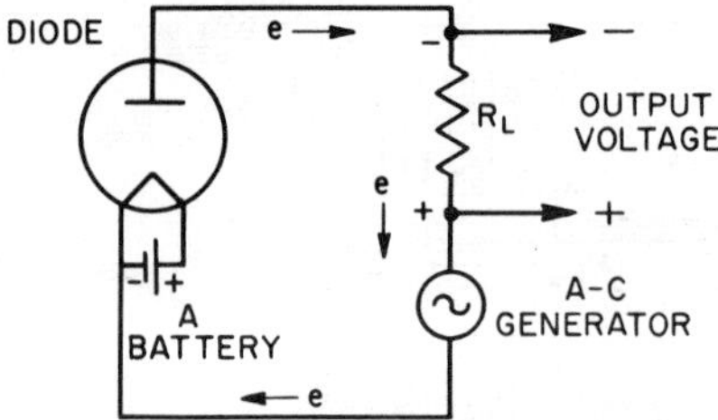

Fig. 9—3.

Circuit showing an alternating voltage being applied to the plate of a diode.

A. *Rectifying action of the diode*

We have seen how making the plate of the diode positive with relation to the cathode attracts to it the electrons emitted by the cathode. These electrons flow through the plate circuit and constitute the plate current. If the plate becomes negative, however, the emitted electrons are repelled, and no plate current flows.

Note, then, that the diode acts as a sort of automatic switch which closes the plate circuit and permits plate current to flow when the plate is positive, relative to the cathode. When the plate becomes negative, the plate circuit is opened and no plate current can flow. In other words, the diode is conductive when its plate is positive and nonconductive when it is negative.

If, as in Figure 9-3, the B battery is replaced by an a-c generator, the plate will be made positive and negative alternately. But plate current will flow, as indicated by the arrows, only when the plate is positive, that is, only during the half-cycles of the generator when a positive charge is being placed on the plate. During the negative half-cycles, no current will flow.

The meter has been replaced by resistor R_L, called the *load resistor.* As the plate current flows through R_L, a voltage drop, with the indicated polarity, is set up across it. This is the *output voltage.* Note that it is a direct voltage.

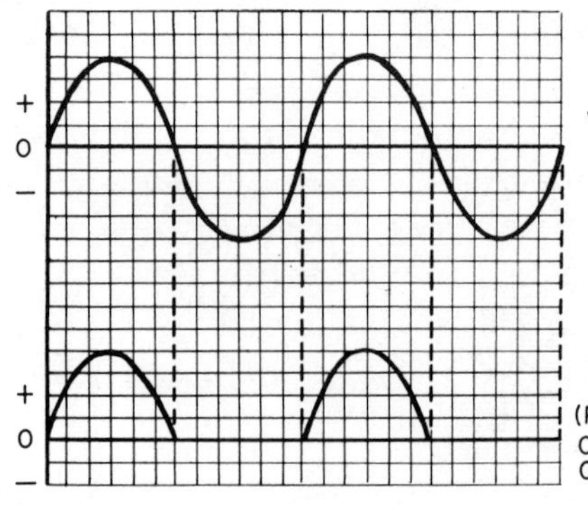

Fig. 9—4.

Graph showing relationship between plate voltage and plate current in the diode.

The action of the diode can be shown graphically, as illustrated in Figure 9-4. The upper portion of the graph indicates the voltage placed upon the plate of the diode by the a-c generator. The lower portion indicates the resulting flow of plate current. Note that, although the input to the tube is an alternating voltage, the resulting plate current is a pulsating direct current. This process of changing an alternating into a direct current is called *rectification,* and is one of the chief functions of diodes.

The rectifying action of the diode can be further demonstrated, using the characteristic curve of Figure 9-2. The alternating-voltage input is plotted against the characteristic curve, producing a pulsating d-c output. (See Figure 9-5.)

It should be noted that Figure 9-5 shows, in reality, two combined graphs. One (the vertical) is of the alternating-voltage input. Here the voltage is plotted against time; each number on the time axis represents a quarter-cycle. The other graph (the horizontal) is that of the plate current of the tube, also plotted against time. The characteristic curve of the tube ties the two graphs together.

Note that the time intervals of both graphs are the same—each number represents a quarter-cycle. At zero, both the input voltage and plate current are zero. At the quarter-cycle mark, the input voltage has risen to its maximum positive value (point A). To locate the corresponding point on the plate-current graph, project a line

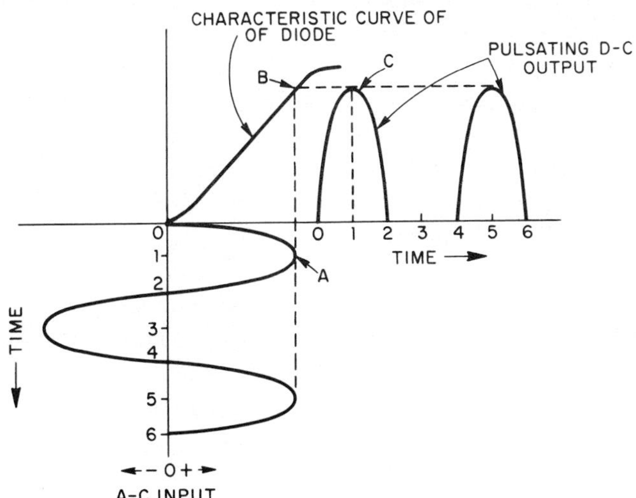

Fig. 9—5. Graph showing the rectifying action of the diode.

vertically from point A until it meets the characteristic curve of the diode (at point B). From this point project a line horizontally until it intersects the quarter-cycle line of the plate-current graph (at point C). In similar fashion the half-cycle, three quarter-cycle, and one-cycle points are plotted.

Note that at the quarter-cycle mark the input voltage and plate current have each risen to their maximum positive values. At the half-cycle mark both the input voltage and plate current are back to zero. During the next half-cycle, the input voltage is negative and the plate current is zero. At the one-cycle mark both are at zero again. The cycle then repeats itself.

B. *Diode characteristics*

The *characteristics* of an electron tube are those electrical features and values which distinguish one tube from another. These characteristics may be shown by means of a graph (as in Figure 9-2) or else they may be tabulated. Some of these characteristics have been described in the last chapter.

(1) *Plate Dissipation.* This is the power lost as heat as a result of the electronic bombardment of the plate.

(2) *Maximum Plate Current.* The maximum average and maximum peak plate currents that may flow in the tube without damage to the plate or cathode.

(3) *Peak Inverse Voltage.* This is the maximum peak voltage that may be applied between the plate and cathode of the tube without danger of breakdown.

(4) *Plate Resistance.* Since resistance is a ratio between voltage and current ($R = E/I$), the resistance of the internal path from the cathode to plate may be found by dividing the plate voltage by the plate current. This value is called the *d-c resistance* of the tube. Its symbol is R_p and it is measured in ohms.

Of greater importance is the ratio between a small change in plate voltage and the corresponding small change in plate current—a ratio that represents the resistance of the path between cathode and plate as a changing or alternating voltage is applied. For this reason, it is called the *a-c plate resistance,* or simply, *plate resistance.* The symbol for plate resistance is r_p and it, too, is measured in ohms.

Vacuum-type diodes generally are employed as rectifiers where relatively small currents at high voltages are required. A typical example is the type 1G3 illustrated in Figure 9-6. The maximum average plate current for this tube is one milliampere and its peak inverse plate voltage is 33,000 volts.

Fig. 9—6. Vacuum-type high-voltage rectifier.

Radio Corporation of America

C. *Gas diodes*

1. HOT-CATHODE DIODES

As we have seen, the space charge repels electrons being emitted by the cathode and so limits the space current. A relatively high plate voltage is required to remove the electrons of the space charge and permit the emitted electrons a free path to the plate. If it were not for the space charge, any small plate voltage could produce the maximum space current.

One method for overcoming the space charge is to insert a very small amount of some gas, such as mercury vapor, into the tube after the air has been removed. Such tubes are known as *gas tubes.* As the electrons emitted by the cathode travel to the plate, they collide with the gas atoms. Since the velocity of these electrons is high, the collisions knock electrons off the gas atoms and thus form positive ions. In turn, the positive ions tend to neutralize the electrons of the space charge, thereby dispelling it. The emitted electrons now can reach the plate more easily.

In manufacture, the pressure of the gas is carefully controlled. The atoms of the gas will not form ions (that is, *ionize*) unless they are struck by electrons with sufficient velocity. Since the velocity of the electrons moving from the cathode to the plate depends upon

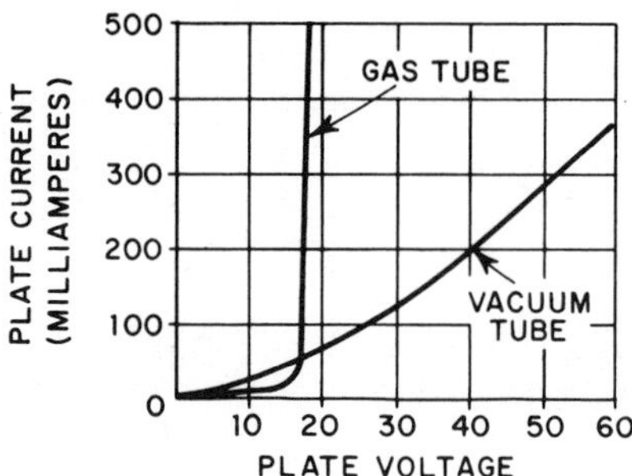

Fig. 9–7.

Plate voltage–plate current characteristic curves showing comparison between a vacuum-type tube and a comparable gas-type tube.

the potential difference between the two electrodes, ionization cannot take place until this potential difference reaches a certain minimum (called the *firing, striking,* or *ignition, potential*). This firing potential depends upon the type of gas employed and its pressure. For typical tubes it ranges from about 10 to 25 volts.

In the vacuum tube, the space charge opposes the flow of electrons from cathode to plate. Hence it affects the internal resistance of the tube. As the plate voltage is increased, the space charge is lessened and the plate current becomes larger. This relationship can be seen from the characteristic curve illustrated in Figure 9-2.

In the gas tube ionization does not take place until the firing potential is reached. Hence the tube acts as a vacuum-type up to this point. Once this potential is reached, the gas begins to ionize. The positive ions tend to neutralize the effect of the space charge, thus decreasing the internal resistance of the tube. Any further increase in the plate voltage has very little effect upon the plate current. This can be seen graphically by examining Figure 9-7 where the characteristic curves of typical gas and vacuum tubes are compared.

Note that after the firing potential of the gas tube is reached, the plate voltage remains fairly constant for different values of plate current. Increasing the plate current increases the degree of ionization of the gas within the tube, thus reducing the internal resistance. Reducing the plate current decreases the degree of ionization, hence increasing the internal resistance. Since the resistance goes down as the plate current goes up, and vice versa, the *IR* drop within the tube remains constant.

The attraction between the positive ions in a gas tube and the negative cathode poses a threat to the latter. Since the ions are relatively heavy, they may bombard the cathode with sufficient force to damage it. Accordingly, care must be taken not to apply plate voltage until the cathode has had enough time to emit sufficient electrons to form a protective space charge around it. This may take up to about 10 minutes, depending upon the tube.

Also, with the space charge dispelled by the positive ions, the internal resistance of the tube drops to a low value and a slight increase of plate voltage may cause enough plate current to flow as to damage the tube. This current flow is held within safe limits by the resistance of the components of the external plate circuit. Care must be taken that this resistance does not become too small, and protective devices, such as fuses, are frequently employed to open the plate circuit if the current becomes too large.

Gas diodes employing heaters or filaments are employed as rectifiers where plate currents up to about 100 amperes may flow. A typical example is the type 857-B illustrated in Figure 9-8. This diode will handle an average plate current of 10 amperes and will withstand a peak inverse plate voltage of 10,000 volts.

Note the structure of the cathode of this tube. Since gas tubes generally are required to furnish higher plate currents than their

Fig. 9—8.

Gas-type rectifier.

Radio Corporation of America

vacuum-type counterparts, the cathode must be capable of a much greater emission of electrons. This means that it must operate at a higher temperature and, as a result, there is danger that its life may be shortened because of excessive evaporation of its active material.

Accordingly, the cathode is surrounded by a cylindrical, metal *heat shield.* This shield reduces heat loss due to radiation and hence the cathode may operate more efficiently at lower temperatures. The inner surface of the shield facing the heat is coated with emitting material and thus can add its emitted electrons to the space current.

2. COLD-CATHODE DIODES

So far, we have been considering the hot-cathode diode. Gas diodes may also be of the cold-cathode type. A typical example of such a tube is the *glow-discharge tube* shown in Figure 9-9. This tube consists of a cathode and plate mounted fairly close together in an envelope containing some inert gas such as argon, neon, or helium. Its symbol is .

As we have seen, the gas in the tube may become ionized if the potential difference between the cathode and plate becomes great enough to reach the firing potential of the gas. The atoms of the gas may also become ionized if exposed to cosmic or ultraviolet rays or radioactive particles in the air. Thus some of the gas atoms of the tube are ionized because of natural conditions. If a voltage is applied to the electrodes, a small current will flow. However, this current will be too small to effect the operation of the tube.

As the applied voltage is increased, the current rises and the electrons produced by natural ionization are speeded up. When the voltage reaches the firing potential of the gas, these electrons have

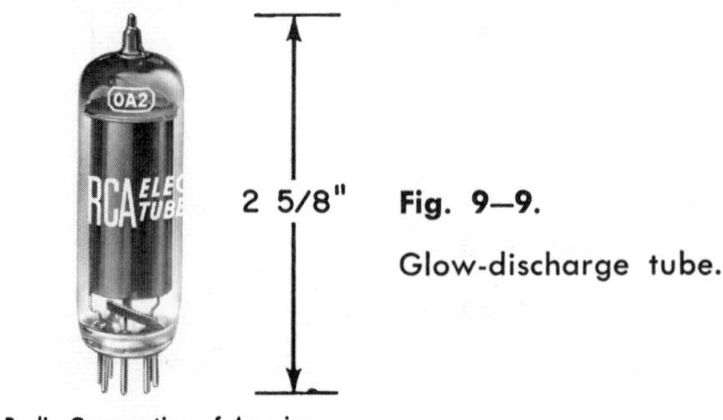

Fig. 9—9. Glow-discharge tube.

Radio Corporation of America

enough velocity so that they can knock electrons off any gas atoms with which they may collide. At the same time the heavier positive ions are attracted to the negative cathode with sufficient force to knock off some of its electrons.

These new electrons are attracted to the positive plate and attain sufficient velocities so as to knock off more electrons from any gas atoms they may encounter on the way. The entire process is cumulative and rapidly reaches a peak (we say the tube "fires" or "discharges"). The current rises rapidly, limited by the resistance of the external circuit to which it is connected.

As we have seen before, the *IR* drop within the tube remains constant at this point (the firing potential) regardless of fluctuations of the current. Because of this, it may act as a *voltage-regulator tube* for special voltage-stabilizer circuits (which will be discussed later).

Voltage regulator tubes are designed to operate within limited ranges, generally at voltages from about 75 to 150 volts and currents in the order of 30 to 40 milliamperes. It gets its name (*glow-discharge*) from the light glow produced as positive gas ions meet up with electrons within the tube and become neutral atoms, giving off light in the process.

QUESTIONS

1. a) Describe the basic structure of the vacuum diode.
   b) Explain what is meant by the *space charge;*
   c) the *space current.*
2. a) Draw the circuit showing how current flows through the tube and its plate circuit.
   b) Indicate the filament and plate circuits.
3. a) Draw a graph showing the relationship between the plate voltage and plate current of the tube.
   b) Explain what is meant by the *saturation point.*
4. a) Explain what is meant by *rectification.*
   b) Explain how the diode may act as a rectifier.
5. Draw the basic circuit of a diode used as a rectifier.
6. Draw a graph showing the rectifying action of the diode.
7. List and explain four *characteristics* of the diode.
8. a) What is meant by the *ionization* of a gas?
   b) What is meant by its *firing potential?*
9. Explain the operation of a *gas diode.*
10. Explain the differences between a vacuum diode and a gas diode.
11. Explain the operation of a *cold-cathode diode.*

10 The Triode

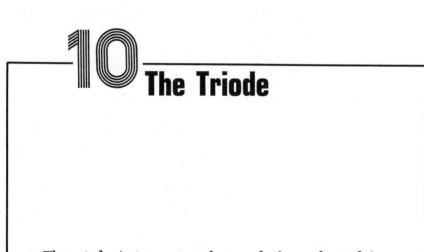

The *triode* (*tri* meaning three; *ode* from electrode) was invented in 1907 when the American scientist, Lee De Forest, inserted a third electrode, a metal screen or *grid,* into the path of the electron stream from the cathode to plate within the diode. By itself, the grid, which consists mainly of open spaces, has little effect on the electron stream. But if the grid is given a positive charge (with respect to the cathode), the stream of electrons is speeded up by the attraction between the grid and the electrons. Some of these electrons strike the wires of the grid, but most of them go right through to the plate (see Figure 10-1A).

Speeding up the electrons means that more will pass a given point in a given length of time. This is the same as saying that more current is flowing. Thus the result of the positive charge on the grid is to increase the space (and plate) current. The more positive the grid, the greater will be the space current (up to the point of saturation).

Should the grid be given a negative charge, the repulsion between it and the electrons would tend to slow down the latter, thus reducing the space current. The more negative the grid, the smaller will be the space (and plate) current (up to the point where the grid becomes negative enough to cut off the space current completely, as shown in Figure 10-1B).

You see, then, that the triode acts as a sort of valve whereby the voltage on the grid controls the plate current.

The grid usually consists of a spiral or mesh of fine wire made of iron, molybdenum, or certain types of alloys. Generally, the tube electrodes are arranged concentrically. The filament or cathode is at the center, the grid surrounds it and, in its turn, the grid is surrounded by the plate (see Figure 10-2). The symbol used for a grid is ——- - . Thus the symbol for the triode becomes . If a filament-type cathode is used, the symbol is .

The diode circuit, you will recall, has two parts—the heater circuit and the plate circuit. The triode circuit has three parts. (See Figure 10-3.) One is the *heater circuit* consisting of the heater or filament and the A battery (or other heater supply). The second is the *plate circuit* consisting of the cathode, the stream of electrons from cathode (through the grid) to the plate, the plate, resistor R_L which is the load resistor (the voltage drop across it is the output voltage), the B battery (or other plate-voltage supply) that makes the plate positive with respect to the cathode, and back to the cathode. The third is the *grid circuit* consisting of the cathode, grid, and the source of grid voltage that is impressed between them.

Note that in the triode, as in all other electron tubes, current flows in a one-way path from cathode to plate within the tube, and through the external plate circuit back to the cathode.

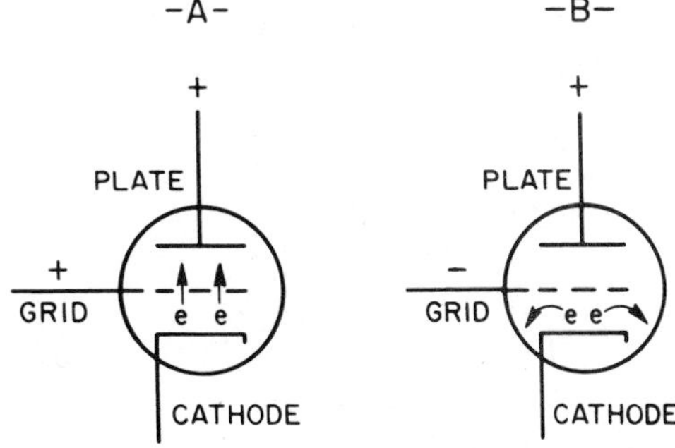

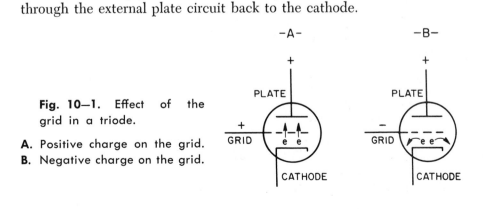

Fig. 10–1. Effect of the grid in a triode.

A. Positive charge on the grid.
B. Negative charge on the grid.

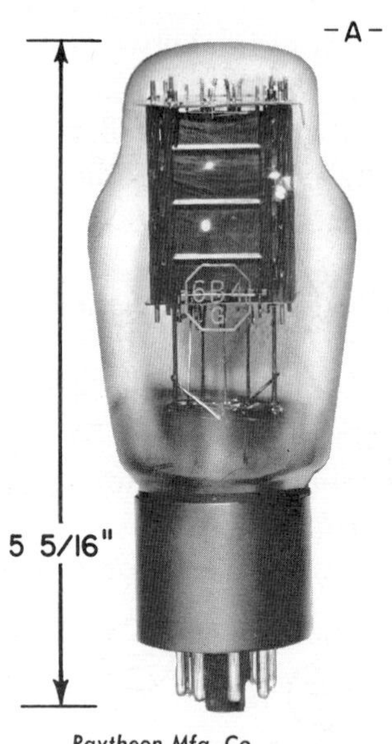

Raytheon Mfg. Co.

Fig. 10–2.

A. Triode, glass envelope, type 6B4-G.
B. Cutaway view showing internal structure.

At this point it might be well to untangle a matter that might lead to confusion. Since we have three circuits for the triode, we have three sets of voltages and currents. The heater circuit presents no difficulty. The heater voltage (E_f) and heater current (I_f) are determined by the tube manufacturer and are kept fixed at his recommended values.

The grid and plate circuits, however, are another matter. Both of these circuits have a-c and d-c components. And the a-c components have average, effective (rms), and instantaneous values. To avoid confusion, a standard system of symbols has been created.

Effective values of voltage are designated by E and effective values of current by I. Instantaneous values are designated by e, for voltage, and i for current.

In the grid circuit we may have a fixed d-c bias furnished by a battery or power supply, whose voltage is designated by E_{cc}. Also in this circuit is an a-c component (the input signal). The *total* value of grid voltage is designated by E_c and the *total* value of grid current by I_c; the *instantaneous total* values are designated by e_c and i_c, respectively. The *effective* voltage of the a-c component is E_g and the effective current is I_g. *The instantaneous effective* voltage and current of the a-c component are e_g and i_g, respectively.

In the plate circuit, the voltage of the B battery or power supply is designated by E_{bb}. The *total* value of plate voltage is designated

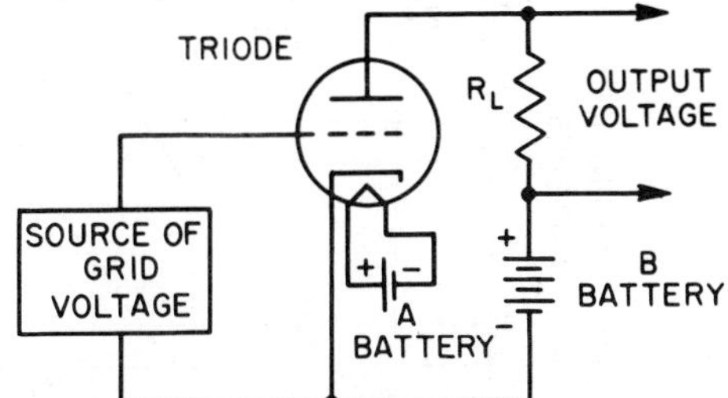

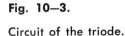

Fig. 10–3.

Circuit of the triode.

by E_b and the *total* value of plate current by I_b; the *instantaneous total* values by e_b and i_b, respectively. The *effective* voltage of the a-c component is E_p and the *effective* current is I_p. The *instantaneous effective* voltage and current of the a-c component are e_p and i_p, respectively.

Other symbols will be explained as we meet them.

A. *Triode characteristics*

1. CHARACTERISTIC CURVES

A study of the triode reveals that there are three values of particular interest to us. These are the *grid voltage* impressed between the grid and cathode of the tube, the *plate voltage* between the plate and cathode of the tube, and the *plate current* flowing in the plate circuit. (The heater voltage, and, therefore, the temperature of the emitter, generally is kept at a constant value, as recommended by the manufacturer of the tube.) The relationship between these values are of interest to us since they give us an insight to the workings of the tube. One method for examining these relationships is by means of characteristic curves.

Let us first examine the relationship in a typical triode between changes in plate voltage and the resulting changes in plate current, keeping the grid voltage constant. Were it not for the action of the grid, we would expect the plate voltage–plate current characteristic curve of this triode to resemble that of the diode illustrated in Figure 9-2. Indeed, if you examine the $E_c = 0$v. curve of the set of curves illustrated in Figure 10-4 you will notice a close resemblance to that of Figure 9-2.

The set of curves of Figure 10-4 was plotted by keeping the grid voltage fixed at a separate value for each curve. Thus, the $E_c = 0$v. curve was plotted with no voltage on the grid. The $E_c = -2$v. curve

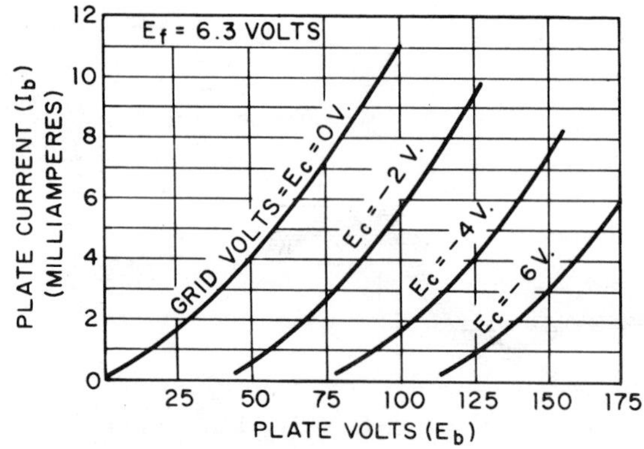

Fig. 10–4. Family of plate characteristic curves for a triode. (Note that these curves have not been plotted up to the saturation point.)

was plotted with the grid 2 volts negative with respect to the cathode. The $E_c = -4$v. curve was plotted with the grid negative by 4 volts. And so on. The reason why only negative values of grid voltage were chosen is because the grid usually is operated in this condition (as you will see later).

A set of curves such as these are called a *family of curves* and, because they show the relationship between plate voltage and plate current, are known as the *plate characteristic curves.* Among other things, they indicate the effect of a signal applied to the grid. If you take a fixed plate voltage, say 100 volts, you will notice that when the grid is 4 volts negative, approximately 1.5 milliamperes of plate current flows. With the grid 2 volts negative, the plate current is about 5.5 milliamperes. With no voltage on the grid, the plate current is about 11 milliamperes. Thus, if the signal applied to the grid varies its voltage between −4 volts and zero, the plate current will vary between 1.5 and 11 milliamperes.

In a similar manner, we can examine the relationship between changes in grid voltage and the resulting changes in plate current, the plate voltage being kept constant. The graph so obtained is called the *mutual characteristic curve* because it shows the relationship between the grid voltage, which is in the input circuit of the tube, and the plate current, which is in the output circuit.

The family of curves illustrated in Figure 10-5 was plotted by keeping the plate voltage fixed at a separate value for each curve.

Thus, the E_b = 100v. curve was plotted with 100 volts on the plate; the E_b = 150v. curve with 150 volts on the plate; and so on. As previously indicated, only negative values of grid voltage were chosen because the grid usually is operated in this condition.

Consider the curve marked E_b = 100v. With the grid approximately 6 volts negative (with respect to the cathode) the plate current is zero. This is the point of *cutoff* where all the electrons emitted by the cathode are blocked by the negative charge on the grid. As the negative charge on the grid is reduced, the attraction of the positive plate can pull some of the electrons through the grid and plate current starts to flow, reaching a value of about 0.5 milliampere when the grid is −5 volts. At −4 volts the plate current is about 1.25 milliamperes; at −3 volts approximately 2.5 milliamperes of current flows. From here on the rise in plate current is more rapid as the grid is made less negative, reaching a value of about 11 milliamperes when the charge on the grid becomes zero. Making the grid positive causes still more current to flow, but this condition is not desirable for several reasons, including the fact that the tube may be damaged.

Note that as the plate voltage is increased, the cutoff point becomes more negative. Thus, when the plate voltage is 100 volts, the cutoff point is at approximately −6 volts. With the plate voltage at 150 volts, the cutoff point is about −9 volts. At 200 volts it is at −12.5 volts, and so on. The more positive the plate, the more negative the grid must be to cut off the flow of electrons from the cathode.

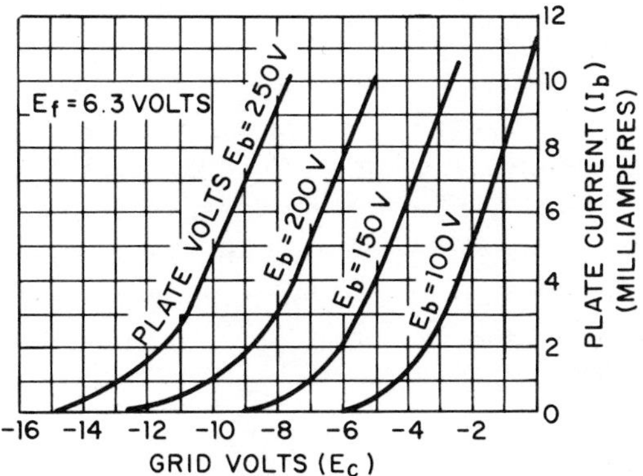

Fig. 10–5. Family of mutual characteristic curves for a triode.

2. TUBE PARAMETERS

The triode operates by the application of a grid, or input, voltage between the grid and cathode, thereby producing a corresponding flow of plate current. To understand the operation of a tube, however, it is not enough to know what happens to the plate current when a certain grid voltage is applied. Since the input voltage generally is a changing rather than constant value, we must know how the tube reacts to small changes.

Another way of looking at this can be seen if you observe the characteristic curves illustrated in Figure 10-4. Note that these curves are not straight lines. Therefore, the plate voltage-plate current relationship is not the same for all portions of the curves. By choosing small values of voltage and current changes, we operate on a small portion of the curve which, accordingly, acts as a straight line. Hence the curvature of the graph is overcome.

These small voltage and current characteristics are called the *tube parameters* and include the *amplification factor, plate resistance,* and *transconductance.*

a. Amplification factor

If you examine the plate characteristic curves illustrated in Figure 10-4, you will find that a 20-volt increase in plate voltage (say, from 65 volts to 85 volts on the $E_c = 0$v. curve) produces an approximately 2.5-milliampere rise in plate current. Examining the mutual characteristic curves for the same tube (Figure 10-5) we find that a 1-volt increase in grid voltage (from -2 volts to -1 volt on the $E_b = 100$v. curve, for example) produces the same 2.5-milliampere rise in plate current. The grid voltage, then, is twenty times as effective as the plate voltage when it comes to controlling the plate current.

This should not be surprising. We would expect the grid to have a greater effect on the flow of electrons than the plate since the former is closer to the source of emission—the cathode. The ratio of the plate-voltage change to grid-voltage change (the plate-current change being the same in both cases) is called the *amplification factor.* That is,

$$\text{amplification factor} = \frac{\text{change in plate voltage}}{\text{change in grid voltage}},$$

when the plate-current change is the same in both cases.

The amplification factor of a triode is determined mainly by the mechanical construction of the tube. The nearer the grid is to the cathode, the greater is its effect on the stream of electrons flowing to the plate, and the greater is the amplification factor of the tube. Also, the finer the mesh of the grid, the greater is the effect of the charged grid upon the electron stream, and, again, the greater the amplification factor. If, on the other hand, the open spaces in the grid are wide, the electrons are able to rush to the plate without being very much affected by the grid charge. This condition, accordingly, makes for a smaller amplification factor.

It should be noted that the electrodes of a tube act as plates of a capacitor, thus permitting the transfer of electric energy from one circuit to the other. As the electrodes are brought closer together or the mesh of the grid becomes finer, the interelectrode capacitances of the tube are increased. Since this usually is undesirable (as you will see later), triodes generally are constructed with low amplification factors.

The space charge within the triode exerts a further limiting effect. As the cloud of electrons surrounds the grid, its effectiveness is reduced and thus the amplification factor is lowered.

The amplification factor of triodes generally ranges in value up to about 100, with most tubes lying between 10 and 40. The triode used in our example has an amplification factor of 20. In electrical formulas, the amplification factor is represented by the Greek letter μ (pronounced *mew*).

b. Plate resistance

The electrons flow from the cathode, through the space charge, past the charged grid, and on to the plate of the tube. There they lose their energy of motion, which is changed to heat. Now, when electrons move through a resistor, they also produce heat. Accordingly, we may look upon the loss of energy within the tube as being analogous to the loss due to resistance and may consider the path between the cathode and plate as offering resistance to the flow of electrons.

Since, in the electron tube, we are more concerned with the *variations* or *changes* in voltage and currents than with the voltages and currents themselves, we are interested in the resistance of the tube which impedes the flow of changing current. The resistance of the path between the cathode and plate to changing current is called the *plate resistance* and is designated by the symbol r_p.

Plate resistance is expressed in ohms and may be calculated by dividing a small change in plate voltage by the corresponding change in plate current, the grid voltage being kept constant. (Note the similarity to Ohm's law.) If, in the graph illustrated in Figure 10-4, you consider, for example, the plate-current change in the $E_c = 0$v. curve from 8 to 9 milliamperes, you will note that it corresponds to a plate-voltage change from approximately 80 to 87.5 volts. Thus a 1-milliampere change in plate current corresponds to a 7.5-volt change in plate voltage.

The plate resistance of this tube then should be

$$r_p = \frac{\text{change in plate voltage}}{\text{change in plate current}} = \frac{7.5}{0.001} = 7{,}500 \text{ ohms.}$$

In most triodes the plate resistances range in values between several hundred and 100,000 ohms.

c. Transconductance

Not only may a small change in plate voltage produce a corresponding change in plate current, but a small change in grid voltage may also produce this small plate-current change. The ratio between the small change in plate current and the small grid-voltage change producing it (the plate voltage being held constant)—that is, the plate-current change divided by the grid-voltage change—is called the *grid-to-plate transconductance* or, more simply, *transconductance.* (The term *mutual conductance* sometimes is used.) The symbol for transconductance is g_m.

You will recall from Ohm's law, resistance is equal to the quotient of voltage divided by current ($R = E/I$). But to compute transconductance, we divide *current* by *voltage* and so obtain the reciprocal of resistance. Accordingly, the unit for transconductance is the *mho,* which is the unit for resistance (ohm) spelled backwards.

If you examine the mutual characteristic curves illustrated in Figure 10-5, you will note that a 1-volt change in grid voltage (from -2 to -1 volt on the $E_b = 100$v. curve, for example) corresponds to a 2.5-milliampere change in plate current. The transconductance of this tube then becomes

$$g_m = \frac{\text{change in plate current}}{\text{change in grid voltage}} = \frac{0.0025}{1} = 0.0025 \text{ mho.}$$

In practice, the mho is too large a unit for easy handling. Accordingly, we use the term *micromho* (μmho), which is one millionth of a mho. Our transconductance thus becomes 2,500 micromhos. Since 2,500 micromhos correspond to a plate-current change of 2.5 milliamperes, for each 1,000 micromhos of transconductance a grid change of one volt produces a plate-current change of one milliampere. The g_m for triodes may range between several hundred to about 10,000 micromhos.

Since all three tube parameters are functions of the grid voltage, plate voltage, and plate current, they are interdependent. The relationship between them may be expressed in the formula

$$\mu = g_m \times r_p.$$

If you refer back to Figure 10-3, you will notice that the voltage output is the result of the voltage drop produced by the plate current flowing through the load resistor (R_L). Tubes designed to supply a high output voltage are called *voltage amplifiers.* A typical type is the 6J5 triode with a plate current of 9 milliamperes, an amplification factor of 20, a plate resistance of 7,700 ohms, and a transconductance of 2,600 micromhos.

As you shall learn later in the book, there are times when we are more concerned with the power output than the voltage output. Since Power $= I^2 \times R$, this means that we desire a relatively high plate current to flow through the plate circuit. Hence we employ *power tubes* with larger emitting surfaces for a greater electron emission and more open grids for these electrons to get through.

The result, then, is a tube with a lower amplification factor and plate resistance. A typical example is the 6B4 triode with a plate current of 60 milliamperes, an amplification factor of 4.2, and a plate resistance of 800 ohms. However, since $g_m = \mu / r_p$, the transconductance is 5,250 micromhos.

B. *Gas triodes*

In our discussion of the gas diode we saw that as an increasing voltage is applied to the plate of the tube very little plate current flows until the potential difference between the cathode and plate reaches the firing potential of the gas (that is, the gas ionizes). At this point the resulting positive ions neutralize the space charge and the plate current rapidly rises to a maximum. To stop this flow, the plate voltage must be reduced to zero or else the plate must be made negative.

Now let us see what is the effect of placing a grid between the cathode and plate. If the grid is made negative enough to cut off the flow of electrons from the cathode, there is no plate current and the gas is not ionized. As the grid is made slightly less negative, some of the electrons get through it and are attracted to the positive plate. However, because the electron flow within the tube is small, the gas still is un-ionized. When the negative charge on the grid is reduced to a certain critical point, enough electrons get through to ionize the gas. At this point the plate current rises rapidly to a maximum.

Once the gas ionizes, some of the positive ions are attracted to the still-negative grid. There they form a kind of sheath around the grid which prevents it from having any further effect on the flow of electrons through the tube. Since there no longer is any retarding effect, the positive plate attracts all the available electrons to it. The tube then behaves just as the gas diode. You see, therefore, that the grid acts as a trigger to start the flow of current.

The vacuum triode acts as a valve whereby a varying grid voltage may cause the plate current to vary in like degree. The gas triode may be considered as a sort of switch by means of which a small voltage applied to the grid can turn on the flow of relatively large plate currents. Note that in both the vacuum triode and gas triode a small grid voltage may control a large plate current.

We call gas triodes *thyratrons*, the symbol for which is the same as for any other type of triode, except that a dot is added in the envelope to show that it contains gas.

In the thyratron a sort of tug-of-war is set up between the attracting force of the positive plate and the repelling effect of the negative grid. Thus, for every positive value of plate voltage there is a negative value of grid voltage sufficient to restrict the flow of electrons to the point where the gas cannot ionize. This relationship is shown in the graph illustrated in Figure 10-6 which is known as the *control characteristic curve*.

In this graph, negative grid voltage is plotted against positive plate voltage. The dotted curve is a typical control characteristic curve. Note that when the plate voltage is equal to 50 volts, a negative grid voltage of 2 volts can prevent the gas from ionizing. When the plate voltage is increased to 100 volts, the grid voltage must be reduced to approximately —2.5 volts. At 200 volts on the plate, the grid voltage must be —3 volts. And so forth.

The pressure of the gas within the thyratron is quite critical. The greater this pressure (within certain limits), the more easily the gas will ionize. That is, it will ionize at a lower plate voltage for a given grid voltage or, conversely, at a more negative grid voltage for a given plate voltage.

There are a number of factors, such as aging, operating conditions and temperatures, etc., that cause the gas pressure within a tube to vary. Further, there are differences in pressure between tubes of a single type owing to slight differences in manufacture. Accordingly, the tube manufacturers furnish, not a single-line control characteristic curve, but a *critical area* (indicated by the shaded area of Figure 10-6) within which the tube may normally function.

The thyratron frequently is operated with alternating voltage applied to its plate. During the half-cycle when the plate is negative, no current will flow through the tube, of course. During the positive half-cycle there is a possibility that current may flow. But this flow can take place only if (a) the plate is more positive than the cathode by at least a certain amount sufficient to start and support ionization (generally, about 15 volts) and (b) the charge on the grid is not more negative than the cathode by a definite amount which is dependent upon the plate voltage.

This can be seen from the graph illustrated in Figure 10-7 where is plotted the positive half-cycle of the alternating voltage applied to the plate and the *critical grid-voltage curve* indicating the grid voltages required to keep the gas from ionizing at all voltage values of the half-cycle.

Thus, at 100 plate volts the critical grid voltage is —2.5 volts; at 200 plate volts, —3.0 volts; at 300 plate volts, —3.25 volts; and at the peak 400 plate volts, —3.5 volts. If at any time during the

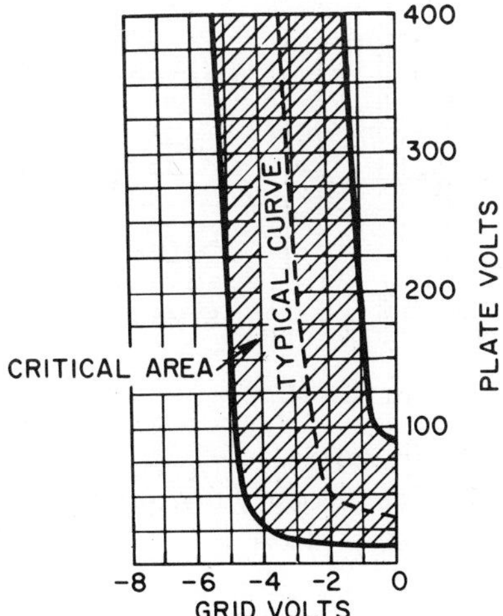

Fig. 10–6.

Control characteristic curve of thyratron.

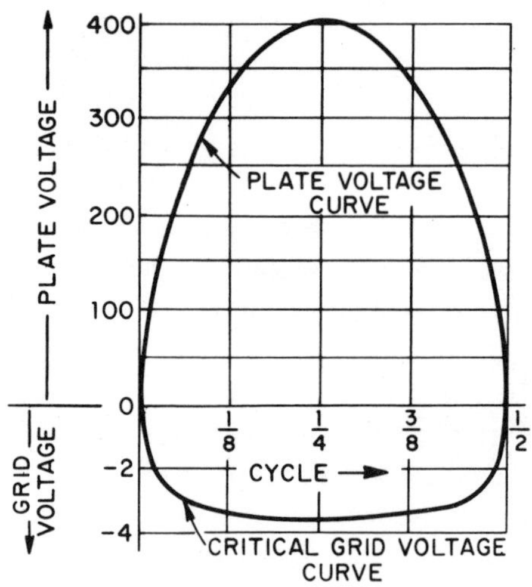

Fig. 10–7.

Graph showing plate-voltage and critical grid-voltage curves of thyratron.

half-cycle, the grid should become less negative (or, what is the same thing, more positive) than the critical voltage indicated for the plate voltage at that point, the gas will ionize, the grid will lose control, and plate current will flow for the rest of that half-cycle. At the next (negative) half-cycle, the plate current will be cut off.

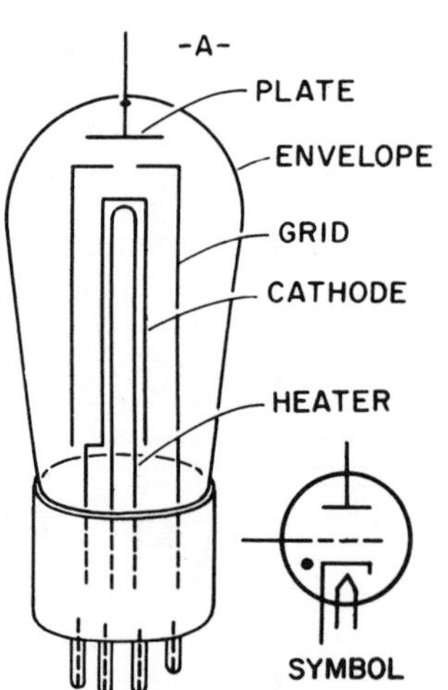

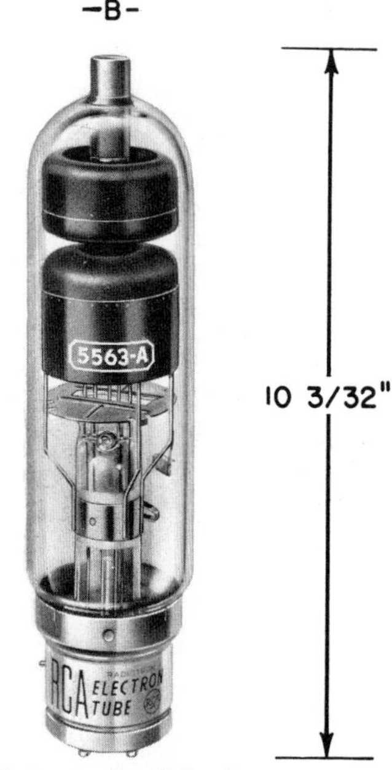

Radio Corporation of America

Fig. 10–8.

A. Thyratron.
B. Commercial thyratron, type 5563-A.

The thyratron is illustrated in Figure 10-8. The envelope generally is of glass for the low-power types and may be of metal for the high-power types. Its cathode may be a filament or of the indirect-heater type. It may contain a heat shield.

The grid generally consists of a cylindrical, metal can, completely surrounding the cathode with one or more holes in the top through which the electrons may pass. The plate usually is a block of nickel or graphite. In some tubes the grid cylinder may be extended to surround (but not touch) the plate.

Various types of gases, such as mercury vapor, xenon, argon, or hydrogen may be used in thyratrons. Plate currents ranging from a few hundred milliamperes to 100, or more, amperes may be controlled.

QUESTIONS

1. Describe the basic structure of the vacuum triode.

2. a) Draw the basic circuit of the triode.
b) Indicate its three circuits.

3. Explain what happens to the plate current as
a) the grid is made negative with respect to the cathode;
b) the grid is made positive with respect to the cathode;
c) an alternating voltage is placed between the grid and cathode.

4. Explain the two points that limit the flow of plate current through the tube.

5. Explain the relationship between the grid voltage, plate voltage, and plate current in a triode.

6. Explain what is meant by the *amplification factor* of the triode.

7. Explain what is meant by the *plate resistance* of the triode.

8. Explain what is meant by the *transconductance* of the triode.

9. Explain the relationship between the amplification factor, plate resistance, and transconductance of the triode.

10. Explain the differences in function and structure between a *voltage-amplifier* tube and a *power* tube.

11. Explain the operation of the *thyratron*.

12. Explain the differences between a vacuum triode and a gas triode.

11 Multigrid and Multiunit Tubes

A. *The tetrode*

We have seen how, in the triode, the voltage on the grid may control the plate current. Generally, we desire this process to be a one-way affair—that is, we do not want the flow of plate current to affect the grid potential.

One of the chief causes of this undesirable effect is the interelectrode capacitance of the tube, especially the grid-to-plate capacitance. The grid and plate act as two plates of a capacitor to pass electric energy from the plate to the grid circuit. You will recall that the danger of too great an interelectrode capacitance kept the amplification factor of the triode low.

It was to reduce the grid-to-plate interelectrode capacitance that the *screen-grid tube* was developed. In this tube, an additional grid, or screen, is inserted between the original grid (now called the *control grid*) and the plate. (See Figure 11-1.) This type of tube is called a *tetrode* (*tetra* meaning four; *ode* from electrode). The symbol for the tetrode is [tetrode symbol]. The control grid is the one nearest the cathode.

The function of the screen grid is to shield the control grid from the electrostatic field around the plate, hence reducing the capacitance between the two. However, merely placing a screen between the control grid and plate of the tube would have no effect upon the capacitance between these two electrodes. To be effective, the screen grid must be connected to the cathode which, of course, is connected to the negative terminal of the B battery. At the same time, on the other hand, the screen grid must be at a positive potential if it is not to interfere with the flow of electrons to the plate.

This dilemma is resolved by connecting the screen grid to a positive point on the B battery (although generally at a lower potential than the plate). Hence the screen grid is positive, with respect to the cathode, as far as the d-c, or *static*, operation of the tube is concerned. At the same time, the screen grid is also connected to the cathode through a capacitor. (See Figure 11-2.)

Since, in operation, the tube handles changing voltages (which have an a-c component), the screen grid thus is at the cathode potential as far as the a-c component, which can pass through the

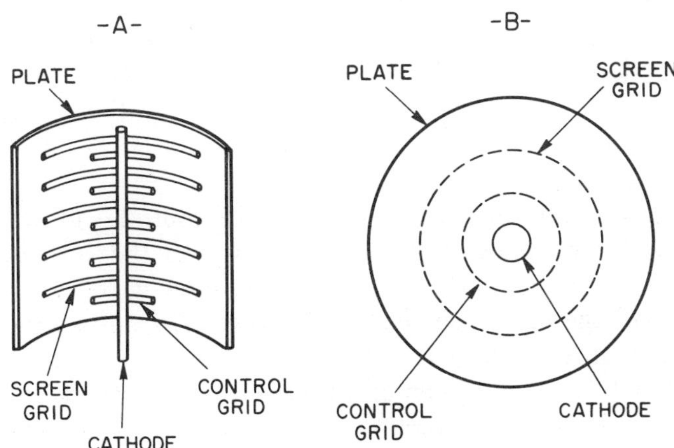

Fig. 11—1. The tetrode.

A. Cutaway side view.
B. Top view.

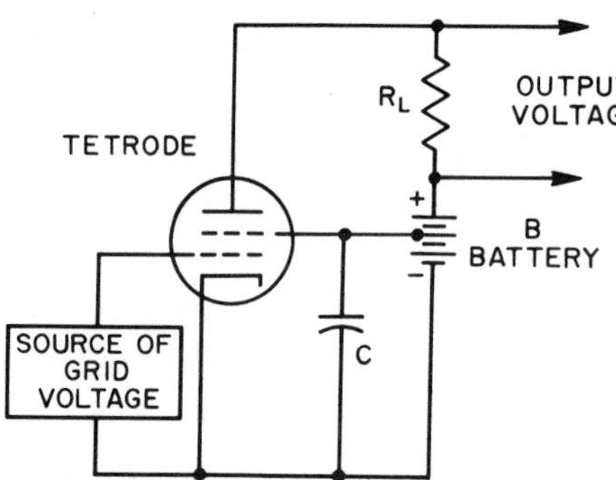

Fig. 11–2.

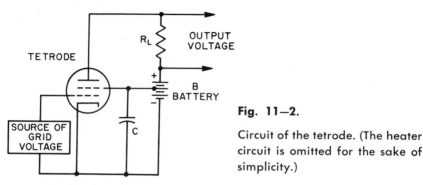

Circuit of the tetrode. (The heater circuit is omitted for the sake of simplicity.)

capacitor, is concerned. That is, the screen grid is *dynamically* coupled to the cathode. Hence, the shielding is effective and the plate-to-grid interelectrode capacitance is reduced. How effective the screen grid is can be seen by comparing the plate-to-grid interelectrode capacitance of the 6J5 triode (3.4 $\mu\mu$f) with that of a type 24-A tetrode (0.007 $\mu\mu$f).

The screen grid is operated at a positive potential, though somewhat less than that of the plate. Thus, the screen grid acts as the plate to pull over electrons emitted by the cathode. Some of these electrons will strike the screen grid, causing a screen current to flow. But the majority of the electrons will continue through the mesh of the screen grid to the plate.

1. TETRODE CHARACTERISTICS

In the triode, the plate current (I_b) is equal to the space current flowing within the tube. But in the tetrode, we have two positive electrodes—the plate and the screen grid—to attract electrons. The space current, therefore, is the sum of the plate current and the screen-grid current.

To differentiate the screen grid from the control grid, we use the subscript 1 for values of the control grid and the subscript 2 for values of the screen grid. The voltage on the control grid becomes E_{c1}, the voltage on the screen grid becomes E_{c2}, and so forth. The space current is equal to $I_b + I_{c2}$.

Note that since the screen grid is closer to the cathode than is the plate, it is chiefly the attraction of the screen grid, rather than the plate, that pulls over the emitted electrons. Since there is very little variation in the screen-grid voltage, this grid exerts a constant pull. Therefore, variations in plate voltage have little effect on the electron stream.

We may plot a graph to show the relationship between variations in space current ($I_b + I_{c2}$) with variations in control-grid voltage (E_{c1}), keeping the screen-grid voltage (E_{c2}) constant. Note that in this respect, the screen grid acts as an anode, and so the cathode, control grid, and screen grid form a kind of triode. As is to be expected, the curve would resemble that for a triode (see Figure 10-5).

When we plot variations in plate current (I_b) and the corresponding variations in screen-grid current (I_{c2}) against variations in plate voltage (E_b), we obtain some interesting results. (See Figure 11-3.) In this graph, the control grid is kept constant at zero volts and the screen grid at 90 volts. The plate current–plate voltage curve appears as a solid line. The screen-grid current—plate voltage curve is shown as a dotted line.

When the plate voltage is at zero, there is no plate current, the entire space current being attracted to the positive screen grid. As the plate voltage increases, some of the electrons pass through the mesh in the screen grid and are pulled to the plate; hence the rise in plate current (as indicated by the rise to point A of the plate curve). Simultaneously, the screen-grid current is reduced in like degree, as indicated by the drop to point A¹ of the screen-grid curve. Note, however, that the screen-grid current is still greater than the plate current.

As the plate voltage continues to rise, more electrons are pulled to the plate. But as the voltage is increased, so is the velocity of the electrons which strike the plate with increasing force. As a result, other electrons are knocked off the atoms of the plate. This phenomenon is known as *secondary emission.*

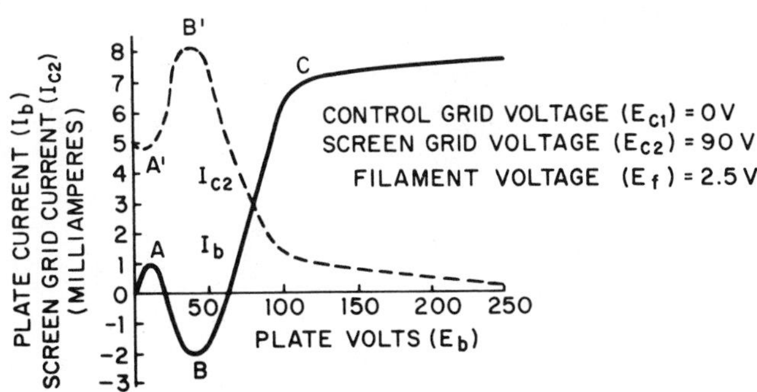

Fig. 11—3. Characteristic curve of the tetrode.

Since the plate still is less positive than the screen grid, most of the secondary electrons are attracted to the screen grid rather than back to the plate. This is shown by the rise in the screen-grid curve to point B^1. The loss of secondary electrons means a loss of plate current as indicated by the dip in the plate curve to point B. Indeed, so many electrons may be lost that the plate current actually may become negative, that is, current flows to the plate from the external circuit.

As the plate voltage begins to approach that of the screen grid, however, more and more of the secondary electrons are recaptured by the plate, and the plate current starts to rise again, the screen-grid current falling in like degree.

When the plate voltage exceeds that of the screen grid, the major portion of the space current flows to the plate. At a certain point (point C of the plate curve), the plate current reaches a level where any further increase in plate voltage has very little effect.

In the tetrode, the control grid reigns supreme. No longer does it have to share its control over the electron stream with the plate. The slightest variation of voltage on the control grid will vary the electron stream reaching the plate without any interference caused by a varying electrostatic field produced by a varying plate voltage. As a result, the amplification factor shoots up. Compare, for example, the amplification factor of the 6J5 triode ($\mu = 20$) with that of the 24-A tetrode ($\mu = 630$).

Since a relatively large change of plate voltage causes very little change in plate current, the plate resistance (r_p) of the tetrode, too, is much larger than for a triode (7,700 ohms for the 6J5 triode; 600,000 ohms for the 24-A tetrode). However, since transconductance (g_m) is equal to μ/r_p, the transconductances of tetrodes generally are somewhat less than those of comparable triodes (2,600 μmhos for the 6J5 triode; 1,050 μmhos for the 24-A tetrode).

The loss of plate current due to secondary emission is a serious drawback to the tetrode for most applications. Accordingly, it has largely been replaced by the pentode (which we shall consider a little later).

2. GAS TETRODES

The gas triode suffers from several drawbacks. Because its grid is so large, it is subject to bombardment of electrons, gas ions, and particles evaporated from the hot cathode. These produce a flow of grid current which, though small, is sufficient to produce an appreciable voltage drop in an external grid circuit whose resistance is very high. This voltage drop may interfere with the triggering action of the grid.

Accordingly, a *shield grid* may be inserted so as to shield the control grid from both the cathode and plate, thus forming a tetrode. (See Figure 11-4.) In a typical example, the shield grid consists of a long tube containing two baffle plates, each of which has a hole through its center. The control grid is a small ring located between the two baffles. Connections are made to the heater, cathode, and shield grid through prongs in the base of the tube. Plate and control-grid connections are made through metal caps sealed in the tube's envelope. The shield grid generally is connected to the cathode.

Because the control grid is small and shielded, it is less subject to bombardment. Further, the shielding reduces the control grid-plate interelectrode capacitance of the tube. As a result of all this, the grid current is reduced to a very low value.

In some of these shielded-grid thyratrons the shield grid is not connected to the cathode, but left free. Positive or negative voltages then may be applied between it and the control grid, which alters the characteristics of the tube to the end that the control grid can more readily trigger its action.

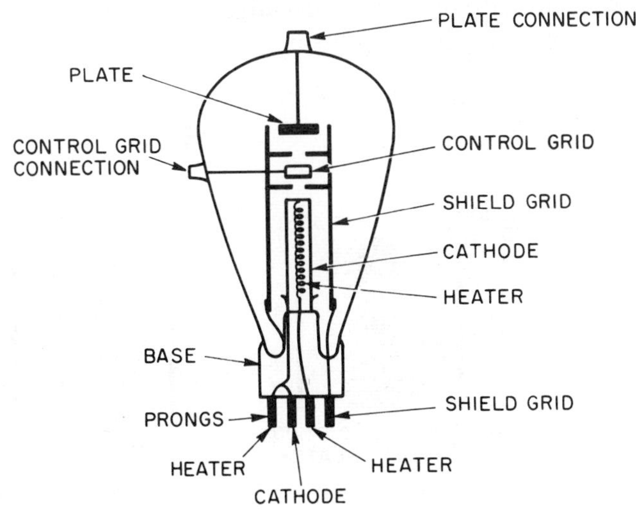

Fig. 11–4. Diagram of four-element thyratron. (Type FG-98.)

B. *The pentode*

The screen grid of the tetrode, you will recall, is a mixed blessing. On the one hand, it makes possible the reduction of the interelectrode capacitance of the tube. On the other hand, it acts as a collector for the electrons knocked off the plate by secondary emission, producing the dip in the characteristic curve illustrated in Figure 11-3.

To get around this latter effect, another grid, called the *suppressor grid,* is inserted between the screen grid and the plate (Figure 11-5). The suppressor grid usually is connected to the cathode of the tube and thus is negative with respect to the plate. As secondary electrons are knocked off the plate, the suppressor grid repels them back to the plate. Although the suppressor grid does not prevent secondary emission, it does overcome the effect caused by it. Accordingly, plate-voltage variations may be quite large, the plate voltage may even drop below the screen-grid voltage, without any serious loss in plate current.

This type of tube is called a *pentode* (*pent* meaning five; *ode* from electrode). The symbol for the pentode is . The grid nearest the cathode is the control grid; the middle grid is the screen grid; the grid nearest the plate is the suppressor grid.

In the symbol illustrated above, the suppressor grid is shown to have an external connection, usually a separate prong in the base. In some pentodes the suppressor grid is connected internally to the cathode. The symbol then becomes .

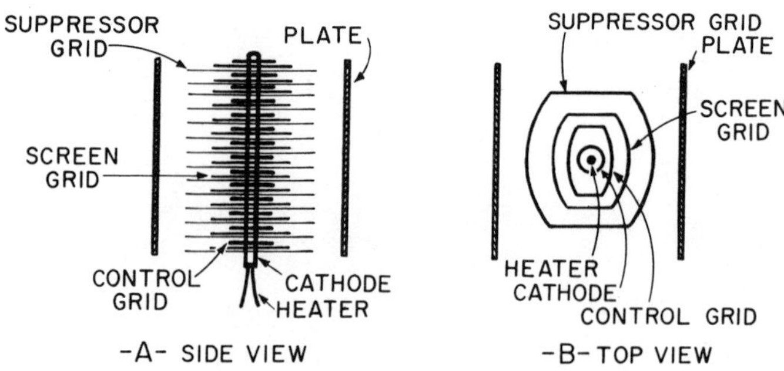

Fig. 11—5. The pentode.

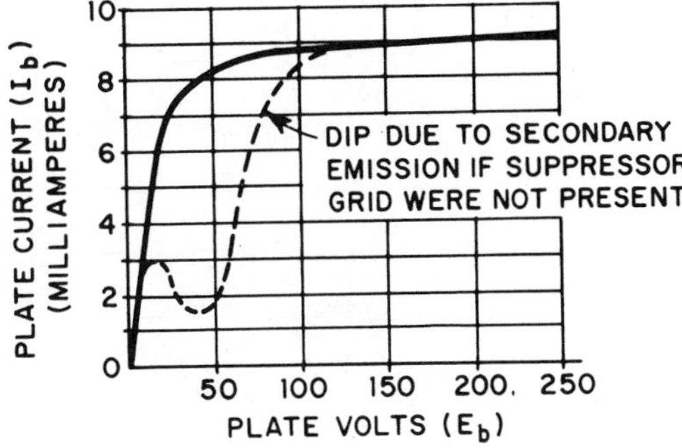

Fig. 11–6.

Plate voltage–plate current characteristic curve of the pentode.

1. PENTODE CHARACTERISTICS

The characteristic curves of the pentode resemble those of the tetrode, except that the dip in plate current resulting from secondary emission is not present. A typical plate voltage–plate current curve is illustrated in Figure 11-6. To obtain this curve, the control-grid voltage (E_{c1}) was held at zero, the screen-grid voltage (E_{c2}) at 90 volts, and the suppressor-grid voltage (E_{c3}) at zero, since the suppressor grid was connected to the cathode. As in the tetrode, the plate voltage has little effect on the plate current, and the curve levels off to a fairly steady value after a certain point is attained.

The control-grid voltage–plate current curve resembles that of the triode. (See Figure 11-7.) Here the plate voltage (E_b) is held constant at 250 volts and the suppressor grid, because it is tied to the cathode, is at zero volts (E_{c3}). The family of curves is obtained by using a separate value of screen-grid voltage (E_{c2}) for each curve.

The amplification factor of the pentode may be very high, some tubes having a μ of 1,500, or more. Plate resistance (r_p), too, is high; 1,500,000 ohms is a common value. The transconductance (g_m) generally is lower than that of a comparable triode, though some pentodes have transconductances as high as 9,000 micromhos.

Pentodes, too, may be constructed as power tubes with larger emitting surfaces and more open grids to permit the flow of greater plate currents. The type 6F6 is a typical power pentode with an amplification factor of about 200, a plate resistance of 78,000 ohms, and a transconductance of 2,550 micromhos.

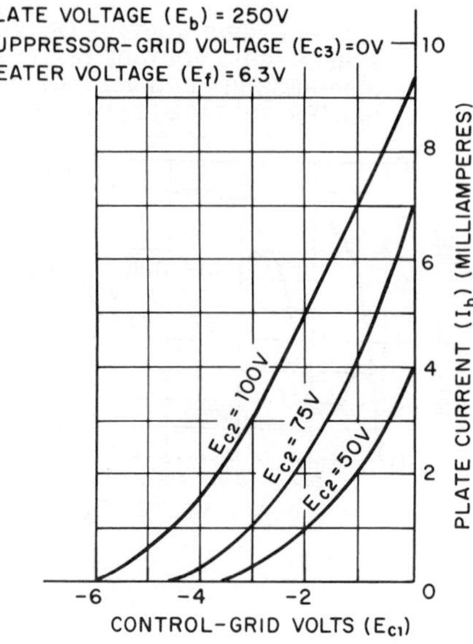

Fig. 11—7.

Family of control-grid voltage—plate current characteristic curves of the pentode.

2. REMOTE-CUTOFF TUBES

The *cutoff point,* you will recall, is the point at which the negative potential on the control grid will repel all electrons and permit none to flow through to the plate. In the ordinary pentode, the tube approaches this point rapidly, as can be seen from the characteristic curves illustrated in Figure 11-7. Such tubes are called *sharp-cutoff pentodes.*

Under certain conditions it is desired that the cutoff point be approached more gradually. Tubes of this type are called *remote-cutoff pentodes.* The difference between the sharp-cutoff and the remote-cutoff tubes is shown graphically in Figure 11-8.

In the sharp-cutoff pentode, the control grid consists of a spiral, or helix, of wire surrounding the cathode. The spacings of this spiral are uniform and close in order to provide the tube with a high

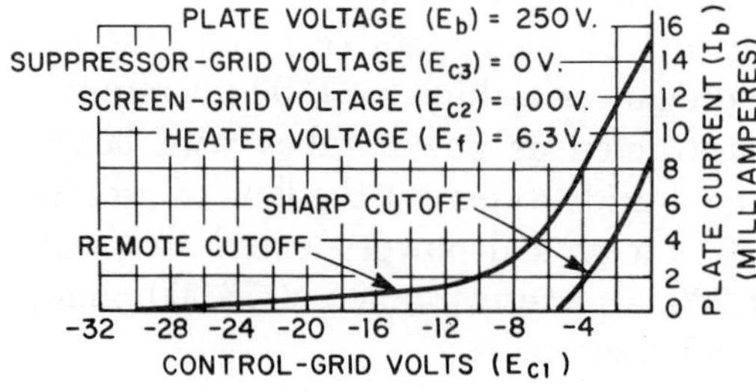

Fig. 11—8.

Graph showing the difference between the characteristic curves of a sharp-cutoff tube and a remote-cutoff tube.

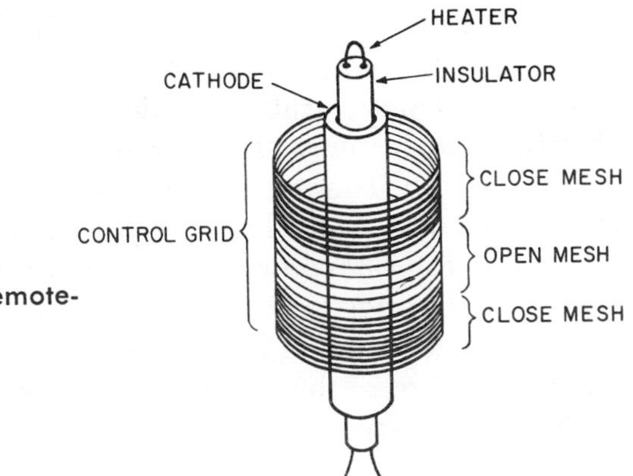

Fig. 11–9.

The control grid of a remote-cutoff tube.

amplification factor. Thus the control grid becomes negative enough to cut off the electron flow at a relatively low voltage.

In the remote-cutoff pentode the mesh of the control grid is closer at the ends and more widely spaced at the center (Figure 11-9). Thus, when the control grid is negative enough to cut off the flow of electrons at the ends, it is not negative enough to stop them from flowing through the more widely-spaced center. The grid must be made still more negative before the entire electron flow will be cut off.

As you have learned, the mesh of the control grid is a factor that determines the μ of a tube. In the remote-cutoff tube this mesh is not uniform, being closer together at the ends and more widely spaced at the center. Hence this type of tube is also known as a *variable-μ*, or *supercontrol, pentode*.

C. *The beam power tube*

When considering the power tube, we are concerned with the amount of available plate current. In the triode, the flow of electrons from the cathode to plate is made to pass through the control grid. In the pentode, there are three grids in its path—the control, screen, and suppressor grids. Of course, the two latter grids serve a useful function, producing a more efficient and higher power tube than the triode. Nevertheless, they have their disadvantages. The screen grid, for example, is an obstacle in the path of the electron stream and, since it carries a positive charge, attracts some of these electrons. This means fewer electrons to reach the plate.

The *beam power* tube was designed to overcome some of the objectionable features of the pentode. It has no suppressor grid. The wires of the control and screen grids are so positioned that they are in a line with each other. (See Figure 11-10.) Thus the electrons emitted by the cathode may travel through the spaces between the wires of both grids in sheets or beams on to the plate of the tube. Very few electrons strike the screen grid, resulting in a lower screen-grid current than for a comparable pentode.

The space between the screen grid and plate is comparatively large. In this space is a set of specially-formed metal *deflecting plates* that are connected to the cathode. Because they are at the same potential as the cathode, they tend to slow down the electrons and concentrate them into narrower beams that strike the plate.

Because they are slowed down, the electrons form a sort of space charge in front of the plate. Thus, electrons that are knocked off

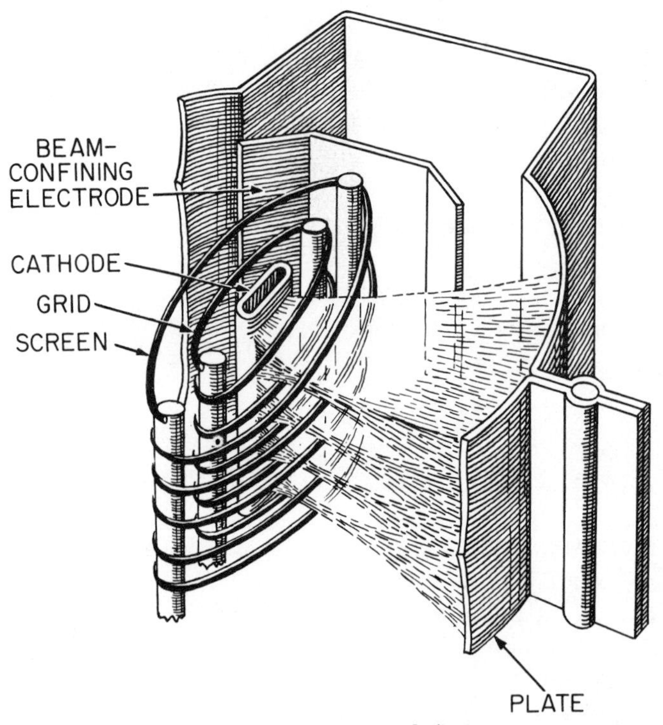

Fig. 11—10. Cutaway view of the beam power tube.

the plate by secondary emission will run into this space charge and be repelled back to the plate. Hence the space charge performs the function of the suppressor grid.

The beam power tube is, in reality, a tetrode that acts as a pentode. For a given emission of electrons, more of the space current reaches the plate, and the plate current of the tube is greater than for a comparable pentode. Although its amplification factor is smaller, so is its plate resistance, and its transconductance is larger. The symbol for the beam power tube is  . Note that the deflecting plates are connected internally to the cathode.

D. *Tubes with more than three grids*

At this point, it may be well to review the basic principle underlying all electron tubes. The cathode emits a stream of electrons which follow a one-way path to the positively-charged plate. Grids, if present, vary the rate at which the electrons flow through the tube. The effect a grid has upon the stream of electrons depends upon its voltage—a positive voltage will speed up the electrons; a negative voltage will slow them down. A varying voltage will cause the electron stream to fluctuate in step.

Thus far, we have discussed tubes having up to three grids—that is, five electrodes (pentode) including the cathode and plate. There are special tubes that have even more electrodes, generally additional grids. Thus a *hexode* (*hex* means six) has six electrodes consisting of a cathode, plate, and four grids. A *heptode* (*hept* means seven) has seven electrodes. An *octode* (*oct* means eight) has eight.

Of special interest in radio receivers is a heptode that is known as a *pentagrid mixer,* or *converter.* (See Figure 11-11.) Here we have two control grids, #1 and #3. (In multigrid tubes the grids are numbered from the cathode to plate. Thus, the grid nearest the cathode is #1. The next is #2, and so on, the highest numbered grid lying next to the plate.) Grids #2 and #4 are joined internally and form a shield and screen grid around grid #3.

The cathode, grid #1, and grid #2 form a sort of triode with grid #2, which is given a positive charge, acting somewhat as an anode. The control voltage is applied to grid #1. Since grid #2 is open-meshed, however, the electron stream, which is fluctuating in step with the variations in voltage on grid #1, passes through the openings and on to the real plate. Thus the cathode, grid #1, and grid #2 act as a composite "cathode" to emit a fluctuating stream of electrons.

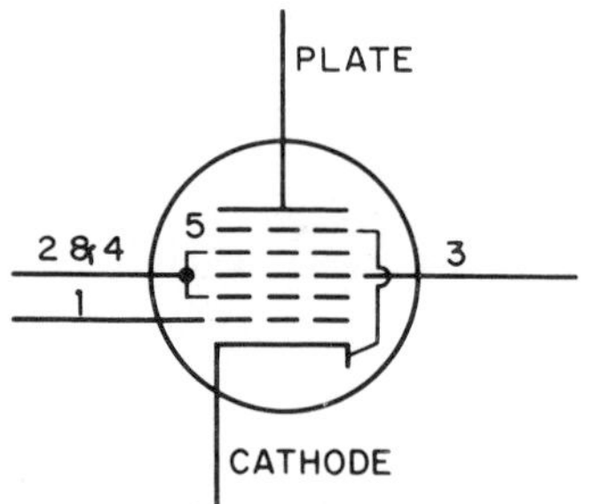

Fig. 11–11.

The schematic diagram of the pentagrid converter tube.

From here on, the tube acts as a pentode. Grid #3 is the second control grid. Grid #4 (which is joined to the positive grid #2) acts as the screen grid, and grid #5 (which is joined to the cathode) acts as the suppressor grid. Hence the electron stream is controlled by the combined actions of grids #1 and #3.

E. *Multiunit tubes*

Two or more tubes may be combined in a single envelope, and we have what are called *multiunit* tubes. All the tubes in the envelope may even share the same cathode, but they differ from the multielectrode tube in one important way. Whereas in the multielectrode tube only one stream of electrons is acted on by all the electrodes, in the multiunit tube the stream of electrons flowing from the cathode divides into two or more parts, each part flowing through its own unit of electrodes.

These tubes are constructed so that the electron stream of one unit is not affected by the electrodes making up any other unit, but proceeds to flow from the cathode through its various grids (if any) to its own plate. Thus we have, in effect, two or more tubes in the same envelope.

QUESTIONS

1. a) Describe the basic structure of the tetrode.
   b) Explain the action of the *screen grid.*
2. Draw and explain the basic circuit of the tetrode.
3. Compare the characteristics of the tetrode with those of the triode and explain the differences.
4. Explain the purpose of the *shield grid* in the gas tetrode.
5. a) Describe the basic structure of the pentode.
   b) Explain the action of the *suppressor grid.*
6. Compare the characteristics of the pentode with those of the triode and explain the differences.
7. Explain what is meant by
   a) a *sharp-cutoff tube;*
   b) a *remote-cutoff tube.*
8. Explain the action of the *remote-cutoff grid.*
9. a) Describe the basic structure of the beam power tube.
   b) Explain the action of the *beam-forming plates.*
10. Explain the difference between *multigrid* tubes and *multiunit* tubes.

Other Types of Tubes

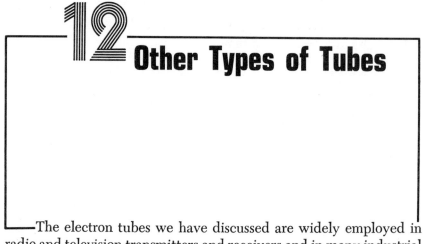

The electron tubes we have discussed are widely employed in radio and television transmitters and receivers and in many industrial applications. In addition, there are a number of tubes used for special purposes. Some of these will be considered in this chapter.

A. *The ignitron*

As previously stated, vacuum diodes are employed as rectifiers to change alternating to direct currents where the currents involved are small. Where they are larger, gas diodes are used. But where the currents are very large, a tube using a mercury-pool cathode, called the *ignitron,* is indicated. Such tubes are used, for example, as rectifiers in the welding process.

The plate, or anode, usually is a graphite block. The envelope may be a glass bulb, but the tubes that handle the larger currents generally have an envelope consisting of an air-tight steel tank from which the air has been evacuated, surrounded by a water jacket to dissipate the heat produced. (See Figure 12-1.)

Electron emission from the mercury-pool cathode is started by a spark between the mercury pool and an auxiliary electrode called the *ignitor.* This spark produces a hot *cathode spot* on the surface of the mercury and it is from this spot that the electrons are emitted.

Electron emission takes place only when the anode is positive—that is, during the positive half-cycle of the alternating current. During the negative half-cycle the cathode spot is extinguished. Accordingly, the ignitor must produce a cathode spot during each positive half-cycle. This necessity for producing a new cathode spot each cycle is a distinct advantage since it offers an opportunity for controlling the tube (as we shall see later).

The basic circuit of the ignitron is illustrated in Figure 12-2. The main circuit consists of the a-c supply, the ignitron tube, and the load (indicated here by R_L). The ignitor circuit contains the ignitor and a small diode rectifier, usually of the gas type. During the negative half-cycle of the a-c input, the plates of the ignitron and the diode are both negative. Hence, both tubes are inoperative.

Fig. 12—1. A. Illustration of the construction of the ignitron.
B. Ignitron.

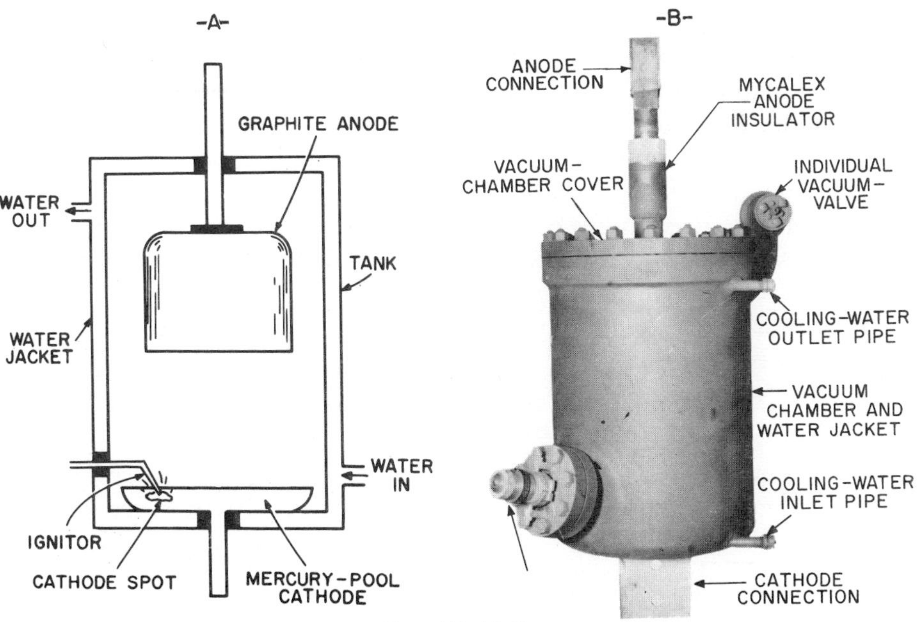

General Electric Company

During the positive half-cycle, both plates are positive. Nevertheless, the ignitron is inoperative because there is, as yet, no electron emission from the cathode. On the other hand, as the plate of the diode becomes positive, electrons flow through the tube from its cathode. This flow of current places a positive charge on the ignitor.

Because of the potential differences between the positive ignitor and the negative mercury pool, a spark is produced which establishes the cathode spot on the surface of the mercury. Electrons flow from this spot to the positive anode of the ignitron. Hence there is a flow of current through the ignitron during each positive half-cycle and a pulsating direct current is applied to the load.

Ignitrons come in a large variety of sizes. Some of the larger ones are able to handle currents whose average values are hundreds of amperes and whose peak values, which are applied only for a small fraction of the a-c cycle, are thousands of amperes. Where even higher currents are involved, several ignitrons may be used together. The voltage of the a-c supply is relatively low, a maximum of about 2,400 volts and generally much less.

B. *The phototube*

In our discussion of types of cathodes used in electron tubes (see Chapter 8, subdivision B) we stated that some tubes employ cathodes that emit electrons when struck by light. This phenomenon is called *photoelectric emission.* Tubes that use this type of emission are called *phototubes.*

Basically, these tubes consist of a cathode coated with some material which emits electrons when struck by light and a positively-charged anode which attracts these electrons. The whole is enclosed in a glass envelope. If the envelope contains a vacuum, the tube is

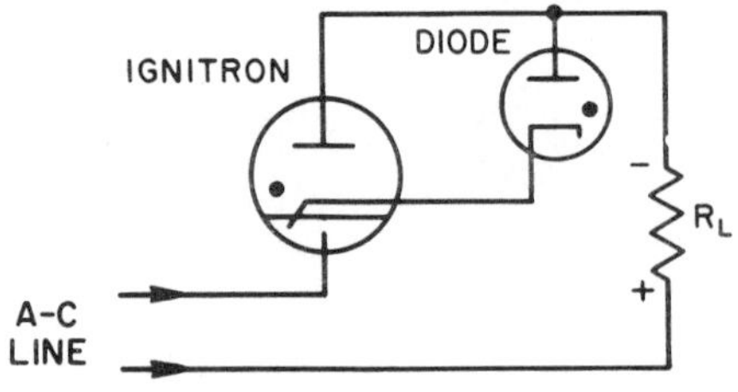

Fig. 12–2.

Basic circuit of the ignitron.

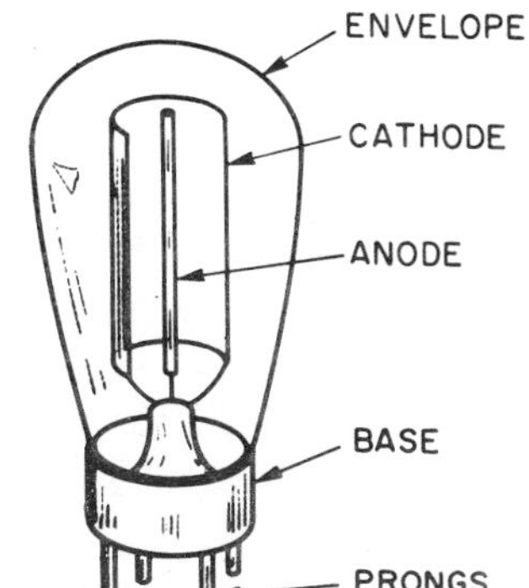

Fig. 12–3.

The phototube.

known as a *vacuum phototube* whose symbol is . If, instead of a vacuum a little gas is added, it is called a *gas phototube* whose symbol is .

The cathode of the phototube generally consists of a half-cylinder of metal coated on its inner surface with the photosensitive material. The anode usually is a thin metal rod lying along the central axis of the cathode. (See Figure 12-3.) To keep out unwanted light, the inner surface of the envelope may be masked, except for a small, circular window facing the inner surface of the cathode through which the light may enter. Or else a shield with a similar window may be placed over the entire tube.

Some tubes have the electrodes arranged vertically, permitting the light to enter from the side. (See Figure 12-4A.) Such a tube is

Radio Corporation of America

Fig. 12–4.

A. Side-on phototube.
B. Head-on phototube.

known as a *side-on* type. Others have the electrodes arranged so that the light enters from the top of the tube. (See Figure 12-4B.) These are called *head-on* types.

Since the phototube is operated by light waves, we must learn a few facts about waves in general and light waves in particular. Waves, as you may have observed at the seashore, travel in a rhythmical pattern. Each *cycle* of the water wave, for example, consists of a crest and an adjacent hollow, and each cycle is followed by a similar cycle.

The speed at which the waves travel is called the *speed of propagation.* The number of cycles passing by a fixed point in a given period of time is the *frequency* of the waves. The distance between a point in one cycle and the corresponding point in the next cycle is known as the *wavelength.* The speed of propagation, the frequency, and the wavelength bear a definite relationship to each other. This may be stated in the formula:

$$\text{wavelength} = \frac{\text{speed of propagation}}{\text{frequency}}.$$

Visible light is but a small portion of a whole spectrum of *electromagnetic waves* that include, in order of ascending frequency, radio waves, infrared (or heat) rays, visible light rays, ultraviolet rays, X-rays, and gamma rays.

Electromagnetic waves travel at a speed of 186,000 miles, or 300,000,000 meters, per second. The frequencies of visible light waves are extremely high. Accordingly, their wavelengths are very small, a minute fraction of an inch. For ease in handling such small values, two units of length are employed. One is the *micron* which is equal to 1/1,000,000 of a meter. The other is the *angstrom unit* of which there are 10,000,000,000 to the meter. (A meter is about 39 inches.)

Visible light covers a range of wavelengths lying, approximately, between 4,000 and 8,000 angstrom units. Visible light, however, is made up of a combination of light of different colors. Thus *violet* light has wavelengths lying between 4,000 and 4,500 angstrom units. Between 4,500 and 5,000 angstrom units we have *blue* light. Between 5,000 and 5,500 angstrom units we have *green* light; between 5,500 and 6,000 angstrom units, *yellow* light; between 6,000 and 6,300 angstrom units, *orange* light; and between 6,300 and 8,000 angstrom units, *red* light.

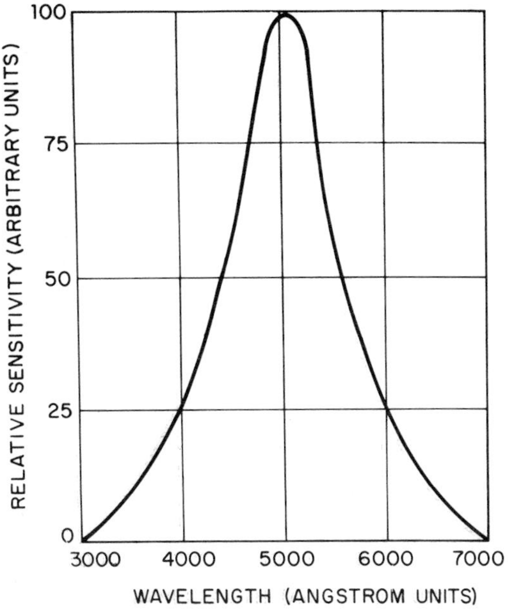

Fig. 12–5.

Sensitivity graph for S-4 material.

There are a number of different types of photosensitive materials used for coating the cathodes of phototubes, each sensitive to a different range of wavelengths. For example, a combination of antimony and cesium, known as the S-4 type, is sensitive to light whose wavelengths lie in a range between 3,000 and 7,000 angstrom units. Other combinations of materials are sensitive to light of different wavelengths.

A sensitivity graph for the S-4 material is illustrated in Figure 12-5. Sensitivity is indicated in arbitrary units ranging from zero to 100. As you can see, this material is not equally sensitive to all light lying within its range. From this graph you will notice that it is most sensitive to light having a wavelength of about 5,500 angstrom units, or, in the green-yellow range.

The electron emission of the photosensitive material of the cathode also depends upon the amount of light, measured in units called *lumens*, that falls upon it. The more light, the more electrons that will be emitted, up to the point of saturation.

In Figure 12-6 you see the anode voltage–anode current curves for a typical vacuum phototube at two values of light (0.1 lumen and 0.5 lumen). Note that the units for current are in microamperes. Even at zero anode potential some of the emitted electrons strike the anode, resulting in a small anode current. As the anode voltage is

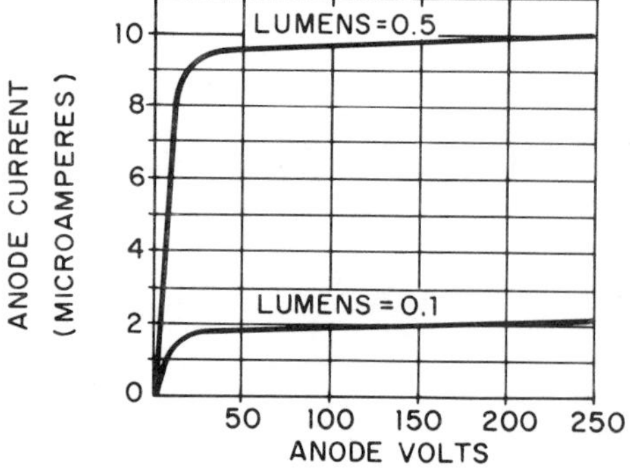

Fig. 12–6.

Anode voltage—anode current curves for typical vacuum phototube.

increased, more of the electrons are attracted, causing a rapid rise in anode current. At a fairly low anode voltage virtually all the electrons are attracted, and further increasing the anode voltage produces practically no increase in anode current.

You can see from the two curves that the anode current is dependent upon the amount of light striking the cathode. As expected, the more light, the larger is the anode current.

The anode voltage–anode current curve for a typical gas phototube at two values of light (0.04 lumen and 0.1 lumen) is illustrated in Figure 12-7. Note that at low anode voltages the tube behaves as a vacuum type—the anode current rises rapidly and then levels off.

At about 25 volts, however, the potential difference between the cathode and anode is great enough to ionize the gas within the tube. The electrons so produced are added to the anode current. Also, the positive ions are attracted to the cathode, bombarding it with enough force to release additional electrons by secondary emission. The anode current rises rapidly. The action is cumulative and unless

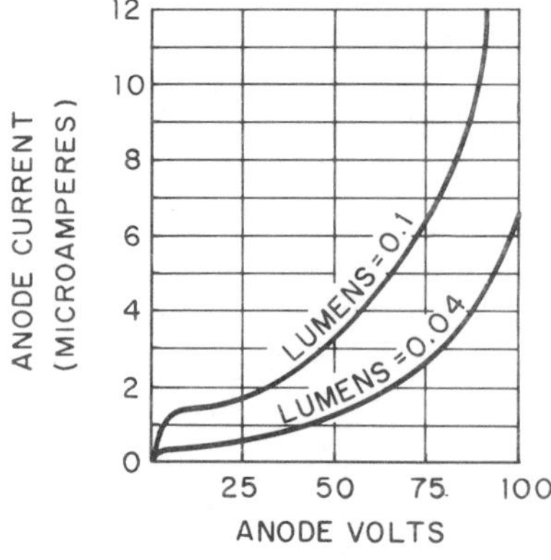

Fig. 12–7.

Anode voltage—anode current curves for typical gas phototube.

some protective device (such as a resistor between the anode of the tube and its voltage supply) is employed, the current may become large enough to damage the tube.

The phototube may also be regarded as a sort of resistor. When the cathode is dark, the tube is nonconductive, its resistance approaching infinity. When the cathode is illuminated, the tube becomes conductive and its resistance drops to a low value, depending upon the characteristics of the tube and the amount of light it receives.

C. *The multiplier phototube*

The phototube, even the gas-type, has a very small output at low levels of illumination. To overcome this defect, the *multiplier phototube* has been developed. In this tube the electrons emitted by the cathode are made to strike in succession a series of positive plates, called *dynodes,* before they end up at the anode.

As each electron strikes a dynode, it knocks off a number of new electrons (generally about five or so) by secondary emission. Thus a single electron emitted by the cathode will produce, say, five electrons at dynode #1. These five electrons will produce about 25 electrons at dynode #2. And so forth. If we employ enough dynodes (and multiplier phototubes generally have between nine and fourteen dynodes), several million electrons will reach the anode for each electron emitted by the cathode. The result, then, is a tube that produces a relatively high output at very low levels of illumination.

A typical multiplier phototube is illustrated in Figure 12-8A. Although some are of the side-on type, most of these tubes are of the head-on type. The dynodes are grouped so that they may fit into a rather small envelope and are shaped so as to focus the electrons to the following dynode.

Light entering the window in the envelope passes through a metal grill and strikes the photoemissive cathode. The function of this grill is to establish an electric field that helps control the paths of the electrons. The symbol for a dynode is —(ı, and the symbol for the entire tube is illustrated in Figure 12-8C.

In order to attract electrons with enough force to produce secondary emission, each dynode must be sufficiently more positive than its

–A–

Radio Corporation of America

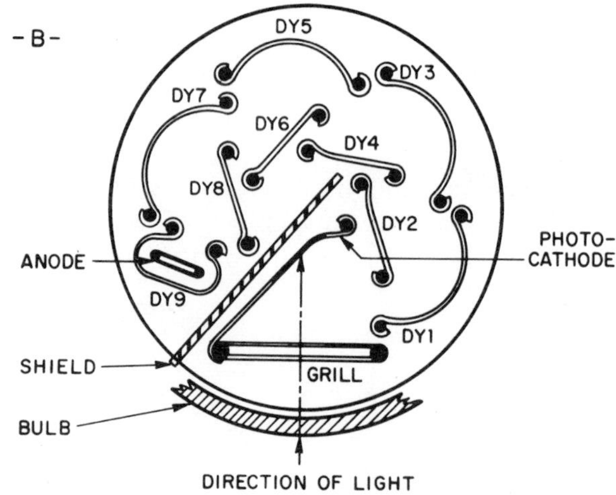

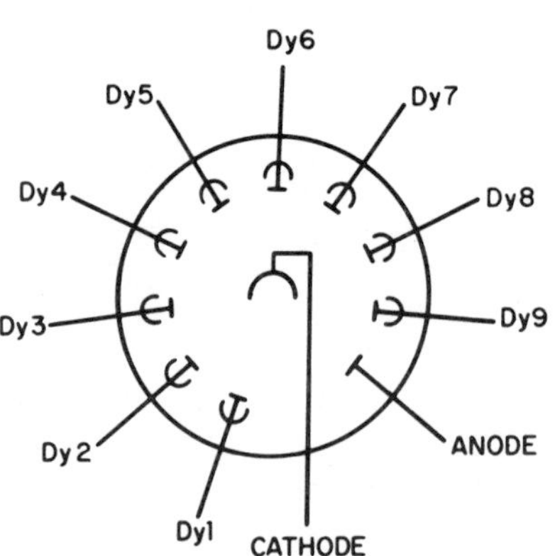

Fig. 12–8.

A. The multiplier phototube.
B. Arrangement of dynodes.
C. Symbol for multiplier phototube.

predecessor. Generally, a potential difference of about 100 to 250 volts is employed. The anode is most positive of all. Hence the potential difference between the anode and cathode may be several thousand volts.

Theoretically, there should be no limit to the number of dynodes employed in a tube. Practically, however, the number is limited because of the difficulty in keeping the electron streams focused and preventing interaction between them, as the number of dynodes is increased. Also, since each dynode operates at a higher voltage than its predecessor, the total voltage may become prohibitively large if too many dynodes are used.

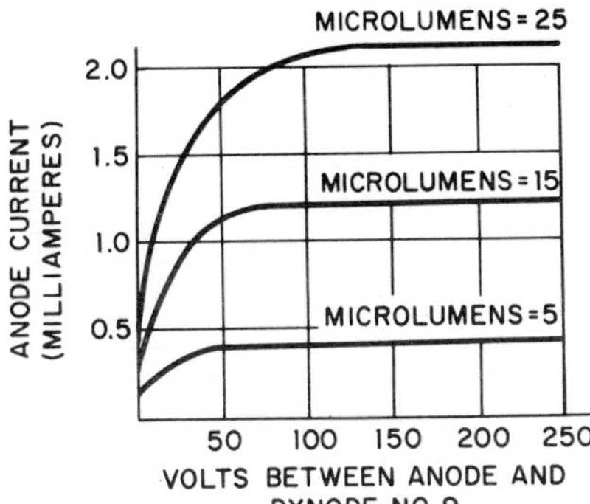

Fig. 12—9.

Anode curves for a typical multiplier phototube.

Several anode curves for a typical multiplier phototube are illustrated in Figure 12-9. The anode current is measured in milliamperes, but note that the illumination is measured in *microlumens* (a microlumen is 1/1,000,000 of a lumen).

D. *The electron-ray tube*

So far, we have studied electron tubes where the stream of electrons emitted by the cathode results in a current flowing through the plate circuit. There are other types of tubes where the electron stream produces light. A common example is the ordinary fluorescent lamp. Another is the *electron-ray tube* that is employed as a visual indicator to replace the meter in certain applications.

The type 6E5 tube is typical. This tube is a vacuum type with the cathode mounted in a vertical position. Around this cathode is placed a funnel-shaped anode, or *target*, tapering down (Figure 12-10). The inner surface of the target is coated with certain

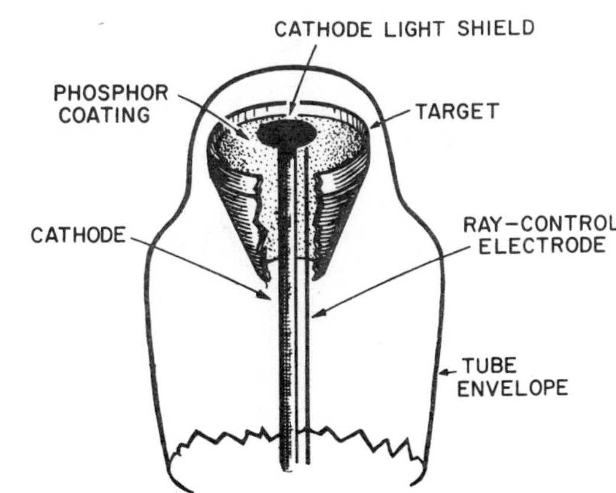

Fig. 12—10.

Construction of the electron-ray tube.

chemicals, called *phosphors,* that glow when struck by an electron stream. Electrons emitted by the heated cathode strike the inner surface of the target, producing light which appears as a ring when viewed looking down at the top of the tube.

Between the cathode and the target, a vertical *deflector,* or *ray-control,* electrode, consisting of a thin wire, is inserted. If this electrode is at the same potential as the anode, or target, it will have little effect on the electron stream, and the glow will be an uninterrupted ring (except for a very thin "shadow" produced where the control electrode obstructs the passage of electrons from the cathode to target). (See Figure 12-11A.)

But if the potential of the control electrode is less positive than that of the target, an electric field is set up between it and the target. This field repels electrons flowing toward the target, and thus the portion of the target in line with the control electrode and its electric field will be cut off from electrons and, therefore, will not glow. The greater the potential difference between the target and the control electrode, the greater will be the dark portion of the target (Figure 12-11B).

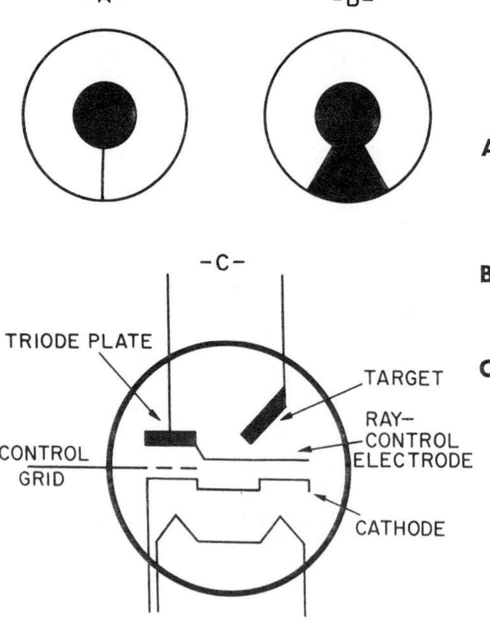

Fig. 12–11.

A. Top view of electron-ray tube when ray-control electrode is at the same potential as the target.

B. Ray-control electrode is at a lower potential than the target.

C. Symbol for the 6E5 electron-ray tube.

Fig. 12—12.

Circuit for the electron-ray tube.

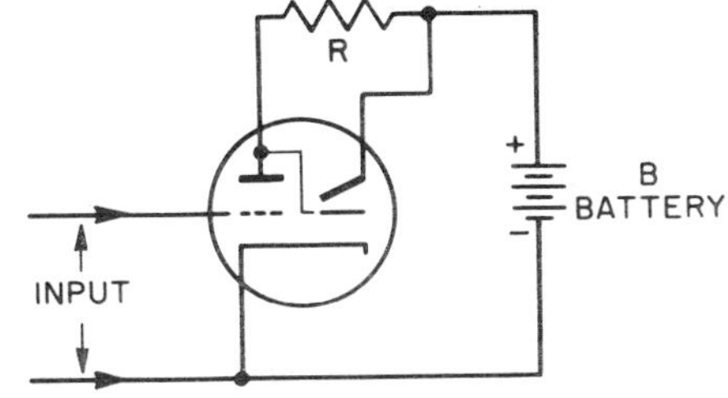

The dark spot in the center of the ring of light is caused by the *cathode light shield* that is so placed as to make the amount of deflection more noticeable. The tube generally is mounted in a horizontal position so that the amount of "shadow" may be seen. The angular spread of this "shadow" is from zero degrees (when the potentials on the control electrode and target are the same) to about 90 degrees (maximum potential difference). Because of the resemblance between the top view of the tube and the human eye, it sometimes is called a *magic-eye tube*.

In the 6E5 tube, a triode is incorporated in the same envelope with the electron-ray assembly. The symbol for the tube is shown in Figure 12-11C. Note that the control electrode is connected internally to the plate of the triode and that the cathode is common to both the triode section and the electron-ray assembly.

The circuit for the electron-ray tube is illustrated in Figure 12-12. Resistor R (generally about one megohm) is connected between the triode plate and the target electrode. If a sufficiently negative voltage is applied to the grid of the triode, practically no plate current flows through the triode section and there is no voltage drop across R. Hence the plate of the triode and the target electrode are at the same potential and the ring of light appears as in Figure 12-11A.

As the negative voltage on the triode grid is reduced, plate current flows through the triode section and a voltage drop appears across R. As a result of this voltage drop, the potential of the triode plate becomes smaller than that of the target electrode. Since the control electrode is connected to the triode plate, the former acquires the same potential as the latter. Thus the control electrode becomes negative compared to the target electrode.

Accordingly, an electric field is set up between the control electrode and the target which obstructs the flow of electrons from cathode to target, thus forming a "shadow." The smaller the negative voltage on the triode grid, the larger will be the plate current through the triode section, the greater the potential difference between the control electrode and target, and the wider the "shadow." When the grid voltage is reduced to zero, the light ring appears as in Figure 12-11B. We thus have a convenient method for obtaining an indication of the grid voltage.

E. *The cathode-ray tube*

One of the most ingenious devices invented by man is the *cathode-ray tube* used in television and radar receivers and in the cathode-ray oscilloscope. As the electron-ray tube just discussed, its operation depends on the fact that if a phosphor is struck by an electron beam, the phosphor will glow.

Basically, the tube consists of an *electron gun* which produces a stream of electrons focused into a fine, needle-like beam, a *high-voltage anode* consisting of an open-ended metal tube that attracts and speeds up the electron beam but permits it to pass through, a *deflection system* which causes the beam to move from side to side and up or down, and a *phosphor screen* against which the beam impinges, producing a spot of light where it strikes. The whole is enclosed in a glass envelope from which all air is evacuated. (See Figure 12-13.)

The electron gun consists of several elements mounted together as a single unit. First, there is an indirect-heater *cathode* which emits the stream of electrons. This cathode is so constructed that the electrons are emitted from its end facing the screen of the tube, rather than from its sides.

The electron stream then passes through the *control grid* consisting of a metal cylinder enclosing the cathode. There is a small opening in the center of the top of this cylinder through which the electrons may pass. The effective size of this opening depends upon the potential difference between the grid and the cathode. The greater the negative potential of the grid, the more it will repel the electrons of the stream and, hence, the smaller the effective size of the opening will be.

Since the intensity of the glow at the screen depends upon the quantity of electrons in the beam striking it, you can see that by varying the potential of the control grid we may control the intensity of the spot of light on the screen. If the control grid be made negative enough, the electron beam will be cut off.

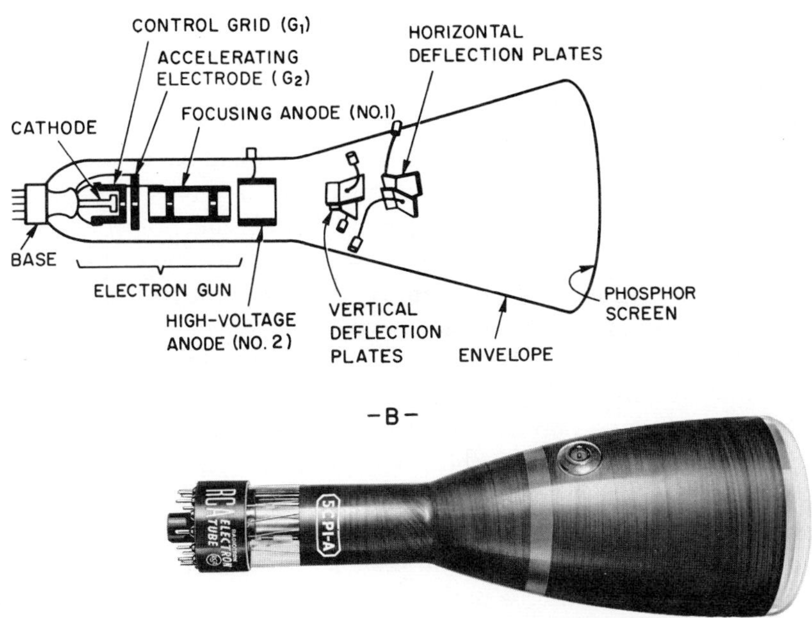

Radio Corporation of America

Fig. 12–13. **A.** Cross-sectional view of cathode-ray tube employing an electrostatic deflection system.
B. Cathode-ray tube.

The next electrode is the *accelerator grid* consisting of a metal cylinder or disk with a small opening through which the electron stream may pass. A high, fixed positive potential is applied to this grid and, as a result, the electrons streaming through the aperture of the control grid are strongly attracted and speed through the opening in the accelerator grid at high velocity. This grid generally is designated as G_2 to differentiate it from the control grid which is designated as G_1. (In some tubes the G_2 grid is omitted.)

Following the accelerator grid is the *focusing anode* (anode #1) which consists of a metal cylinder open at both ends. Two metal disks, each with a small opening at its center, are set inside the cylinder. A fairly high, fixed positive potential is placed on this anode. As a result of this potential and because of the electric field created between this anode and the more positive high-voltage anode which follows it, the electron stream is sped up and focused into an extremely thin beam. The focusing anode acts on the electron beam in a manner somewhat similar to the action of a glass focusing lens upon a beam of light. All the electrodes of the electron gun are connected to prongs set into the base of the tube.

The *high-voltage anode* (anode #2) consists of an open-ended cylinder through which the electron beam may pass. As its name implies, high positive voltage (generally thousands of volts) is applied to this anode through a button set in the envelope. Its function is to speed up the electron beam on its way through the deflection system to the screen.

Before the electron beam reaches the phosphor screen it must pass through the *deflection system* that is able to move the beam in the vertical plane (up or down) and in the horizontal plane (from side to side). In this way the beam can be made to strike any spot on the screen. There are two types of deflection systems; the *electrostatic* type and the *electromagnetic* type.

In the electrostatic deflection system two sets of metal deflection plates are mounted in the neck of the tube in such a manner that the electron beam must pass between the plates of each set. The plates of one set are mounted parallel to the ground. However, because they move the electron beam in the vertical plane, they are known as the *vertical deflection plates.* The plates of the second set are mounted at right angles to those of the first set and, because they move the electron beam in the horizontal plane, are known as the *horizontal deflection plates.* Connections to the plates are made through four buttons set in the side of the envelope.

Since the electrons of the beam carry a negative charge, they will be attracted to a positively-charged plate and repelled from a negatively-charged one. Look at Figure 12-14. Only the vertical deflection plates are shown here. Normally, the electron beam passes unhindered through these plates to strike the center of the screen (Figure 12-14A).

Now assume that a voltage is applied to the plates so that the upper plate is charged negatively and the lower plate is charged positively, as shown in Figure 12-14B. The electron beam will be repelled from the negative upper plate and will be attracted to the positive lower plate. Thus the beam is deflected down and will strike the screen below its center. The greater the potential difference between the plates, the more the beam will be deflected down.

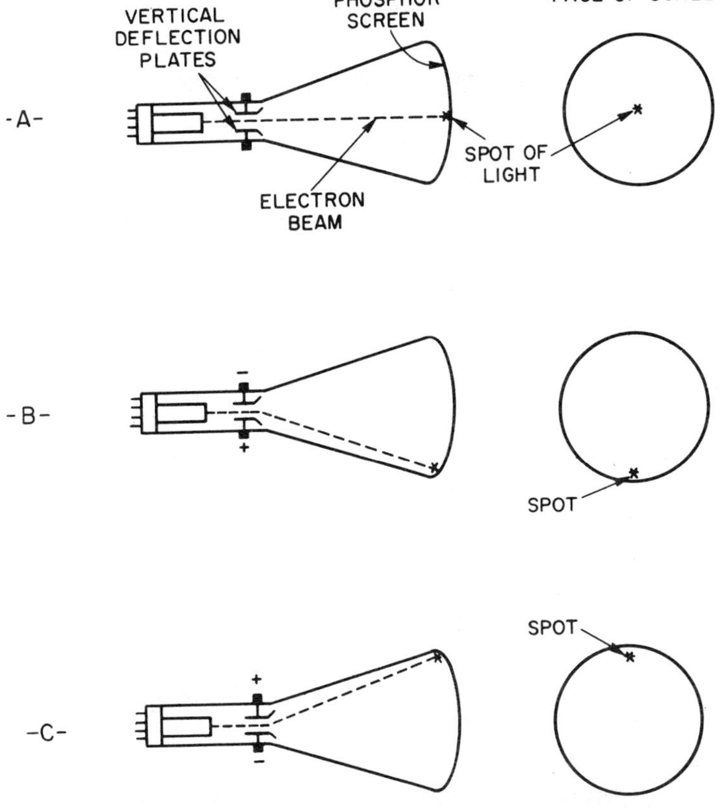

Fig. 12—14. How the electrostatic deflection system works.
A. No charge on the vertical deflection plates.
B. Negative charge on top plate, positive charge on bottom plate.
C. Positive charge on top plate, negative charge on bottom plate.

If the charges on the plates are reversed (Figure 12-14C), the beam will be deflected upward. Thus, by controlling the voltage on the plates, we can cause the electron beam to move up and down vertically, producing spots of light where the beam strikes the screen. Since the phosphor screen is a thin deposit on the inner face of the wide end of the envelope, we can see the light through the glass.

It should be pointed out that when we say that one plate of a set is positive and the other negative, we mean that one is positive *relative to the other.* For example, suppose the upper plate has a potential of +10 volts and the lower one a potential of −10 volts. The difference of potential between the plate is 20 volts, and the electron beam will be deflected toward the upper plate to a degree depending upon this difference of potential.

Now assume the upper plate receives a potential of +120 volts and the lower plate a potential of +100 volts. The upper plate still is the positive one (relative to the lower plate) and the potential difference, as before, is 20 volts. Accordingly, the electron beam will be deflected toward the upper plate in the same degree.

In a similar manner, the spot of light can be made to move horizontally from side to side by the action of the horizontal deflection plates. Thus, by the simultaneous action of both the vertical and horizontal deflection plates, the spot of light can be moved all over the surface of the screen. The ends of all the deflection plates are flared out to prevent interference with the electron beam.

Now, suppose we apply a voltage whose waveform is as shown in Figure 12-15 to the horizontal deflection plates. This is known as a *sawtooth* waveform, for obvious reasons. Note that the voltage starts at zero, rises relatively slowly to its maximum positive value, and then drops rapidly to its maximum negative value. Then the cycle is repeated. The time it takes the voltage to rise from its maximum negative value to its maximum positive value is known as the *trace,* or *sweep,* time. The time it takes for the voltage to fall from its maximum positive to its maximum negative value is known as the *retrace,* or *flyback,* time.

See what is happening to the spot of light in the meantime. Starting from the center of the screen (at zero volts), it moves horizontally relatively slowly to the right of the screen as the voltage rises, reaching the extreme right as the voltage attains its maximum

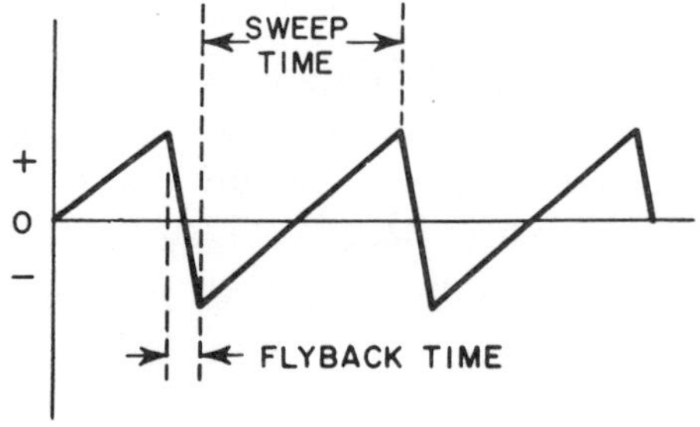

Fig. 12–15.

Waveform of the sweep voltage.

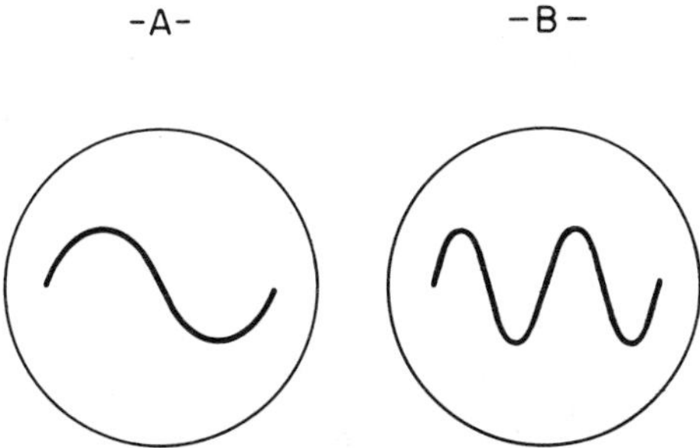

Fig. 12—16. Traces seen on the face of the cathode-ray tube.
A. Sinusoidal voltage at the vertical deflection plates at same frequency as sawtooth voltage at the horizontal deflection plates.
B. Sinusoidal voltage at the vertical plates at twice the frequency of sawtooth voltage at the horizontal plates.

positive value. Then it moves rapidly to the left, reaching the extreme left side as the voltage attains its maximum negative value. Then the cycle is repeated. Thus the spot sweeps from side to side in a straight line in step with the cycles of the sawtooth voltage. You can see why this sometimes is called the *sweep voltage.*

Let us see what happens if we apply a sinusoidal alternating voltage to the vertical deflection plates. When the voltage is at zero, the spot of light will be at the center of the screen. As the voltage rises to its maximum positive value, the spot of light will move up vertically to the top of the screen. As the voltage falls to its maximum negative value, the spot will move vertically to the bottom of the screen. Thus the spot will move up and down in a straight line in step with the frequency of the alternating voltage.

Now what happens if we apply the sawtooth voltage to the horizontal deflection plates at the same time we apply the sinusoidal voltage to the vertical deflection plates? As the sawtooth voltage is sweeping the spot horizontally from side to side, the sinusoidal voltage is moving it up and down. If the frequencies of both voltages are the same, the spot will trace across the screen a single sine wave (Figure 12-16A). If the frequency of the sinusoidal voltage is twice that of the sawtooth voltage, the spot will sweep through two sine waves (Figure 12-16B). And so on.

If the frequencies involved are too great for the eye to follow, we cannot determine the movements of the spot of light across the screen. But the phosphors employed will continue to glow for a short time after they are struck by the electron beam. Hence we can see the path of the spot (called the *trace*) across the screen.

You can see how the cathode-ray tube can be used as a voltmeter to measure the voltage applied to the vertical deflection plates. All we need to know is the number of volts required to produce a vertical deflection of some unit length and measure the vertical height of the trace. We can also use it to measure the frequency of the vertical voltage. If we know the frequency of the sweep voltage and a single cycle appears in the trace, then the frequency of the vertical voltage is the same as that of the sweep voltage. If two cycles appear, the vertical voltage has twice the frequency of the sweep voltage, and so forth.

The tube may also be used to show traces of voltages with irregular waveforms. Because the electron beam is so light it has practically no inertia and so can be moved very rapidly from spot to spot on the screen. The waveforms of voltages with frequencies running into the millions of cycles per second can be depicted. Tubes using electrostatic deflection are generally employed in cathode-ray oscilloscopes.

Since the electron beam constitutes a flow of current, it is surrounded by a magnetic field. Accordingly, we may employ an external electromagnet whose magnetic field will react with the one around the electron beam to produce a deflection of the beam. This method is called the *electromagnetic deflection* system and is widely employed in television and radar receivers.

Look at Figure 12-17. Here you see a cross-sectional view of the neck of the tube. In the center is the electron beam flowing out of the page toward you. This electron beam is surrounded by a magnetic field, as represented by the circle of dotted lines around the beam. The arrows on these dotted lines indicate the direction of these magnetic lines of force.

Two electromagnets are placed above and below the neck of the tube. The coils of these electromagnets are connected in series and their windings so arranged that as a direct current flows through them, as indicated, a north pole appears above the tube and a south pole below it. The magnetic field between the two poles is indicated by the dotted lines connecting them, and the direction of the magnetic lines of force is indicated by the arrows on the dotted lines.

Note that to the right of the electron beam the two magnetic fields augment each other, as indicated by the arrows all in the same direction. On the left side of the beam opposing arrows show that the two fields oppose and weaken each other. Because the resulting field is stronger on the right than on the left, the electron beam is deflected to the left.

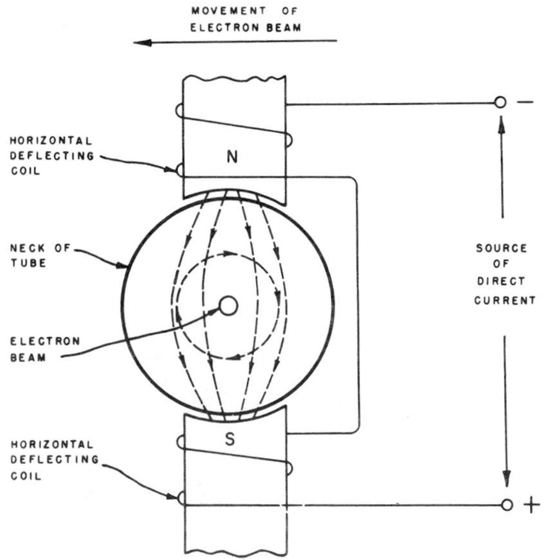

Fig. 12—17. How electromagnetic deflection coils can deflect the electron beam.

The greater the current flowing through the coils, the stronger their magnetic field will be, and the more the electron beam is deflected. By reversing the current flow through the coils, the direction of their magnetic field is reversed, and so is the direction of deflection of the beam. These coils are the *horizontal deflection coils.*

A similar set of two coils, placed to the left and right of the neck of the tube, causes the electron beam to be deflected up and down; they are called the *vertical deflection coils* (see Figure 12-18A). In practice, both the horizontal and vertical deflection coils are enclosed

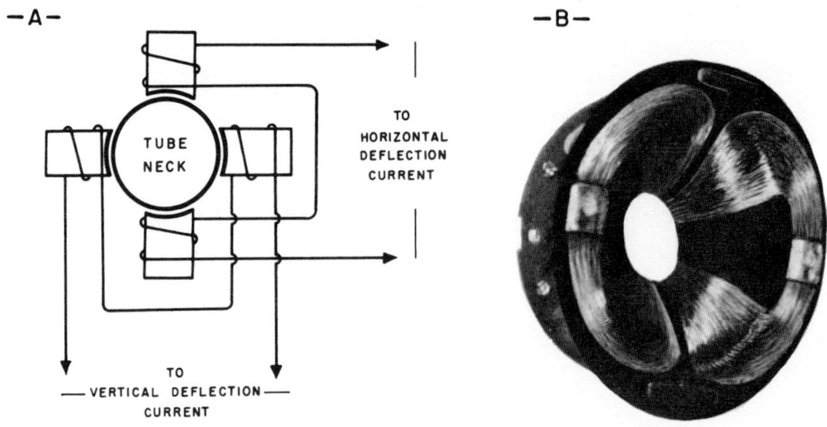

Fig. 12–18. **A.** How the vertical and horizontal deflection coils are placed around the neck of the tube.
B. A deflection yoke.

in a common form, called the *yoke,* which fits over the neck of the tube. By properly positioning the yoke on the neck of the tube and controlling the amount of current flowing through each set of coils, the electron beam can be deflected to strike any desired spot on the screen.

The screen consists of a thin coating of phosphors deposited upon the inner face of the wide end of the envelope. These phosphors *fluoresce,* or glow, when struck by the electron beam. In addition, they continue to glow for a short period of time after the electron beam passes. This is called *phosphorescence.* The color of the light produced and the duration of the afterglow depends upon the composition of the phosphors. The colors generally are green, blue, or white and the duration of afterglow is short, medium, or long.

In the cathode-ray oscilloscope a phosphor that produces a green trace with a medium afterglow (designated as P1) generally is required. Where the voltages to be examined are of extremely short duration and do not repeat cyclically (these are called *transients*), a long-persistent phosphor (P2) is required. Where we wish to photograph the trace produced on the screen, a P5 phosphor that produces a blue trace (which can photograph well) of short duration is employed. In television receivers a P4 phosphor, which produces a white light, generally is used.

Radio Corporation of America

Fig. 12–19. Cathode-ray tube used in the television receiver.

For use in the cathode-ray oscilloscope, a tube with a round face five or seven inches in diameter generally is employed. A larger diameter tube usually is used in the radar receiver. For television purposes, a tube with a rectangular face is most common (Figure 12-19). The size of this tube is measured by the length of the diagonal of the face and tubes are available in sizes up to 27 inches.

Very often, the inner surface of the flare of the envelope is coated with a conductive graphite coating called *aquadag*. This coating is connected internally to the high-voltage anode and thus has a high positive charge. Its purpose is to attract, and thus remove from the tube, electrons knocked off from the screen by secondary emission as it is struck by the high-velocity electron beam.

QUESTIONS

1. a) **Describe the basic structure of the ignitron.**
   b) **Explain the action of the *ignitor*.**
2. **Draw and explain the basic circuit of the ignitron used as a rectifier.**
3. a) **Describe the basic structure of the phototube.**
   b) **Explain its action.**
4. **Explain the relationship between the speed of propagation, frequency, and wavelength of electromagnetic waves.**
5. a) **What two factors of the light striking a phototube affects its anode current?**
   b) **Explain.**

6. What is the effect upon the resistance of a phototube
   a) when its cathode is dark;
   b) when its cathode is illuminated?
7. a) Describe the basic structure of the multiplier phototube.
   b) Explain its action.
8. a) Describe the basic structure of the electron-ray tube.
   b) Explain its action.
9. Describe the basic structure of the cathode-ray tube.
10. In the cathode-ray tube, explain the action of
    a) the *control grid;*
    b) the *focusing anode;*
    c) the *high-voltage anode.*
11. Explain the action of the *deflection plates* in a cathode-ray tube using an electrostatic deflection system.
12. Explain the action of the *yoke* in a cathode-ray tube using an electromagnetic deflection system.
13. Explain how the *trace* is formed upon the phosphor screen of the cathode-ray tube.
14. What is meant by
    a) the *sweep time;*
    b) the *retrace time?*
15. Explain how a sine curve may be traced upon the phosphor screen of a cathode-ray tube.
16. Explain how the cathode-ray tube may be used to determine the frequency of an unknown voltage.

Section 3

How the Electron Tube Is Employed

In spite of the bewildering array of circuits employing electron tubes, it may be of some comfort to find that, for the most part, the tubes are used for one of three purposes. One is as a *rectifier* that changes alternating to direct current. The second is as an *amplifier* for small voltages or currents. The third is as an *oscillator* that generates alternating currents.

13 Rectifiers and Power Supplies

We already have learned how the diode acts as a rectifier, changing alternating current to direct (Chapter 9). Inasmuch as most of the power mains are of the a-c type, one of the chief functions of the diode is as a rectifier for the power supplies that operate from these power mains and furnish the direct currents and voltages required by many electronic devices and circuits. The vacuum tube diode has been largely replaced by the semiconductor diode, and the following discussion and circuitry can be applied to either. The transition took place because the semiconductor diode doesn't require any heater supply and offers fewer losses as well as greatly reduced size and cost. (See Chapter 29.)

A. *Half-wave rectifier*

The basic rectifier circuit is illustrated in Figure 13-1A. Transformer T is a step-down type that reduces the line voltage to a value suitable for the heater of the rectifier. During those half-cycles of the line voltage when the plate of the diode is positive, relative to its cathode, current flows through the circuit (including the load, shown here as resistor R_L) in the direction indicated by the arrows.

During the alternate half-cycles when the plate is negative, the tube is nonconductive and there is no flow of current. Because current flows through the load only during half the cycle, this type of circuit is called a *half-wave rectifier circuit.*

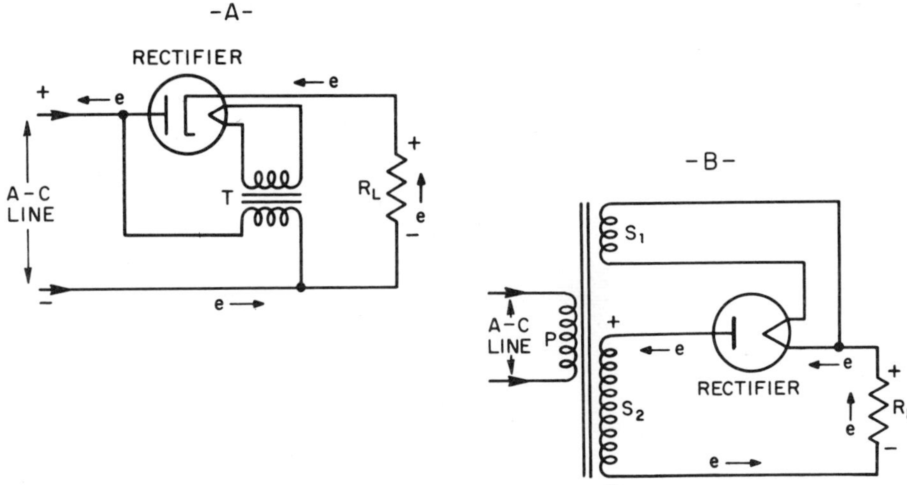

Fig. 13—1. **A.** Basic circuit of the half-wave rectifier.
B. Half-wave rectifier, using a step-up transformer.

Where we wish a voltage higher than that supplied by the line, a step-up transformer may be employed, as illustrated in Figure 13-1B. (Note that a filament-type diode is shown here instead of the indirect-heater type of Figure 13-1A. Of course, either type may be used.) Secondary S_2 is the step-up winding of the transformer. Secondary S_1 is a step-down winding for the filament of the tube.

As in the circuit of Figure 13-1A, when the plate is positive (relative to the cathode), current flows through the load in the direction indicated by the arrows. When the plate is negative, there is no flow of current.

B. *Full-wave rectifier*

We may supply direct current to the load during each half of the cycle by using two diodes and a transformer whose secondary winding is center-tapped. Look at Figure 13-2A. Assume a half-cycle when the top of the secondary winding is positive and the bottom is negative. Because of its position half-way between the ends of the winding, the center-tap is negative, relative to the top end of the winding, and positive, relative to the bottom end.

Only diode D_1 will pass current since its plate is positive. The plate of diode D_2 is negative and no current flows through it. The current flow in the circuit is indicated by the arrows.

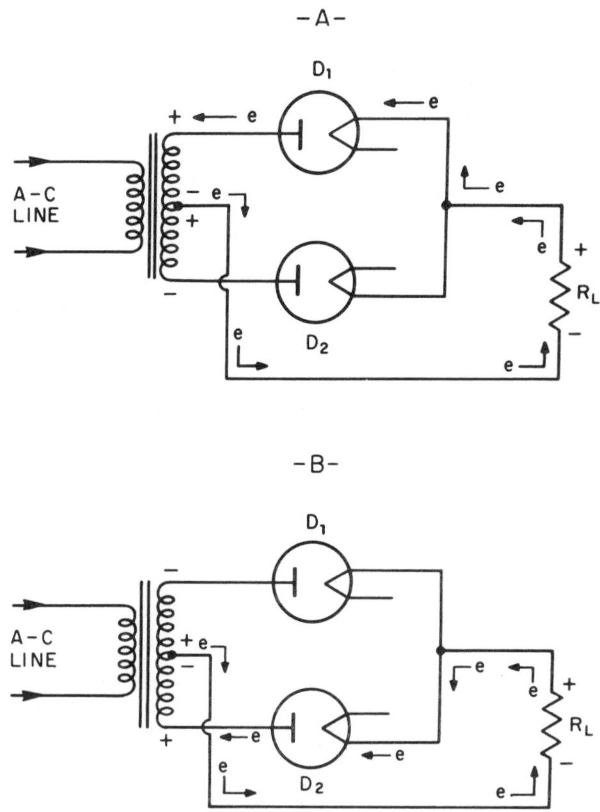

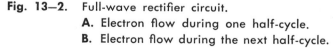

Fig. 13–2. Full-wave rectifier circuit.
A. Electron flow during one half-cycle.
B. Electron flow during the next half-cycle.

During the next half-cycle, the polarity of the transformer is reversed (Figure 13-2B). The top end of the winding is negative and the bottom is positive. The center-tap is positive, relative to the top end of the winding, and negative, relative to the bottom end.

Now D_2 is operative and no current flows through D_1. Again the arrows indicate the direction of current flow in the curcuit. Note that in both instances the direct current flowing through the load is in the same direction. We call this a *full-wave rectifier circuit.*

A graphic comparison between half-wave and full-wave rectifiers is shown in Figure 13-3. The waveform of the alternating voltage applied to the plates of the diodes is illustrated in Figure 13-3A. In Figure 13-3B we see the waveform of the current flowing through the load of the half-wave rectifier. Note that this current consists of a series of direct-current pulses, one for each positive half-cycle of the alternating voltage. The negative half-cycles are blocked out by the nonconductive tube.

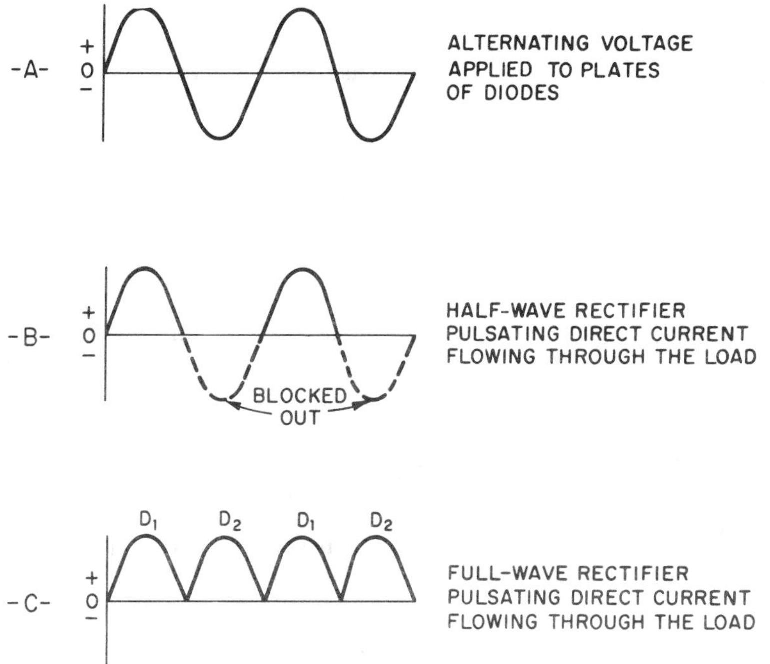

Fig. 13—3. Graph showing the relationship between the input voltage and the output voltages of the half-wave and full-wave rectifier circuits.

Radio Corporation of America

Fig. 13—4.

Duodiode, type 5Y3-GT, for use in full-wave rectifier circuit.

In Figure 13-3C we see the waveform of the current flowing through the load of the full-wave rectifier. Again, this current consists of a series of direct-current pulses. But this time there is one pulse for each half-cycle of the alternating voltage.

The full-wave rectifier circuit has a number of advantages over the half-wave type. For one, each tube carries only half the load current, whereas the half-wave rectifier tube has to carry the entire load. Also, since there are two output pulses per cycle for the full-wave rectifier and only one pulse per cycle for the half-wave type, the output of the full-wave type is easier to filter (as shall be shown a little later).

The half-wave rectifier circuit, on the other hand, has a number of advantages over the full-wave type. Since, in the full-wave circuit, only half the secondary winding of the transformer is used at a time, this winding must have twice as many turns as a winding used in a half-wave circuit to produce a similar voltage. Also, the full-wave rectifier requires two tubes.

This latter disadvantage may be overcome, however, by enclosing both diodes in one envelope. Such a tube is called a *full-wave rectifier tube*, or *duodiode*, an example of which is illustrated in Figure 13-4.

C. *Bridge rectifier*

Another full-wave circuit is the *bridge rectifier* shown in Figure 13-5. Here, four diodes are arranged in such a way that when the top of the secondary winding is negative and the bottom positive (Figure 13-5A), electrons will flow to junction A. These electrons will make the plate of D_3 negative; hence this tube will be nonconductive. But they also will make the cathode of D_1 negative with respect to its plate, hence D_1 becomes conductive and electrons can flow through it to junction B.

In the same manner, D_2 becomes nonconductive and the electrons are forced to flow on to the bottom end of the load resistor R_L (which thus becomes negative). Then they flow through R_L to junction D.

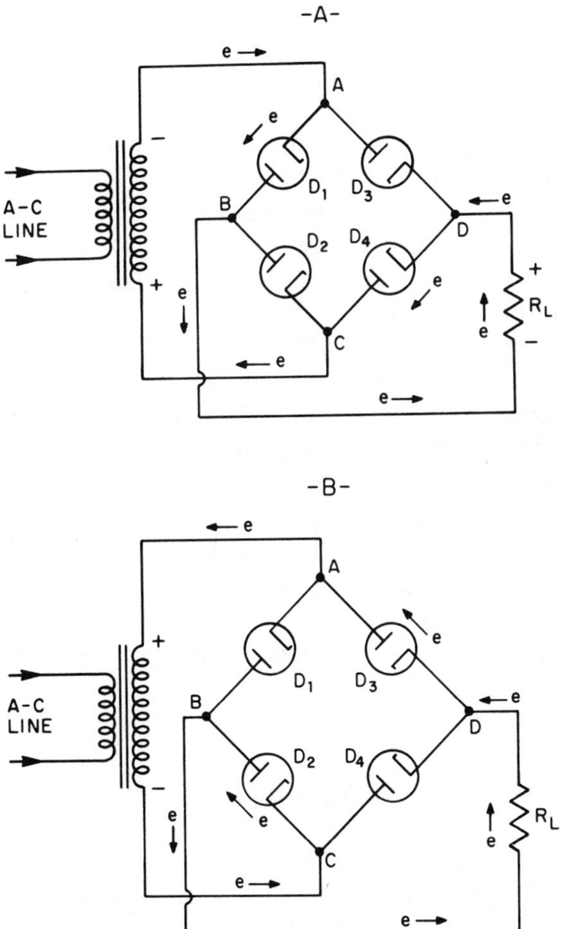

Fig. 13—5. Bridge rectifier circuit.
A. Electron flow during one half-cycle.
B. Electron flow during the next half-cycle.

Since D_3 is nonconductive, the electrons flow through conductive D_4 to junction C. Since D_2 is nonconductive, the electrons go on to the positive end of the secondary winding.

During the next half-cycle (Figure 13-5B), the polarity of the secondary winding reverses. Electrons flow from the negative end to junction C. D_4 is nonconductive and the electrons flow through D_2 to junction B. D_1 is nonconductive and the electrons flow through R_L to junction D. Since D_4 is nonconductive, the electrons go through D_3 to junction A, and then to the positive side of the secondary. Note that the output is full-wave and that during each half-cycle current is flowing through the load resistor in the same direction.

The bridge circuit has a number of advantages over the center-tapped full-wave circuit previously described. Since the entire secondary winding is being used during each half-cycle, the bridge rectifier delivers about twice the voltage of a comparable center-tapped full-wave rectifier. Also, since there always are two non-conductive tubes in series (see Figure 13-5), the inverse voltage across a tube in the bridge circuit is half that across a tube in a comparable full-wave circuit.

The disadvantage of the bridge circuit is that four tubes and three separate filament or heater windings are needed. This can be seen from an examination of Figure 13-5. You will notice that the cathodes of D_3 and D_4 are tied together. Hence, one heater winding may be used for both tubes. But the cathodes of D_1 and D_2 are at a different potential from each other and from those of D_3 and D_4. Thus, separate heater windings must be used for D_1 and D_2, else there might occur a breakdown between the cathodes and heaters of the tubes. However, the need for any heater windings can be avoided if semiconductor rectifiers are used. Such rectifiers will be described in Chapter 27.

D. *Filter circuits*

The output from the rectifier, regardless of type, is a pulsating direct current. This current contains two components—one a steady direct component, the other an alternating component (see Chapter 5, subdivision G). For many electronic applications only a steady direct current is required. The function of the *filter circuit,* then, is to remove the alternating component from the output of the rectifier.

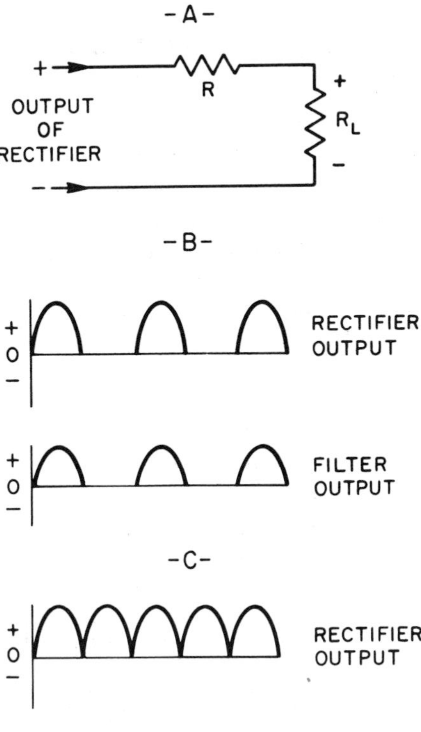

Fig. 13–6.

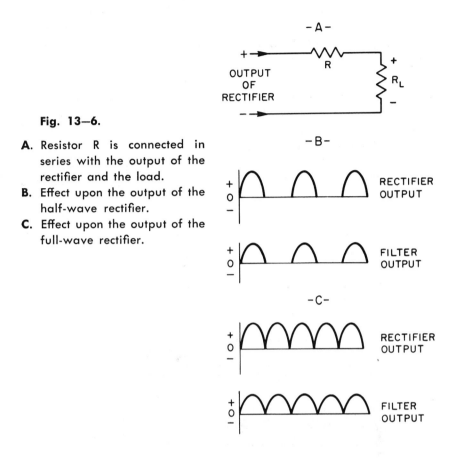

A. Resistor R is connected in series with the output of the rectifier and the load.
B. Effect upon the output of the half-wave rectifier.
C. Effect upon the output of the full-wave rectifier.

The effect of placing a resistor (R) in series with the load (R_L) is shown graphically in Figure 13-6. Since the resistor will impede equally the flow of the steady d-c component and the a-c component, the net result is merely to reduce the amplitudes of the d-c pulses. The effect upon the output of the half-wave rectifier is illustrated in Figure 13-6B. The effect upon the full-wave rectifier is shown in Figure 13-6C.

The action of a large capacitor across the output from the rectifier is shown in Figure 13-7. Figure 13-7A shows the circuit with C as the capacitor and resistor R_L serving as the load. If the resistance of R_L is large, it will permit only a small current to flow through it. Hence the load it places upon the rectifier output is light. If, on the other hand, the resistance is low, the current flow through it will be large, and the load on the rectifier output is heavy.

Figure 13-7B shows what happens in the case of a half-wave rectifier. As a pulse comes from the rectifier, the voltage across C rises nearly as fast as the pulse. When the output from the rectifier drops, however, the voltage across C does not fall to zero. Instead, the capacitor slowly discharges across R_L. Before this discharge is complete, the next pulse comes along and again C is charged up to peak voltage. The capacitor thus may be considered as a sort of tank, storing energy during the rectifier pulses and discharging it through the load between pulses.

If the load is light (that is, if R_L is large), the amount of discharge between pulses is slight and the voltage applied to the load tends to

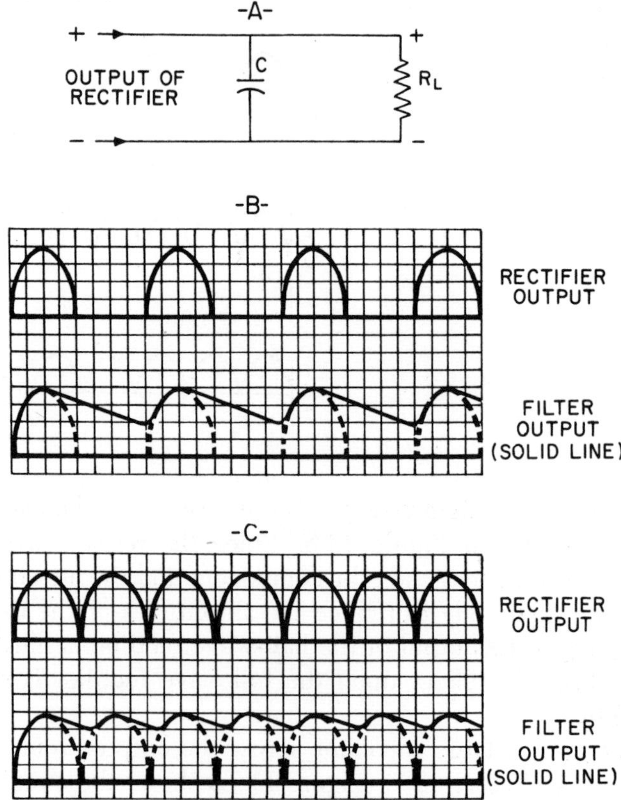

Fig. 13—7. **A.** The capacitor-filter circuit.
B. Waveforms showing filtering action with half-wave rectifier.
C. Waveforms showing filtering action with full-wave rectifier.

have a fairly constant amplitude. If, however, the load is heavy (small R_L), the amount of discharge between pulses is greater and the amplitude of the voltage applied to the load varies greatly. Hence, using just a capacitor as a filter is not suitable for circuits where appreciable current is needed.

In Figure 13-7C the action of the capacitor on the full-wave rectifier is shown. Note that the amplitude variations of the voltage applied to the load here are less than for a comparable half-wave rectifier. Thus, full-wave rectification makes for easier filtering.

Now let us consider the action of a large inductor in series with the output from the rectifier (Figure 13-8A). We know that as

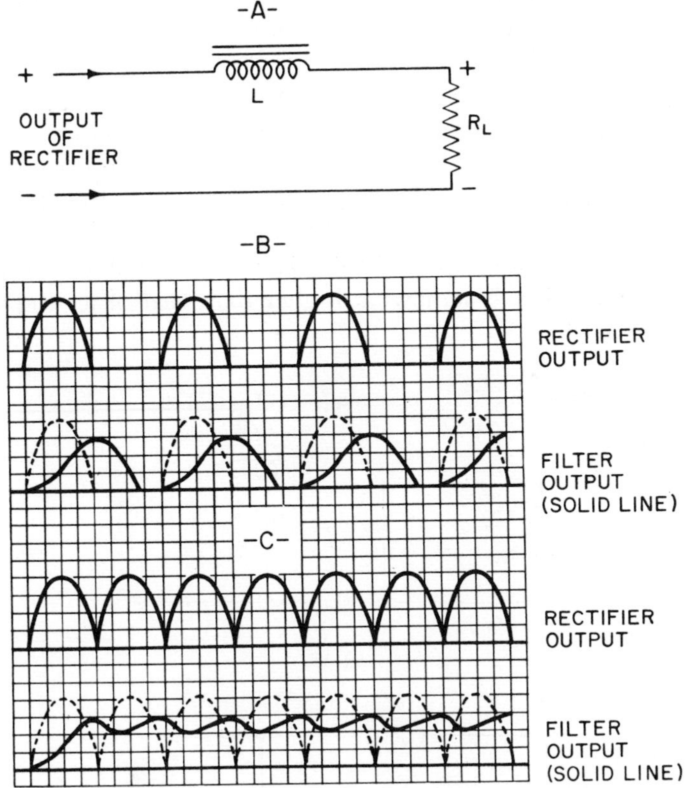

Fig. 13–8. **A.** The inductor-filter circuit.
B. Waveforms showing filtering action with half-wave rectifier.
C. Waveforms showing filtering action with full-wave rectifier.

current flows through an inductor, the inductance resists any change in the amplitude of the current flowing through it, and thus tends to prevent the current from rising or dying down.

If the current is supplied by a half-wave rectifier, the output from the filter would appear as in Figure 13-8B. Because of the ohmic resistance of the inductor, the amplitudes of the pulses are reduced. The filtering action, however, is slight.

The filter output when a full-wave rectifier is used is shown in Figure 13-8C. Again, the amplitudes of the pulses are reduced. But the filtering action is much better, resembling that of the capacitor-type filter.

In practice, the power supply generally employs both capacitor and inductor filters. There are several arrangements of such circuits. The filter capacitor C may be placed across the output of the rectifier as in Figure 13-9A. Such a filter is known as a *capacitor-input filter.* Another arrangement is to have the series inductor come first (Figure 13-9B). Such a filter is known as an *inductor-input* filter.

In both types, the capacitor acts to store and release electric energy and the inductor acts to oppose any change in the magnitude of the current. The resulting output is a fairly steady direct current, nearly free of any a-c component. The output voltage from the capacitor-input filter is somewhat higher than that of a comparable inductor-input filter.

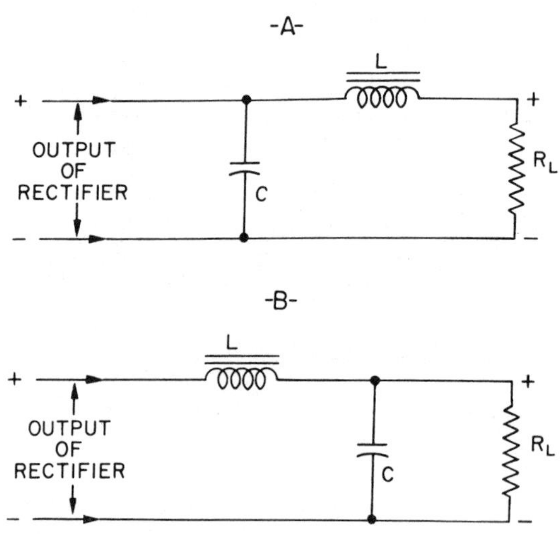

Fig. 13—9. **A.** Capacitor-input filter circuit.
B. Inductor-input filter circuit.

Additional capacitors and inductors may be added to remove any of the a-c component that might have gotten past the first portion of the filter. A common type of filter circuit is shown in Figure 13-10A. Note that this is the capacitor-input circuit of Figure 13-9A, with another capacitor (C_1) added. This capacitor acts to bypass any of the a-c component that may get through C and L. A filter of this type sometimes is called a *π-section* filter (π is the Greek letter *pi*).

Another variation is the circuit shown in Figure 13-10B. This, you will note, consists of two inductor-input sections (see Figure 13-9B).

Capacitors used in power-supply filter circuits are called *filter capacitors.* Where voltages up to about 400 volts are concerned, electrolytic capacitors, with capacitances up to 100 microfarads and higher, frequently are employed. Where higher voltages are involved, paper and oil-filled capacitors, with values up to about 12 microfarads, are used.

The iron-core inductors used in power-supply filter circuits are called *filter choke coils,* and thus the inductor-input filter is most commonly known as the *choke-input filter.* Inductances ranging from about one to 30 henrys are used frequently. The wire of the choke coil must be heavy enough to pass the current required by the load. Also, the resistance of the winding, must be low, otherwise the

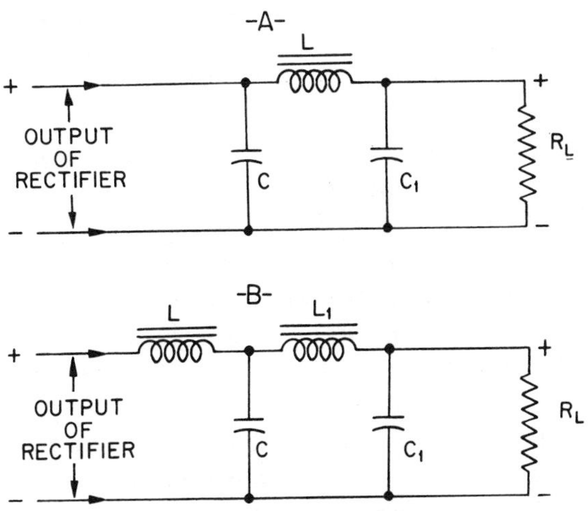

Fig. 13—10. **A.** π-section filter circuit.
B. Two-section inductor-input filter circuit.

voltage drop across it may be so large that not enough voltage will be left for the load. Resistances of from about 100 to 1,000 ohms are common, the choke coils used to pass the heavier currents having the lower resistances.

Where the current requirement of the load is small, a resistor of several thousand ohms may be used instead of the choke coil and the filtering is accomplished only by the large filter capacitors. Since the current is low, the voltage drop across the resistor is not too great. The result is a cheaper power supply. Also, the absence of the choke coil removes a possible source of stray magnetic fields. This is desirable in certain electronic applications.

E. *Regulated power supplies*

The voltage output of the ordinary power supply is not constant, but changes with fluctuations in the load, in the power supply, and in the line voltage. For many applications such voltage variations are not serious. But for some applications these voltage variations are undesirable. A number of *voltage-regulator* circuits have been developed to overcome these variations and thereby keep the voltage output of the power supply at a constant level.

Simplest, perhaps, is the circuit employing glow-discharge tubes (see Chapter 9 subdivision C,2) as voltage regulators. The basic circuit is illustrated in Figure 13-11A. Essentially, the glow-discharge tube acts to keep a constant potential across its terminals. The load (R_L) then is connected across these terminals and so is kept at constant potential.

The value of this potential depends upon the type of tube employed, varying from 75 volts for the type 0A3 tube to 150 volts for the 0D3. The tube will keep this voltage constant only within its current range which varies from a minimum of 5 milliamperes to a maximum of about 40 milliamperes. If the current through the tube is exceeded, the tube will lose its regulating ability. It is the function of resistor R to limit the current through the tube to its operating range.

Should a voltage greater than a single tube can regulate be required, a number of tubes may be connected in series, as illustrated in Figure 13-11B. Here three different types of tubes are employed —the 0A3 rated at 75 volts, the 0C3 at 105 volts, and the 0D3 at 150 volts. Hence the voltage across all three is 330 volts. If intermediate

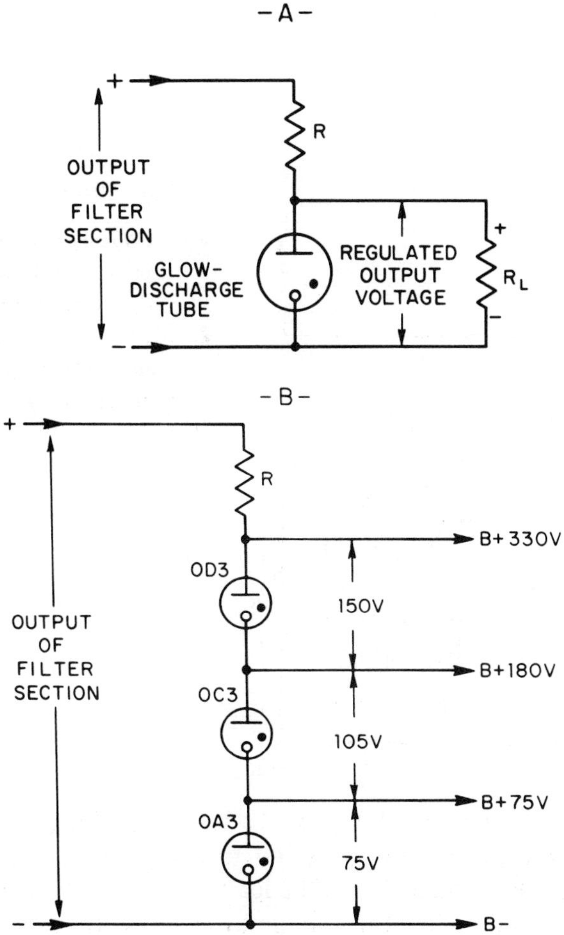

Fig. 13–11. **A.** Regulated power supply using a glow-discharge tube. **B.** Regulated power supply using three glow-discharge tubes connected in series.

voltages are required, they may be obtained from the junctions between the tubes. The current range, however, is the same as for a single tube—that is, between about 5 and 40 milliamperes.

The glow-discharge tubes are limited, however, because they can operate only at a fixed voltage and a limited current. For this reason they cannot be used where greater currents are involved. When larger currents are required, it is customary to use a voltage-regulating circuit employing a *grid-controlled regulator tube.*

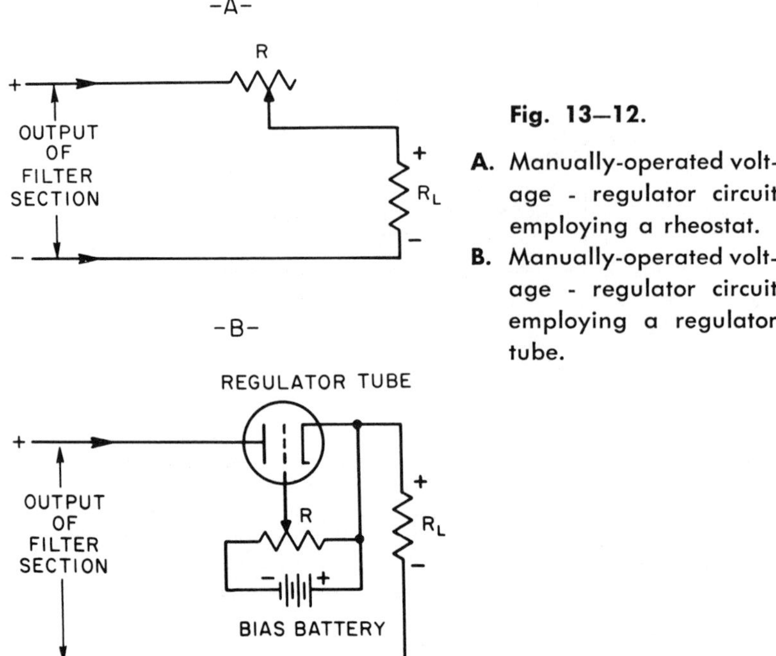

Fig. 13–12.

A. Manually-operated voltage - regulator circuit employing a rheostat.

B. Manually-operated voltage - regulator circuit employing a regulator tube.

Examine the circuit of Figure 13-12A. The output of the filter section of the power supply is fed to the load through the series rheostat R. The voltage across the load is equal to the voltage output of the power supply less the voltage drop across R, which is adjusted so that the voltage across the load is at its normal operating voltage.

Now assume that the voltage across the load drops because of either a greater current demand by the load or a reduced voltage output of the power supply. Reducing the resistance of R reduces the voltage drop across it, and the voltage across the load is raised to its normal operating value. If the voltage across the load becomes too high, increasing the resistance of R would bring it back to normal again. This is an example of a manually-operated voltage regulator.

The same result can be obtained by means of a triode connected in place of rheostat R, as illustrated in Figure 13-12B. As you know, the path through the tube offers a resistance to the flow of current. This is known as the d-c plate resistance (R_p). If the grid is made negative, the flow of current is reduced and the plate resistance is increased. If the grid is made positive (or less negative), the current flow is increased and the plate resistance is reduced.

The effect of varying the potential of the grid, then, is to vary the resistance offered by the tube to the flow of current from the power supply to the load. The more negative the grid, the larger is the resistance of the tube; the less negative the grid, the smaller is the resistance.

The potential of the grid (known as the *grid bias*) is varied by means of potentiometer R which is across the bias battery. Hence, if the load voltage drops from its normal value, adjusting R so that a smaller negative bias is placed upon the grid reduces the plate resistance of the tube and the load voltage rises to its normal value. If the load voltage rises above its normal value, adjusting R to place a larger negative bias on the grid increases the plate resistance and the load voltage drops back to normal. The triode operated in this manner is called a *regulator tube.* Thus we have another example of a manually-operated voltage regulator.

The bias on the grid of the regulator tube can be controlled automatically, as illustrated by the circuit of Figure 13-13. Note that resistor R_1 is in the plate circuit of the *control tube.* Hence the voltage drop across R_1 depends upon the plate current in that tube, and the polarity of this voltage drop is as indicated. The larger the plate current of the control tube, the greater will be the voltage drop across R_1, and the more negative will be the grid bias of the regulator tube. Hence, the larger will be its plate resistance.

Fig. 13—13. Automatic voltage-regulator circuit.

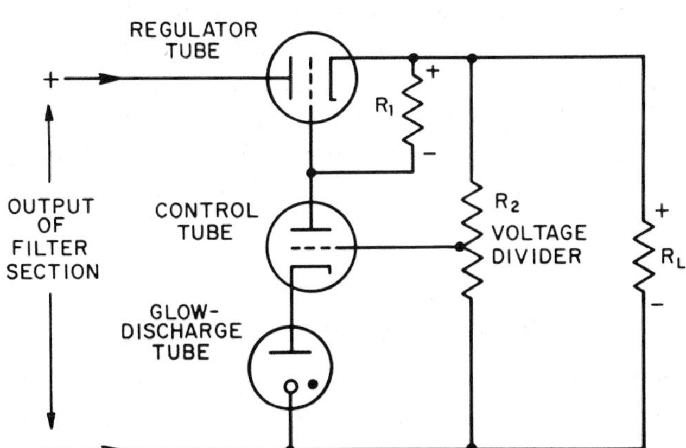

The flow of current through the control tube depends, in turn, upon the bias voltage on its grid. This bias is established by the difference in potential between its cathode (which is held at a constant potential by means of a glow-discharge tube) and its grid. The grid obtains its potential from a tap on a voltage divider (R_2) which is connected across the output of the power supply in parallel with the load (R_L). This tap is established at some point where a normal operating voltage will appear across the load.

Now, assume that the voltage across the load increases. The voltage across the voltage divider also increases and the potential of the tap, too, rises. Thus the potential of the grid of the control tube rises. Its cathode, however, will remain at constant potential because of the glow-discharge tube. Thus, the grid will become more positive with respect to the cathode and more current will flow through the control tube and its plate resistor R_1. The plate of the control tube will become more negative, and so will the grid of the regulator tube. This will increase the plate resistance of the regulator tube, there will be a larger voltage drop across it, and the voltage across the load will be dropped back to its normal value.

If the voltage across the load tends to decrease from its normal value, the grid of the control tube becomes less positive, less current flows through the control tube, and this, in turn makes the grid of the regulator tube less negative. As a result, the plate resistance of this tube decreases and the voltage across the load rises to its normal value.

The regulator tube must be capable of passing the current required by the load. Where one tube cannot carry this current safely, two or more tubes may be connected in parallel.

Such a voltage regulator is very effective in keeping the output at a constant voltage. It also aids the filtering action of the power supply by ironing out any variations that may be due to the presence of any of the a-c component that leaked through the filter section. Its chief drawbacks are that it requires three extra tubes and that the power supply must make up for the voltage drop across the regulator tube or tubes.

F. *Voltage multipliers*

Where the desired voltage output of the power supply is larger than that of the line voltage, a step-up transformer may be employed. Another method is to use a *voltage-multiplier* circuit.

The operation of this circuit depends upon alternately charging two or more capacitors to a value close to the peak line voltage and then permitting them to discharge in series. The total output voltage thus becomes equal to the sum of the voltages appearing across the individual capacitors. Diode tubes (or semiconductor rectifiers, which will be discussed in Chapter 27) automatically perform the required switching operations.

There are a number of variations of these voltage-multiplier circuits but, most common, the *series,* or *cascade, voltage doubler,* whose basic circuit is illustrated in Figure 13-14, is employed. Let us start with the half-cycle when a positive charge is placed on the plate of D_1. This diode then is conductive and capacitor C_1 is charged up with the indicated polarity. D_2 is nonconductive, since its plate is negative.

The charge on C_1 is very close to the *peak* value of the line voltage. This peak value, you will recall, is 1.41 times the *effective,* or *rms,* value (see Chapter 5, subdivision C,1).

During the next half-cycle D_2 becomes conductive and D_1 nonconductive. Capacitor C_1 adds its potential to the line voltage and C_2 is charged through D_2 to twice the peak line voltage. Actually, the output is somewhat less, since the load is constantly draining away the charge on the capacitors and thus they never reach the peak-line-voltage condition. Capacitor C_3 is an additional filter capacitor.

The voltage output of this circuit depends in large part upon the values of C_1 and C_2. The larger they are, the closer to twice the peak line voltage will be the total voltage output. Values of 100 microfarads, or more, are common.

Both capacitors should be the same value, else some of the alternating current may leak through. However, since C_1 is charged to peak line voltage, its working-voltage rating need be only enough to handle safely the peak line voltage. On the other hand, capacitors C_2 and C_3 must have working-voltage ratings sufficient to handle

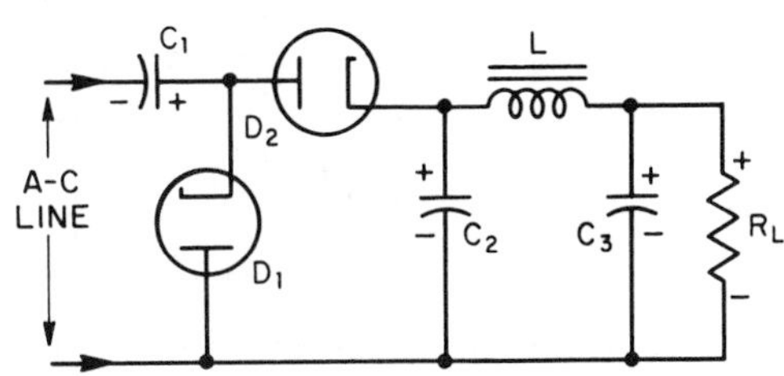

Fig. 13–14.

The series voltage-doubler circuit.

safely *twice* the peak line voltage. Note that the voltage-multiplying circuit will operate only on alternating current.

It is obvious that several voltage doublers may be connected so that their combined outputs produce a voltage that is a larger multiple of the line voltage. However, such circuits are not ordinarily employed.

G. Controlled rectifiers

So far, all the rectifiers we have discussed have been *free-running*, that is, the tube becomes conductive when its plate becomes positive with respect to its cathode and remains conductive throughout the entire half-cycle during which this condition prevails. In the thyratron we have the possibility of controlling the portion of the half-cycle during which current flows through the tube. Since the *average* output of the rectifier depends upon the duration of the current flow, we thus have a means for controlling this output. Such a rectifier is called a *controlled rectifier.*

As we have learned, the thyratron is nonconductive unless its plate voltage is positive enough to initiate ionization and its grid voltage is less negative than the critical grid voltage for any specific value of plate voltage (refer to Chapter 10, subdivision B). If, at any time when the plate is sufficiently positive, the grid voltage becomes less negative than the critical value, the thyratron becomes instantaneously conductive and current continues to flow through the tube as long as the plate is positive.

One method for controlling the thyratron is illustrated in Figure 13-15. The voltage on the grid is controlled by the setting of potentiometer R_1 across the grid battery. Alternating anode voltage is

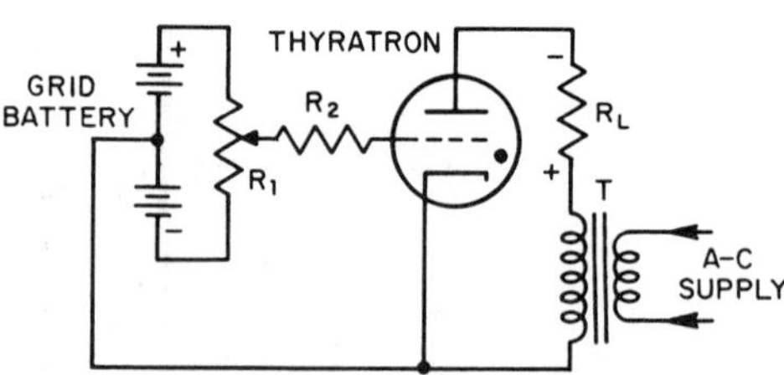

Fig. 13—15. Circuit of one method for controlling the thyratron.

applied by means of transformer T. R_2 is a resistor that limits the flow of grid current. R_L is the load.

When potentiometer R_1 is adjusted to its midpoint, the grid voltage is zero. The resulting action is illustrated graphically in Figure 13-16A. As the anode voltage becomes positive, current starts flowing, the grid loses its control, and current flows for the full positive half-cycle. The duration of current flow is indicated by the heavy portion of the anode-voltage curve.

As the slider of R_1 is moved toward the negative end of the potentiometer, the grid bias becomes negative. The resulting action is illustrated in Figure 13-16B. The grid-bias line crosses the critical grid-voltage curve at point #1 and current starts flowing through the tube at that instant of the positive half-cycle which corresponds to point #1 (point A).

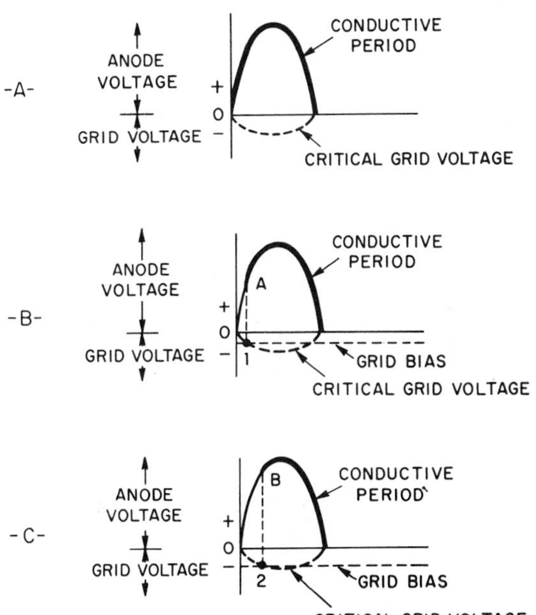

Fig. 13—16. Graphs indicating the flow of current through the thyratron during the positive half-cycle of the anode voltage. (The heavy portions of the curves indicate the period of current flow.)

A. Flow at zero grid bias.

B. Flow when small negative bias is placed on the grid.

C. Flow when larger negative bias is placed on the grid.

Current flows for the remainder of the half-cycle. Note that the duration of the current flow (as shown by the heavy portion of the curve) is less than in the previous case. Thus the average current output of the thyratron is less.

In Figure 13-16C we see the graphic representation of what happens when the slider of R_1 is adjusted to make the grid bias more negative. The grid-bias line crosses the critical grid-voltage curve at point #2 and current starts flowing at point B of the anode-voltage curve. The duration of current flow is still less, and so is the average current output of the tube.

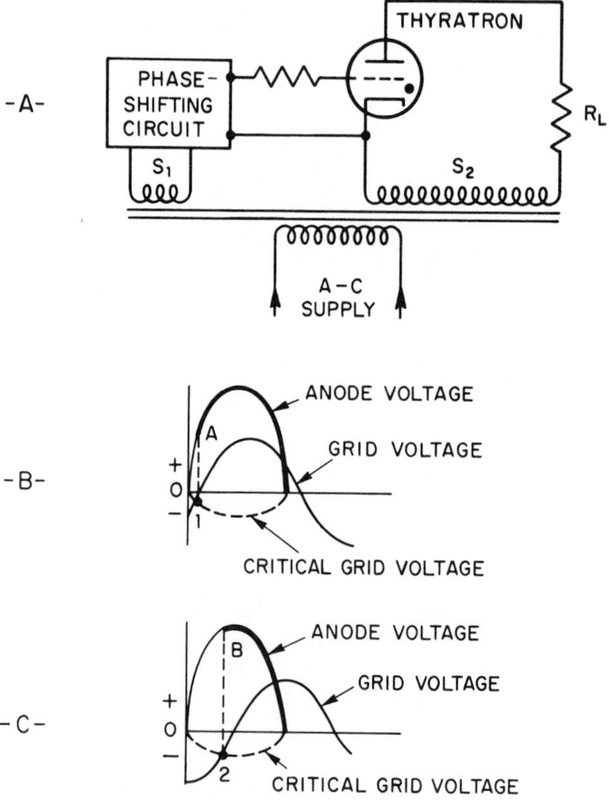

Fig. 13—17. **A.** Circuit of phase-shifting method for controlling the thyratron.
B. Flow when grid voltage lags slightly behind the anode voltage.
C. Flow when grid voltage lags more behind the anode voltage.

Another method for controlling the thyratron is illustrated in Figure 13-17A. Secondary S_2 of transformer T supplies the alternating voltage to the anode of the tube. Secondary S_1 supplies alternating voltage to the grid of the tube. However, the phase of this grid voltage, relative to that of the anode voltage, can be shifted by means of the *phase-shifting circuit* between secondary S_1 and the grid.

Figure 13-17B shows graphically the effect of shifting the phase of the grid voltage so that it lags slightly behind that of the anode voltage. Current starts flowing through the tube at a point (point A) in the positive half-cycle which corresponds to point #1 where the grid-voltage curve crosses the critical grid-voltage curve, and continues to flow for the rest of the positive half-cycle. Because this point occurs early in the half-cycle, the average current is relatively large.

In Figure 13-17C we see what happens when the phase-shifting circuit is adjusted so that the grid voltage lags further behind the anode voltage. The grid-voltage curve crosses the critical grid-voltage curve at point #2, and current starts flowing at a point (point B) in the positive half-cycle corresponding to point #2. Because this point occurs later in the half-cycle, the average current is less.

QUESTIONS

1. **Draw and explain the basic circuit of a diode used as a half-wave rectifier.**
2. **Draw and explain the basic circuit of the full-wave rectifier.**
3. **Explain the advantages and disadvantages of the half-wave and full-wave rectifiers.**
4. **Draw and explain the basic circuit of the bridge rectifier.**
5. **a) Explain the function of the filter circuit of a power supply.**
   b) Draw and explain the circuit of a *π-section filter*.
6. **Draw and explain the basic circuit of a regulated power supply using a *glow-discharge tube*.**
7. **Draw and explain the basic circuit of an automatic voltage-regulated power supply.**
8. **Draw and explain the basic circuit of the series voltage doubler.**
9. **Explain how the average plate current of a thyratron rectifier may be controlled by**
   a) varying the grid voltage;
   b) producing a phase shift in the grid voltage.

14 Amplifiers

We have seen that if a varying voltage is applied to the input (generally between the cathode and control grid) of the electron tube, it will cause a current, varying in like degree, to flow in the output circuit of the tube (generally between the plate and cathode). By inserting a resistor or reactor in the output circuit, the current flowing through it will produce a voltage drop. This voltage drop, which is the output signal, varies in step with the input voltage and may be many times as great in amplitude. Hence, the input voltage is amplified. An amplifier used this way is a *voltage amplifier.*

Where the output of the amplifier is to operate some power-consuming device such as a loudspeaker, relay coil, or transmitter antenna, we are more concerned with the power output of the amplifier than with its voltage output. The electron tube and its accompanying components then are redesigned for greater power-handling ability. Such an amplifier is called a *power amplifier.*

Here again, the semiconductors in the form of transistors have taken over most applications. We must still understand vacuum tube amplifiers for some applications, as well as for repair and maintenance work.

Amplifiers may be placed in several categories, depending upon the frequencies of the currents and voltages they are called upon to handle. The human ear is sensitive to sound frequencies up to about 15,000 cycles per second. Accordingly, voltages and currents whose frequencies range up to about 15,000 cycles per second are known as *audio-frequency* (*a-f*) voltages and currents, and amplifiers that operate with such voltages and currents are called *audio-frequency amplifiers.*

Voltages and currents whose frequencies lie beyond the audio range are known as *radio-frequency* (*r-f*) voltages and currents, and amplifiers that operate with such voltages and currents are called *radio-frequency amplifiers.* Amplifiers that handle voltages and currents whose frequencies lie in both the audio- and radio-frequency ranges as, for example, certain amplifiers used in television, are called *wide-band,* or *video, amplifiers.*

Finally, there is the *direct-coupled amplifier* that can handle both direct and alternating voltages and currents.

The basic circuit of the electron-tube amplifier is illustrated in Figure 14-1. The varying input signal, represented by an a-c generator (–(∼)–), is applied as a voltage between the cathode and control grid of the tube. As a result of the input signal, a current, called the *plate current,* varying in like degree, is made to flow through the output circuit consisting of the path from cathode to plate within the tube, through the output, or *load,* resistor (R_L), through the B battery, and back to the cathode.

Because of this current flow, a voltage drop is produced across R_L. This voltage, which is the output signal, varies in step with the variations of the input signal.

The relationship between the voltage of the input signal applied between the cathode and control grid (the *grid voltage*) of the tube and the current it causes to flow through the output circuit (the

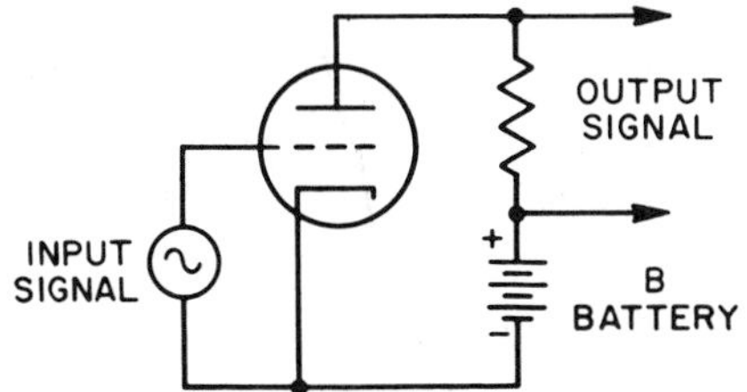

Fig. 14–1.

Basic circuit of the electron-tube amplifier.

plate current), is illustrated graphically in Figure 14-2A. From this you will note that if the grid is negative (relative to the cathode) by 6 volts or more, the tube is cut off and no plate current may flow.

As the grid is made less negative, the plate current gradually increases until the negative charge on the grid is 3 volts. Reducing the negative charge on the grid further causes the plate current to rise in a straight line.

With this in mind, let us suppose that we apply an input signal between the cathode and the control grid of the tube. (For the sake of simplicity, we will assume that this signal is an alternating voltage, sinusoidal in waveform and having an amplitude of 1.5 volts.) See Figure 14-2B.

When the input voltage is zero (point A) the plate current is about 11 milliamperes. You can ascertain this by projecting point A vertically until it intercepts the characteristic curve at the *operating point,* which corresponds to 11 milliamperes on the plate-current scale. Point A″ of the output-current curve corresponds to point A of the input-signal curve and lies on a line projected horizontally to the right from the operating point on the characteristic curve.

As the input voltage rises to a peak of +1.5 volts during the positive half-cycle of the input signal (point B), you would expect the plate voltage to rise in like degree. However, as the grid becomes positive it forms a sort of diode with the cathode that rectifies the incoming signal. As a result, there is a flow of current in the grid circuit, and a voltage is developed that opposes that of the incoming signal. Consequently, the grid voltage is reduced to about +0.75 volt (point C). Point C″ on the output-current curve may be found by projecting point C upward until it intercepts the characteristic curve (at point C′) and then horizontally to the right.

At point D of the input-signal curve the grid voltage again drops to zero. Point D″ of the output-current curve can be determined as before. At point E of the input-signal curve the grid voltage falls to −1.5 volts. At this point the grid is negative relative to the cathode. Hence there is no rectification and no flow of grid current. At point F the grid voltage is back to zero.

If you compare the output-current curve with the input-signal curve, you will notice that the waveform of the former has been distorted. The distortion is due to the fact that, at one portion of the cycle, the grid became positive. Since such distortion of the signal generally is not desirable, such an amplifier is not suitable.

The grid can be prevented from becoming positive by placing a *bias* upon it of, say, —3 volts. (See Figure 14-2C.) Now the operating point is at —3 volts and this is the voltage on the grid when the input signal is zero. This signal, then, will swing the grid from —3 volts (at zero signal) to —1.5 volts (at the peak of the positive half-cycle of the input signal), back to —3 volts as the input signal drops to zero, to —4.5 volts (at the peak of the negative half-cycle), and then back to —3 volts at zero signal. At no time does the grid become positive.

The output-current curve can be plotted as in the previous case. But note that, although the positive half-cycle of this curve resembles in waveform the positive half-cycle of the input signal, the negative half-cycle of the output-current curve is distorted. This is because the negative half-cycle of the input signal operated upon the *curved* bottom portion of the characteristic curve.

The proper solution is to choose a bias so that the grid voltage never becomes positive and the entire input signal operates on the *straight-line* portion of the characteristic curve. (See Figure 14-2D.) Here the operating point is chosen at —1.5 volts. At zero signal the grid voltage is —1.5 volts. The input signal swings the grid to zero (at the peak of the positive half-cycle of the signal), to —3 volts (at the peak of the negative half-cycle), and back to —1.5 volts (at zero signal). You will notice that the waveform of the output-current curve resembles that of the input signal.

For a specific tube, the grid bias recommended by the tube manufacturer should be employed. This, and other pertinent information may be found in the handbooks supplied by the various tube manufacturers.

One method for inserting grid bias is illustrated in Figure 14-3A. A C battery of suitable voltage is placed in the grid circuit so that the grid is made negative with respect to the cathode, thus establishing the operating point. Because the bias remains constant, this method of biasing is called *fixed bias*.

Another fixed-bias method, where a power supply is employed, is illustrated in Figure 14-3B. Two resistors, R_1 and R_2, are connected in series and form a voltage divider across the output from the power supply. The value of R_2 is such that the voltage drop across it is equal to the bias voltage. Point C—, the most negative point, is connected in series with the signal source to the grid of the tube. The junction between R_1 and R_2 is connected to the cathode.

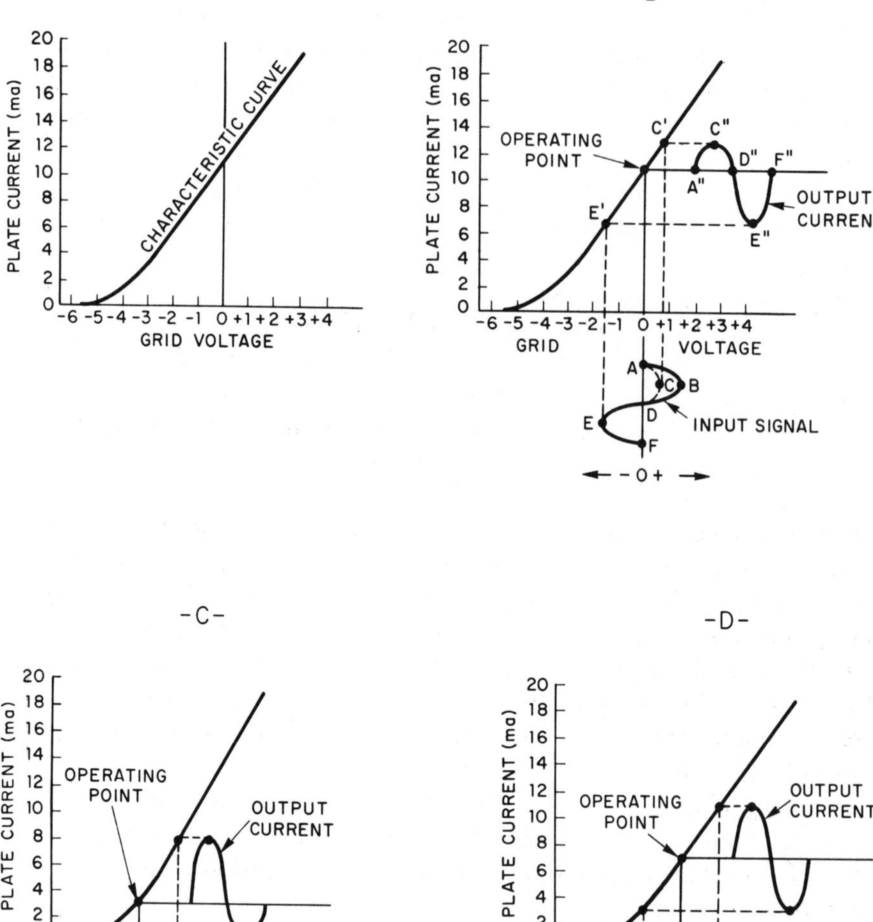

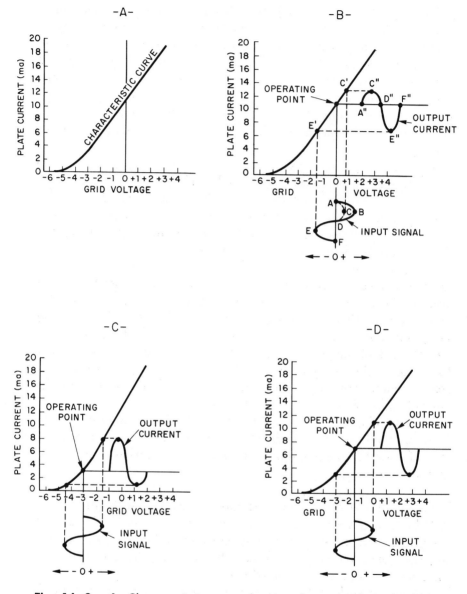

Fig. 14–2. **A.** Characteristic curve showing the grid voltage and plate current relationship.

B. Graph showing distorted output as the grid becomes positive. (Grid bias too small.)

C. Graph showing distorted output when grid bias is too large.

D. Graph showing undistorted output with proper grid bias.

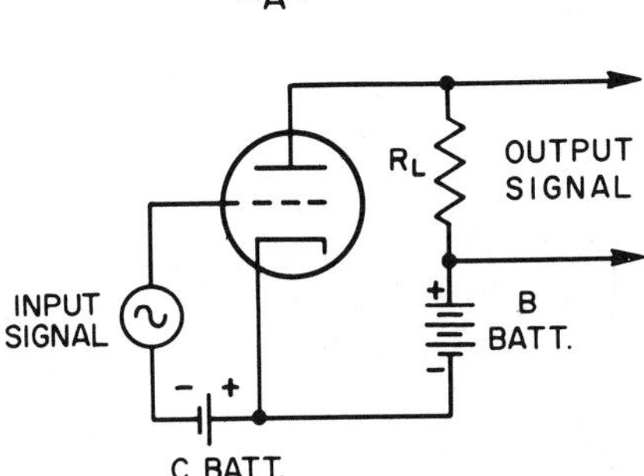

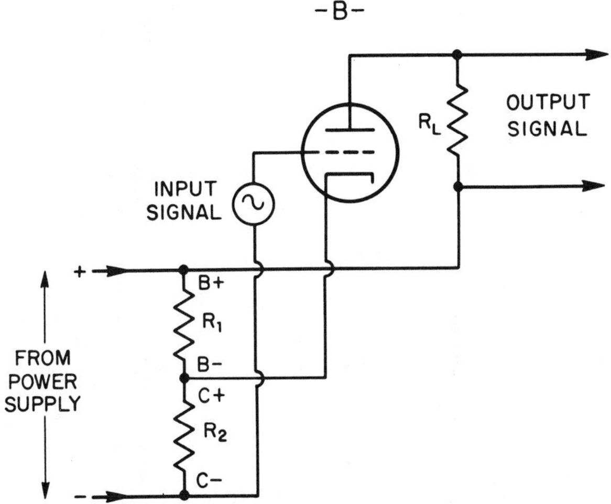

Fig. 14—3. Circuits illustrating fixed bias.
A. Grid bias obtained from C battery.
B. Grid bias obtained from power supply.

Thus the grid is negative, with respect to the cathode, by the voltage drop across R_2. Point B+, the most positive point, is connected to the plate of the tube through R_L. You can see that the voltage relationship here is similar to that of Figure 14-3A.

Still another method for obtaining grid bias is illustrated in Figure 14-4A. During the positive half-cycle of the input signal the grid becomes positive. Accordingly, it forms a sort of diode with the cathode of the tube and rectified current flows in the grid circuit, charging the grid capacitor C_g with the indicated polarity. During the negative half-cycle of the input signal there is no rectifying action and the charge on C_g starts to leak off through the grid resistor R_g to the cathode.

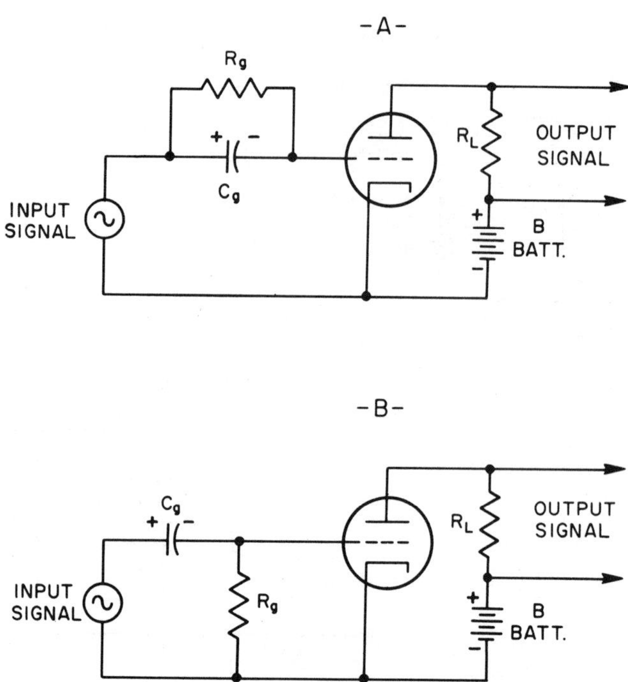

Fig. 14—4. Circuits illustrating grid-leak bias.

Because R_g has a high resistance, the charge on C_g leaks off slowly and is renewed as the next positive half-cycle of the signal comes along. Depending upon the amplitude of the signal and the values of C_g and R_g, a proper negative bias thus is placed upon the grid of the tube.

Since this bias is obtained from the signal itself, this method is called *self-bias*. It sometimes is designated as *grid-leak bias*. The difficulty with this method is that, if the input signal is removed, the bias, too, is lost. As a result, the operating point is shifted further up on the characteristic curve of the tube, producing a rise in plate current (see Figure 14-2). This rise in plate current may be great enough to wreck the tube. Accordingly, some type of protective device frequently is incorporated into the circuit to prevent the plate current from rising to too high a value.

A variation of the above circuit is illustrated in Figure 14-4B. You will note that it is, essentially, the same as the one shown in Figure 14-4A.

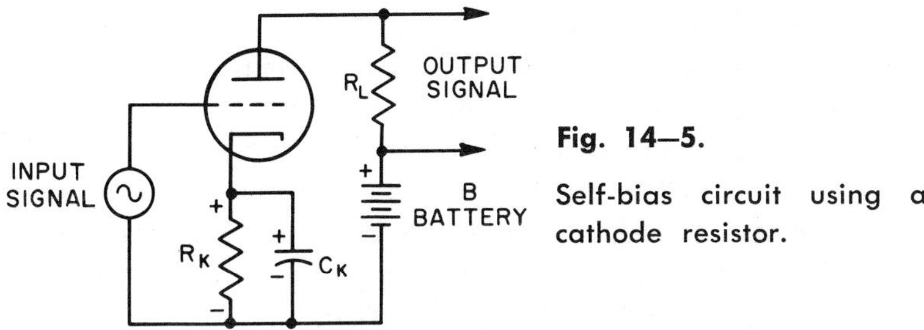

Fig. 14—5.

Self-bias circuit using a cathode resistor.

Another self-bias method, one that is most frequently employed in amplifiers, is the *cathode-resistor* circuit illustrated in Figure 14-5. Resistor R_k of suitable value is inserted in the cathode circuit of the tube. Plate current then flows from the cathode to the plate of the tube, through R_L and the B battery, through R_k, and back to the cathode of the tube.

This flow of current produces a voltage drop across R_k with the indicated polarity. The amount of the plate current and the resistance of R_k determine the magnitude of this voltage drop. Since the positive side of R_k is connected to the cathode and the negative side to the grid, the voltage drop across R_k thus becomes the grid bias for the tube. As the plate current and grid bias are dependent upon each other, there is no danger of the plate current rising to too high a value. The greater the plate current may rise, the larger will be the negative grid bias, thus tending to reduce the plate current.

The current flow through the plate circuit is, of course, a direct current. But because it follows the variations of the input voltage, it is a fluctuating direct current. Hence the voltage drop across R_k, too, will fluctuate in like degree.

Normally, the grid bias of the tube should be some steady value. Accordingly, the *cathode bypass capacitor* C_k is placed across the cathode resistor to filter out the fluctuations. Where the fluctuations are at radio frequencies, a cathode capacitor of 0.001 μf may be sufficient. But where the fluctuations are at audio frequencies, a value of 5 μf, or more, may be required.

It is interesting to examine the phase relationships between the various voltages in the amplifier. Figure 14-6A shows what happens in an amplifier as the signal voltage impressed on the grid of the tube increases in value. (Since the signal here is represented as the voltage output of an a-c generator, the rise in signal voltage is indicated by the + sign at the symbol for the generator.)

The negative grid bias of the tube is partially overcome, and the grid, therefore, has a less negative charge. This is the same as saying that the grid swings toward the positive. The plus symbol on the grid in this figure, therefore, does not represent the polarity of the grid (which rarely is allowed to become positive), but rather indicates the direction in which the grid is swinging.

Note that, in any instant, the potential of the plate of the tube is less than that of the B battery by the voltage drop across the load resistor R_L. Thus, if no current flows through the plate circuit, the voltage drop across R_L is zero, and the potential of the plate is the same as that of the B battery. However, as the plate current increases, so does the voltage drop, and the plate potential becomes smaller.

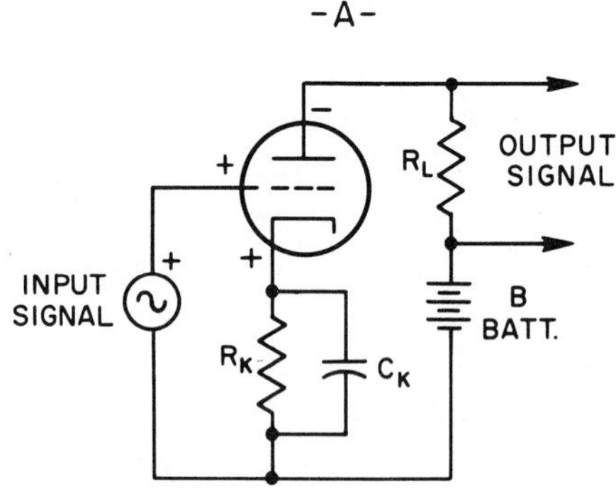

Fig. 14–6.

Circuits showing phase relationships in an amplifier.

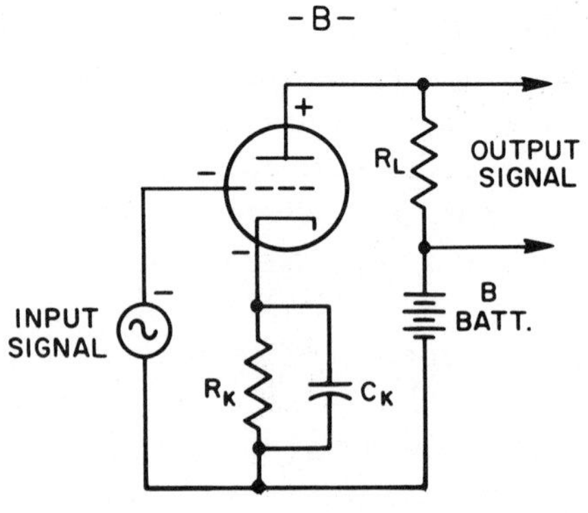

As the grid becomes less negative, more plate current flows, and the plate potential becomes less positive. The minus sign on the plate in this figure indicates this fact rather than the actual voltage on the plate which, of course, is positive. Similarly, the current flowing through the cathode resistor R_k becomes greater, thus producing a greater voltage drop across this resistor, which tends to make the cathode more positive.

As the signal voltage decreases in value (represented by the − sign at the a-c generator in Figure 14-6B), the grid becomes more negative. This cuts down the plate current and also the voltage drop across the load resistor. Thus, the plate potential becomes more positive. At the same time, the lesser plate current produces a smaller voltage drop across the cathode resistor, and thus the cathode becomes less positive (or more negative).

From these facts we may gather:

1. The grid and plate are out of phase. As the grid goes toward the positive, the plate goes toward the negative, and vice versa.
2. The grid and cathode are in phase, both going toward the positive and negative together.
3. The signal-input and output voltages are 180 degrees out of phase. (This 180-degree phase difference applies only if the load is resistive. If, however, the load contains components that are inductive or capacitive, the inductive and capacitive reactances will cause the phase difference to vary from the 180-degree value.)

At this point it might be well to consider certain features found in electronic circuit diagrams. In Figure 14-7A you see the familiar circuit of the electron-tube amplifier. The power supply may be a B battery or one of the power supplies operating from the a-c mains discussed in the previous chapter.

Note that the grid circuit, cathode circuit, and plate circuit join at a common point which is connected to "ground" (whose symbol is ⏚). The reason for this "ground" connection is to establish a common reference point from which to measure the various potentials of the circuit. This "ground" may actually be the ground through the grounded side of the power main feeding the power supply. Often, it is the metal chassis upon which the various components are mounted.

The more usual representation of this circuit appears as shown in Figure 14-7B. Note that the two circuits are identical. Instead of connecting all the portions of the circuit at one point and then

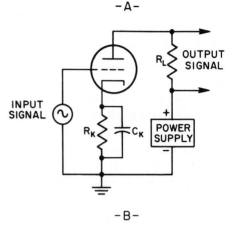

Fig. 14–7.

A. Circuit of the electron-tube amplifier.
B. The more conventional representation of the same circuit.

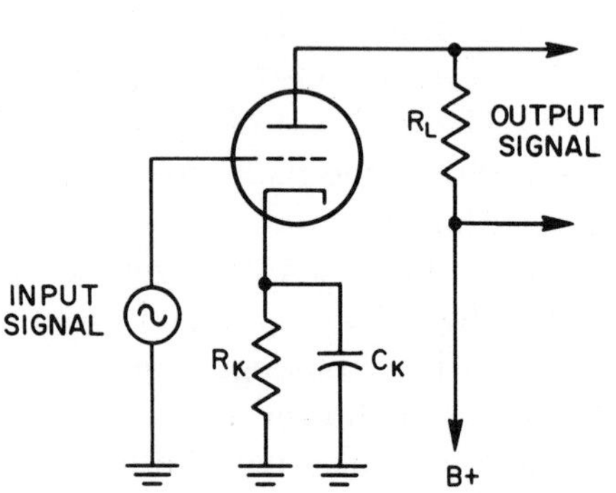

connecting this point to ground (as in Figure 14-7A), each portion is shown grounded. Also, instead of depicting the entire power supply, only the positive terminal is shown (B+). The negative terminal of the power supply is presumed to be connected to ground.

Where a pentode is employed, the circuit appears as illustrated in Figure 14-8. Resistor R_1 is a *dropping resistor* that reduces the B+

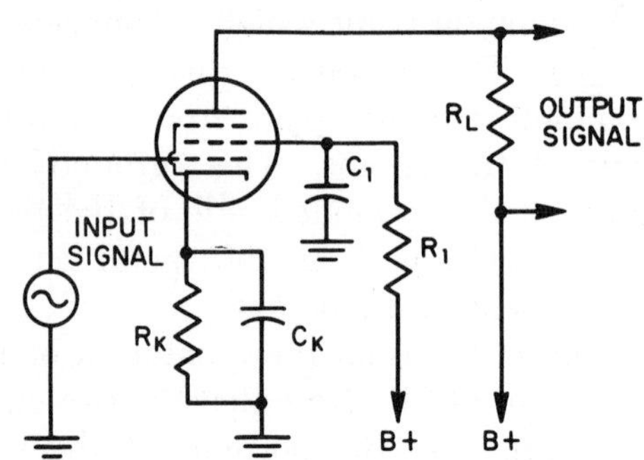

Fig. 14–8.

Electron-tube amplifier employing a pentode.

voltage to the proper level for the screen grid. Capacitor C_1 connects the screen grid to ground, as far as the a-c component of the signal is concerned. Note that the suppressor grid is connected internally to the cathode.

QUESTIONS

1. Explain the difference between a *voltage amplifier* and a *power amplifier*.

2. Explain the classification of amplifiers on the basis of the frequencies of the signal voltages.

3. Draw and explain the basic circuit of an amplifier employing a triode.

4. a) Explain the function of *grid bias*.
b) What is the effect of making the grid bias too low?
c) What is the effect of making the grid bias too high?

5. a) Draw a graph of a triode used as an amplifier, showing the relationship between the input signal and the characteristic curve of the tube.
b) What is meant by the *operating point?*

6. Draw and explain the circuit of a triode amplifier using *fixed bias*.

7. Draw and explain the circuit of a triode amplifier using *grid-leak bias*.

8. Draw and explain the circuit of a triode amplifier using a *cathode resistor* and *cathode bypass capacitor* for bias.

9. In an amplifier, explain the phase relationship between
a) the grid voltage and the plate voltage;
b) the grid voltage and the cathode voltage;
c) the signal-input and output voltages.

10. Draw and explain the circuit of an amplifier employing a pentode.

15 Audio-frequency Amplifiers

A. Voltage amplifiers

Of itself, the ordinary electron tube is capable of functioning in circuits where the frequencies of the voltages and currents involved may range from zero (direct current) to millions of cycles per second. What limits the frequency range of a specific amplifier are the limitations of the other components associated with it. Thus a capacitor which will offer an infinite impedance to direct current may act as a virtual short circuit at high frequencies. On the other hand, an inductor which offers a very low impedance to low-frequency currents will exhibit an extremely high impedance at high frequencies.

The electron tube is used with associated components to form a *stage of amplification.* Such a stage is shown in Figure 15-1A. The input signal voltage is represented by the symbol e_g, the load resistor by R_L, and the resulting plate current by the symbol i_p. As the plate current flows through the load resistor, a voltage drop (e_p),

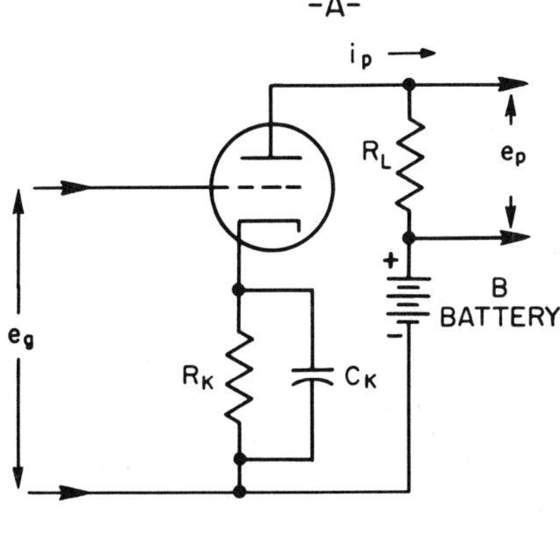

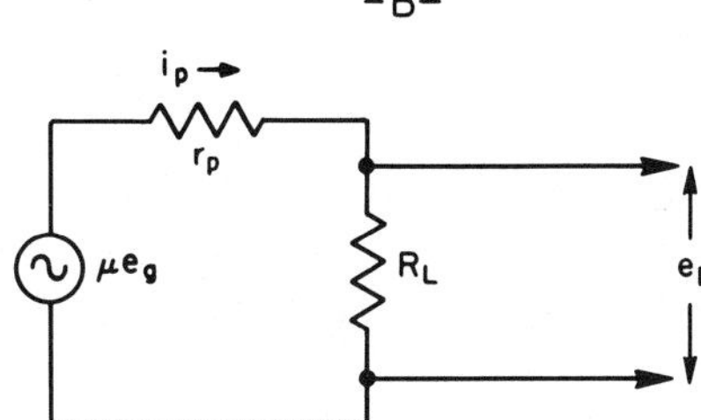

Fig. 15—1.

A. A stage of amplification.
B. Its equivalent circuit.

which is the output voltage of this stage, occurs across R_L. Note that all voltages and currents discussed here (unless otherwise stated) are the a-c components produced by the signal, rather than the d-c components produced by the battery.

The equivalent of the circuit shown in Figure 15-1A is the circuit of Figure 15-1B. It may be assumed that if the a-c generator, whose voltage output is equal to the input signal voltage multiplied by the amplification factor of the tube (μe_g), is connected in series with a resistor (r_p) equal to the plate resistance of the tube and the load resistor (R_L), the current (i_p) flowing through the load resistor and the output voltage (e_p) developed across it will be the same as in the original circuit. We may consider the voltage amplification of this stage as the ratio of the output voltage to the input voltage, or e_p/e_g.

By Ohm's law, we see that

$$i_p = \frac{\mu e_g}{r_p + R_L}.$$

Also, the voltage drop across R_L is equal to its resistance multiplied by the current flowing through it, or $e_p = i_p \times R_L$. Substituting for i_p, we get

$$e_p = \frac{\mu e_g}{r_p + R_L} \times R_L, \text{ or } \frac{\mu e_g \times R_L}{r_p + R_L}.$$

As previously stated,

$$\text{Voltage Amplification} = \frac{e_p}{e_g}, \text{ or}$$

$$\frac{\dfrac{\mu e_g \times R_L}{r_p + R_L}}{e_g}, \text{ or } \frac{\mu R_L}{r_p + R_L}.$$

We may state, therefore, that

$$\text{Voltage Amplification} = \frac{\text{Amplification Factor} \times \text{Load Resistance}}{\text{Plate Resistance} + \text{Load Resistance}}.$$

Note that the voltage amplification of the stage is not the same as the amplification factor of the tube, but is determined by the combined effect of the amplification factor, plate resistance, and load resistance. If the load resistance is made increasingly large, the voltage amplification of the stage approaches the amplification factor of the tube as a limiting factor.

From the above, you can readily see that as we increase the load resistance, we increase the voltage amplification of the stage. This can be confirmed by examination of the circuit shown in Figure 15-1A. Since the output voltage results from the voltage drop across the load resistor, the greater the value of this resistor, the greater the voltage drop, and thus the greater will be the output voltage. Hence, it is indicated that the resistance of the load resistor of the voltage amplifier should be as large as possible.

But see what happens to the plate voltage as the load resistor is increased in value. The d-c voltage on the plate produced by the B battery is cut down by the voltage drop across the load resistor, which is connected between the plate and the B battery. Thus, increasing the load resistor reduces the plate voltage. Of course, the B battery may be increased to compensate for this larger voltage drop, but to do so might require voltages not conveniently obtained.

For this reason, a compromise value for the load resistor usually is chosen. The plate load resistor generally has a value from about 100,000 to 500,000 ohms, or, roughly, up to about ten times the plate resistance of the triode. It is suggested that data issued by the tube manufacturer be consulted for the recommended values to be used.

We might consider a stage of voltage amplification using the type 6SF5 triode, as an example. The amplification factor of the tube is 100, and its plate resistance is 66,000 ohms. If the recommended load resistor of 500,000 ohms is used, the voltage amplification of this stage becomes

$$\text{Voltage Amplification} = \frac{100 \times 500{,}000}{66{,}000 + 500{,}000}, \text{ or } 88.3.$$

In the case of pentodes, the plate resistance usually is so large compared to the load resistance that the latter may be neglected in the denominator of the formula for voltage amplification. The formula then becomes

$$\text{Voltage Amplification} = \frac{\mu R_L}{r_p}.$$

Since, as we have seen previously (Chapter 10, subdivision A), the transconductance (g_m) is equal to μ/r_p, the above formula may be written as

$$\text{Voltage Amplification} = g_m \times R_L.$$

In our illustrations of the amplifier we have depicted the input section with an a-c generator supplying the input signal. Actually, the circuit should be as illustrated in Figure 15-2. Generally, we are interested in the variations (the a-c component) of the signal. Accordingly, capacitor C (which is known as the *coupling capacitor*) blocks out the d-c component of the signal, but passes the a-c component to the grid of the tube. Its value must be such as to offer

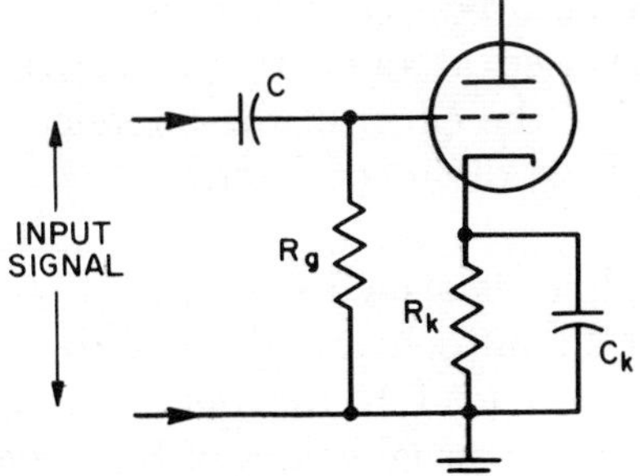

Fig. 15–2.

Input circuit of the amplifier.

very little impedance to the signal at the frequencies involved. For audio-frequency operation a coupling capacitor of from about 0.001 to 0.01 microfarad generally is employed.

As the electron stream from the cathode to the plate flows through the tube, some of the electrons strike the grid. Such electrons are relatively few, but since they are blocked from flowing to the ground (through the signal source) by capacitor C, they accumulate on the grid. Also, should the signal make the grid positive at any instant, more electrons are attracted and trapped on the grid. This accumulation of electrons places an unwanted negative bias on the grid.

Accordingly, a path is provided for them to leak off to ground through the *grid resistor* (R_g). If the resistance of R_g is too high, the electrons will leak off too slowly. If the resistance is too low, the resistor will place too large a load on the source of the input signal. Values of about 0.5 to 1.0 megohm are customary.

At this point, it may be well to consider the load the tube itself places upon the signal source. When the grid is negative with respect to the cathode (as is normal), its input impedance, theoretically, is infinite (that is, it acts as an open circuit). Actually, it is not an open circuit since the grid and cathode form the plates of a small capacitor across the signal source. The capacitive reactance of this capacitor depends upon the frequency of the signal.

If the frequency is high, the capacitive reactance is low and, hence, a large load is placed upon the signal source. If the frequency is low, the capacitive reactance is high and, therefore, a light load is placed upon the signal source. This load, of course, reduces the signal voltage.

At audio frequencies, triodes with relatively large interelectrode capacitances may be used. But at the higher radio frequencies, pentodes, with much lower interelectrode capacitances, generally are employed. (Triodes may be used at radio frequencies if we employ special circuits to neutralize the effect of the interelectrode capacitance of the tube. This will be discussed later.)

The output circuit of the amplifier is shown in Figure 15-3. In Figure 15-3A you see the output circuit containing a resistive load (R_L). The voltage drop across this load is the output voltage. Values of about 0.1 to 0.25 megohm are usual values of R_L, depending upon the type of tube employed as well as other factors. (The value specified by the tube manufacturer should be used.)

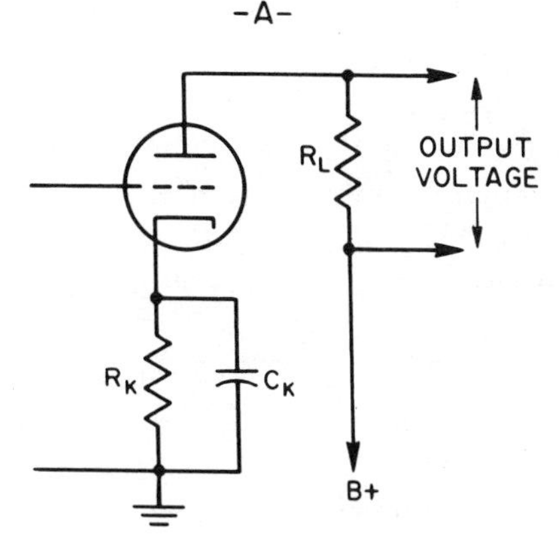

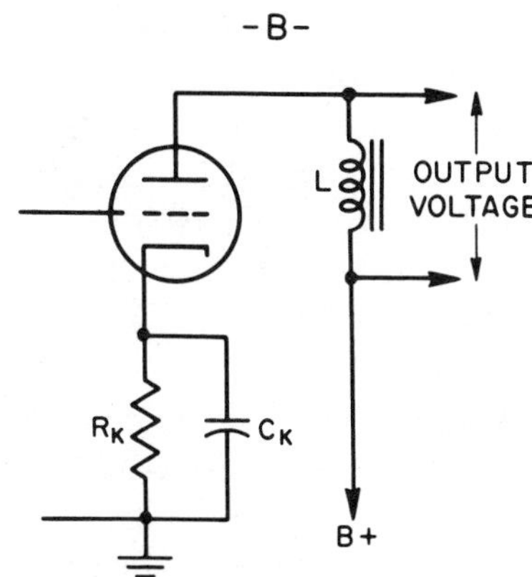

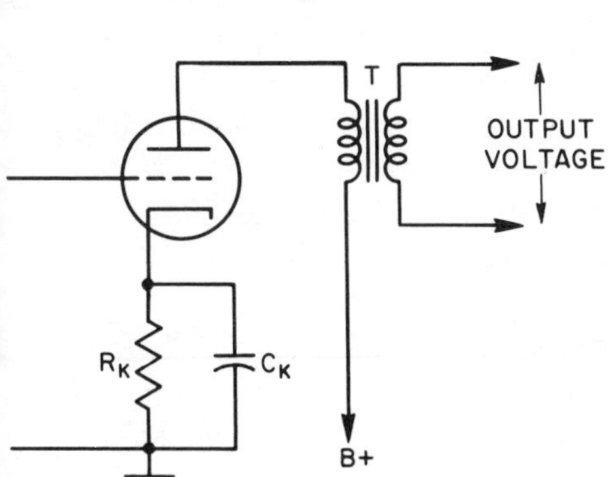

Fig. 15—3. Output circuit of the amplifier.

A. Circuit using a resistive load.
B. Circuit using an inductor as a load.
C. Circuit using a transformer as a load.

Since the voltage drop across R_L is subtracted from the voltage of the power supply, the voltage applied to the plate of the tube is considerably reduced. One method to overcome this difficulty is illustrated in Figure 15-3B. Here an iron-core inductor (L) is substituted for R_L. The ohmic resistance of its winding is relatively low, hence the voltage applied to the plate is higher. But the inductive reactance to the a-c component of the signal is high, hence the output voltage is large.

The difficulty with this system is the fact that L offers a varying inductive reactance to currents of varying frequencies. The higher the frequency, the greater will be the reactance; the lower the frequency, the less the reactance. Since the current flowing in the plate circuit varies in frequency (but always in the audio range for a-f amplifiers), the output voltage will depend, not only on the magnitude of this current, but its frequency as well. This produces a certain amount of distortion of the output signal.

Another method is illustrated in Figure 15-3C. Here the load is a step-up transformer (T). The primary winding is in the plate circuit. The output voltage is taken from across the secondary winding. Because of the step-up turn ratio, the output voltage thus is increased.

The transformer method suffers from the same frequency-discrimination defect as does the inductor method. Hence the resistor method of Figure 15-3A generally is employed. It has the additional advantages of being cheaper, lighter, and has no magnetic field that may interfere with the signal.

While on the subject of the output circuit, we may examine an interesting variation illustrated in Figure 15-4. Note that the output voltage is taken from across the cathode resistor R_k. Since this resistor is part of the plate circuit, the plate current flows through it. And since the cathode bypass capacitor is omitted, the signal variations are present. The plate resistor is omitted and capacitor C_1 is used to bypass signal variations to ground, thus keeping a steady potential at the plate.

Since the resistance of R_k is low, the output voltage, too, is low. The voltage gain of this stage (output voltage/input voltage) is less than one. Because the grid and cathode are in phase (see

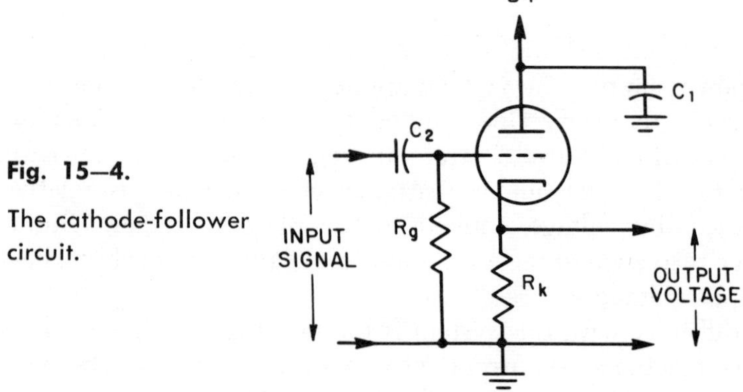

Fig. 15—4.

The cathode-follower circuit.

Figure 14-6), the input and output voltages, too, are in phase. It becomes obvious, then, why this type is called a *cathode-follower circuit.*

You may wonder why such a circuit is desirable. But note that in the ordinary amplifier circuit the output is taken from across a high-resistance source (R_L). In the cathode-follower circuit it is taken from a low-resistance source (R_k). Thus the circuit acts as a sort of transformer with a high-impedance input (the grid circuit) and a low-impedance output (the cathode circuit). Under certain circumstances such an impedance-changing device is desirable.

B. Coupling devices

Frequently, where the input signal is very small, two or more stages of amplification may be used in *cascade.* That is, the output signal of one stage is fed to the input of the following stage for further amplification.

A typical two-stage amplifier is illustrated in Figure 15-5. The two stages are coupled by means of the resistor-capacitor network consisting of R_{L1}, R_{g2}, and C_2. Hence this circuit is called a *resistance-capacitance-coupled amplifier* or, more simply, a *resistance-coupled amplifier.*

Another method for coupling the stages is illustrated in Figure 15-6. Here the coupling device is the step-up transformer T_1. Hence this circuit is called a *transformer-coupled amplifier.*

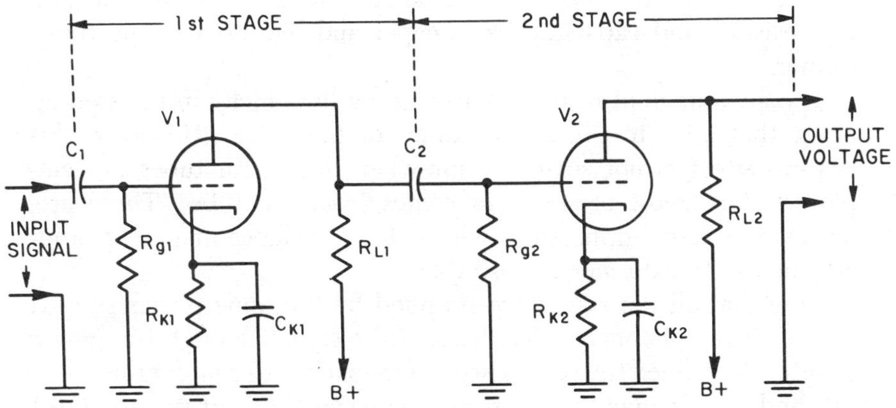

Fig. 15—5. Two-stage, resistance-coupled amplifier.

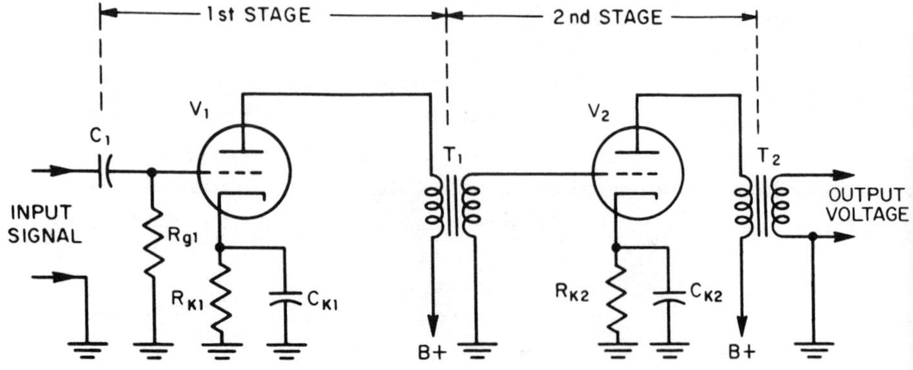

Fig. 15–6. Two-stage, transformer-coupled amplifier.

Too many stages cannot be coupled together because of the danger of leakage, or *feedback*, of the signal voltage from one stage to the preceding stages. This feedback interferes with the signal and thus generally is not desirable. Customarily, resistance-coupled amplifiers do not have more than three stages and transformer-coupled amplifiers two.

Most frequently, amplifiers employ resistance coupling. For one, the *R-C* coupling network exhibits less discrimination against currents of different frequencies than does the coupling transformer. Hence there is less distortion of the output signal. Also, the transformer is surrounded by a magnetic field that may interfere with the signal. The *R-C* network has no such magnetic field. Further, the resistors and capacitor are cheaper and lighter than the transformer.

The advantage of transformer coupling lies chiefly in the step-up effect that adds to the amplification of the stage. However, this step-up effect is not so important where high-gain tubes are employed since most of the gain comes from the tube. There are, however, certain applications where the impedance-matching property of the transformer is desirable.

Note that all the stages are supplied by the same power supply. Hence there is danger of feedback of the signal through the power supply. To reduce the possibility of feedback, a *decoupling network* (R_2 and C_4) frequently is inserted between the load resistor (R_L) and the power supply in each stage (Figure 15-7). Should any of

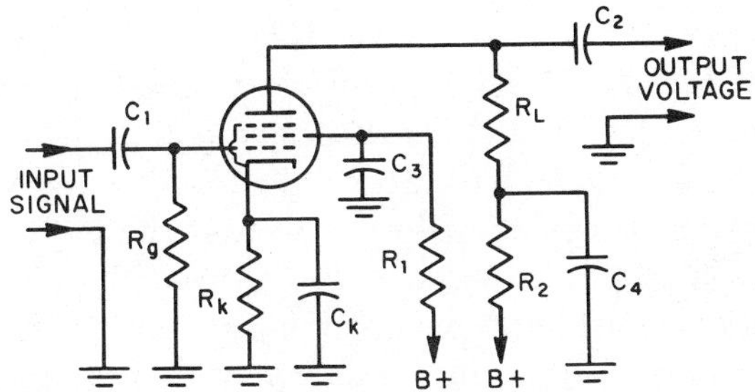

Fig. 15–7. Pentode amplifier using a decoupling network.

the signal voltage from a succeeding stage attempt to enter through the power supply, it would be reduced (*attenuated*) by R_2 and by-passed to ground through C_4.

The value of R_2 is much lower than that of R_L (values of about 5,000 to 10,000 ohms are common). The value of C_4 should be large enough to offer a low reactance at the lowest frequencies handled by the amplifier.

C. *Power amplifiers*

We may obtain a clear understanding of the function of the power amplifier, perhaps, if we consider a typical application such as the public-address system by means of which a speaker on the platform is heard by every person in a large auditorium. As the speaker addresses the microphone, the sound waves are converted into a fluctuating current whose variations conform to the variations of the sound waves. (How this is accomplished will be discussed in Chapter 23.)

This fluctuating current places a similarly-varying voltage (the input signal) at the grid of the voltage-amplifier tube. The output of this stage is an amplified version of the input signal. If necessary, one or more additional stages of voltage amplification may be employed to bring the signal voltage up to an even greater amplitude.

The amplified signal voltage now is applied to the grid of the power-amplifier tube. The output of this stage is the plate current whose variations correspond to those of the signal voltage. Hence we have here an amplified version of the fluctuating current flowing from the microphone. The plate current flows through the coil of the loudspeaker, producing sound waves that are similar, but of much greater power, to the sound waves leaving the speaker's mouth.

The load of the power stage generally is not a resistor but some power-consuming device such as the voice coil of a loudspeaker or the coil of a relay. As you have seen (Chapter 7, subdivision D, 1), maximum transfer of power will take place if the impedance of the load matches the output impedance of the tube (the plate resistance r_p). For practical considerations involving reduction of distortion of the signal, however, a slight mismatch is required. (The tube manufacturer's recommendation should be followed.) Note that in the voltage amplifier impedance-matching is not essential since the important thing is voltage, not power, output.

Power amplifiers are designed to handle power ranging from a few milliwatts to hundreds of kilowatts. A typical power stage employing a beam power tube is illustrated in Figure 15-8. If the impedance of the load matches the plate resistance of the tube, the circuit of 15-8A may be employed. If, however, the impedance of

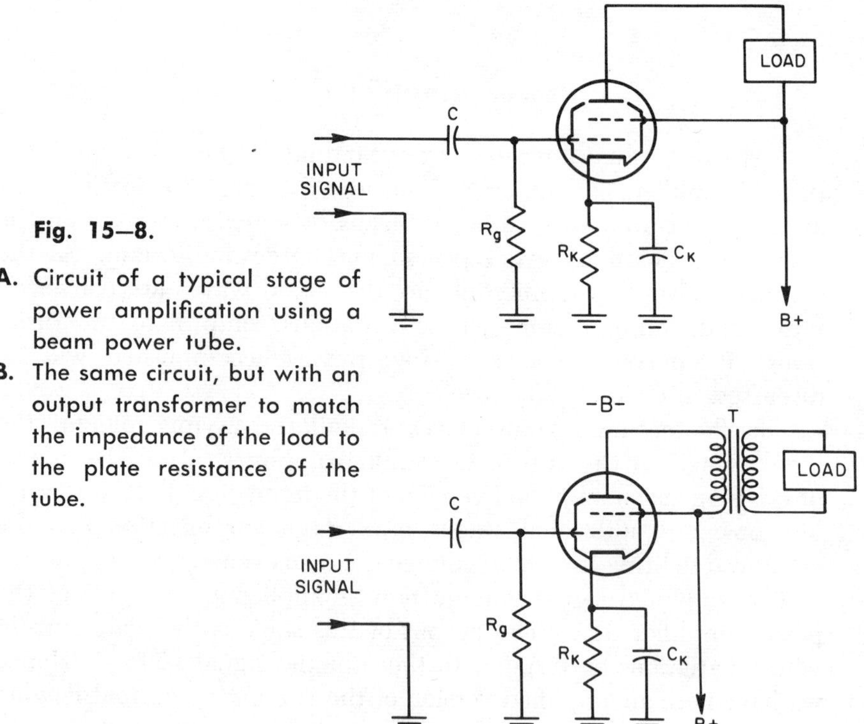

Fig. 15—8.

A. Circuit of a typical stage of power amplification using a beam power tube.

B. The same circuit, but with an output transformer to match the impedance of the load to the plate resistance of the tube.

the load is some other value, transformer T, called the *output transformer*, may be used as an impedance-matching device between the load and the tube, as illustrated in Figure 15-8B.

1. PUSH-PULL CIRCUIT

It is desired, sometimes, to obtain more power output from an amplifier than a single tube is capable of delivering. Of course, this demand may be met by using a larger tube. Another solution frequently employed is to use the *push-pull* circuit illustrated in Figure 15-9.

The output voltage from the final voltage-amplifier stage (V_1) is transformer-coupled to the power-amplifier stage by means of transformer T_1, which is known as an *input transformer.* The secondary winding of this transformer is center-tapped, each end of the winding connecting to the grid of one of the power tubes (V_2 and V_3). The center tap is connected to ground. Cathode resistor R_k furnishes the grid bias to both tubes.

The primary winding of transformer T_2 (the *output transformer*) is also center-tapped. Each end of that winding goes to the plate

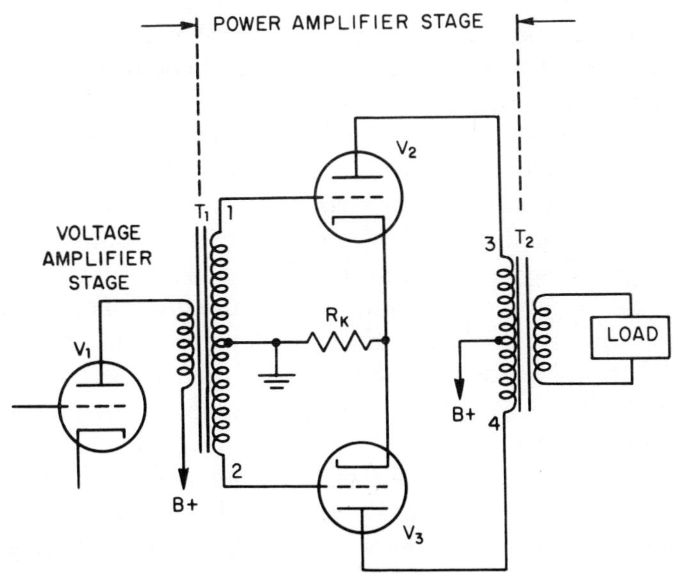

Fig 15—9. Circuit of a push-pull stage of power amplification.

of one of the power tubes, and the positive terminal of the power supply (B+) is connected to the center tap. The load is in the circuit of the secondary winding of T_2.

The fluctuating plate current flowing through the primary winding of T_1 sets up an alternating voltage across its secondary. Assume an instant when point #1 of the secondary winding is positive; point #2 then is negative. The positive voltage on the grid of tube V_2 speeds up the electron flow from its cathode to its plate, on to point #3 of the primary winding of T_2, through the upper half of that winding, on to the B+ terminal of the power supply, and (not shown in the diagram) through the power supply and through R_K to the cathode.

As the current flows through the upper half of the primary winding of T_2, a varying voltage is set up across the secondary winding of that transformer (owing to the expanding magnetic field).

Now let us see what is happening in tube V_3 at this same instant. The negative voltage on its grid slows down the flow of electrons to the plate and, as a result, its plate current is reduced. This means that the current flowing through the lower half of the primary winding of T_2 falls off.

But as this current falls off, it, too, sets up a varying voltage across the secondary winding of T_2 (owing to the collapsing magnetic field). The two voltages across the secondary winding of T_2 are in the same direction and, therefore, reinforce each other. Thus we can get about twice the power output of a single tube.

The push-pull circuit offers another advantage. As explained previously (see Chapter 14), we generally seek to operate the amplifier with such a grid bias that the operating point falls upon the straight-line portion of the characteristic curve of the tube and with an input-signal voltage whose maximum amplitude does not drive the grid positive on the positive swing of the signal or beyond cutoff at the negative swing (see Figure 15-10A). Thus plate current flows at all times and the waveform of the plate current resembles that of the signal voltage. An amplifier operated this way is called a *Class A amplifier.* Voltage amplifiers generally are operated in Class A.

If we increase the amplitude of the input-signal voltage we will obtain a larger plate current. Because we do not wish the signal to drive the grid positive, we may accommodate the larger signal by increasing the negative grid bias, thus shifting the operating point more to the negative. See Figure 15-10B.

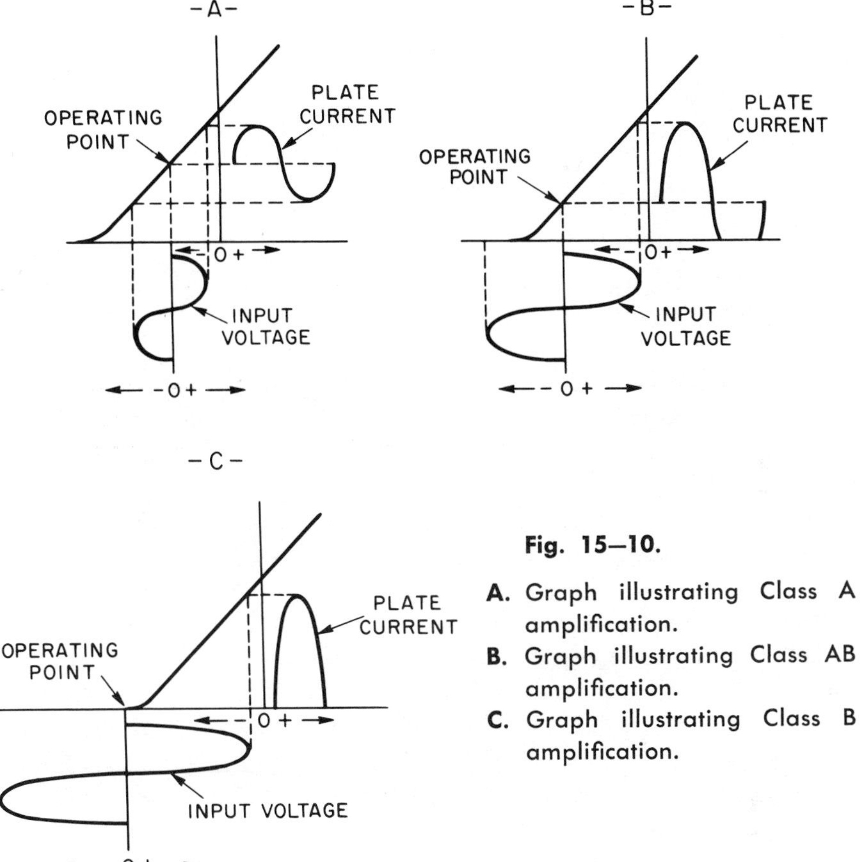

Fig. 15—10.

A. Graph illustrating Class A amplification.

B. Graph illustrating Class AB amplification.

C. Graph illustrating Class B amplification.

But see what happens on the negative swing of the signal. The grid is driven beyond cutoff for a certain portion of the cycle, plate current stops flowing during that period, and the waveform of the plate current is distorted. An amplifier in which the signal voltage and grid bias are such that plate current in the tube flows for appreciably more than half, but less than the entire cycle, is called a *Class AB amplifier*.

If we increase the grid bias still more until it reaches the cutoff value of the tube, we can accommodate a still larger input-signal voltage (with a still larger plate current). See Figure 15-10C. Now the plate current flows only during one-half of each cycle. This type of amplifier is called a *Class B amplifier*.

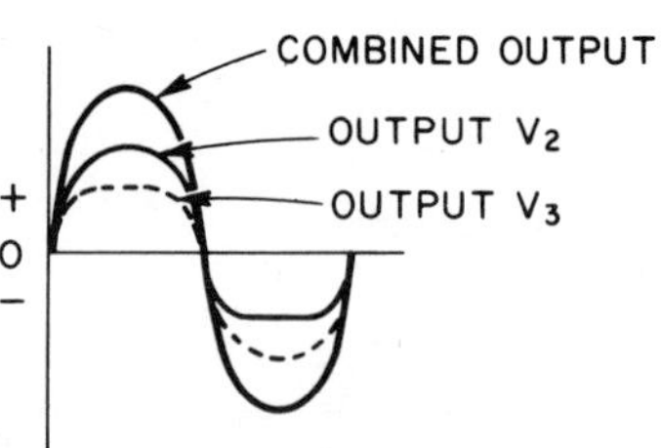

Fig. 15—11.

Graph showing the action of the push-pull circuit.

The increased plate current resulting from Class AB and Class B operation is desirable, but the distortion of the signal is not. It is here that the push-pull circuit comes to the rescue.

You will recall that both tubes of the push-pull stage simultaneously deliver their outputs to the center-tapped primary winding of the output transformer. These two outputs are added together to give the total output of the stage of amplification.

Look at Figure 15-11. The curve with the light solid line represents the waveform of the output of tube V_2. Note that during the first half-cycle, the output from tube V_2 is the undistorted signal. During the next half-cycle the signal is distorted.

The dotted curve represents the waveform of the output from tube V_3. Now the first half-cycle is distorted; the next half-cycle is undistorted. But since the two outputs are added together, the *total* output for each half-cycle (shown by the heavy solid-line curve) is the same and, therefore, the overall effect is to produce an output signal whose waveform closely resembles that of the input signal. Thus the push-pull circuit permits the use of an input signal of greater amplitude with the resulting greater plate current.

Note that in Figure 15-9 the cathode bypass capacitor was omitted. This is possible because, since the plate current of one tube is increasing as that of the other tube is decreasing, the flow of current through the cathode resistor is essentially constant at all times and, hence, no bypass capacitor is required.

2. PHASE INVERTER

From our discussion of the push-pull circuit, we may gather two facts concerning the signal input to the grids of the tubes. First, the signal voltage applied to each grid of the tubes must be approximately equal in amplitude. Secondly, since the grid of one tube swings toward the positive as the grid of the other swings negative, these signal voltages must be *180 degrees out of phase* with each other. This is accomplished by the center-tapped secondary winding of the input transformer (see Figure 15-9).

The same result may be accomplished with a stage of resistance-coupled amplification, called a *phase inverter.* See Figure 15-12. You will recall that in an amplifier with a resistive load the plate and cathode are 180 degrees out of phase with each other (see Chapter 14). If we were to obtain our input signal voltage for the push-pull stage from these two points, we would then have the necessary 180-degree out-of-phase relationship.

Note that the upper portion of the diagram is a straight resistance-coupled amplifier with its load resistor (R_{L1}), coupling capacitor (C_2), and grid resistor (R_{g2}). The signal is applied to the control grids of the push-pull beam power tubes (V_2 and V_3) from points A and B. These points are 180 degrees out of phase with each other. Since the same plate current flows through R_{L1} and R_{L2}, if the values of these resistors are equal, the amplitudes of the signal voltages applied to the grids of the push-pull tubes, too, will be equal. Thus the requirements of the push-pull circuit are met.

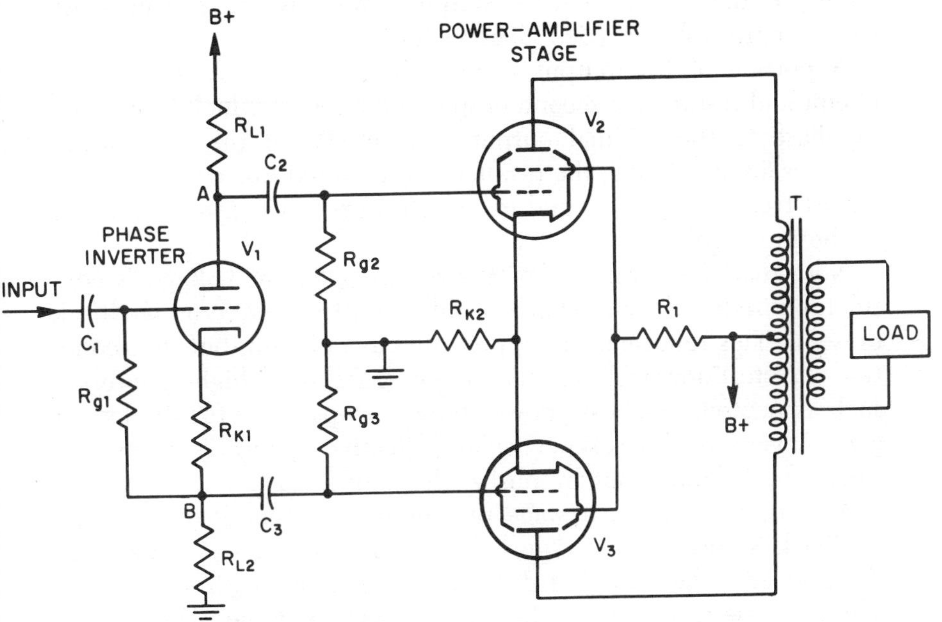

Fig. 15—12. The phase-inverter circuit.

The cathode resistor R_{k1} of the phase-inverter tube (V_1) is unbypassed. This omission (as we shall see later) is desirable since it reduces distortion in the signal. The screen grid and plates of the power tubes connect to the same B+ point. The dropping resistor (R_1) is used to insure that a lower B+ voltage appears on the screen grids than on the plates.

There are many variations of the phase-inversion circuit illustrated here. The basic principles for all, however, remain the same, that is, signals of equal amplitude and opposite phase are fed to the control grid of each of the push-pull tubes.

3. INVERSE FEEDBACK

The problem of distortion of the signal is a serious one and various methods have been devised for solving it. In audio-frequency amplifiers a process known as *inverse feedback* is frequently employed. In essence, the process consists of feeding back, or superimposing upon the input signal, a portion of the output signal. How it operates is illustrated graphically in Figure 15-13.

Assume that the input voltage (A) has a waveform which resembles a sine curve and that the output voltage (B) developed in the plate circuit is distorted as shown, owing to distortion in the amplifier stage. Note that the output voltage is opposite in phase to the input voltage. (See Chapter 14.)

A portion of this output voltage (C) is fed back to the grid circuit and results in a second output voltage (D) which is opposite in phase to the original output voltage. When the two output voltages are combined, the resultant (the dotted-line curve of E) is almost free of distortion, and its waveform resembles the sine curve of the input voltage.

Note that the amplitude of the resulting output voltage is considerably reduced from what it would be if there were no feedback present. This represents an appreciable loss of gain, but the reduction of distortion is well worth it. Besides, the use of high-gain tubes, such as pentodes or beam power types, compensates for the loss of gain. Inverse feedback is particularly effective against distortion that cannot be balanced out by push-pull circuits.

A typical circuit of a stage of amplification employing inverse feedback is illustrated in Figure 15-14. The a-c (signal) voltages at the grid and plate are 180 degrees out of phase with each other. Thus, inverse feedback is achieved if a portion of the plate voltage is fed back to the grid.

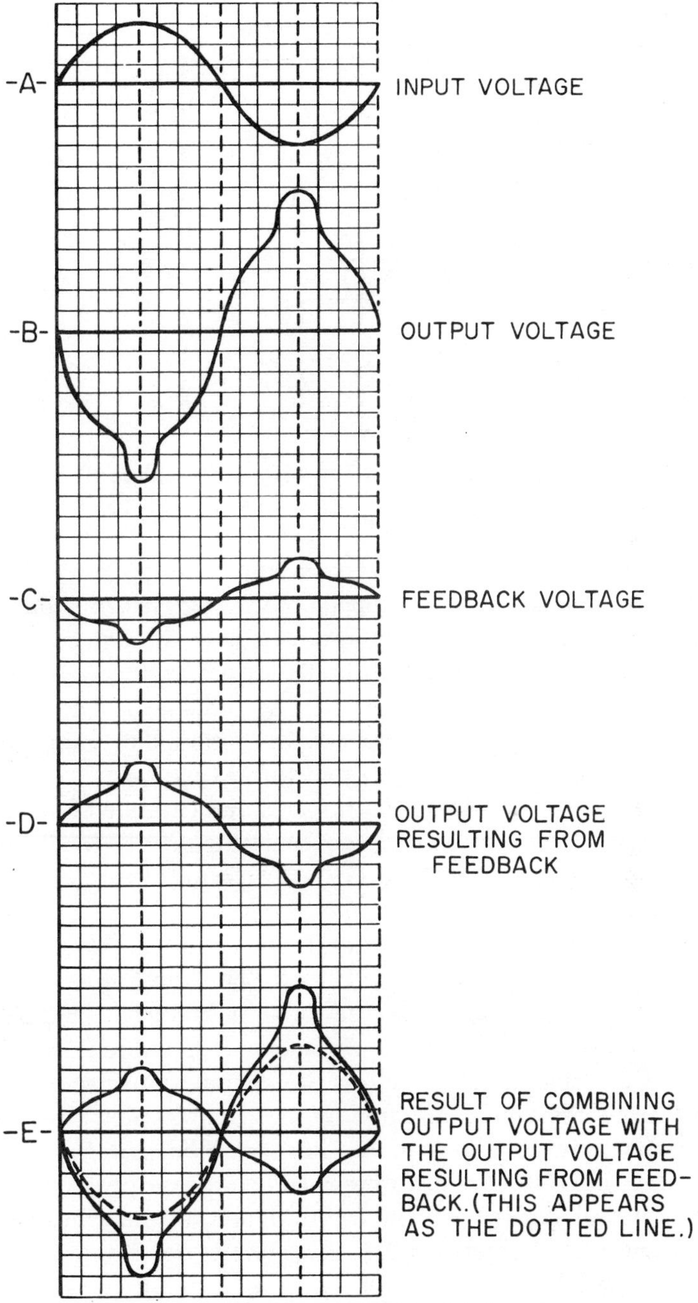

Radio Corporation of America Receiving Tube Manual

Fig. 15–13. Graph showing how inverse feedback eliminates distortion.

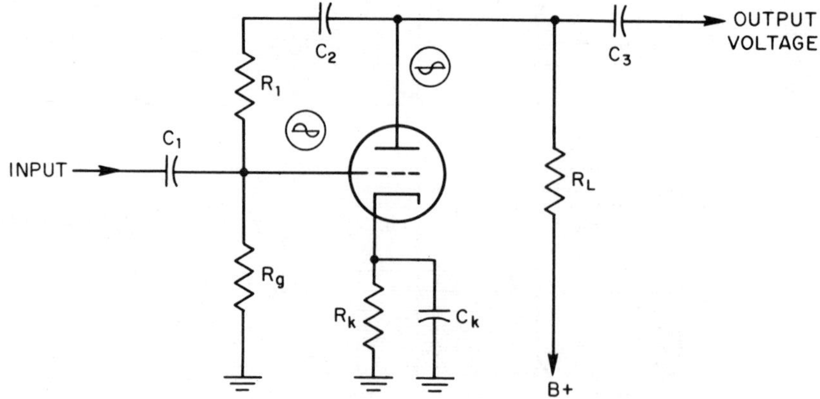

Fig. 15—14. Inverse-feedback circuit for a single stage of amplification.

Capacitor C_2 is a blocking capacitor used to keep the direct voltage of the B supply from getting to the grid. Its impedance must be low, so that the output voltage may flow through it with small loss. Resistor R_1 and grid resistor R_g form a voltage divider which determines how much voltage is to be fed back. The amount of feedback is a compromise between the amount of gain we desire and the amount of reduction of distortion we wish to achieve.

As you know, the voltages on the grid and cathode of an amplifier are in phase. Hence we may obtain the same inverse-feedback effect if the plate voltage were fed back to the cathode, instead of the grid of the tube.

Only distortion that originates in the components included in the inverse-feedback circuit is reduced. Thus, in Figure 15-14, only the distortion originating in the tube is reduced. Figure 15-15 illustrates an inverse-feedback circuit, or *loop,* which includes two stages of amplification, as well as the output transformer T.

Note that the signal has undergone a triple phase reversal between the input to the grid of the first tube and the secondary winding of the output transformer. Hence the output voltage is 180 degrees out of phase with the input signal and may be fed to the grid or cathode of the first tube. Resistor R_1 determines how much voltage is fed back. Capacitor C_3 is a blocking capacitor which prevents any d-c components in the cathode circuit from being shunted by the low d-c resistance of the secondary winding of the output transformer.

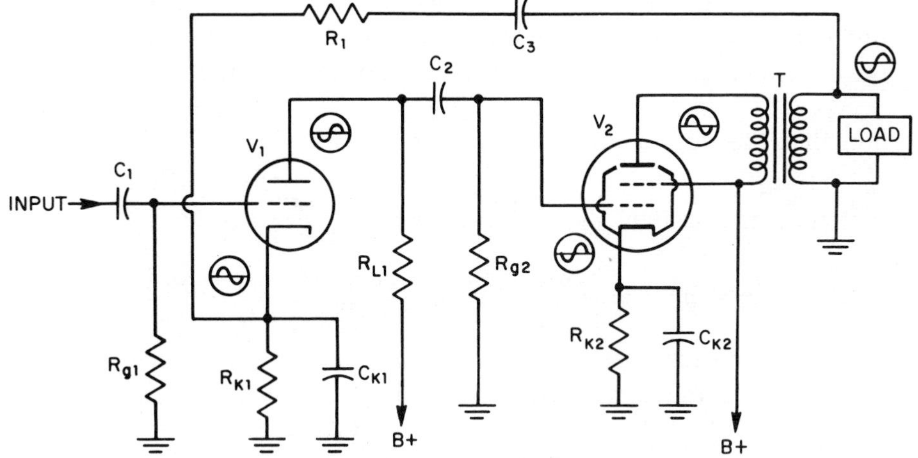

Fig. 15—15. Inverse-feedback circuit covering two stages of amplification including the output transformer.

Theoretically, any number of stages of amplification may be included in the inverse-feedback loop. Practically, however, certain complications creep in when we go beyond two stages which, generally, is the maximum number of stages included in the feedback loop.

The circuits we have illustrated are of the *voltage-feedback* type. There is another type of inverse feedback produced by the flow of plate current, called the *current-feedback* type. This feedback is produced by omitting the cathode bypass capacitor from across the cathode resistor of the stage.

When discussing the use of the cathode bypass capacitor (see Chapter 14) we saw that the function of this capacitor is to bypass the fluctuations (a-c component) of the plate current which normally would flow through the cathode resistor. Since the cathode and plate are 180 degrees out of phase, if there were no cathode capacitor to bypass it, the a-c component of the plate current flowing through the cathode resistor would produce an inverse-feedback effect similar to that produced by voltage feedback. Sometimes both voltage and current feedback are employed in the same amplifier.

4. POWER AND THE DECIBEL

The power output of the amplifier may be calculated by multiplying the voltage in the output circuit by its current. If we consider the equivalent circuit of the amplifier illustrated in Figure 15-1B, we find that the output voltage (e_p) is equal to $\frac{\mu e_g \times R_L}{r_p + R_L}$. The current flowing through R_L (i_p) is equal to $\frac{\mu e_g}{r_p + R_L}$. Thus:

$$\text{Power} = e_p \times i_p = \frac{\mu e_g \times R_L}{r_p + R_L} \times \frac{\mu e_g}{r_p + R_L} = \frac{R_L \times (\mu e_g)^2}{(r_p + R_L)^2}.$$

Since, as we have seen, maximum power output is obtained when the plate resistance (r_p) is equal to the load resistance (R_L), we get

$$\text{Maximum Power} = \frac{r_p \times (\mu e_g)^2}{(2r_p)^2} = \frac{(\mu e_g)^2}{4r_p}.$$

The input voltage (e_g) discussed here is the *effective* (rms) value. If we wish to consider *peak* voltages, we should keep in mind that when we consider sinusoidal a-c phenomena, the effective voltage is equal to the peak voltage divided by the square root of two, that is, by 1.41. (Refer to Chapter 5, subdivision C,1.) Thus our formula becomes

$$\text{Maximum Power} = \frac{(\mu e_{g\ max.})^2}{8r_p}$$

where $e_{g\ max.}$ is the peak input voltage.

In actual practice, the load resistance may be less than the plate resistance (as in the case of pentodes) or several times as large as the plate resistance (as in triodes). The value of the load resistance varies with the characteristics of the tube, and in all cases the tube manufacturers' recommendations should be followed. These recommendations take into consideration the permissible distortion as well as the power output of the tube.

In addition to the power output, we frequently are interested in the relationship between the power input and the power output of an amplifier. In order to understand how two power levels may be compared, let us briefly examine sound power.

Fortunately for us, the human ear does not hear sounds in their direct power ratio. Thus, we can listen to ordinary conversation quite comfortably and yet be able to hear thunder, which is many times as loud, without having our eardrums broken. This is because the human ear is a *logarithmic* device.

The *common logarithm* of a number is the number of times 10 must be multiplied by itself to equal that number. Thus, the logarithm of 100 (that is 10×10, or 10^2) is 2. The logarithm of 1,000 ($10 \times 10 \times 10$, or 10^3) is 3. The logarithm of 100,000 (10^5) is 5. In mathematics, we write this relationship as

$$\log_{10} 100{,}000 = 5.$$

In comparing two powers, we use the unit *bel,* which is the logarithm of the ratio of these two powers. Thus, in comparing the power of the ordinary conversation with that of thunder (which is taken to have 100,000 times the power of ordinary conversation), we say that the increase in sound is equal to

$$\log_{10} \frac{\text{Power of Thunder}}{\text{Power of Conversation}}, \text{ or } \log_{10} \frac{100{,}000}{1}, \text{ or 5 bels.}$$

A more convenient unit is the *decibel* (abbreviated as db), which is one tenth of a bel. Thus, the increase in sound from ordinary conversation to thunder would be equal to

$$10 \log_{10} \frac{100{,}000}{1}, \text{ or 50 decibels.}$$

The same method is employed in measuring the increase in power of electronic devices. Thus, the increase in power in an amplifier would be

$$\text{Gain in Decibels} = 10 \log_{10} \frac{\text{Power Output}}{\text{Power Input}}.$$

From this, we can obtain the following table:

Power ratio of 1	=	0 db gain
Power ratio of 10	=	10 db gain
Power ratio of 100	=	20 db gain
Power ratio of 1,000	=	30 db gain

and so on.

When the power output is greater than the power input, we say that we have a decibel gain, or *decibel up,* and indicate it as +db. When the power output is less than the power input, we say that we have a decibel loss, or *decibel down,* and indicate it as −db.

When several amplifiers are hooked together to work into one another, we multiply the amplifications. Thus, if three amplifiers, each having an amplification gain of 100, are connected together, we get a total gain of 100 × 100 × 100, or 1,000,000. But in the decibel system, we add the decibel gains. Thus, we get +20db+20db +20db, or +60db. Similarly, if we connect two amplifiers, one of which has a gain of +30db and the other a loss of −10db, the net result would be +30db−10db, or +20db.

Doubling the power produces a gain of +3db. Thus, if the volume control of an amplifier is turned up so that the power rises from, say, 4 watts to 8 watts, we say the gain is up +3db. If, conversely, the power output is reduced from 4 watts to 2 watts, the gain is down −3db. Again, if the original 4 watts is increased to 8 watts, the gain is up +3db. Increasing the power output to 16 watts produces another gain of +3db and the total gain is +6db. At 40 watts the power has been increased ten times and the total gain is +10db. And so on.

You will notice that when we say that an amplifier has a gain of +10db, it has no meaning in terms of actual power output, meaning merely that the power output is ten times as great as the power input. For this reason radio engineers have chosen a reference level of 6 milliwatts (0.006 watt). If we say now that the output is, say +30db, we indicate an output of 30db up from 6 milliwatts. Thus:

$$\text{Power Output} = +30\text{db} = 1{,}000 \times 0.006 \text{ watt} = 6 \text{ watts.}$$

Although the six-milliwatt reference level is standard in radio, telephone and sound engineers may use still other reference levels. Care should be taken to avoid confusion.

QUESTIONS

1. **For the triode, state and explain the formula for voltage amplification in terms of *amplification factor, plate resistance,* and *load resistance.***
2. **For the pentode, state and explain the formula for voltage amplification in terms of**
   a) ***amplication factor, plate resistance,* and *load resistance;***
   b) ***transconductance* and *load resistance.***
3. **In a stage of a-f amplification, explain the function of**
   a) **the *coupling capacitor;***
   b) **the *grid resistor;***
   c) **the *plate resistor.***

4. **Draw the circuit showing**
   a) ***resistance-capacitance coupling;***
   b) ***transformer coupling.*** **Explain the advantages and disadvantages of each type.**
5. **Draw the circuit of a pentode amplifier using a decoupling network and explain its function.**
6. **Explain why, in a power amplifier, the impedance of the load must be matched to that of the tube.**
7. **Draw and explain the basic circuit of two triodes used in a *push-pull* stage of a-f amplification.**
8. a) **Explain, in terms of grid bias, Class A, Class AB, and Class B amplification.**
   b) **Explain the advantages and disadvantages of each type.**
9. **Explain why Class B may be employed for a stage of a-f push-pull amplification.**
10. a) **Explain the phase relationship between the signal voltages that are applied to the grids of the tubes of a stage of push-pull amplification.**
    b) **Explain the function of a *phase-inverter stage.***
11. a) **Explain the function of *inverse feedback* in the a-f amplifier.**
    b) **Draw and explain the circuit of a stage of a-f amplification using inverse feedback.**
12. **In the case of inverse feedback, explain the difference between *voltage feedback* and *current feedback.***
13. **Explain what is meant by a *decibel.***

16 Radio-frequency, Video, and Direct-coupled Amplifiers

A. *Radio-frequency amplifiers*

There are two main functions for the radio-frequency amplifier. In receivers, it is used to amplify a weak radio-frequency signal until it is strong enough to operate a loudspeaker or a cathode-ray tube. In the transmitter and certain industrial applications, it is used to amplify the small r-f signal generated by an oscillator (which will be discussed later) until it is powerful enough to operate some power-consuming device such as an antenna.

Audio-frequency amplifiers are required to amplify the entire a-f band which covers a range of approximately 15 kilocycles (15 kc). Amplification is, theoretically, uniform over the entire audio range.

The entire radio-frequency range, on the other hand, covers thousands of megacycles (mc). Since no one r-f amplifier is able to cover the entire frequency range, the range is broken up into various bands, each used for a specific purpose. In the *amplitude-modulated* (AM) broadcast band, the frequency spread, as allocated by the Federal Communications Commission, is from 535 to 1,605 kc, a spread of 1,070 kc. The frequencies of transmitting stations within this band are spaced from 10 to 15 kc apart to prevent interference with each other.

In the *frequency-modulated* (FM) broadcast band, the frequency spread is from 88.1 to 107.9 mc, a spread of 19,800 kc. Transmitting stations are spaced 200 kc apart.

Two bands of frequencies have been allocated for television broadcasting. One, the *very-high-frequency* (VHF) band, covers a range of 162 mc from 54 to 216 mc. The other, the *ultra-high-frequency* (UHF) band, covers a range of 420 mc from 470 to 890 mc. Stations in both bands are spaced 6 mc apart.

Within a specific band, each transmitting station is assigned the frequency at which it is to operate. However, each station transmits, not only at this frequency, but rather a relatively narrow band of frequencies lying at either side of this assigned frequency. Thus, for example, a station lying in the AM broadcast band which is assigned a certain frequency will transmit a signal whose frequencies encompass a band extending 5 to 7.5 kc on either side of the assigned frequency.

The r-f amplifier, then, is adjusted to cover, not the entire AM broadcast band (which covers a range of 1,070 kc) simultaneously, but rather a band of about 15 kc, corresponding to the spread of a single station. We say that the *bandwidth* of the r-f amplifier is 15 kc. The amplifier is adjusted (*tuned*) to one station at a time.

In the FM broadcast band where each station has a frequency spread of 200 kc, the bandwidth of the r-f amplifier is about 150 kc. In the television bands the r-f amplifier has a bandwidth of 6 mc.

The r-f amplifier is designed to operate only in a specific band. Hence we have an r-f amplifier for the AM broadcast band, another type for the FM broadcast band, and still another type for the television bands. These amplifiers differ from each other because of the frequencies at which they operate and their bandwidths.

The r-f amplifier, then, serves a dual purpose. One is as a *bandpass filter* (see Chapter 7, subdivision E) which passes signals from the desired station and rejects all others. The other, of course, is to amplify these r-f signals.

This rules out the use of resistance coupling between stages since the chief virtue of this coupling method is its almost uniform amplification over a wide range of frequencies. It also rules out the use of the laminated iron-core a-f transformer for coupling r-f stages.

For one, these transformers do not have the required bandwidth. Further, the effect of the stray capacitance of their many turns, though fairly small at audio frequencies, is large enough at radio frequencies to reduce the amplification to almost zero. Besides, the eddy-current losses that would occur in the laminated iron cores of these transformers at radio frequencies would be tremendous (see Chapter 7, subdivision D,1).

The device most commonly employed to couple stages of r-f amplification, is the *tuned radio-frequency transformer,* as illustrated in Figure 16-1. In this circuit r-f transformer T_1 is the *input* transformer and T_2 is the *output* transformer. The secondary winding of T_1 is tuned to the frequency of the incoming signal by means of variable capacitor C_1. The secondary of T_2 is tuned to the same frequency by means of variable capacitor C_2.

You will recall that in our discussion of the resonant circuit (see Chapter 7, subdivision C,1) we learned that a parallel-resonant circuit offers a very high impedance to a current at the resonant frequency, but a low impedance to currents at other frequencies. Thus, if C_1 of Figure 16-1 is adjusted to tune the secondary winding of T_1 to resonance at the frequency of the desired signal, a relatively large voltage will appear across the secondary (and, hence, at the grid of the tube) for signals of this frequency. For all other signals the voltage will be very low.

Similarly, if C_2 is adjusted to tune the secondary winding of T_2 to resonance at the frequency of the desired signal, the winding will exhibit a large impedance for signals of this frequency and a very low impedance for other signals. The large impedance of the secondary winding of T_2 is reflected back to the primary winding (see

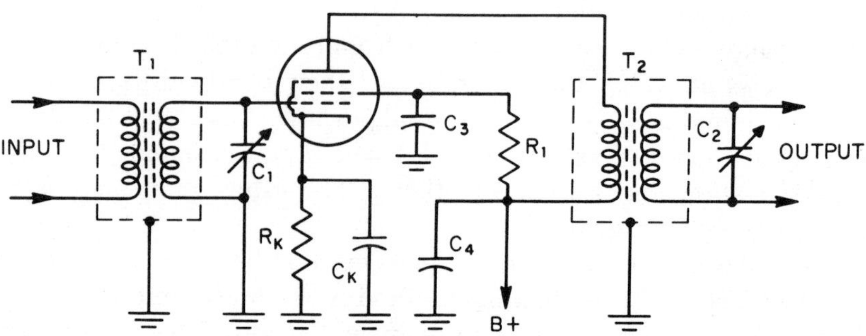

Fig. 16—1. Stage of tuned r-f amplification.

Chapter 7, subdivision D,2) which is in the plate circuit of the tube, a condition that makes for high voltage amplification. At all other frequencies the amplification is very low.

The shunting effect of the grid-to-cathode capacitance of the tube, too, is minimized. This small capacitance is in parallel with the relatively large capacitance of variable capacitor C_1, and thus has but a small additive effect. This is especially true if pentodes, which have small interelectrode capacitances, are employed.

A serious difficulty encountered in r-f amplifiers is the prevention of feedback from the output of a stage to its own input or to another stage. The danger of feedback is greater for r-f than for a-f amplifiers because of the higher frequencies involved.

One cure for feedback is the use of bypass capacitors between the B+ leads and ground (C_4 of Figure 16-1). These capacitors permit the signal to flow to ground without the necessity of passing through the power supply. Since this power supply is common to all the stages, one source of feedback thus is eliminated.

Another method is proper design which separates the grid and plate leads. Still another method is the use of nonmagnetic *shielding*, usually of aluminum or copper, which isolates the grid and plate circuits of a stage of r-f amplification from each other as well as each complete stage from the others. These shields are well grounded to the chassis. Even the wires that connect the components of the grid and plate circuits often are encased in flexible copper-braid tubing which, in turn, is grounded.

Usually, instead of shielding the entire stage of amplification, it is deemed sufficient merely to shield the r-f transformer by enclosing it in a grounded metal can, usually of aluminum. This shielding is indicated by the dashed-line boxes around the transformers in Figure 16-1.

The tubes, too, are encased in grounded metal shields, if they are of the glass type. The metal-type tubes are self-shielding, and a ground connection is made to the pin of the base which is connected to this outer shield.

However, eddy currents are generated in these shields by the magnetic fields around the windings of the transformers. The power for these currents must come from the enclosed components. Thus, eddy currents represent a loss of electric energy at the expense of the signal. For this reason, the shielding is used only where needed and the shield cans are not mounted too close to the r-f transformers.

Since the laminated iron core employed by a-f transformers cannot be used by the r-f transformer, the latter frequently employs an air core. The difficulty with the air-core type, however, is that there is very little magnetic linkage between the primary and secondary windings. Accordingly, modern r-f transformers usually employ a core of powdered iron which increases the inductance and linkage without excessive eddy-current losses.

The windings of the r-f transformer may be tuned by the ordinary air-type variable capacitor. Thus, by adjustment of this capacitor the transformer may be tuned to resonance at different frequencies (or different bands of frequencies). Of course, the transformer may be tuned by employing a variable inductance and a fixed capacitance. Accordingly, many r-f transformers have a powdered-iron core that can be moved in or out to vary the inductance of the windings. The fixed capacitor generally is of the mica or ceramic type.

If more than one stage of r-f amplification is employed (Figure 16-2), each stage must be tuned simultaneously to the frequency of the signal. Although the use of several tuned circuits increases the selectivity of the amplifier (see Chapter 7, subdivision D, 2), the need for manipulating a number of variable capacitors is a nuisance. To overcome this nuisance, the shafts of the variable capacitors are connected together so that they tune simultaneously as one dial is manipulated.

This is called *ganging*. Most commonly, all the variable capacitors are mounted on one shaft. Thus, we have a *two-gang capacitor*, a *three-gang capacitor*, and so forth, depending upon the number of sections ganged together. The dotted lines in the diagram indicate that the various capacitors are ganged together. Where the transformers are tuned by movable powdered-iron cores, these cores may be similarly ganged.

Since all the tuned circuits must tune to the same frequency and since it is virtually impossible to construct two amplifiers that are exactly alike, small semivariable capacitors, called *trimmers*, are usually connected across the larger variable capacitors. The symbol for the trimmer is T .

Since the trimmers are in parallel with the large variable capacitors, they vary the overall capacitance of the tuned circuit slightly, thus compensating for small differences. These trimmers may be of the *compression* type (illustrated in Figure 6-7A). Another type frequently employed consists of a small disk of ceramic with a metal

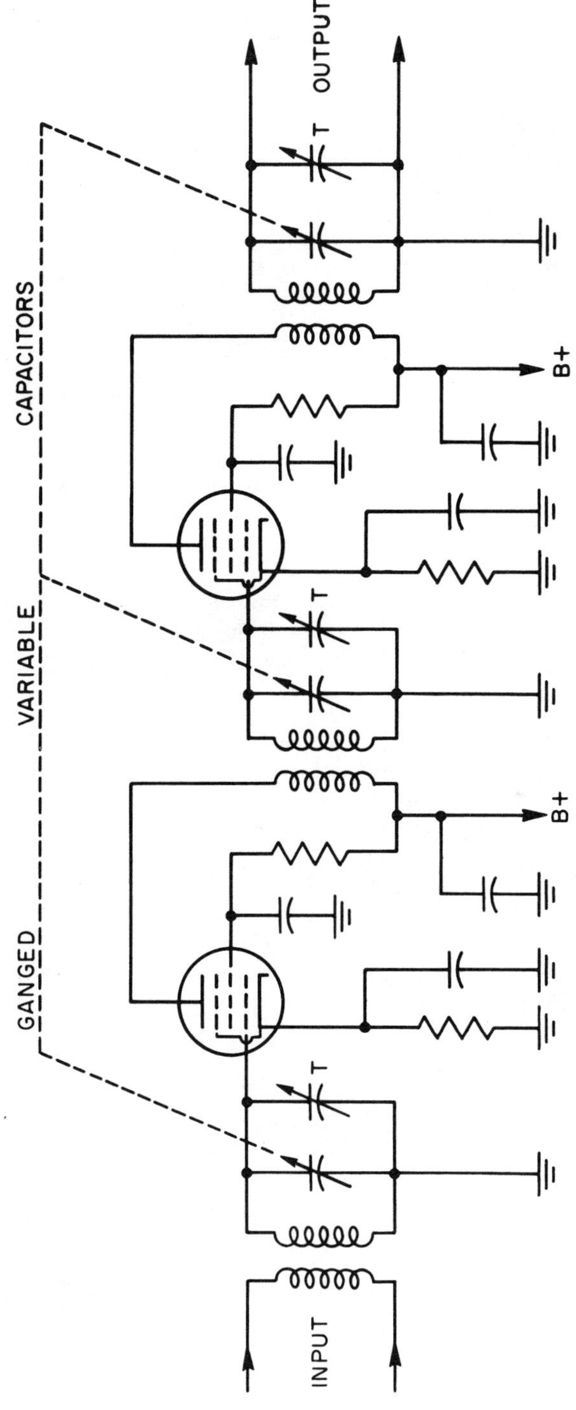

Fig. 16–2. Two stages of tuned r-f amplification. (Shielding is omitted in this diagram for the sake of simplicity.)

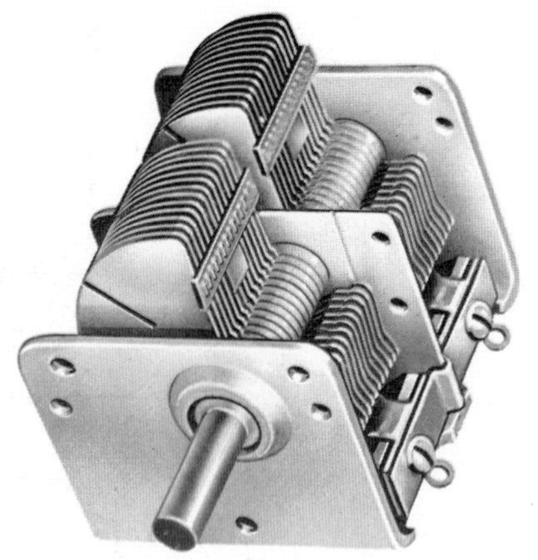

Fig. 16–3.

Two-gang and three-gang variable capacitors.

plate deposited on one side. Another metal plate is so arranged that it can be rotated across the other side of the ceramic disk. Thus, by adjusting the rotatable plate, the capacitance of the trimmer may be varied.

As we shall see later in the book, most radio and television receivers operate by converting the frequency of the received signal to a predetermined frequency, called the *intermediate frequency.* The r-f amplifier, then, is fixed to operate at this frequency, rather than being tunable over the entire band.

Such an amplifier is called an *intermediate-frequency* (*i-f*) *amplifier* (see Figure 16-5) and its transformers are known as *intermediate-frequency transformers.* Since it no longer is necessary to tune these transformers over the entire spread of the band, it becomes feasible

Fig. 16–4.

Ceramic trimmer capacitor.

Centralab Div. of Globe-Union, Inc.

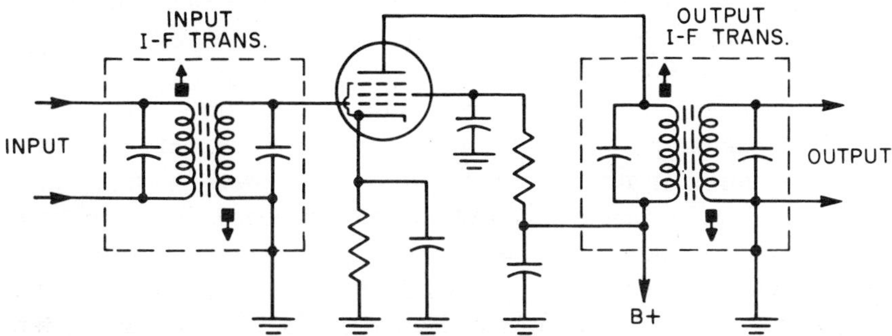

Fig. 16–5. Stage of i-f amplification.

to tune both the primary and secondary windings to the intermediate frequency. Thus, by adding tuned circuits the selectivity of the receiver is increased.

To compensate for small variations between the i-f transformers, each winding has a movable powdered-iron core that can be moved in or out, thus varying the inductance slightly. (See Figure 16-6.) The symbol indicating the adjustable core is ⇡. In some of the earlier transformers the inductances were fixed and the windings were tuned by semivariable trimmers.

We have stated that we seek to avoid feedback from the plate to grid circuits of the r-f amplifier. This is particularly troublesome where triodes are employed since such tubes have a relatively large plate-to-grid interelectrode capacitance by means of which the feedback may take place. Accordingly, by a process called *neutralization*,

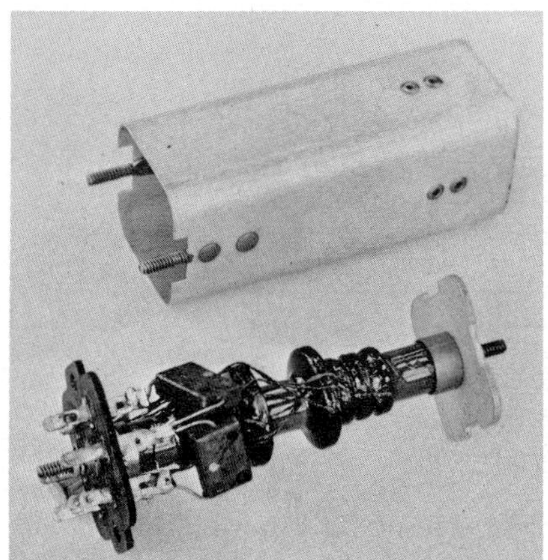

Fig. 16–6.

I-f transformer, showing construction.

a portion of the voltage from the output circuit of the stage is fed back to the grid circuit to cancel out the grid voltage caused by feedback. (See Figure 16-7.)

Feedback occurs from the plate to grid circuits through the plate-to-grid capacitance, which is indicated by the imaginary capacitor C_{GP}. We can neutralize this feedback by impressing upon the grid a voltage equal in magnitude and opposite in phase. Thus, the two voltages will "buck" each other out.

The two ends of the primary winding of the output transformer (A and B) are of opposite phase. By correct location of the B+ tap we can obtain an opposite-phase voltage equal to the feedback voltage. If this opposite-phase voltage is fed to the grid through the *neutralizing capacitor* C_N, the two voltages will cancel out. The neutralizing capacitor should equal the plate-to-grid interelectrode capacitance of the tube (a few micromicrofarads).

Partly because of the necessity for neutralization of the triode, pentodes generally are employed in r-f amplifiers. However, a certain amount of "noise" is introduced to the signal by all tubes. And the larger the number of grids in the tube, the greater is the noise. Thus pentodes are noisier than triodes.

It was to capitalize on the lower noise level of the triode that the *grounded-grid* amplifier circuit was developed. In the conventional amplifier, the input signal is applied between the grid and cathode, and the output signal is obtained between the plate and cathode. The grounded cathode then is common to both the input and output circuits.

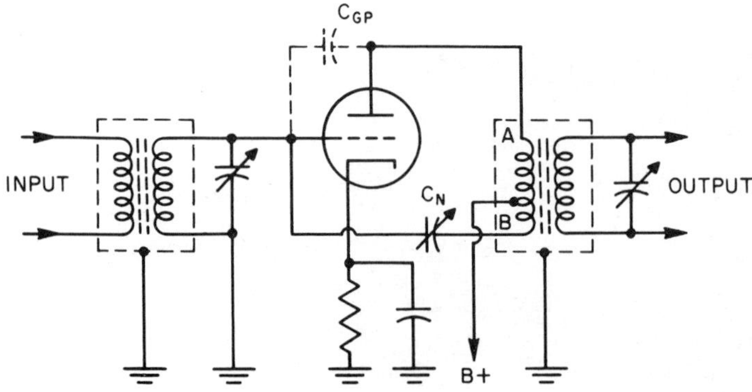

Fig. 16–7. Stage of neutralized tuned r-f amplification.

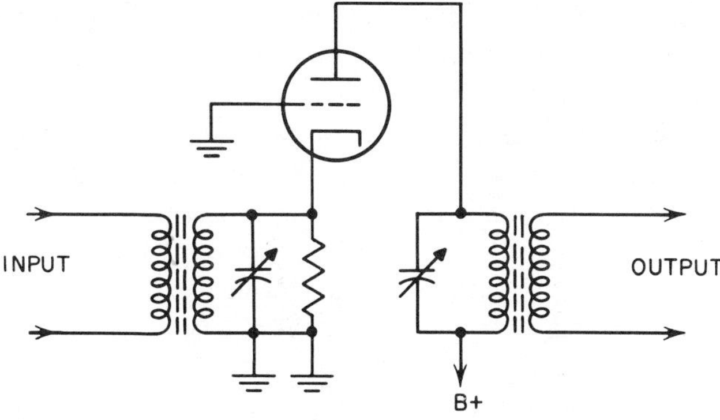

Fig. 16—8. The grounded-grid amplifier.

In the grounded-grid amplifier (see Figure 16-8) the input signal is also applied between the grid and cathode, but this time the grid is grounded. Now the output is obtained between plate and grid, which is common to the input and output circuits.

The grounded-grid amplifier has a number of advantages besides the use of a lower-noise triode. At very high frequencies, the triode, in a conventional grounded-cathode circuit, is not employed because the feedback between output and input circuits inherent to such an arrangement makes for an unstable amplifier. However, in the grounded-grid amplifier, the grounded grid acts as a shield between the input and output circuits, thus reducing the feedback. Hence triodes may be employed at these frequencies.

The grounded-grid amplifier has the added advantage that it need not be neutralized. The grid-to-plate currents flowing because of the interelectrode capacitances in the grounded-cathode circuit cause feedback since they flow between the input and output circuits. In the grounded-grid circuit, however, grid-to-plate currents do not flow through the input circuit. Hence, no feedback results. Accordingly, the amplifier need not be neutralized.

1. CLASS C AMPLIFIERS

The transmitter and the signal it transmits will be discussed later in the book. At this point, let us consider a few details. The oscillator generates an r-f current of constant frequency and amplitude. This is called the *carrier current.* At the same time, the information we wish to convey (say, sound) is applied to the microphone and is converted to an a-f *sound current* whose waveform conforms to the variations of the sound waves.

The carrier current is amplified by r-f amplifiers and the sound current by a-f amplifiers. When both have reached their proper levels, the a-f sound current is impressed upon the r-f carrier current by a process called *modulation.* Depending upon the type of modulation employed, the amplitude of the carrier current or its frequency is made to vary in step with the variations of the sound current. Note that the modulated carrier current now has two components; the a-f component and the r-f one.

This modulated carrier current, flowing through the antenna system of the transmitting station, sets up a radio wave that has the same waveform as the current. When the radio wave is intercepted by the antenna of the receiving station, it induces a current in the receiver's antenna system. The waveform of the induced current resembles that of the radio wave and, hence that of the original modulated carrier current of the transmitter.

At the receiver, the signal is amplified by the r-f amplifiers and, by a process called *demodulation,* the a-f component (which carries the desired information) is separated from the r-f carrier component. The a-f component then is amplified by the a-f amplifiers and applied to the loudspeaker, which converts it to sound.

Note that in all the stages of the receiver, the signal carries the waveforms of the sound current. Hence the input signal must always be reproduced faithfully in amplified form. This applies to both the r-f and a-f amplifiers. It follows that all the amplifier stages must operate in Class A, except for the a-f power output stage which may operate in Class AB or Class B if a push-pull circuit is employed.

We have seen how, by selecting the grid bias, we may determine the operating point on the characteristic curve of the tube (see Chapter 14). In Class A operation the grid bias is such that the plate current flows for the entire cycle. In Class AB operation plate current flows for more than half, but less than for the full cycle. The waveform of the output current is distorted, but the amplitude is greater. In Class B operation the plate current flows for only half the cycle, but the amplitude is still greater.

By choosing a still more negative grid bias, the plate current can be made to flow for less than a half-cycle, but its amplitude is even greater. (See Figure 16-9.) This is called *Class C* operation. Note that the signal voltage drives the grid positive at its positive peaks. The resulting flow of grid current reduces the amplitude of the signal voltage somewhat, as indicated by the dotted curves.

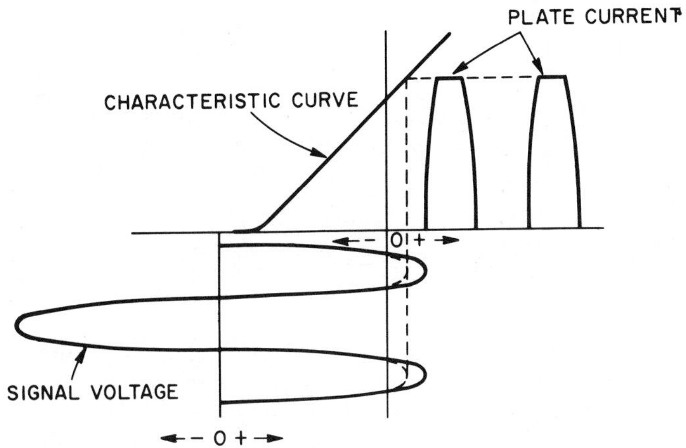

Fig. 16–9. Graph illustrating the Class C amplifier.

The waveform of the plate current is greatly distorted, so much that even the push-pull circuit cannot restore it to the original waveform. Hence Class C amplifiers are not used in the receiver.

Returning to the transmitter, the oscillator generates a relatively weak r-f carrier current of constant frequency and amplitude. This current can be amplified by means of Class C r-f amplifiers up to the point where modulation is applied.

Consider the typical r-f amplifier circuit illustrated in Figure 16-10. The r-f carrier current generated by the oscillator at the required frequency is fed to the grid of the tube whose grid bias, as determined by the value of the cathode resistor R_k, causes it to operate as a Class C amplifier. The *tank circuit,* consisting of C_1 and L_1 in the plate circuit, is tuned to resonance at the oscillator frequency.

Oscillations of plate current are set up in the tank circuit (see Chapter 7, subdivision C,1). To keep these oscillations going it is not necessary to feed power to the tank circuit during the complete cycle of oscillation. It is enough that a short pulse, in step with the oscillations, of course, be given to the oscillating current.

The effect is the same as if you would push a child in a swing. You do not have to push during the whole swing. It is enough to give a slight push during a small portion of the swing to keep the child going.

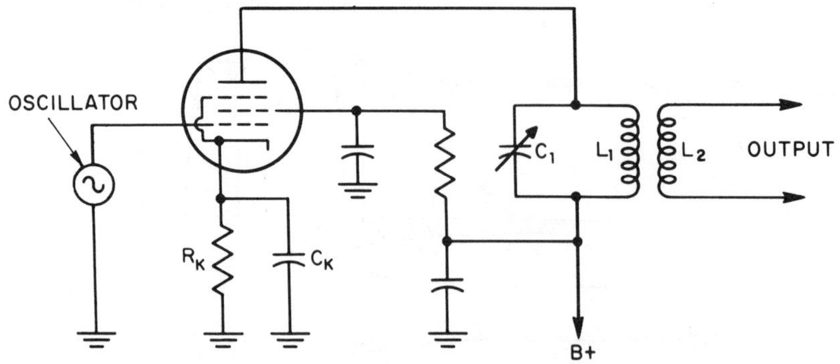

Fig. 16—10. Typical r-f amplifier circuit.

So, you see, all we need are short pulses of plate current in step with the oscillations in the tank circuit. And, as shown in Figure 16-9, such plate-current pulses are produced by the Class C amplifier. The effect, then, is to produce in the plate tank circuit an amplified version of the carrier current generated by the oscillator.

Of course, after the modulation has been added to the carrier current, Class C amplification can no longer be employed. Accordingly, the final r-f amplifier of the transmitter may be a power stage operating in Class A or, more frequently, a push-pull stage operating in Class B. (See Figure 16-11.) Note that this circuit resembles the a-f push-pull circuit. There are, however, some differences.

The input is to a tuned circuit consisting of the center-tapped inductor L_2 and variable capacitors C_1 and C_2. These capacitors are, in reality, two sections of a *split-stator* capacitor so arranged that as the capacitance of one section is increased, the capacitance of the other section is decreased in like degree. Each section tunes half of L_2.

The plate circuit consists of a similar center-tapped inductor L_3 and a split-stator capacitor C_3 and C_4. Fixed bias, obtained from the power supply, generally is employed for this stage. Note the decoupling circuits consisting of RFC_1 and C_5 in the grid-bias circuit and RFC_2 and C_7 in the B+ circuit. These decouplers are employed to eliminate feedback through the power supply. RFC_1 and RFC_2 are *r-f choke coils,* consisting of a number of turns of wire and are used instead of the customary resistors. These choke coils offer a high impedance to signals at radio frequencies, but a low ohmic resistance to the direct currents from the power supply.

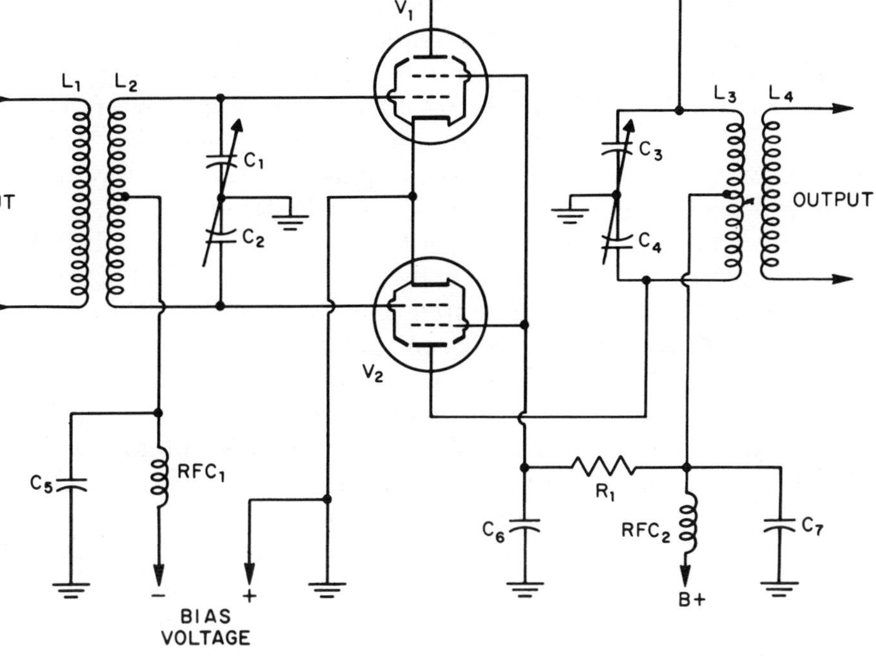

Fig. 16—11. R-f push-pull stage.

As has been indicated, the r-f carrier current is generated by the oscillator. Hence, where a high-frequency carrier current is required, the oscillator must be capable of operating at this high frequency. However, the higher the frequency, the more difficult it is to keep the oscillator stable. It would be desirable to operate the oscillator at a lower frequency and to multiply the frequency of its output to the desired frequency. The Class C amplifier offers such a solution.

Let us return to our analogy of the swing. Assume it takes one second for the swing to make one back-and-forth oscillation. We can keep the swing going, therefore, if we apply a push every second. That is, we apply one push each cycle.

Suppose now that the swing is speeded up so that it makes two back-and-forth oscillations each second. If we still apply a push a second, we can keep the swing going. Only this time we apply one push every two cycles. Similarly, we can keep the swing going if we apply the push every three cycles, and so forth. The only requirement is that the oscillations must be a whole-number multiple of the pushes.

With this in mind, turn to the *frequency-multiplier* circuit illustrated in Figure 16-12. The input is to the tuned circuit consisting of L_2 and C_1 which is tuned to resonate at the frequency of the oscillator. The tube is biased to operate as a Class C amplifier. The plate current then consists of a series of pulses in step with the frequency of the incoming signal, that is, one pulse for each cycle.

The plate tank circuit consisting of L_3 and C_2 is tuned to a whole-number multiple of the frequency of the incoming signal. Let us say, for example, that the plate tank frequency is twice that of the incoming signal. Thus it will receive one pulse of plate current every second cycle and hence its oscillations will be maintained. The output, accordingly, will be twice the frequency of the incoming signal.

Since the input circuit is at a different frequency than that of the output circuit, feedback presents no problem. Accordingly, a triode may be employed and no neutralization is required.

By tuning the plate tank circuit to a higher whole-number multiple of the input frequency, the output may have a still higher frequency. Thus we may have frequency-triplers and frequency-quadruplers. It generally is not advisable to multiply the frequency too many times in a single stage. Where a great deal of frequency-multiplication is needed, we may feed a stage of frequency-multiplication into a second similar stage.

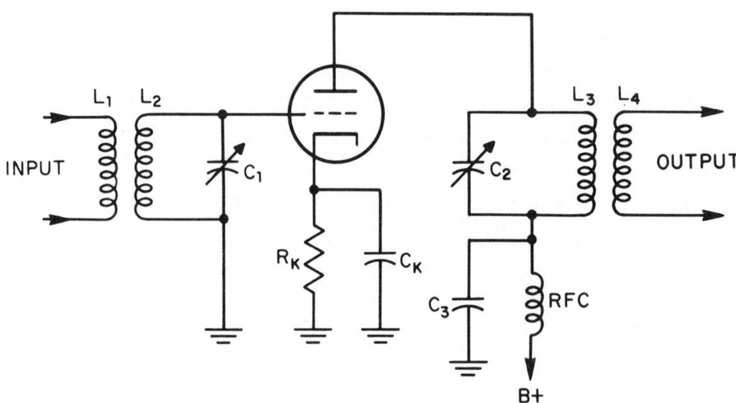

Fig. 16—12. Circuit of frequency multiplier.

B. Video amplifiers

Just as *audio* pertains to the audible, or sound, signal in the ordinary radio receiver, so *video* pertains to the visible, or picture, signal in the television receiver. Hence, amplifiers designed to amplify these picture signals are called *video-frequency amplifiers.*

Since these video-frequency signals cover a frequency range of about 30 to 4,000,000 cycles per second (under present standards), the video-frequency amplifier must operate over a very wide range, or band, hence it also is known as a *wide-band amplifier.* It is very important that amplification be uniform throughout this band.

These requirements immediately suggest a circuit similar to the resistance-coupled a-f amplifier. However, amplification at low frequencies falls off because of the increased impedance of the coupling capacitor, and amplification at the high frequencies falls off because the interelectrode capacitances of the tubes and the capacitances of the connecting wires act to shunt the load resistor, hence lowering its resistance.

One obvious remedy for the drop in amplification at low frequencies is to increase the size of the coupling capacitor. However, although this capacitor may be increased to have a value of about 0.1 microfarad (as compared to a maximum value of about 0.01 microfarad for the a-f amplifier), increasing this capacitor beyond this value will increase the stray capacitances which, in turn, will adversely affect amplification at high frequencies.

We may compensate for loss of amplification at low frequencies by connecting resistor R between the load resistor (R_L) and the plate supply, as illustrated in Figure 16-13. This resistor is bypassed to ground by the capacitor C. At high frequencies, the plate current

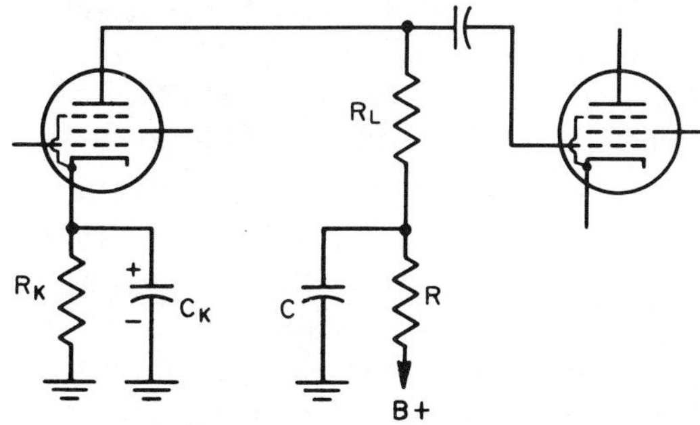

Fig. 16—13. Circuit for low-frequency compensation.

will flow through C (whose value is fairly high) rather than through R, and thus the effective plate load will be the resistor R_L. But at low frequencies, the impedance of C is quite high. Therefore the effective plate load will be increased by approximately the value of R. This results in increased voltage amplification.

At audio frequencies, the value of the cathode-resistor bypass capacitor (C_k) usually is about five microfarads. But since the low frequencies encountered in video-frequency amplification are somewhat lower than those of audio-frequency amplification, this bypass capacitor must have a much greater value if loss of amplification is to be avoided. Accordingly, electrolytic capacitors of about 100 microfarads frequently are employed.

At high frequencies, the plate-to-cathode capacitance of the preceding tube, the grid-to-cathode capacitance of the following tube, and the capacitance of the connecting wires act to shunt the load resistor thereby reducing its effective resistance. This, in turn, reduces the output voltage of the amplifier.

To meet this difficulty, tubes with small interelectrode capacitances (pentodes) and high transconductances are used in video amplifiers. The transconductance (g_m), you will recall, is the ratio between the change in plate current and the change in grid voltage producing it (see Chapter 10, subdivision A, 2). Thus, given a certain change in grid voltage, a tube with a greater transconductance will produce a greater change in plate current than a tube with a smaller transconductance. Or, to put it in a different way, for a given output voltage to be produced by a given change in grid voltage, the tube having a greater g_m will need a smaller load resistor than a tube with a smaller g_m (since $e_p = i_p \times R_L$).

Since the shunting effect of the interelectrode capacitances at high frequencies is to reduce the impedance of the load, tubes with high g_m are used. The output voltage of the video amplifier thus is maintained in spite of the drop in load resistance.

For example, compare the g_m of an ordinary pentode, such as the 6SJ7 (1,650 micromhos), with the g_m of the pentode section of the 6EB8 (12,500 mircomhos) which frequently is used in video amplifiers. (The 6EB8 tube is a dual type, consisting of a pentode and triode in the same envelope.) Since a one-volt change in grid voltage will produce a one-milliampere change in plate current for each 1,000 micromhos of transconductance, a one-volt change of grid voltage will produce a 1.65-milliampere change of plate current in the 6SJ7 tube, and a 12.5-milliampere change of plate current in the

6EB8 tube. Or, to put it differently, the load resistor of the 6SJ7 tube need be about 7½ times as great as that of the 6EB8 tube to produce the same output voltage.

In addition to reducing the load resistance of the preceding tube, the interelectrode and lead capacitances act to shunt, and so reduce, the input impedance to the following tube. This tends to lower the grid voltage and thereby reduce the amplification. By adding a small inductor (L), called a *peaking coil,* as shown in Figure 16-14A, the coil and the shunting capacitances form a parallel-resonant circuit which, at high frequencies, places a high impedance across the input of the second tube. At low frequencies, the inductor has practically no effect on the circuit.

Another method for compensating for the loss of amplification at high frequencies is shown in Figure 16-14B. Here the peaking coil (L) is connected in series with the coupling capacitor. At high frequencies, this inductor forms a low-impedance series-resonant circuit with the shunting capacitances, causing a larger voltage to appear at the grid of the second tube. As before, at low frequencies the inductor has virtually no effect on the circuit. Sometimes, two peaking coils are used, one in shunt and the other in series. In addition, video amplifiers must be designed for short leads and a minimum of capacitance between stages.

Even with the use of tubes having high g_m and the other compensating devices, the gain per stage is low. The gain has been sacrificed to achieve adequate amplification over the wide video band. A number of stages are necessary, therefore, to obtain appreciable amplification.

C. *Direct-coupled amplifiers*

In most instances, the input signal to the amplifier is an a-c phenomenon. Accordingly, this input signal can be applied through a coupling capacitor or transformer. Where the input signal is a steady direct current or voltage, however, such coupling devices cannot be employed. The signal then must be applied to the control grid of the amplifier tube either directly or through some device, such as a resistor or coil, that can pass direct currents. Such amplifiers are called *direct-coupled amplifiers.*

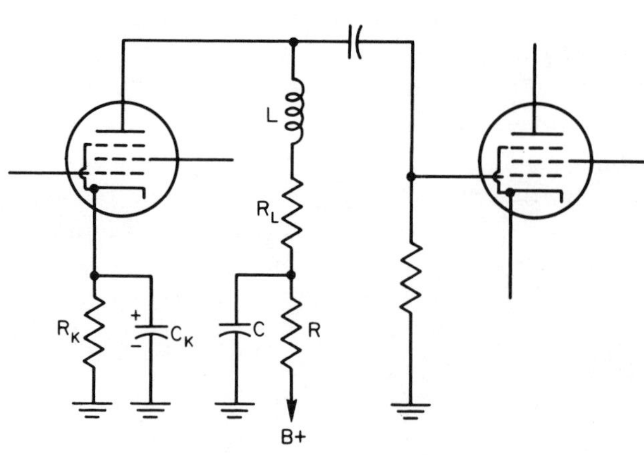

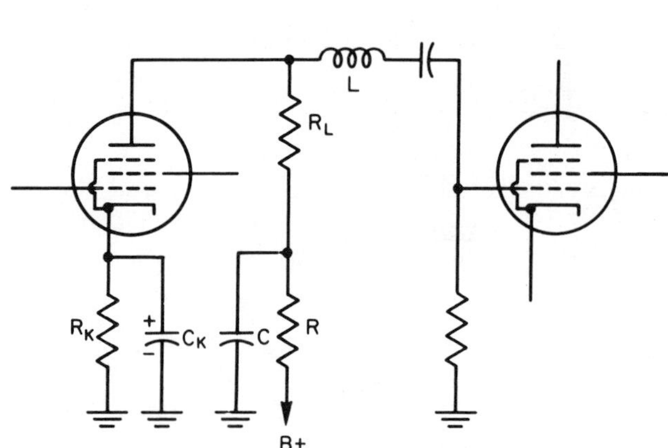

Fig. 16—14. Circuits for high-frequency compensation.
A. Peaking coil in shunt.
B. Peaking coil in series.

The circuit of the basic direct-coupled amplifier is illustrated in Figure 16-15. The bias battery establishes the operating point of the tube. With no signal applied, a certain amount of plate current flows and a voltage drop (the output voltage) is established across the load resistor R_L.

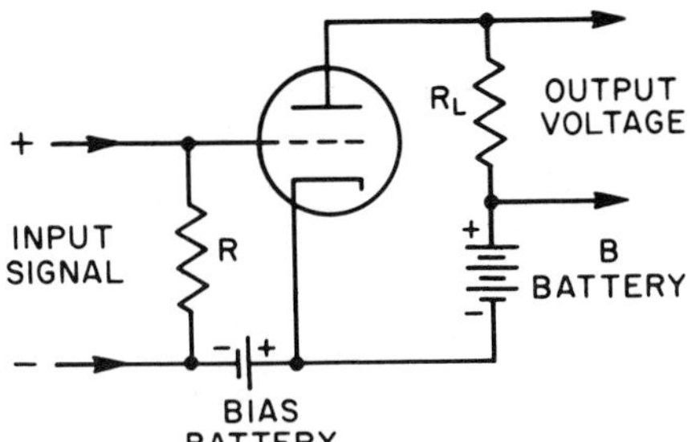

Fig. 16–15. Basic circuit of the direct-coupled amplifier.

The input signal, applied with the indicated polarity, tends to neutralize the charge placed on the grid by the bias battery. The larger the input signal, the less negative the grid of the tube becomes, and the greater the plate current and the output voltage. Since the grid of the tube must not become positive, the maximum voltage of the input signal cannot exceed the voltage of the bias battery.

Let us assume that the stage has a gain of 10. Then a 1-volt signal applied to the input produces a 10-volt change in the output voltage. It is in this way that it acts as an amplifier.

It is possible to connect several direct-coupled amplifiers in series. The ordinary resistance-capacitor method of coupling, previously described, cannot be used, since the low-frequency response is limited by the impedance of the coupling capacitor. For direct currents (zero frequency), this impedance becomes infinitely great.

A method for the direct coupling of the output of one tube to the grid of the second must be employed (see Figure 16-16). As shown here, the circuit will not operate because of the high positive voltage placed by battery B_1 on the grid of the second tube. To get around this difficulty, a second battery (battery B_2 of Figure 16-17) must be

Fig. 16–16. Direct coupling between amplifiers. (This circuit will not operate.)

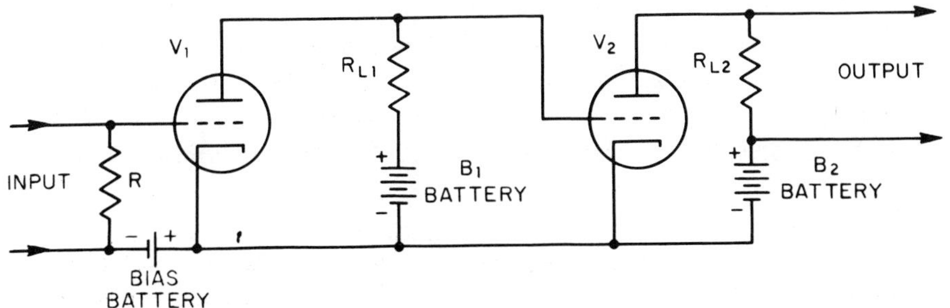

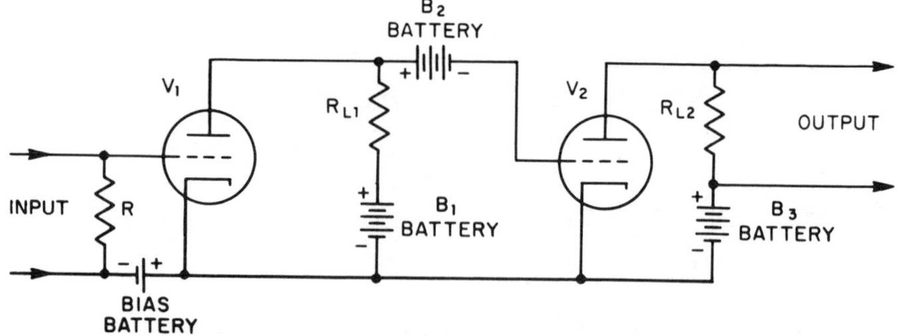

Fig. 16–17. Direct coupling between amplifiers using a secondary battery (B_2) to buck out the voltage of the first.

placed in such a way as to buck out the positive voltage to the grid of the second tube placed by battery B_1. In addition, this second battery must also supply a negative bias on that grid. However, because of the voltage drop across the load resistor (R_{L1}), the voltage required of battery B_2 may actually be less than that of battery B_1.

Another method of presenting the circuit shown in Figure 16-17 is illustrated in Figure 16-18A. Here, all the various batteries are combined into one large tapped battery. Note that the voltage of this large battery has to be approximately twice that of the ordinary B battery normally employed in an amplifier. Note also that the cathode of the second tube is at a fairly high potential. If a common heater supply is used for both tubes, the insulation between the second cathode and its heater must be sufficient to withstand this voltage. To avoid breakdown, separate heater supplies for each tube are sometimes used.

If a power supply is used instead of batteries, a voltage divider may be employed to tap off the various potentials, as shown in Figure 16-18B. As before, the total voltage of this power supply is approximately twice that used by the conventional a-c amplifier. The various taps on the voltage divider must be selected carefully for proper balance. Since any small variation in any part of the circuit unbalances the complete circuit, this type of amplifier has a tendency to become unstable. More than two stages are rarely employed.

In addition to amplifying steady direct voltages, the direct-coupled amplifier may be used to amplify signals containing very

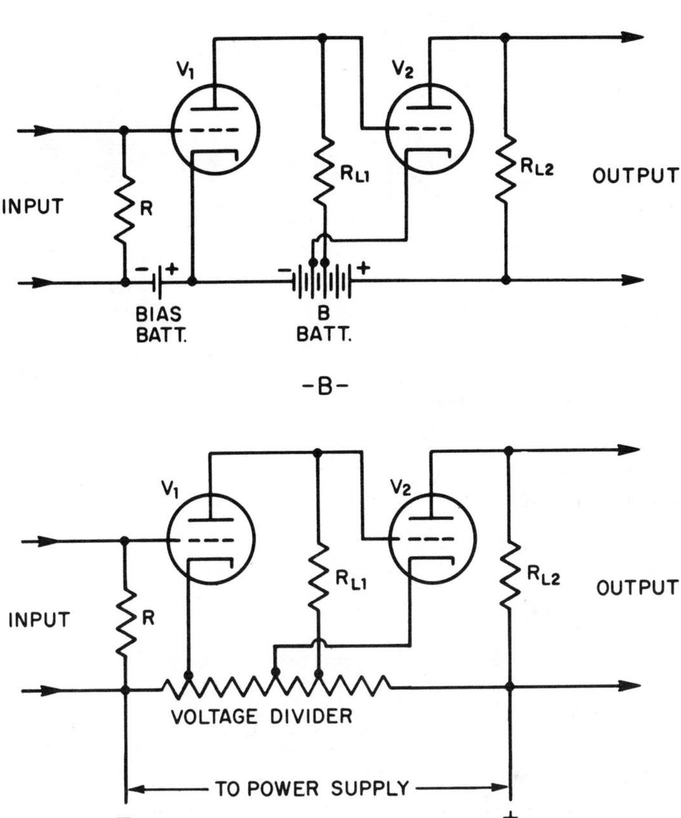

Fig. 16—18. Direct-coupled amplifier.
A. Using a common tapped battery.
B. Using a voltage divider across the power supply.

slow variations in voltage. It also may be used to amplify audio- and radio-frequency signals just as any a-c amplifier.

An interesting application of direct coupling to r-f amplification is the *cascode* amplifier, illustrated in Figure 16-19, which frequently is employed in the tuner section of television receivers. The signal is fed to the grid of V_1 which is a conventional grounded-cathode amplifier. Variations in the signal voltage at the grid of V_1 appear as amplified variations in voltage at its plate (although opposite in phase to the voltage applied to the grid).

The plate of V_1 is coupled directly to the cathode of V_2 through the r-f choke coil RFC. Triode V_2 is a grounded-grid amplifier whose

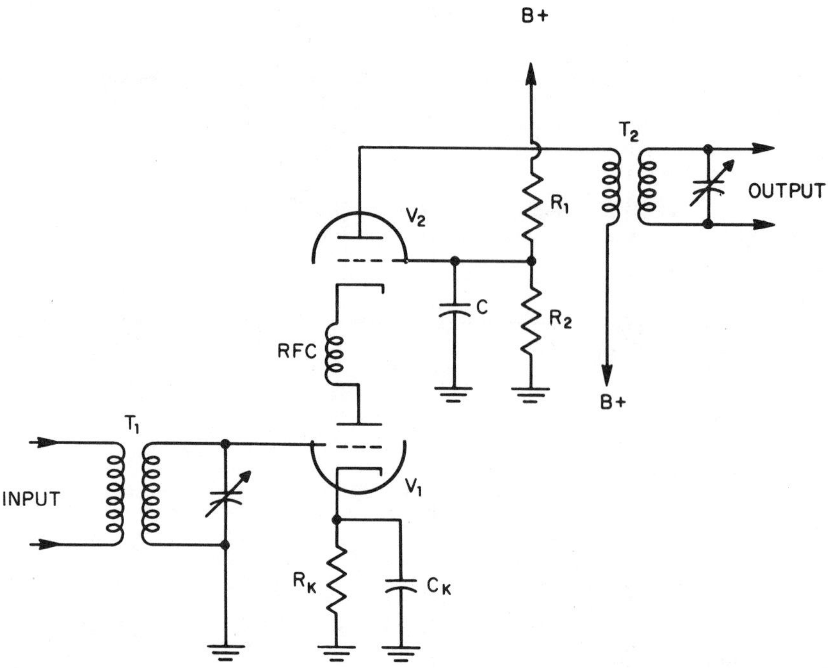

Fig. 16–19. The cascode r-f amplifier.

grid is grounded, as far as the signal is concerned, through capacitor C. The amplified signal-voltage variations that appear at the plate of V_1 thus are applied to the cathode of V_2. Hence they appear in further amplified form at its plate.

Note that the cathode of V_2 is positive, because of its junction with the plate of V_1. The grid of V_2, accordingly, must receive a slightly less positive potential than its cathode if the tube is to function as an amplifier. Bias for V_2, therefore, is obtained from the B+ supply through a voltage divider formed by R_1 and R_2.

In effect, then, we have here two stages of r-f amplification connected in series, or in cascade. Hence the overall amplification of the cascode stage is relatively high, especially if high-*mu* twin triodes are employed. Because triodes are used, the noise introduced by this stage is low.

QUESTIONS

1. Explain how the function of a radio-frequency amplifier differs from that of the audio-frequency amplifier.

2. Explain how the circuit of the radio-frequency amplifier differs from that of the audio-frequency amplifier.
3. Explain why resistance coupling and iron-core transformer coupling are not suitable for r-f amplification.
4. Draw and explain the basic circuit of a stage of tuned r-f amplification.
5. Explain two methods employed for reducing feedback in r-f amplifiers.
6. a) Explain the process of *neutralization* used when a triode is employed as an r-f amplifier.
   b) Explain why a pentode is more suitable than a triode as an r-f amplifier.
7. a) Explain Class C amplification in terms of grid bias.
   b) Explain why Class C cannot be used for a-f amplification.
   c) Explain the advantages and disadvantages of Class C amplification.
8. Draw and explain the circuit of the grounded-grid amplifier.
9. Draw and explain the circuit of the r-f frequency multiplier.
10. a) Explain the need for low-frequency compensation in the video amplifier.
    b) Explain how it is accomplished.
11. a) Explain the need for high-frequency compensation in the video amplifier.
    b) Explain how it is accomplished.
12. a) Draw and explain the basic circuit of a two-stage direct-coupled amplifier.
    b) Explain its advantages and disadvantages.
13. Draw and explain the circuit of the cascode amplifier.

17 Oscillators

It becomes necessary, sometimes, to generate alternating currents at very high frequencies. Where the frequencies are low, up to several thousand cycles per second, the ordinary rotating-armature type of generator may be employed. But where the frequencies are high, ranging up to millions, and even billions, of cycles per second, such generators cannot be used. The incredibly high speed of rotation necessary for such frequencies would soon cause the generator to fly apart by centrifugal force. As we shall see in this chapter, however, electron tubes, together with associated components, may be made to generate these currents. We call such generators *oscillators.*

There are a great many different types of oscillator circuits. Since it is impossible to discuss all of them in a single chapter, we shall confine ourselves to those most frequently employed.

A. *The L-C oscillator*

In Chapter 7, subdivision C, 1, you learned that if a current is started flowing in a closed loop containing inductance and capacitance, the current will flow back and forth (*oscillate*) in that circuit until its energy is dissipated by the resistance it encounters. We should examine this matter more closely.

Let us start with the simple oscillatory circuit illustrated in Figure 17-1A. Assume that capacitor C is charged with an excess of electrons (negative charge) on its upper plate and a deficiency (positive charge) on its lower plate. As a result, an electrostatic field is set up across it which causes the electrons to move in a clockwise direction through the circuit. As the electrons move through coil L, they produce a magnetic field around it.

As the capacitor discharges itself, the flow of current slows down and tends to cease. The cessation of current, in turn, produces a collapse of the magnetic field around the coil. As the field collapses, the lines of force cut across the loops of the coil, inducing an electromotive force across them. This electromotive force is of such polarity as to cause the electrons to continue moving in the same clockwise direction through the circuit. Thus, instead of equalizing the distribution of electrons, an excess now appears on the lower plate of the capacitor, and the capacitor is charged again, though now in the reverse manner.

As the magnetic field expends itself, the electromotive force driving the electrons in a clockwise direction disappears. Now the electrostatic field across the charged capacitor starts to drive the electrons in a counterclockwise direction through the circuit. This process, which is called the *flywheel effect*, continues with electrostatic-field energy being converted into magnetic-field energy, and back again.

Theoretically, once started, the back-and-forth oscillations of the electrons should continue indefinitely. However, the resistance of

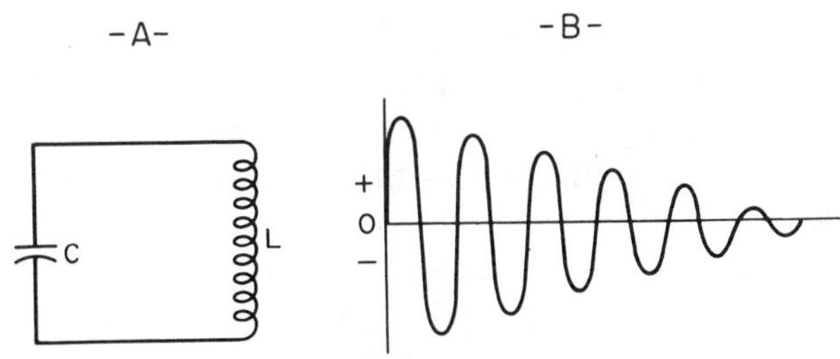

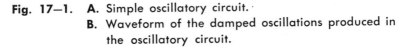

Fig. 17—1. **A.** Simple oscillatory circuit.
B. Waveform of the damped oscillations produced in the oscillatory circuit.

the circuit absorbs a portion of the energy during each cycle, converting it to heat. Accordingly, the amplitude of the current is a little lower each cycle until, finally, the electrons come to rest.

The waveform of the current flowing in the circuit during the oscillations is illustrated in Figure 17-1B. The clockwise flow of electrons corresponds to the positive half-cycles and the counterclockwise flow to the negative half-cycles. Note that the amplitude of the current is greatest for the first cycle and gradually dies away with each succeeding cycle. Oscillations with this type of waveform are called *damped* oscillations. Note, too, that the waveform is sinusoidal. This is characteristic of *L-C* oscillations.

As you have learned, the frequency of these oscillations is determined by the values of L and C. A fair approximation of this frequency may be obtained from the formula:

$$f = \frac{1}{2\pi\sqrt{L \times C}}\,.$$

where f is the frequency in cycles per second, L is the inductance in henrys, and C is the capacitance in farads. The value of π (pi) is 3.14.

As we have stated, resistance in the circuit produces losses which soon stop the oscillations. But if we had some method for replacing the electric energy lost in the circuit, the oscillations would continue indefinitely.

Such replacement of lost electric energy is accomplished by taking advantage of the amplifying ability of the electron tube. As you know, the energy developed in the plate circuit of the tube is greater than that impressed on the grid circuit. (Of course, the power supply furnishes the additional energy.) If the *L-C* oscillatory circuit (called the *tank circuit*) were to be connected to the input of an electron tube, the amplified energy would appear in the plate circuit. If a sufficient quantity of this plate energy were to be fed back to the tank circuit (the grid circuit) to compensate for the energy lost through resistance, oscillations in that circuit could be maintained indefinitely.

The basic circuit of the oscillator is illustrated in Figure 17-2. The oscillator voltage, which is applied to the grid of the tube, is developed by the L_1-C_1 tank. The output current appears in the plate circuit and flows through the *feedback coil* (L_2), thus feeding back a voltage by transformer action to the tank circuit to overcome the losses and keep the oscillations going.

The amount of feedback can be controlled by varying the number of turns of the feedback coil and by varying its coupling to coil L_1 of the tank circuit. Also, the feedback voltage must be in phase with the voltage of the grid circuit. Otherwise it would oppose it and soon cause the oscillations to die out. Since the grid voltage is 180 degrees out of phase with the plate voltage, the feedback must be accompanied by a 180-degree phase shift. This phase shift is accomplished by the feedback coil.

Thus we have an oscillator in which the direct current supplied by the B battery is converted into an alternating current whose frequency is determined by the values of L_1 and C_1. We can vary the frequency of this alternating current by changing the values of L_1 or C_1, or both.

In our discussion of the Class C amplifier (see Chapter 16, subdivision A, 1) we saw how oscillations in the tank circuit can be kept going by applying a series of pulses in step with the oscillations. Hence our oscillator tube is operated in Class C. Values of grid leak (R_g) and grid capacitor (C_g) are selected to bias the tube properly. The function of C_2 is to form a path for the a-c components of the plate current which avoids having them flow through the B battery.

A question might arise at this point. How are the oscillations started in the circuit illustrated in Figure 17-2?

To answer this question, consider what happens at the moment we start heating the cathode of the tube. When the cathode is cold, no electrons are emitted and, therefore, there is no plate current. But as the cathode heats up, more and more plate current flows. Thus, a variation is produced in the plate current.

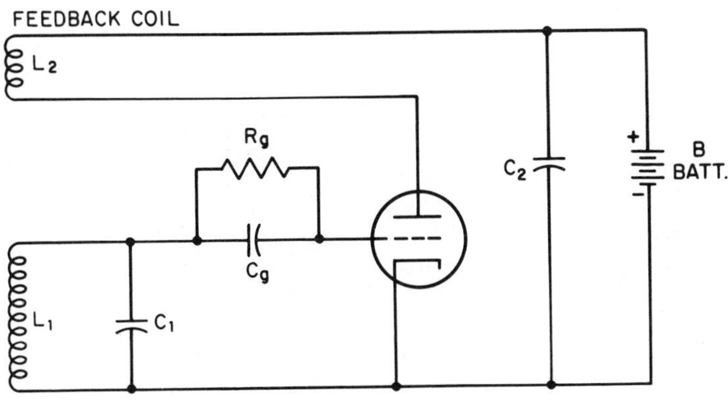

Fig. 17—2. Basic oscillator circuit.

This variation, no matter how slight, is reproduced in the grid circuit by means of feedback, and is now amplified by the tube, producing a still greater variation in the plate current. This process progresses rapidly, and soon the circuit is oscillating at its maximum value.

Suppose that the oscillations in the tank circuit start to die down and reduce in amplitude. The grid voltage is reduced. But as the grid voltage dies down (approaches zero), the plate current increases (see Figure 16-9). In turn, the feedback is increased, which increases the amplitude of the oscillations. Thus, oscillation is prevented from dying out.

Now suppose the oscillations continue to build up. Is there a limit they can reach? This limit is determined by the characteristic curve of the tube. You will recall that the upper end of the curve bends and flattens out into what is known as the *saturation point* of the tube (see Figure 9-2). Increasing the grid voltage beyond this point will produce little, if any, increase in plate current. Thus, the feedback reaches its maximum limit.

Since an increase in current in the tank circuit is accompanied by increased losses due to resistance and since the feedback is fixed at its maximum value, the amplitude of the oscillations in the tank circuit, too, will be fixed at a certain maximum level. Of course, a more powerful tube, with the saturation point further removed, may be used if oscillations of greater amplitude are desired.

The B battery can be connected in the circuit by one of two methods. One is the *series-feed* method shown in Figure 17-2. Here the battery is in series with the feedback circuit. Where it is necessary or desirable to keep the B battery out of the feedback circuit, the *shunt-* or *parallel-feed* method may be used. (See Figure 17-3.)

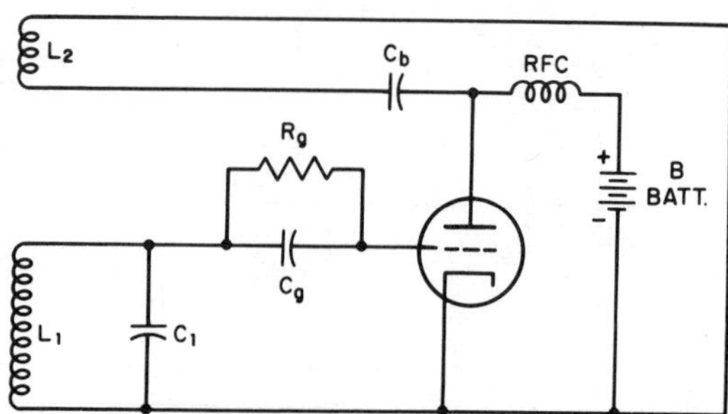

Fig. 17–3. Oscillator employing shunt, or parallel, feed.

The r-f choke coil (RFC) is used to keep the alternating currents out of the B battery. Capacitor C_b is a blocking capacitor used to keep the direct current out of the oscillating circuit, and it must be large enough to offer negligible impedance to the alternating currents at the frequencies at which they are generated.

There are several methods for obtaining the output from the oscillator. Two most frequently employed methods are illustrated in Figure 17-4. In Figure 17-4A, a coupling capacitor (C_2) is connected to the "hot" (ungrounded) end of the tank coil L_1. If a

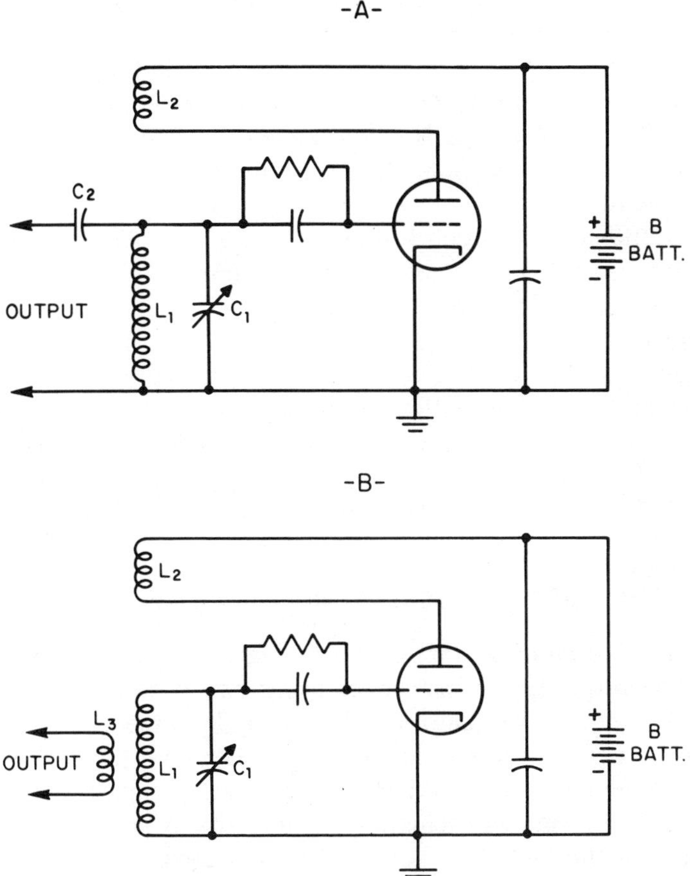

Fig. 17—4. How output is obtained from the oscillator.
A. By means of a coupling capacitor.
B. By means of an inductively-coupled coil.

lesser output is desired, capacitor C_2 may be attached to taps on the winding instead of the top. Coil L_2 is the feedback coil.

Note that capacitor C_1 is variable, thus permitting the frequency of the oscillator to be varied. In some instances, C_1 is fixed and coil L_1 has a movable powdered-iron core which can be moved in or out, thus varying the inductance of L_1 and, hence, the frequency of the oscillator.

In Figure 17-4B, output is obtained by means of a coil (L_3) inductively-coupled to the tank coil L_1. The amount of output may be varied by varying the degree of coupling between L_1 and L_3.

Note that the feedback must not only compensate for the losses in the oscillatory circuit, but also for the energy drained away by the requirements of the load upon the oscillator.

The process of feeding energy from the plate to the grid circuit in phase is called *regeneration.* Hence the oscillator whose circuit is illustrated in Figure 17-2 is called a *regenerative oscillator.* Since the tuned tank circuit is in the grid circuit, it also is known as a *tuned-grid oscillator.*

A variation of this circuit is illustrated in Figure 17-5A. Note that the circuit is similar to the one shown in Figure 17-2, except that an alternate method for connecting the grid leak R_g is employed (see Chapter 14).

Sometimes, the tuned tank circuit is placed in the plate circuit, as in Figure 17-5B. The action is the same as before—some of the energy is fed back to the grid circuit and amplified by the tube, and the amplified energy compensates for the losses in the tank circuit. We call this circuit a *tuned-plate oscillator.*

1. THE HARTLEY OSCILLATOR

The *Hartley oscillator,* shown in Figure 17-6, is a variation of the regenerative type. Instead of using two separate coils inductively coupled, this oscillator employs a single tapped coil. The bottom portion (from the tap down) is the plate coil; the entire coil is the grid tank coil and, together with the tuning capacitor C, determines the frequency of oscillation. The feedback is in proper phase by virtue of the fact that the plate is connected to one end of the coil and the grid to the other end, the cathode being connected to the tap.

Assume an instant when the top of the coil is negative and the bottom positive. The grid becomes negative, cutting down the flow

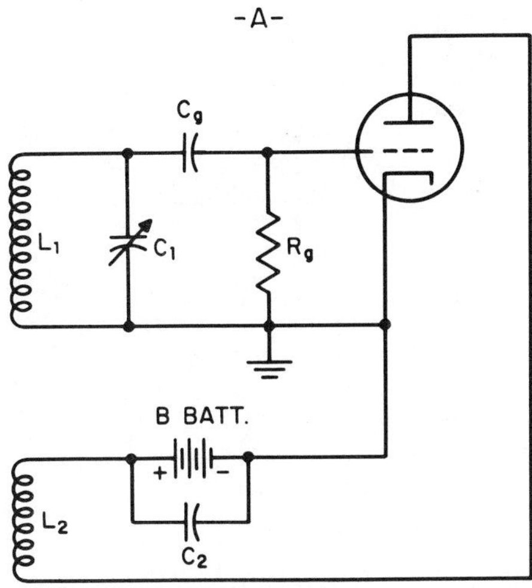

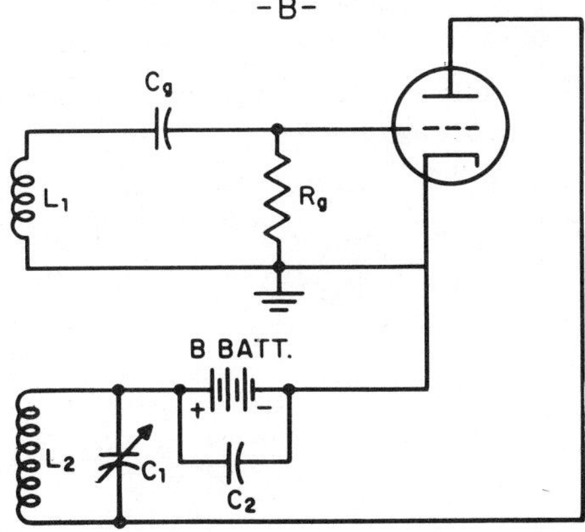

Fig. 17–5.

A. Tuned-grid oscillator.
B. Tuned-plate oscillator.

of plate current. This causes the plate to become more positive, and thus a positive charge of proper phase is fed back to the bottom of the coil. The amount of feedback is controlled by adjusting the tap on the tank coil.

2. THE COLPITTS OSCILLATOR

The *Colpitts oscillator,* shown in Figure 17-7, is very similar to the Hartley oscillator. Instead of using a tapped tank coil, the tank capacitor is tapped. This is accomplished by connecting two variable

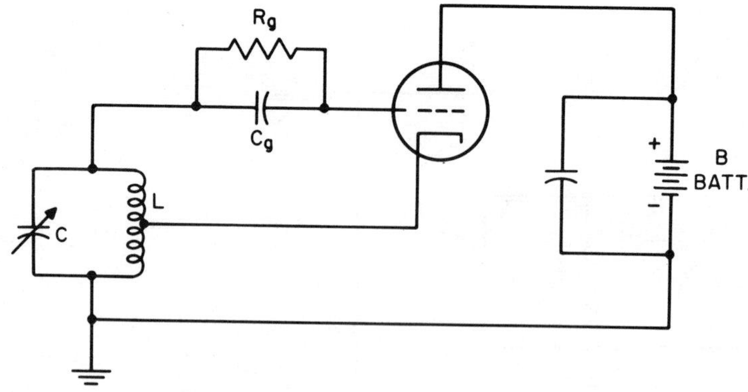

Fig. 17–6. The Hartley oscillator.

capacitors, C_1 and C_2, in series and using the junction between them as a tap. Feedback is controlled by varying the ratio between the two capacitances. One capacitor is increased while the other is decreased by the same amount, and this keeps the overall capacitance, and hence the frequency, constant. Feedback is of the proper phase, as explained in the section on the Hartley oscillator.

Frequency changes often are obtained by varying the value of L. The reason is that it is quite difficult to vary the overall capacitance of C_1 and C_2 in series without disturbing the ratio between them that is essential for proper feedback.

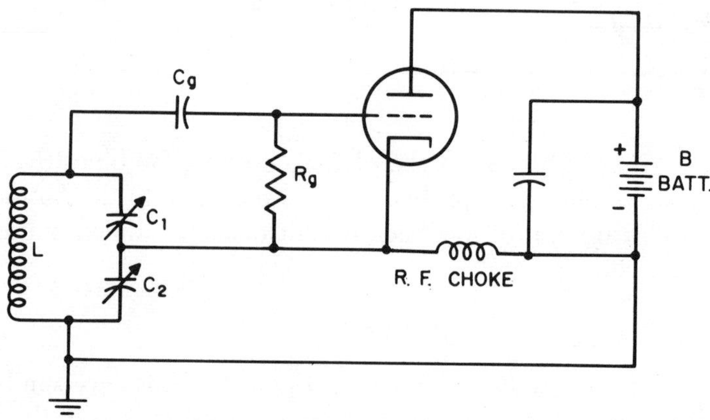

Fig. 17–7. The Colpitts oscillator.

3. THE TUNED GRID–TUNED PLATE OSCILLATOR

Feedback from the plate to the grid may take place within the tube itself by virtue of the interelectrode capacitances between the elements. Although these interelectrode capacitances exist within tubes in all kinds of oscillators, in certain types, they become the chief source of feedback.

One type of oscillator employing interelectrode capacitance for feedback coupling is the *tuned grid–tuned plate oscillator* shown in Figure 17-8. Here, one of the two resonant circuits is in the grid circuit and the other is in the plate circuit. There is no inductive coupling between coils L_1 and L_2. Feedback occurs through the grid-to-plate capacitance of the tube (shown as C_{GP}).

Both tank circuits are tuned to the same frequency and, in practice, the oscillator operates at a frequency that is slightly lower than the natural frequency of either tank circuit. If the two tank circuits differ somewhat in their natural frequencies, the tank circuit having the higher efficiency will determine the frequency of the oscillator. Feedback may be adjusted by varying the tuning of either the grid or the plate circuit.

4. THE CRYSTAL OSCILLATOR

An interesting variation of the tuned grid–tuned plate oscillator is the *crystal oscillator* illustrated in Figure 17-9. Certain crystalline substances, such as Rochelle salt and quartz, possess the property of converting electric energy to mechanical energy, and vice versa. If a mechanical stress is applied to the crystal, an electrostatic field

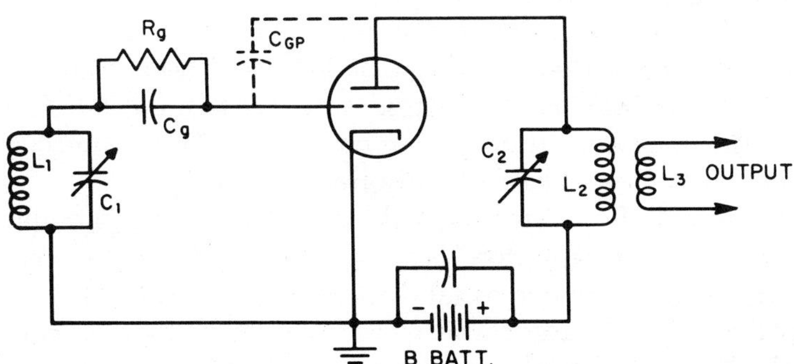

Fig. 17–8. The tuned grid–tuned plate oscillator.

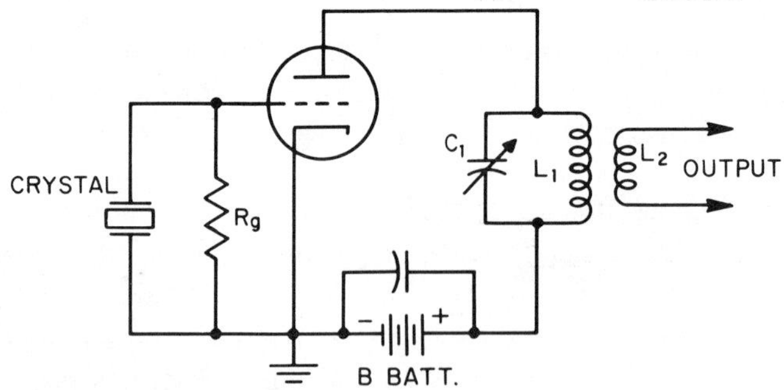

Fig. 17–9. The crystal oscillator.

appears between its faces. Conversely, when a voltage is applied to electrodes on two parallel faces of the crystal, a mechanical strain occurs in the crystal. This is known as the *piezoelectric effect.* The symbol for the crystal and its electrodes is ⏚ .

Electric energy applied to the two parallel faces of the crystal produces a mechanical strain in the crystal. This strain, in turn, produces an electrostatic field, which, in turn, again produces a strain. This process repeats itself. At the natural period of the mechanical vibrations of the crystal, the two actions may be made mutually self-sustaining by bringing in sufficient electric energy to replace that which is lost as heat during each cycle. The effect of the crystal, then, is to produce an oscillating voltage whose frequency is determined by the natural frequency of the crystal. This frequency, in turn, is determined by the mechanical structure of the crystal. Quartz crystals generally are employed because of their great mechanical strength and such crystals can be cut with a natural frequency that may be millions of cycles per second.

Note the similarity to the *L-C* oscillating circuit where an electrostatic field is converted to a magnetic field, and back again. Consequently, the crystal and its electrodes can replace the *L-C* tank circuit in the oscillator. It acts as an extremely efficient tuned circuit and produces a remarkably steady frequency output.

In Figure 17-9, resistor R_g is the grid leak. Note the apparent absence of a grid capacitor. But you will recall that the crystal is sandwiched between two metal electrodes. The whole, then, acts as a capacitor whose charge furnishes the bias for the tube. Frequently, a cathode resistor is also employed to furnish a protective bias, should the crystal cease oscillating.

The circuit shown here is only one of many types of crystal-oscillator circuits. Because the output frequency of such an oscillator is extremely stable, it is used to prevent frequency variations in transmitters and in test and laboratory equipment where constant frequency is desired.

5. THE ELECTRON-COUPLED OSCILLATOR

The *electron-coupled oscillator,* whose circuit is illustrated in Figure 17-10, is a type that is frequently employed. Note that, if we consider the screen grid as the anode of a triode consisting of the cathode, control grid, and screen grid, we have, in effect, a Hartley oscillator with the B+ applied in shunt feed. Coil L_1 is not inductively coupled to L_2 or L_3.

As a result of the oscillations in the grid tank circuit, the control grid is made alternately positive and negative. During the positive half-cycles, electrons flow to the screen grid (acting as an anode). During the negative half-cycles, the control grid becomes negative and the electron flow is cut off. The result, then, is a series of pulses, one pulse for each oscillation of the grid tank circuit.

Since the screen grid has open spaces, these pulses continue through to the real plate of the tube which is positive. Thus a series of pulses, in step with the oscillations of the grid tank circuit, is applied to the plate tank circuit consisting of L_2 and C_2. Since this tank circuit is tuned to the same frequency as the grid tank circuit, an amplified version of the oscillating current in the grid tank circuit is

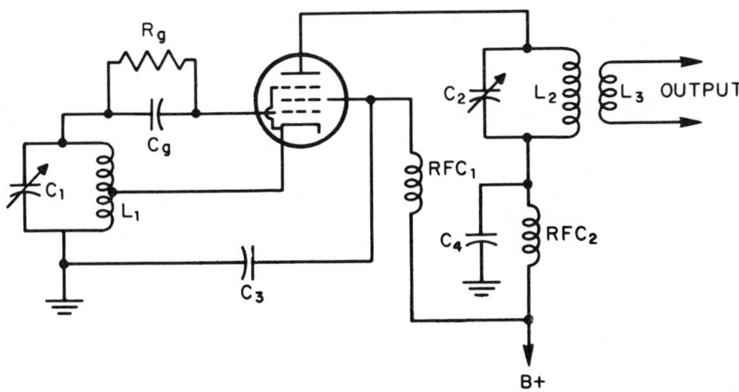

Fig. 17—10. The electron-coupled oscillator.

set flowing. Coil L_3 is the output coil. R-f choke coils RFC_1 and RFC_2 are used to keep the r-f current out of the power supply.

B. *The R-C oscillator*

Assume a battery, resistor (R), and a capacitor (C) are connected in series, as illustrated in Figure 17-11. As the capacitor is charged, an electrostatic field is built up across it. When the charging potential is removed, the electrostatic field will force a current flow (provided there is a closed circuit) in a direction opposite to the charging force. Thus, an alternating current is set up.

The time required for the capacitor to charge up (and discharge) depends upon the values of R and C. (See Chapter 7, subdivision B.) Hence these values determine the frequency of the oscillations. This is the principle upon which *resistor-capacitor* (*R-C*), or *relaxation, oscillators* operate.

The frequencies of *R-C* oscillators are relatively low, from a fraction of a cycle up to several hundred kilocycles per second. *L-C* oscillators, on the other hand, are capable of producing currents and voltages with frequencies in the millions of cycles per second.

The waveform of the output from the *L-C* oscillator is sinusoidal. Relaxation oscillators produce outputs whose waveforms are more varied, such as sawtooth, square-wave, etc., depending upon the circuits involved. There are a number of types of such *R-C* oscillators. Some of the most frequently employed types will be described here.

1. SWEEP-CIRCUIT OSCILLATORS

You will recall that in our discussion of the cathode-ray tube we stated that we used a sawtooth voltage to sweep the spot of light horizontally across the face of the tube (see Chapter 12, subdivision E). The generator of this type of voltage is called a *sawtooth*, or *sweep-circuit, oscillator.*

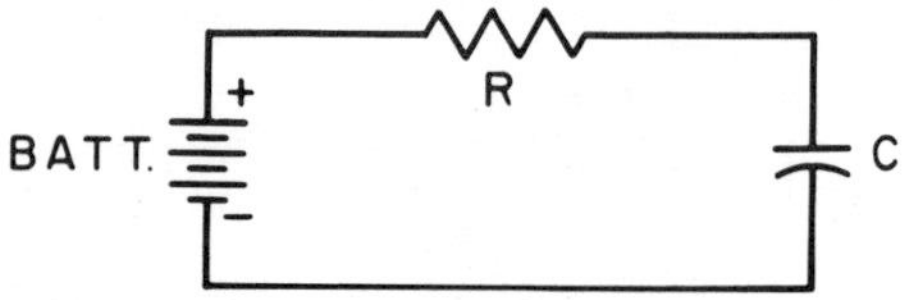

Fig. 17—11.

Basic circuit of the *R-C* oscillator.

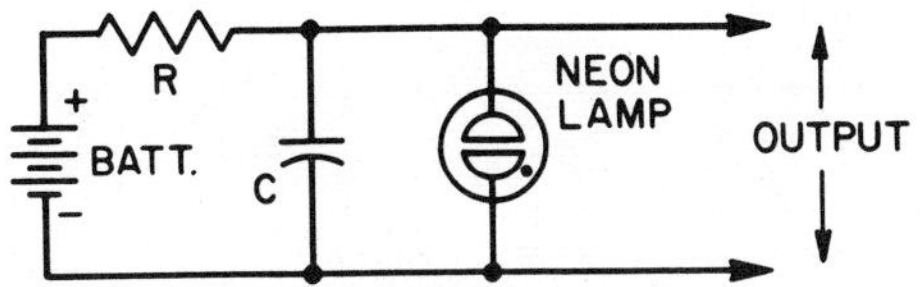

Fig. 17–12. The neon-lamp sweep-circuit oscillator.

The circuit for such an oscillator is shown in Figure 17-12. The battery charges the capacitor (C) through a resistor (R). The rate of charge depends upon the values of capacitance and resistance. The *neon lamp* contains two plates sealed into a glass envelope. The air has been evacuated from the lamp, and in its stead, a small amount of neon gas has been introduced. The lamp is connected across the capacitor.

When the voltage across the capacitor becomes high enough (usually about 50 or 60 volts), the neon gas ionizes and becomes a short circuit which discharges the capacitor almost immediately. When this happens, the voltage across the capacitor (and hence across the neon lamp) drops until the gas ceases to be ionized (the gas becomes *de-ionized,* or *extinguished*). The short circuit across the capacitor is removed and it charges up once again. Then the cycle is repeated.

The waveform of the output voltage (taken from across the neon lamp) is shown in Figure 17-13. The battery potential should be much greater than the ionizing potential of the gas. The frequency of the sawtooth voltage is the number of times the voltage rises and falls per second. This frequency is determined by the values of the resistor and capacitor and may be varied by changing these values.

Another generator of the sawtooth voltage is the *thyratron sweep oscillator* illustrated in Figure 17-14. The thyratron tube differs from the neon lamp in that, whereas the ionizing potential of the neon lamp is a fixed value, the grid of the thyratron controls the electron

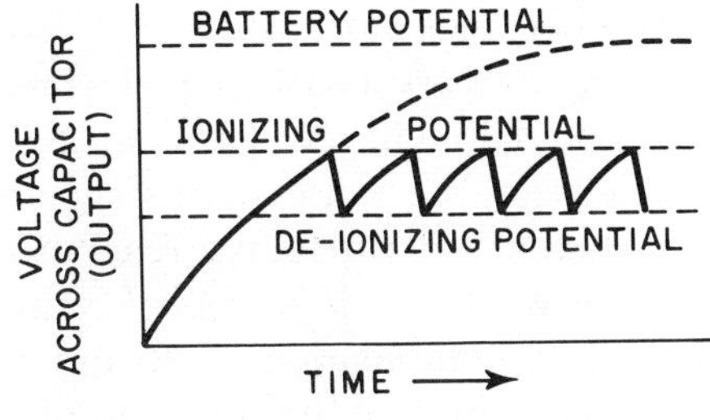

Fig. 17–13. Waveform of the output voltage of the neon-lamp oscillator.

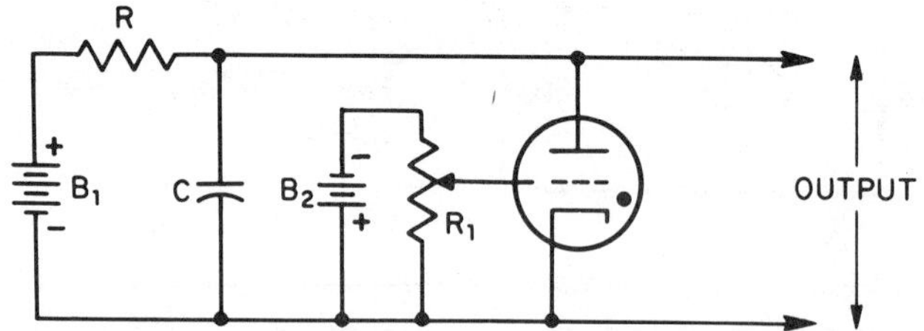

Fig. 17—14. The thyratron sweep-circuit oscillator.

flow through the tube and thus determines its ionizing potential (see Chapter 10, subdivision B). The more negative the grid, the higher is the ionizing potential; the less negative the grid, the smaller the ionizing potential.

This oscillator operates on the same principle as the neon-lamp oscillator except that the frequency may be varied by changing the grid bias on the thyratron as well as by changing the values of R or C. Battery B_1 charges the capacitor and battery B_2 supplies the negative grid bias, as determined by the setting of potentiometer R_1.

A small negative bias means that the tube will fire at a lower potential than if the negative bias were larger. The voltage across the capacitor becomes sufficient to fire the tube sooner and, therefore, less time need elapse to reach the ionizing potential. Thus more rises and falls of voltage across the capacitor will occur in a given length of time; hence the frequency is increased.

2. THE PHASE-SHIFT OSCILLATOR

Another *R-C* oscillator is the *phase-shift oscillator* whose circuit is illustrated in Figure 17-15. As heater current is applied, the cathode starts emitting electrons and the resulting plate current causes a voltage drop across cathode resistor R_k. As a result, the grid becomes negative.

As the grid goes negative, the plate becomes more positive. Feedback from plate to grid is through a network consisting of three capacitors (C) and three resistors (R). You will recall that an *R-C* circuit causes a phase shift up to 90 degrees, the phase angle depending upon the relative values of resistance and capacitive reactance (see Chapter 7, subdivision B). Here we have three such circuits and the values are chosen so that each combination of R and C produces a phase shift of 60 degrees. The effect of the three circuits, then, is to produce a total phase shift of 180 degrees.

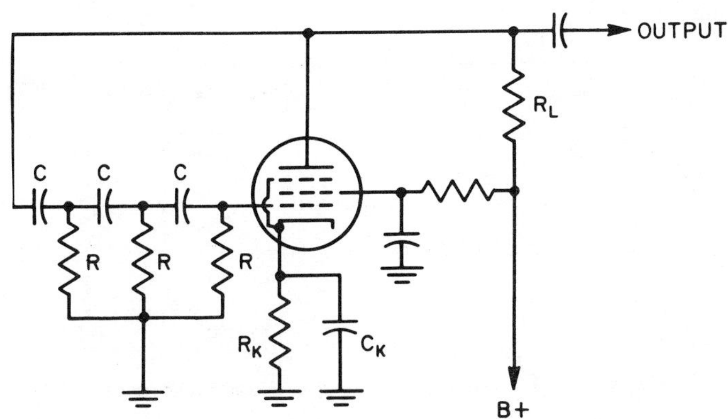

Fig. 17–15. The phase-shift oscillator.

As a result, the positive pulse from the plate undergoes a 180-degree phase shift and appears as a negative pulse on the grid. Thus the grid becomes still more negative. This process continues until the grid becomes negative enough to cut off the tube.

When enough electrons leak off the grid to start the tube operating again, the plate starts going negative (less positive), making the grid positive. This continues until the grid reaches its maximum positive value, which coincides with the saturation point of the tube, and there is no further amplification.

As further amplification ceases, the plate swings toward the positive, making the grid swing toward the negative. As a result, the amplification drops off, making the plate still more positive and the grid more negative. This cumulative process continues till the grid becomes negative enough to cut off the tube. Then the entire process starts again. The frequency at which these changes take place depends upon the values of the three capacitors (C) and three resistors (R).

The waveform of the output voltage generally is quite complicated, but if the amplitude is kept quite low, a fairly good sine wave is produced. Frequency change is accomplished by varying all three capacitors or all three resistors simultaneously. Because of the difficulties presented, this type of oscillator generally is employed to produce a single fixed frequency.

3. THE MULTIVIBRATOR

An *R-C* oscillator that is widely employed is the *plate-coupled multivibrator* whose basic circuit is illustrated in Figure 17-16.

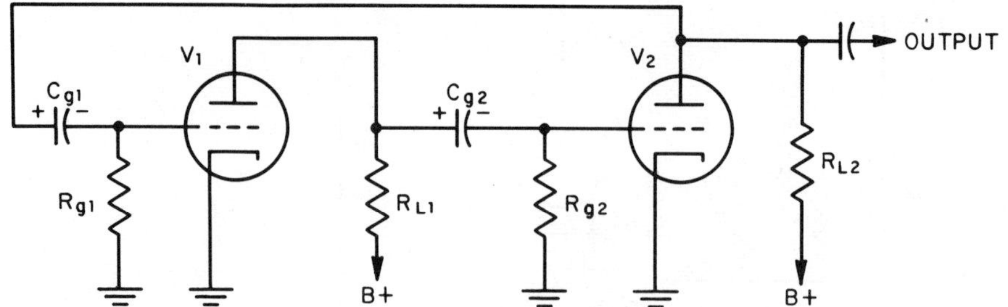

Fig. 17—16. The plate-coupled multivibrator.

Essentially, it consists of two similar resistance-coupled amplifiers, coupled to each other. Thus V_1 and V_2 are two similar triodes (which may be contained in one envelope), R_{g1} is equal to R_{g2}, R_{L1} is equal to R_{L2}, and C_{g1} is equal to C_{g2}.

As plate voltage is applied, both V_1 and V_2 start to conduct and the charges on C_{g1} and C_{g2} build up with the indicated polarities. Since it is virtually impossible to produce two amplifiers that are identical, one of the tubes will conduct somewhat more heavily at the start than the other. Assume that V_1 conducts more heavily.

The increase of plate current of V_1 causes a large voltage drop across R_{L1} and, as a result, the plate voltage of V_1 is reduced. As the plate voltage is reduced, C_{g2} starts to discharge, making the grid of V_2 more negative. Hence, an increase in plate current of V_1 causes a reduction in plate current of V_2.

As the plate current of V_2 is reduced, so is the voltage drop across R_{L2}, causing a voltage rise at the plate. This voltage rise, in turn, swings the grid of V_1 positive, thus further increasing its plate current. The action is cumulative, resulting in V_1 conducting at its maximum while V_2 is cut off. (Normally, the grid of V_1 would go very highly positive, but heavy grid current flows, charging C_{g1} which, in turn, keeps down the grid voltage.)

Tube V_2 is held below cutoff by the negative charge on C_{g2}. As this charge leaks off to ground through R_{g2}, the voltage on the grid of V_2 gradually becomes less negative. When it reaches the cutoff point, V_2 starts to conduct, causing a drop in its plate voltage which, in turn, causes a drop in voltage at the grid of V_1. Now the entire process is reversed and ends with V_2 conducting at its maximum and V_1 cut off.

The negative charge on C_{g1} holds V_1 below cutoff until this charge can leak off through R_{g1}. Then the entire cycle starts again. You can see that C_{g1}, C_{g2}, R_{g1}, and R_{g2} determine the frequency of the multivibrator.

We may further visualize the action of the multivibrator by studying the waveforms of the voltages at both grids and plates. (See Figure 17-17.) At the start of the cycle (point A) the voltage at the grid of V_1 is at its maximum positive value and the grid of V_2 has dropped well below the cutoff point. The heavy flow of grid current in V_1 quickly reduces the grid voltage to zero where it remains while the negative charge on the grid of V_2 is leaking off.

Meanwhile, the plate voltage of V_1 starts at a low positive value and rises somewhat as the grid voltage drops to zero. Then, as the grid voltage remains constant, so does the plate voltage. In tube V_2, the plate voltage rises to the full potential of the power supply as the tube is cut off and is maintained at this level as long as there is no flow of plate current.

At point B the charge has leaked off sufficiently so that the voltage at the grid of V_2 passes the cutoff point. Now V_2 is conducting and the grid voltage of V_1 drops below its cutoff point, cutting off the tube. As V_1 is cut off, its plate voltage rises to the full potential of

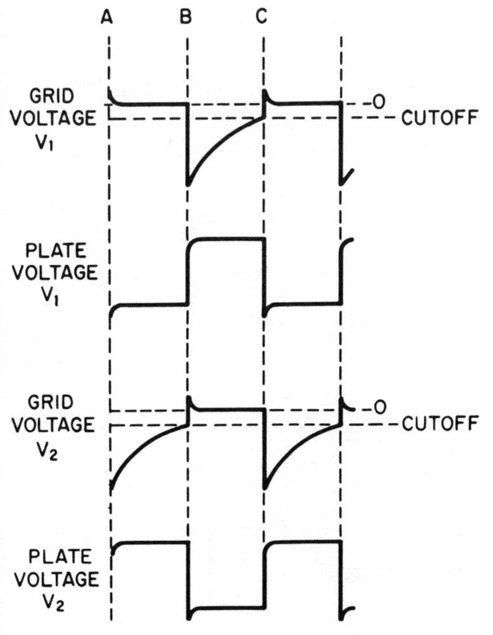

Fig. 17—17.

Waveforms of the voltages at the grids and plates of the multivibrator.

the power supply and is maintained at that level until the negative charge on the grid of V_1 has leaked off sufficiently so that it can start conducting again (at point C) and so repeat the entire cycle. In V_2 the plate voltage drops and is maintained at a low level as long as the tube is conductive.

Note the waveforms of the voltages at the plates of the tubes. Such a waveform is known as a *rectangular,* or *square, wave.*

A feature of the multivibrator is that its frequency may be locked in step (*synchronized*) with a series of pulses applied to either grid. Take V_1, for example. The waveform of its grid voltage under normal (*free-running*) conditions is shown in Figure 17-18A. In Figure 17-18B we see two positive pulses that are applied to its grid. The effect of adding these pulses is shown in Figure 17-18C.

Pulse #1 raises the grid voltage during the period while it still is below cutoff. Since the grid still is below cutoff, nothing happens and the multivibrator continues as usual. Pulse #2, however, is applied just before the grid voltage reaches the cutoff level and brings the grid above cutoff. The tube becomes conductive and so initiates the cycle. Thus the frequency of the multivibrator can be synchronized with a series of positive pulses applied just before the grid reaches its cutoff level. The frequency of these triggering pulses then, must be close to the free-running frequency of the multivibrator. The synchronizing, or *sync,* pulses may be applied to the grid of V_2 as well.

Since a pulse which occurs well before the point at which the tube becomes conductive has no effect on its frequency, sync pulses having a frequency about twice that of the multivibrator may be

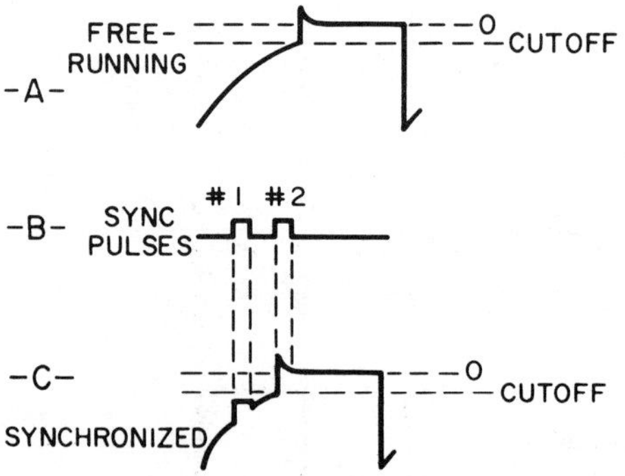

Fig. 17—18.

Graph showing how the multivibrator is synchronized by incoming sync pulses.

employed. Thus every other pulse would trigger the multivibrator and the alternate pulses would have no effect. Similarly, the multivibrator may be triggered by sync pulses which have a frequency that is any whole-number multiple of the free-running frequency of the oscillator.

We thus have a *frequency divider* whereby signals of a higher frequency (the sync pulses) are used to produce signals of a lower frequency (the output of the multivibrator).

A popular variation of this multivibrator is the *cathode-coupled multivibrator* whose circuit is illustrated in Figure 17-19. As plate voltage is applied, both V_1 and V_2 start to conduct. As V_1 conducts, the flow of plate current causes its plate voltage to drop. This drop drives the grid of V_2 negative, decreasing its plate current. The drop in plate current reduces the voltage drop across R_k, which furnishes the grid bias for both tubes.

As the bias on V_1 is thus reduced, its plate current rises, causing a further drop in its plate voltage. This, in turn, makes the grid of V_2 still more negative. The process continues until V_2 is cut off and V_1 is conducting heavily.

Tube V_2 is held cut off until C has discharged sufficiently through R_{g2} so that the grid of V_2 is brought up to its cutoff point and V_2 starts to conduct. The resulting flow of plate current causes a

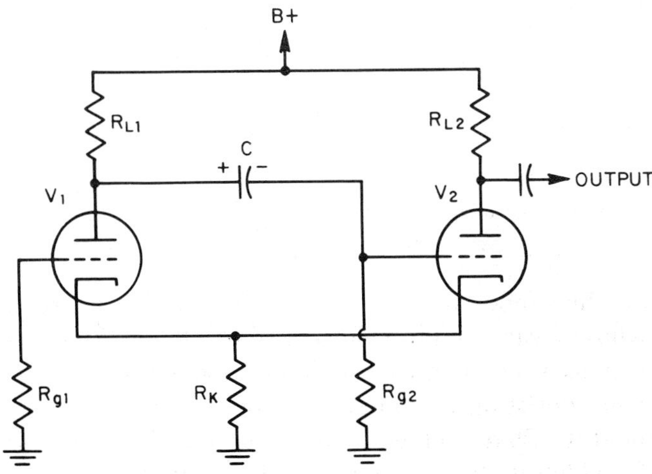

Fig. 17–19. The cathode-coupled multivibrator.

voltage drop across R_k, which thus increases the negative bias on V_1. Its plate current drops, producing a rise in its plate voltage.

The rise in plate voltage of V_1 drives the grid of V_2 more positive as C charges from the B+ supply. The plate current of V_2 increases, more bias is developed across R_k, less plate current flows in V_1, and a still greater positive voltage is placed on the grid of V_2. (As in the case of the plate-coupled multivibrator, the grid of V_2 normally would go very highly positive, but heavy grid current flows, charging capacitor C, which in turn keeps the grid at a low positive value.) This process continues until V_1 is cut off and V_2 is conducting heavily.

Tube V_1 is held at cutoff for the time it takes C to charge. As C becomes charged, the grid of V_2 becomes less positive, resulting in a drop in plate current. This drop of plate current produces a drop in the negative bias developed across R_k that is applied to the grid of V_1. When C has charged up sufficiently, the negative bias on the grid of V_1 has been reduced to the point where V_1 again starts to conduct, and the entire cycle is repeated. (Note that this time when the grid of V_2 goes negative, the charge on C drives it down far below cutoff.)

The waveforms resemble those for the plate-coupled multivibrator. The period during which V_1 is cut off depends upon the time required for C to charge. The period during which V_2 is cut off depends upon the time required for C to discharge. The plate voltage rises to the B+ voltage as the tube is cut off, remains at that voltage during cutoff, and drops to a low value as the tube becomes conductive. It remains at this value during the period of conduction and rises again to repeat the entire cycle. Thus the waveform resembles a series of square pulses.

The cathode-coupled multivibrator, too, may be triggered by sync pulses, just as the plate-coupled type.

4. THE BLOCKING OSCILLATOR

Still another frequently employed *R-C* oscillator is the *blocking oscillator* whose basic circuit is illustrated in Figure 17-20. This you will recognize as a regenerative oscillator. Oscillations are set up in a tank circuit consisting of coil L_1 and the distributed capacitance in and around it. Plate voltage is fed back into the tank circuit by means of L_2 which is closely coupled to L_1. Output is taken from a tertiary winding, L_3.

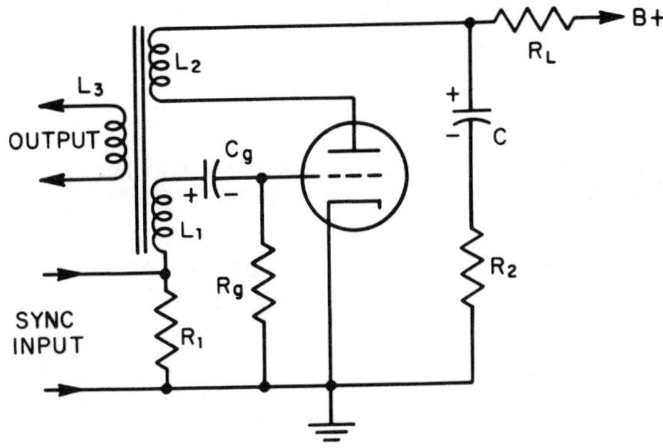

Fig. 17–20. The blocking oscillator.

As the oscillation goes into its positive half-cycle, the positive charge it places on the grid of the tube causes a grid current to flow and, as a result, capacitor C_g charges heavily with the indicated polarity. During the negative half-cycle of the oscillation, the negative charge on C_g adds to the negative voltage of the oscillation, dropping the grid far below the cutoff point.

The losses in the transformer formed by L_1 and L_2 tend to dampen out the oscillations. Accordingly, the next positive half-cycle produces a voltage of lower amplitude. Since this voltage is not sufficient to bring the grid voltage above the cutoff point in the face of the large negative charge on C_g, the tube remains cut off. Hence there is no feedback and the oscillations in the tank circuit cease.

With the cessation of oscillations, the negative charge on C_g starts leaking off through resistor R_g. When enough of the charge has leaked off to raise the grid above the cutoff point, oscillation starts again, and the entire cycle is repeated. The time required for the charge on C_g to leak off sufficiently so that oscillations start up again is determined by the values of C_g and R_g. You see that once every cycle, as the grid goes positive, there is produced a large pulse of plate current.

While the tube is cut off, capacitor C charges, with the indicated polarity, from the B+ supply through resistor R_L. When the tube becomes conductive, C discharges through the tube. (C never discharges fully, since the tube is conductive for a relatively short period of time.) As a result, the waveform of the voltage across C resembles a sawtooth.

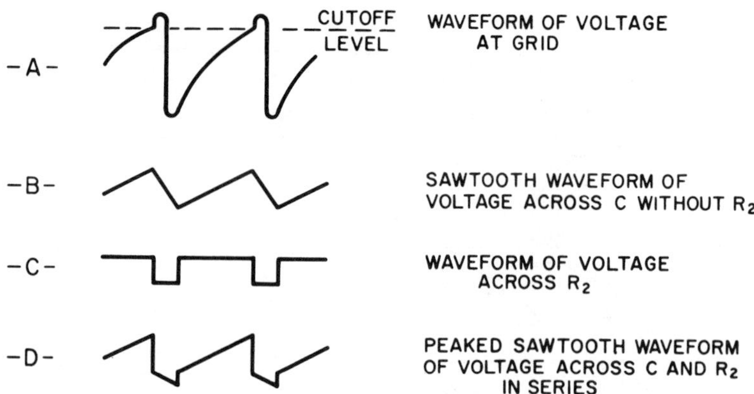

Fig. 17–21. Waveforms of voltages at various points of the blocking oscillator.

So far, we have been discussing what happens to C as though R_2 (called a *peaking resistor*) were not there. While C is charging, the voltage drop across R_2 remains constant. But during the interval when the tube is conductive and C discharges, the voltage across R_2 drops. When the tube is cut off, this voltage rises again.

In Figure 17-21 you see the waveforms of the various voltages we have been discussing. Note that the combination of the voltages across C and R_2 produce a *peaked sawtooth* waveform.

The blocking oscillator, as the multivibrator, can be brought into synchronization by means of a series of positive pulses applied to the grid of the tube. These pulses, at a frequency close to the free-running frequency of the blocking oscillator, must be applied across R_1 just before the tube becomes conductive. And, as the multivibrator, the blocking oscillator may act as a frequency divider where the frequency of the sync pulses is a whole-number multiple of the frequency of the oscillator.

QUESTIONS

1. a) **Explain, in terms of the movement of electrons, the action in an *L-C* oscillatory circuit.**
   b) **Explain the *flywheel effect.***
2. **Draw and explain the basic circuit of the *regenerative oscillator.***

3. **Draw and explain the basic circuit of the *Hartley oscillator.***
4. **Draw and explain the basic circuit of the *Colpitts oscillator.***
5. **Draw and explain the basic circuit of the *tuned grid–tuned plate oscillator.***
6. **Draw and explain the basic circuit of the *crystal oscillator.***
7. **Draw and explain the basic circuit of the *electron-coupled oscillator.***
8. **Explain the basic principle upon which *R-C* oscillators operate.**
9. **Draw and explain the basic circuit of the *neon-lamp sweep-circuit oscillator.***
10. **Draw and explain the basic circuit of the *thyratron sweep-circuit oscillator.***
11. **Draw and explain the basic circuit of the *phase-shift oscillator.***
12. **Draw and explain the basic circuit of the *plate-coupled multivibrator.***
13. **Draw and explain the basic circuit of the *cathode-coupled multivibrator.***
14. **Explain how the multivibrator may be *synchronized* by a series of pulses applied to the grid of either tube.**
15. **Draw and explain the basic circuit of the *blocking oscillator.***

Section 4

The Electron Tube in Industry

The uses of electronics in industry are so widespread that it is impossible to discuss them all in one book, much less in a few chapters. Also, the full understanding of many of these applications requires knowledge of more advanced electronics.

Accordingly, only a few representative examples of industrial electronic devices and applications will be discussed in this section. And these discussions will be simplified so as to present the prin ciples involved, rather than a rundown of all the technical details.

In the main, these electronic devices perform one or more of four functions. These are: *switching,* the turning on or off of some device such as a bell, lamp, or motor; *measuring,* generally some physical quantity such as temperature, force, or weight; *timing,* initiating or terminating some process after a predetermined lapse of time; and *controlling* as, for example, making a motor run faster or slower, or varying the current in a welding machine to form a hotter, or cooler weld.

Basically, most of the applications operate along the following lines. Physical changes, such as changes in quantity, position, or condition are translated into electrical quantities that are proportional to the changes. Generally, these electrical quantities are so small that they must be amplified before they are able to actuate some device such as a meter, motor, or electromagnet.

The devices that translate the physical changes into electrical quantities are called *sensing devices,* or *transducers.* Typical examples are the *phototube,* which translates light energy into electrical energy, and the *thermocouple,* which translates heat energy into electrical energy.

Electron tubes play a large part in industrial electronics, mainly in their roles as rectifiers, amplifiers, and oscillators. Frequently, these tubes acquire new names when used in industry. Thus a high-vacuum, hot-cathode diode is known as a *kenotron.* If this diode is gas-filled, it may be called a *phanotron.* A high-vacuum, hot-cathode tube with one or more grids is called a *pliotron.*

18 Electronic Relays

We are all familiar with the way a switch is used to turn a light or motor on or off. When the contact points touch (switch *closed*) the circuit is completed and current from the power lines flows to the device being controlled. When the contact points are separated (switch *open*), the current is interrupted and the controlled device stops.

The same thing can be done with an electromagnetic switch called a *relay*. One advantage of the electromagnetic relay over the ordinary switch is that the former can be operated by an electric current.

It is the current which operates the relay that is of interest to us. It may come from an illuminated phototube or a heated thermocouple, for example. These are the sensing devices or transducers which, under the influence of light, heat, etc., generate an electrical

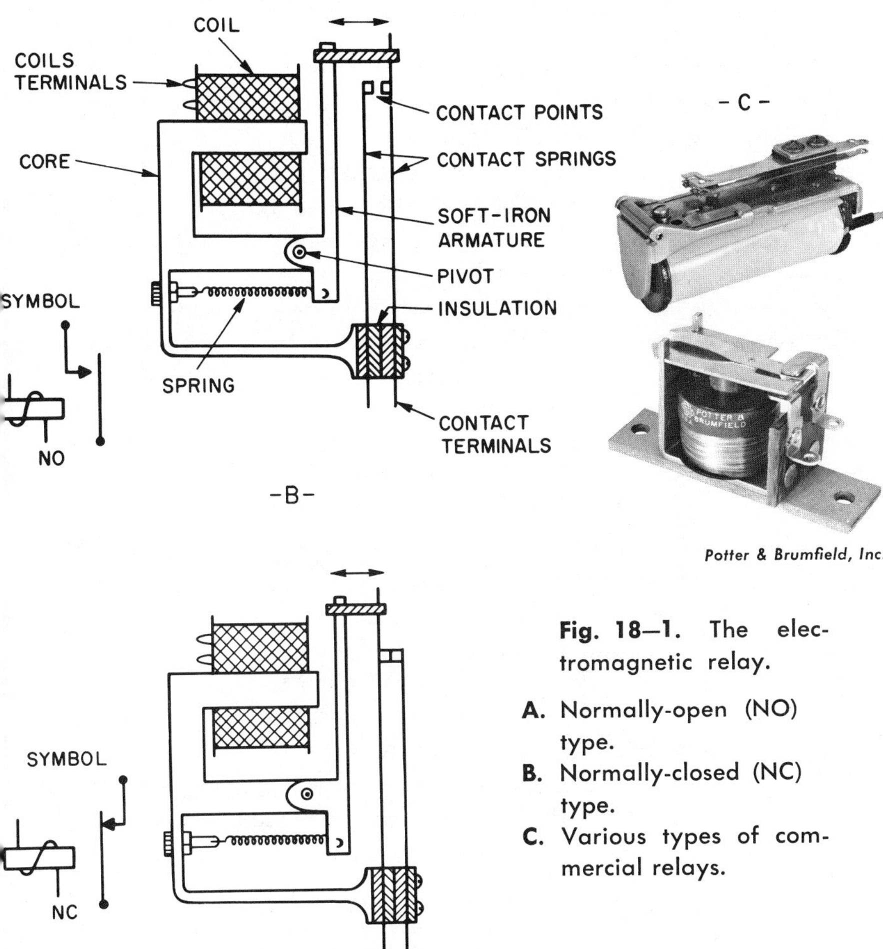

Potter & Brumfield, Inc.

Fig. 18—1. The electromagnetic relay.

A. Normally-open (NO) type.
B. Normally-closed (NC) type.
C. Various types of commercial relays.

signal which energizes the relay. Since these signals generally are quite feeble, an amplifier usually is employed between the sensing device and the relay to bring the signal up to the proper level to operate the relay.

The combination of sensing device, amplifier, and electromagnetic relay is called an *electronic relay*. Note that it acts as a switch that is either open or closed.

A. The electromagnetic relay

The *electromagnetic relay* is one of the most widely used devices of industrial electronics. Basically, it is a switch operated by an electric current. (See Figure 18-1A.) It consists of an electromagnet (coil and core) which, when energized, attracts a soft-iron armature, thus closing a pair of contact points. When the electromagnet is de-energized, the armature is released and a spring pulls it back. This causes the contact points to separate.

The coil may be wound with many turns so that only a few milliamperes of current need flow through it to attract the armature. On the other hand, it may be so constructed as to require an energizing current of one or more amperes.

As illustrated in Figure 18-1A, the contact points are shown as *normally open* (NO), that is, with the coil unenergized, the contact points are separated. When the coil is energized, the contact points touch. In Figure 18-1B, the contact points are shown as *normally closed* (NC), that is with the coil unenergized, the contact points touch. When the coil is energized, the contact points separate.

The basic relay circuit is illustrated in Figure 18-2. Closing the switch in the *control circuit* causes current from the battery to flow through the coil of the relay which thus becomes energized. The soft-iron armature is attracted, closing the relay contacts. As these contacts (which are in the *controlled circuit*) close, current from the a-c power line flows through the lamp, causing it to become illuminated. If the switch now is opened the relay is de-energized,

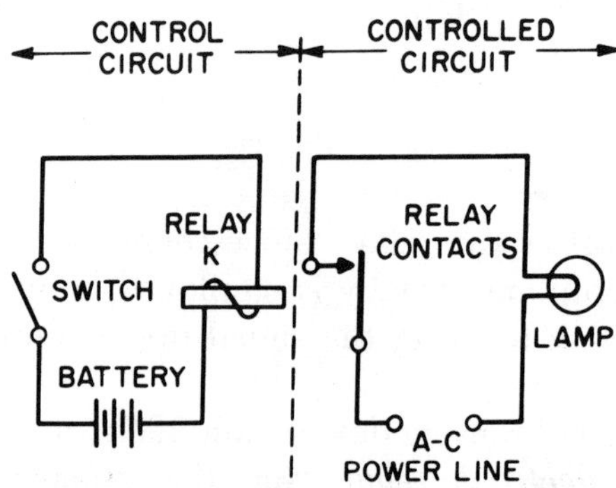

Fig. 18—2. Basic relay circuit.

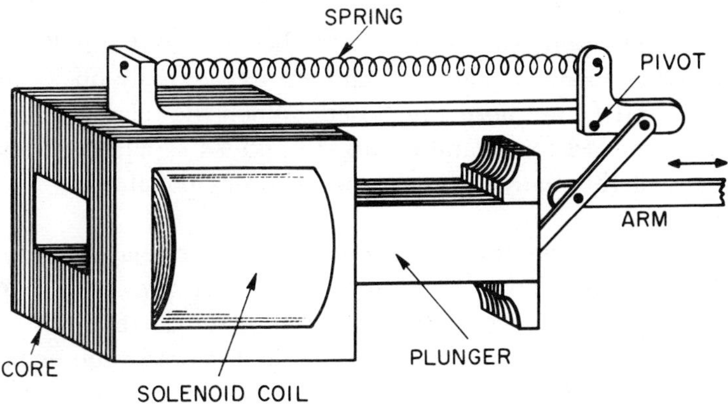

Fig. 18–3. The solenoid, or sucking coil.

the relay spring pulls the contact points apart, and the lamp goes out. The letter symbol for the relay is the letter K.

The lamp is the controlled device. It may as well be an electric motor, bell, or an electromagnetically-operated device such as a valve or counter. One common device is the *solenoid,* or *sucking coil,* illustrated in Figure 18-3. As current flows through the solenoid coil, which is wound on a hollow iron core, a magnetic field is created that pulls, or sucks, the soft-iron plunger into the center of the core. The movement of the plunger operates an arm that actuates a valve, counter, etc. When the flow of current through the solenoid coil ceases, the magnetic field disappears and the spring pulls the plunger out of the core, thus moving the arm back to its original position.

As you have learned, the energizing of the relay closes the contact points, thus completing the controlled circuit. By the same token, the energizing of the relay may cause a pair of normally-closed contact points to be separated, thus opening the controlled circuit. In this way the controlled device may be turned on or off.

The relay may operate as many contact points as is practical. For example, instead of the single-pole, single-throw arrangement illustrated in Figure 18-1A and B, we may have a single-pole, double-throw switch as shown in Figure 18-4A. Contact points #1 and #2 are normally open and contact points #2 and #3 are normally closed. When the relay is energized, contact points #2 and #3 separate and points #1 and #2 touch.

Another arrangement, illustrated in Figure 18-4B, shows two separate single-pole, single-throw switches operated simultaneously by the relay. Contact points #1 and #2 are normally open; contact points #3 and #4 are normally closed. When the relay is energized, points #1 and #2 touch and points #3 and #4 separate. These are but a few of the many different combinations possible.

Under certain conditions, it is desired that the relay remain in its energized position (that is, with the armature pressed up against the electromagnet) even after the energizing current has ceased flowing through the coil. For example, in a certain type of burglar alarm, the relay is energized and a bell is sounded as an intruder interrupts a beam of light that is focused upon a photocell. Obviously, the intruder cannot be depended upon to remain in the path of the light beam long enough for the alarm to be noticed. What is needed is an alarm that, once triggered, will continue to sound, even though the triggering pulse is removed.

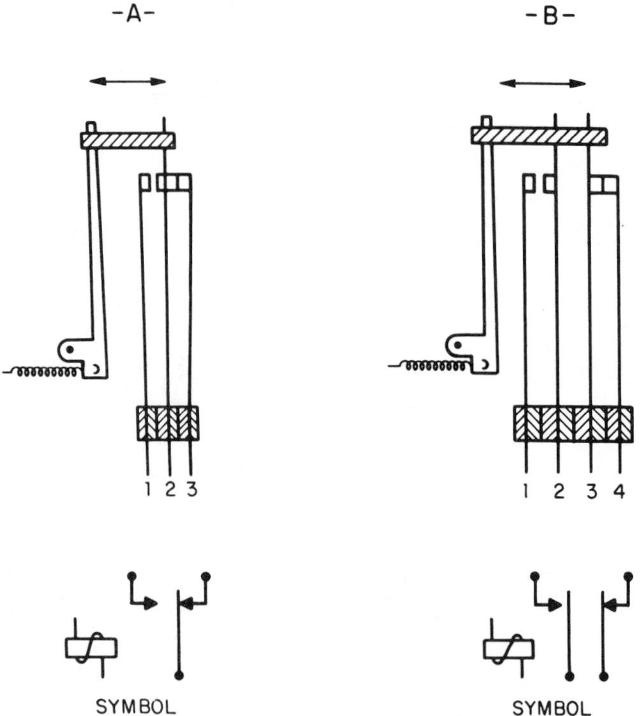

Fig. 18—4. Several contact-point arrangements for the relay.

The relay used for this purpose is a *latching* type. In Figure 18-5A you see a manual-reset latching relay. As the relay is energized, an arm, operated by a spring, slips into place, holding the contacts in position even after current stops flowing through the relay coil. To reset the relay, the button at the front must be pushed in, releasing the contacts.

An electrical-reset type is illustrated in Figure 18-5B. It consists, in effect, of two relays mounted end-to-end so that the latch arm of one engages that of the other. If relay #2 is energized, its armature, and the latch arm which is attached to it, is pulled to the right. This permits the latch arm of relay #1 to fall beneath the latch arm of relay #2, thus holding the #2 armature and contacts in position even after relay #2 is de-energized. To reset the relay, a current, which may come from some other source, is sent through the coil of relay #1. Its armature and latch arm are pulled to the left, releasing the latch arm of relay #2, thus releasing its armature and contacts.

The coil of the relay may be energized by either direct or alternating current. Obviously, the armature will be attracted by a south pole as well as by a north pole. However, if alternating current is used, the rapid alternations will cause the armature to vibrate or *chatter*. As a result, it will not make clean contacts or breaks.

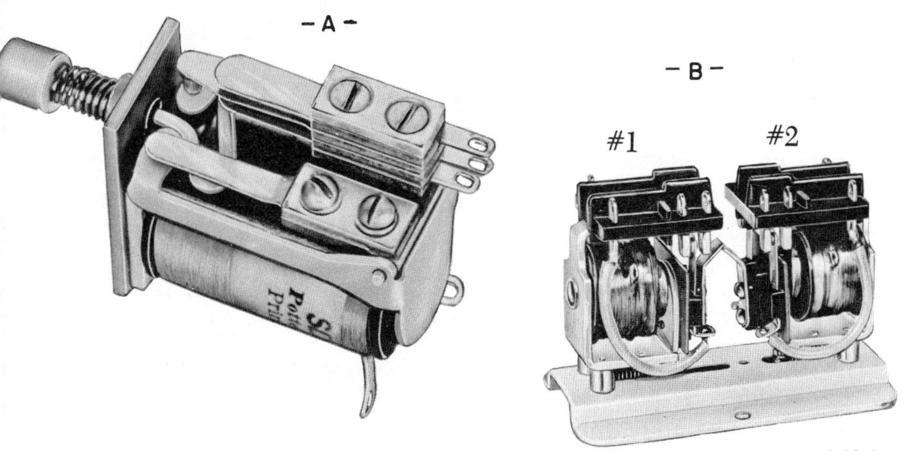

Potter & Brumfield, Inc.

Fig. 18–5. Latching relays.
A. Manual-reset type.
B. Electrical-reset type.

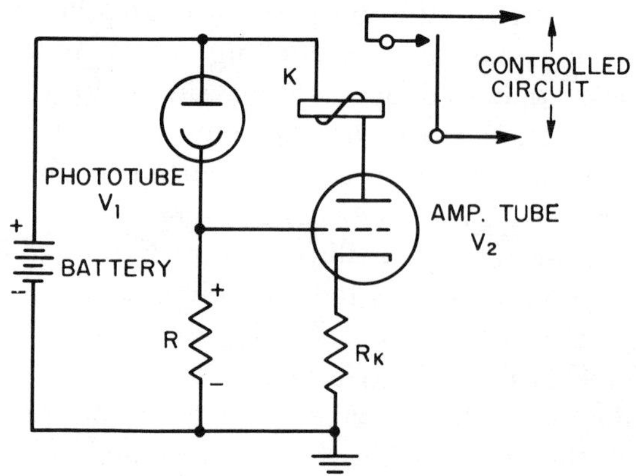

Fig. 18—6. Circuit of a forward-acting photoelectric relay operating from batteries.

There are a number of methods whereby the chatter can be reduced if the relay is supplied with alternating current. One obvious method is to rectify the alternating current and to apply the resulting direct current to the relay. Another method will be discussed a little later when we consider the photoelectric relay.

B. *Electronic relays employing a phototube as a sensing device*

The phototube (see Chapter 12, subdivision B) frequently is used as a sensing device. In the absence of light, this tube offers a very high resistance, approaching infinity. When illuminated, however, the cathode emits electrons which are attracted to a positive anode, the tube becomes conductive, and its resistance drops to some low value.

A typical circuit for the *photoelectric relay*, is illustrated in Figure 18-6. When no light strikes the phototube (V_1), its resistance is very high. The values of the battery and cathode resistor (R_k) of the amplifier tube (V_2) are such that not enough plate current flows through the amplifier tube to energize the relay.

As the phototube is illuminated, it becomes conductive and current flows from the battery, through resistor R and the phototube, and back to the battery. This current produces a voltage drop across R with the indicated polarity. As a result, the grid of V_2

becomes more positive (or, what is the same thing, less negative) with respect to its cathode, causing a greater flow of plate current.

As the illumination of the phototube is increased, the flow of current through it and R increases in like degree, and so does the voltage drop across R. When the illumination reaches a certain level, the voltage drop across R becomes sufficient to reduce the negative bias on the grid of V_2 to the point where enough plate current can flow to energize the relay.

Because the current flow through the phototube and R is only a few microamperes, the resistance of the latter must be quite large to produce a sufficient voltage drop across it. Values for R usually are in the order of megohms.

A photoelectric relay operating in this way is called *forward-acting*. It is used wherever a change from darkness to light is made to actuate some device. For example, metal toys are placed on a moving belt and made to pass in front of a spray gun that coats them with some dark material. Occasionally, the spray gun misses a toy.

The moving belt then carries the toys to a place where an intense light is beamed on them. Since the properly-sprayed toys are dark, they do not reflect any light. The missed toy, however, is shiny and reflects the light to a phototube that normally is not illuminated. The phototube now causes its relay to be energized which, in turn, actuates an electromagnetic device that pushes the defective toy from the belt.

Another application of the forward-acting photoelectric relay is in the automatic garage-door opener. As the headlights of an automobile illuminate a phototube, the relay is energized, turning on a motor that opens the door.

The photoelectric relay can be constructed so that the relay is energized as a light beam shining on the phototube is interrupted or obscured. In this condition the relay is called *reverse-acting*. (See Figure 18-7.)

Note that the phototube and R act as a voltage divider across the battery. When the phototube is illuminated it becomes conductive and its resistance drops to a low value. Because the resistance of R is very high, the voltage drop across it is much greater than that across the tube.

Accordingly, the grid of V_2 becomes slightly positive, with respect to the ground, but quite negative with respect to its cathode. With the amplifier tube in this condition its plate current is not large enough to energize the relay.

When the phototube is darkened, its resistance becomes very high. The battery voltage now is divided so that a greater portion appears across the tube. The positive charge applied to the grid of the amplifier tube is increased, the plate current rises, and the relay is energized.

An example of this type of photoelectric relay is a burglar alarm where a beam of light normally is focused upon the phototube. When an intruder passes through the beam the tube is darkened, energizing the relay and causing a bell to sound or a light to go on. The same set-up can be used as a fire alarm to indicate when smoke obscures the beam of light.

It can also be used to operate an electromagnetic counter as persons or objects pass through the beam, or to operate a motor or similar device that opens a door as an approaching person interrupts the beam. It is widely used as a safety device on industrial machines to stop the machine if the operator inadvertently places his hand in a dangerous area and interrupts the light beam in so doing.

The photoelectric relays illustrated in Figures 18-6 and 18-7 are d-c operated, obtaining their power from some d-c source such as batteries. Frequently, we wish to operate the relay from the a-c power mains. Of course, we may employ a power supply using a rectifier, similar to those described in Chapter 13. However, it is possible to apply the alternating voltage to the anodes of the phototube and amplifier tube. The tubes then can operate only on the alternate half-cycles when the plates are positive.

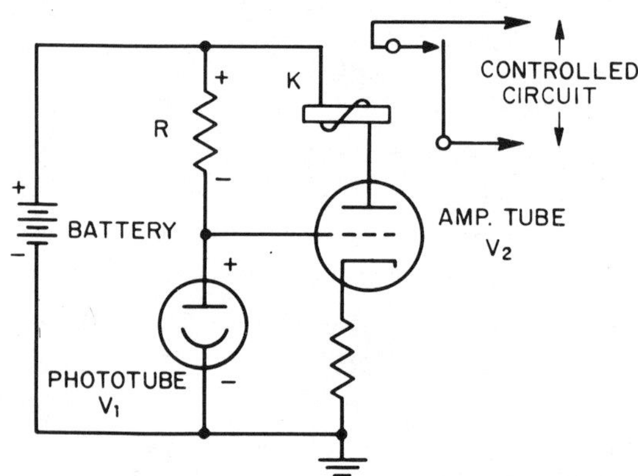

Fig. 18–7. Circuit of a reverse-acting photoelectric relay operating from batteries.

A typical forward-acting circuit for an a-c operated photoelectric relay is illustrated in Figure 18-8. Power is obtained by means of the tapped secondary of a transformer operating from the a-c line. The tap is so arranged that the voltage (E_g) between it and side A of the secondary winding is a small portion of the entire secondary voltage. Note that the phototube and R act as a voltage divider across the secondary voltage (less a small voltage drop across the relay coil). Voltage E_g is applied between the cathode and grid of the amplifier tube.

With the phototube dark, there is no emission and the resistance of the tube is very high, practically an open circuit. Under these conditions, let us start with a half-cycle when side A of the secondary winding is negative and side B positive. The negative voltage is applied to the grid of the amplifier tube, driving it below cutoff and so no plate current flows through the relay.

During the next half-cycle, side A of the winding becomes positive and side B negative. The plate of the amplifier tube becomes negative and, again, no plate current flows through the relay. Thus the relay remains unenergized as long as the phototube is dark.

Now let us see what happens when the phototube is illuminated. The tube becomes conductive and its resistance drops to a low

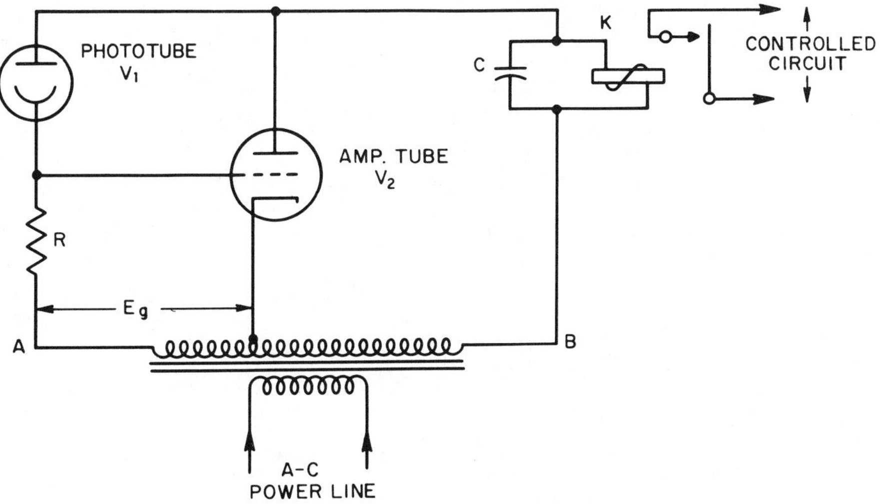

Fig. 18—8. Circuit of a forward-acting, a-c operated, photoelectric relay.

value. Start with a half-cycle when side A of the secondary winding is negative and side B positive. The plates of both the phototube and the amplifier tube are positive with respect to their cathodes. The grid of the amplifier tube is made negative with respect to its cathode by the voltage E_g. However, since the phototube is conductive, the flow of current through R produces a voltage drop of such magnitude and polarity as to overcome the negative voltage on the grid of the amplifier tube and thus raise the tube above cutoff. As a result, plate current flows and the relay is energized.

During the next half-cycle when side A is positive and side B negative, the plate of the amplifier tube becomes negative, thus cutting off the flow of plate current. However, capacitor C, which had charged up during the previous half-cycle, now starts to discharge through the relay coil, thus keeping the relay energized until the next half-cycle comes along. (This is one method by which a relay is kept from "chattering" when a-c operated.)

A typical reverse-acting circuit for an a-c operated photoelectric relay is illustrated in Figure 18-9. With the phototube illuminated, the tube is conductive and its resistance is quite low. Let us start with a half-cycle when side A of the secondary winding of the transformer is negative and side B positive. Since the resistance of the phototube is low, the grid of the amplifier tube becomes quite

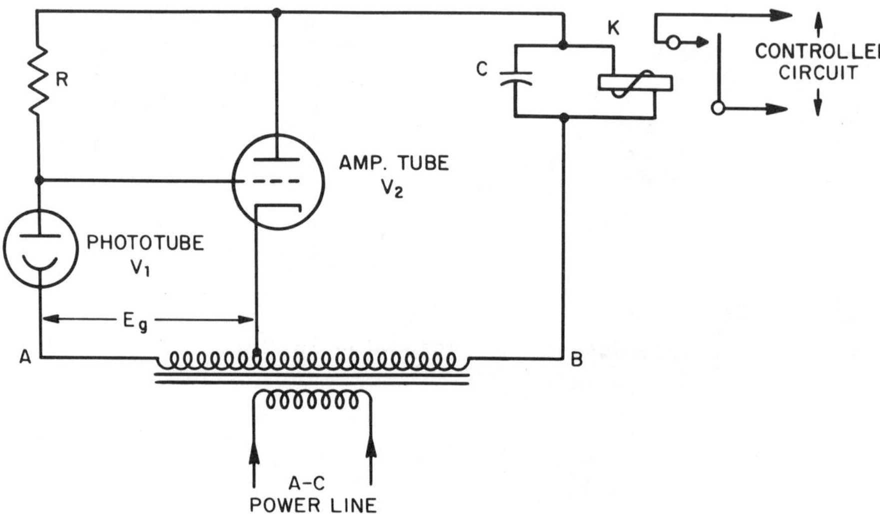

Fig. 18–9. Circuit of a reverse-acting, a-c operated, photoelectric relay.

negative relative to its cathode, as a result of voltage E_g. The tube is cut off and no plate current flows.

During the next half-cycle, side A becomes positive and side B negative. Now the plate of the amplifier tube is negative and, again, there is no flow of plate current. So, as long as the phototube is illuminated there is no plate current and the relay remains unenergized.

When the phototube is darkened, however, emission ceases and its resistance becomes very high. Again, let us start with side A negative and side B positive. Now both the plate of the amplifier tube and the plate of the phototube are positive with respect to their cathodes. As a result of E_g, a small negative voltage is placed upon the grid of the amplifier tube. But, because the phototube is dark, very little current flows through R, thus producing a very small voltage drop across it. As a result, the grid of the amplifier tube becomes positive. Plate current flows and the relay is energized.

During the next half-cycle, side A becomes positive and side B negative. Because its plate has become negative, plate current in the amplifier tube is cut off. However, capacitor C now starts to discharge through the relay coil and the relay remains energized.

The plate current in a high-vacuum amplifier tube generally is quite small. Hence the relay it energizes must be a low-current type. This means that the springs and contact points carried by them must be light. And this, in turn, means that these light contact points cannot be used to switch heavy currents as, for example, the currents required to operate a motor.

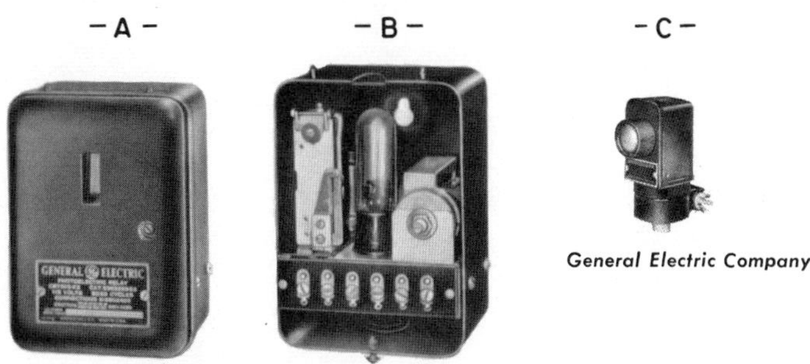

General Electric Company

Fig. 18—10. **A.** External view of photoelectric relay. **B.** Internal view of photoelectric relay. **C.** Light source for photoelectric relay.

Where heavy currents must be switched, we need a heavier relay with heavier springs and contact points. This implies a heavier plate current flowing through the relay coil. To obtain the heavier plate current, a thyratron tube may be substituted for the high-vacuum amplifier tube. The circuit then may be the same as illustrated in Figure 18-8 and 18-9, except that a resistor generally is inserted in series with the control grid of the thyratron to limit the flow of grid current.

QUESTIONS

1. Explain what is meant by an *electronic relay* and explain how it is used.

2. a) Describe the *electromagnetic relay*.
   b) Explain the action of a *self-latching relay*.

3. Draw and explain the basic circuit of an electromagnetic relay, showing what is meant by the *control circuit* and the *controlled circuit*.

4. a) Draw and explain the circuit of a *forward-acting* photoelectric relay operating from batteries.
   b) Give an example of how it is employed.

5. a) Draw and explain the circuit of a *reverse-acting* photoelectric relay operating from batteries.
   b) Give an example of how it is employed.

6. Draw and explain the circuit of an a-c operated *forward-acting* photoelectric relay.

7. Draw and explain the circuit of an a-c operated *reverse-acting* photoelectric relay.

8. a) Draw and explain the circuit of a capacitive relay operating from batteries.
   b) Give an example where such a circuit may be used.

19 Electronic Measurement

Measurement plays an important part in industry. We need to know "how heavy?", "how long?", "how hot?", and lots more. Also, in the control of the various devices we need more than merely a switch to turn them on or off. For example, to prevent overheating, we want a device that will turn an electric oven off after it reaches a predetermined temperature. Or a device that will shut off a a hopper after a certain quantity of material has dropped into a container.

Obviously, measurement is essential for such controls. The sensing devices, or transducers, some of which are the same as those employed in the electronic relay, translate physical changes into electrical quantities that are proportional to the changes. Then these electrical quantities are measured on some device such as a meter whose scale may be calibrated in units of the physical phenomenon producing the electrical values. Or else, when the electrical quantities reach a certain predetermined level, some electrical device such as a motor or an electromagnetic switch or valve is operated.

Fig. 19—1.

Meter-relay.

Assembly Products, Inc.

As an example, consider the instrument illustrated in Figure 19-1. It resembles the ordinary moving-coil meter, except that it also is a relay. The pointer carries the moving contact and when the current flowing through the meter is sufficient to produce a certain predetermined deflection of the pointer, the moving contact touches a fixed contact and the controlled circuit is closed. The position of the fixed contact is determined by means of a screw adjustment on the face of the meter and is indicated by a second pointer. Thus, when the two pointers coincide, the contact points touch.

Since the meter-relay is a delicate instrument, the contact points must be light and so can pass very little current. Where the controlled device requires heavy currents, the meter-relay is used to operate a more rugged electromagnetic relay which, in turn, operates the controlled device.

A. *Electronic voltmeter and recording instruments*

1. ELECTRONIC VOLTMETER

The moving-coil meter movement (see Chapter 4, subdivision E) is a sensitive and efficient device. Some of these movements are able to produce a full-scale deflection with as little as 10 or 20 microamperes flowing through their coils. However, it suffers from two disadvantages that make it unsuitable for measurement in certain electronic circuits.

First, it is not sensitive enough to measure the minute currents that are present in some electronic circuits. Or, even if these currents are sufficient to register on the more sensitive meters, the movements of such meters are extremely delicate and so may easily be damaged.

Second, the meter has a tendency to "load down" certain electronic circuits. For example, assume you are trying to measure the voltage drop across a two-megohm resistor using a voltmeter which, including its multiplier, has a resistance of 100,000 ohms. To measure the voltage drop, the meter is connected in parallel with the resistor. Hence the two-megohm resistance is shunted by another of 100,000 ohms. The total resistance of the circuit thus falls to a little less than 100,000 ohms and the voltage drop across it becomes about one-twentieth of its normal value. The reading on the meter, obviously, will be grossly erroneous.

Both of these disadvantages are overcome by the *electronic voltmeter,* or *vacuum-tube voltmeter* (VTVM), whose basic circuit is illustrated in Figure 19-2. As you see, it consists of a d-c amplifier with an ordinary moving-coil meter connected in its plate circuit. Start with the probes shorted to each other. A certain amount of plate current flows in the plate circuit and is registered on the meter. As a result of this plate current, a voltage drop occurs across the cathode resistor (R_k) and a negative bias is placed on the grid of the tube.

The probes now are separated and the unknown voltage placed across them. If, as a result of this voltage, the grid becomes less negative, more plate current will flow. If the polarity of this voltage is such that the grid becomes more negative, less plate current flows. These changes in plate current are indicated on the meter. Since the amount of plate current is dependent upon the voltage on the grid, the meter can be calibrated to read grid voltage.

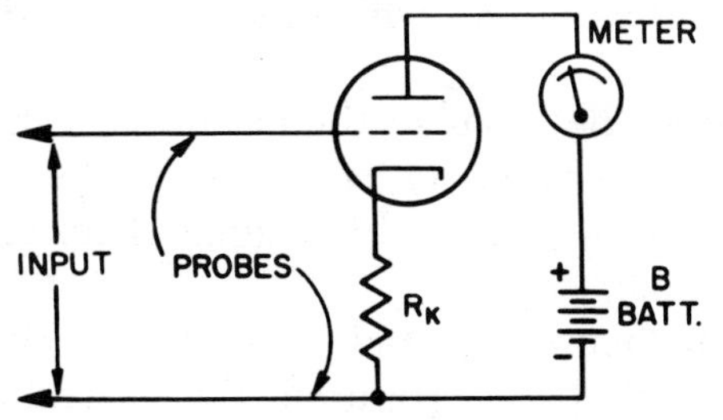

Fig. 19–2.

Basic circuit of the electronic voltmeter.

But notice that the resistance of the input circuit is extremely high, approaching infinity. Hence the loading effect on the circuit under test is very small. Also, because of its amplifying action a very small input voltage to the electronic voltmeter may cause a large flow of plate current. Thus the meter in that circuit need not be extremely sensitive and a more rugged type may be employed. In this way the two disadvantages of the moving-coil voltmeter are overcome.

If larger voltages are to be measured, a *voltage divider* may be used in front of the input to the electronic voltmeter. This device consists of a resistor, or group of resistors connected in series. The unknown voltage is applied to both ends of the divider, hence appearing as a voltage drop across it. By means of suitable taps, a definite, known portion of the unknown voltage drop may be measured and the reading on the meter multiplied by an appropriate multiplication factor. For example, assume the voltage drop across a portion that is one-tenth of the entire resistance of the divider produces a three-volt reading on the meter. Then we would know that the voltage drop across the entire divider must be 30 volts.

The circuit of a practical electronic voltmeter is illustrated in Figure 19-3. The range of the meter is determined by the position of the RANGE switch which connects the grid of the amplifier tube to a suitable tap on the voltage divider.

Note that the meter is in the cathode circuit of the tube instead of the plate circuit. Since these two circuits are in series, the current in both is the same and the meter can register it equally well where it is. The scale of the meter is a zero-center type and zero adjustment is obtained by means of the 10,000-ohm potentiometer in the plate circuit. This potentiometer adjusts the plate voltage on the tube to a point where the plate current, with no input voltage applied to the grid, is sufficient to give half-scale deflection (zero reading) on the meter. The 3,000-ohm rheostat in the cathode of the tube is used to compensate for variations which may be caused by changing tubes.

Because of the voltage divider, the input to the electronic voltmeter no longer approaches infinity. However, since the resistance of this divider is approximately 15 megohms, the loading effect on the circuit under test is quite small.

The instruments illustrated in Figure 19-2 and 19-3 are d-c voltmeters. If alternating voltages are to be measured, they may be

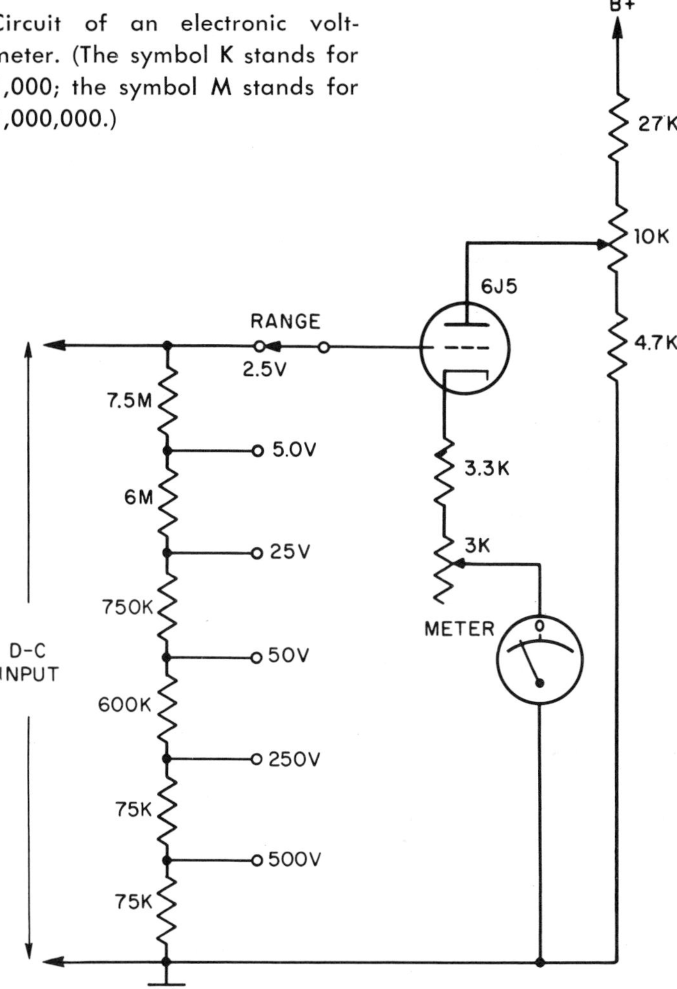

Fig. 19–3.

Circuit of an electronic voltmeter. (The symbol K stands for 1,000; the symbol M stands for 1,000,000.)

Hickok Electrical Instrument Co.

rectified and changed to direct voltages which then may be measured on the d-c voltmeter. Frequently, rectification takes place in an external *a-c probe* that is connected to the d-c electronic voltmeter by means of a shielded cable.

The circuit for such a probe is illustrated in Figure 19-4. Tube V is the rectifier and R_1 is its load resistor. Capacitor C_1 is a block-

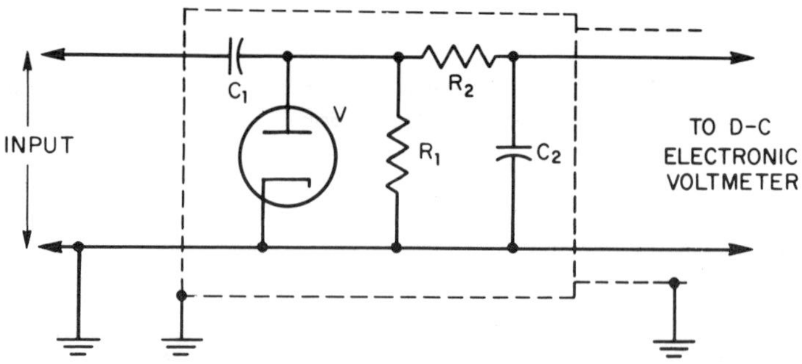

Fig. 19—4. Circuit of a-c probe used with a d-c electronic voltmeter.

ing capacitor to keep any d-c component of voltage under test from entering the instrument. R_2 and C_2 form an *R-C* filter to remove the a-c component from the rectified voltage that is applied to the electronic voltmeter.

2. RECORDING INSTRUMENTS

In many industrial processes it is essential to keep a continuous written record. For example, suppose we wish such a record of the temperature in a furnace. A suitable sensing device, whose electrical output is proportional to the temperature, is mounted in the side of the furnace. (Such a device will be described later.) The output of the transducer is amplified and fed to the input of the recording device. This consists of an electric motor or clockwork mechanism that causes a ruled sheet of paper to pass at a uniform rate beneath a pen. (See Figure 19-5.)

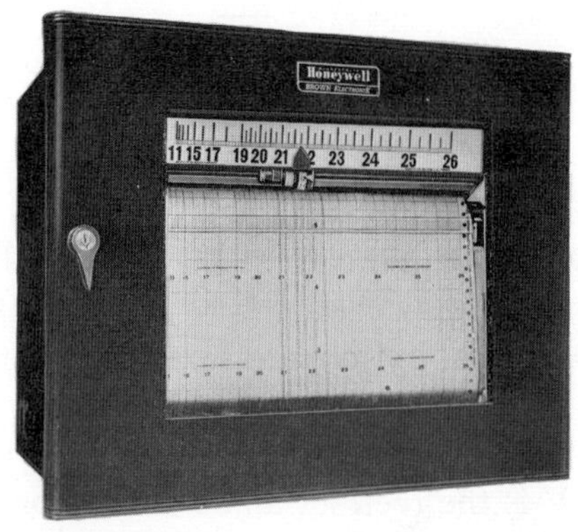

Fig. 19—5. Recording meter.

Minneapolis-Honeywell Regulator Corp.

The movement of the pen is controlled by an electromagnetic device which is actuated by the current from the amplifier. The greater the temperature of the furnace, the greater will be the output from the amplifier, and the greater will be the excursion of the pen over the moving paper. The chart may be divided into units of hours, days, or even weeks. In this way the temperature of the furnace at any given time may be ascertained.

Such recorders are used for a great many purposes other than the recording of temperature. Thus, with suitable transducers, they may record pressure, rate of flow, acidity, in fact, anything that can be measured.

B. *Temperature measurement*

Perhaps the most common temperature-sensing device is the thermometer where a column of mercury sealed in a glass tube expands in direct proportion to the degree of heat applied to it. The height of the column is noted by means of a scale, calibrated in degrees of temperature, that is etched on the surface of the glass tube.

The thermometer can be used as a control device by sealing a metal electrode into the side of the tube at a point corresponding to a desired temperature. (See Figure 19-6.) Another electrode sealed in near the bottom of the thermometer makes contact with the mercury column. Then, as the column reaches the desired point, it touches the upper electrode, closing the circuit and so operating a

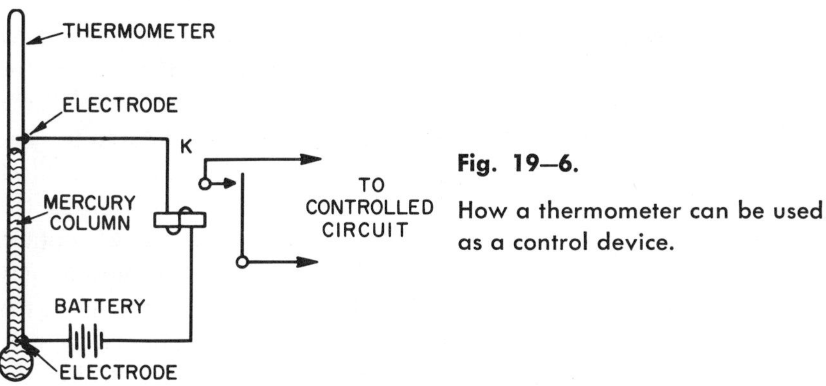

Fig. 19–6. How a thermometer can be used as a control device.

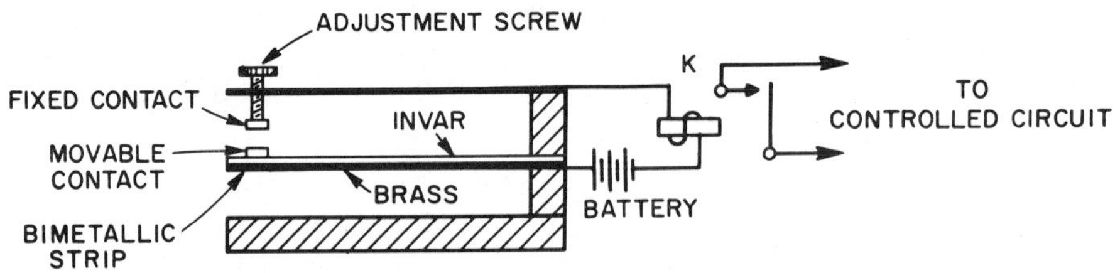

Fig. 19—7. The thermostat.

relay which may actuate some device in the controlled circuit. When the mercury column falls below the desired point, the relay is de-energized.

Another temperature-sensing device is the *thermostat* illustrated in Figure 19-7. The heart of this device is a bimetallic strip consisting of two dissimilar metals, each with a different rate of expansion due to heat, welded together. Commonly employed are *brass,* which has a relatively large rate of expansion, and *invar,* an alloy of nickel and iron which has a small rate of expansion. If, as shown in Figure 19-7, the bimetallic strip is heated, the greater expansion of the brass will cause the free end of the strip to bend upward. When cooled, the strip will return to its normal position.

The bending of the bimetallic strip is proportional to the degree of heat applied to it. Of course, temperature may be indicated by fastening a pointer to the end of the strip and permitting it to move over a suitable scale. This, however, generally is not practical since the thermostat must be close to the source of heat as, for example, inside an oven.

The thermostat may also be used as a control device by fastening a contact point at the end of the bimetallic strip so that, when the heat causes the strip to bend a certain amount, the movable contact touches a fixed contact, thus closing the relay circuit. The temperature at which the contact points will touch can be controlled, within a certain range, by the adjustment screw that brings the fixed contact nearer to or further away from the movable contact.

As an example of how it acts as a control, the thermostat may be mounted within an electric oven. When the temperature in the oven reaches a certain predetermined point, the bimetallic strip bends sufficiently so that the contact points touch, energizing the relay. The oven's heating element, which is in the controlled circuit, is turned off and the oven starts to cool. As the temperature drops, the strip starts to straighten out. When this has proceeded to the point where the contact points are separated, the relay is de-energized and the heating element is turned on again. Then the entire process is repeated, thus keeping the temperature within the oven fairly constant between narrow limits.

An interesting variation of the thermostat shown in Figure 19-7 is the *Klixon switch* illustrated in Figure 19-8A. The bimetallic element consists of a springy disk. Normally, the curvature of this disk is such that the movable contacts fastened to it touch a set of fixed contacts, thus completing the controlled circuit. As the disk is heated, the difference between the expansions of the two metals of which it is composed causes the disk to curve in the opposite direction, thus separating the contacts and opening the circuit.

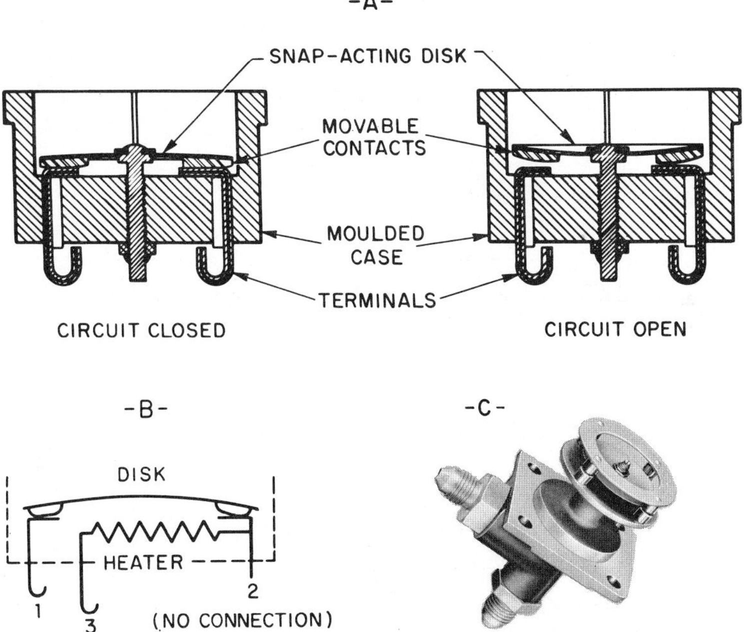

Metal and Controls Div., Texas Instruments, Inc.

Fig. 19–8. **A.** Cross-sectional view of the Klixon switch closed and open.
B. Circuit of the Klixon overload protector.
C. The Klixon valve.

Because of the springiness of the disk, the action is not gradual. The disk snaps from one curvature to the other when its temperature reaches a certain predetermined level. When the temperature falls to another predetermined level, the disk snaps back to its original curvature.

Such switches frequently are mounted inside the housing of electric motors to prevent overheating. The switch is connected in series with the motor and the power line. Normally, the switch is closed. But should the temperature within the motor housing rise to a dangerous level, the switch opens the line circuit, thus stopping the rotation of the motor. When the temperature drops to a safe level, the disk snaps back to its original curvature, and the motor may resume its rotation. Some switches contain a reset button that can be used manually to reverse the curvature of the disk from its "off" position to its "on" position.

The Klixon switch may also be used as a protective circuit-breaker to prevent overloads in an electrical circuit. Thus, as illustrated in Figure 19-8B, a heater is inserted so that the current flows through it and the switch. As the current rises to a dangerous level, the heat produced by the heater is great enough to cause the disk to change its curvature, thus opening the circuit and stopping the flow of current.

The principle of the Klixon switch may also be applied to operate a thermal valve. Thus, the bimetallic disk, by responding to changes in temperature, transmits its force mechanically to open or close an opening rather than to make or break an electrical circuit. (See Figure 19-8C.)

As sensing devices, both the thermometer and thermostat suffer from certain drawbacks. The thermometer is fragile. The thermostat is not too accurate a measuring device. Further, neither are sensitive enough for some of the applications required by industry.

In 1822, Thomas J. Seebeck, a German physicist, found that if two dissimilar metal wires or strips are joined at one end and this junction is heated, a small voltage will appear between the cool, open ends (see Figure 19-9A). The magnitude of this voltage is proportional to the temperature difference between the heated junction and the cool free ends.

Such a device is known as a *thermocouple* and the phenomenon is known as the *Seebeck effect.* (Interestingly, this effect is reversible.

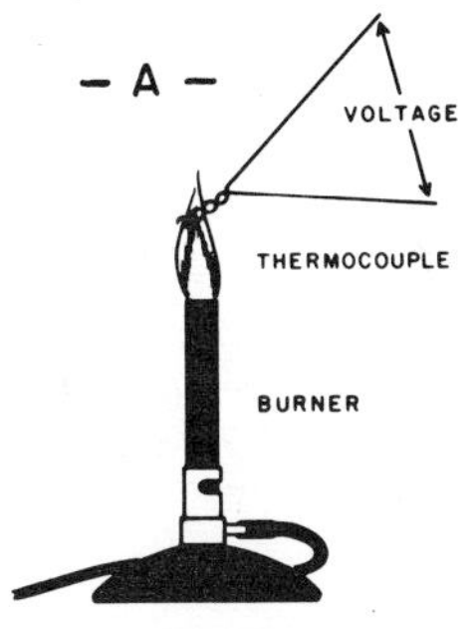

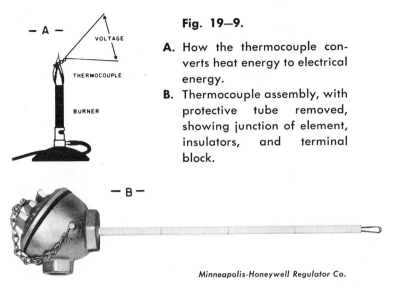

— B —

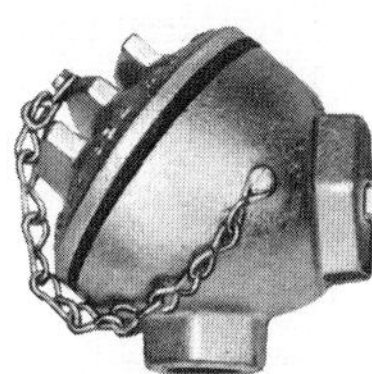

Fig. 19–9.

A. How the thermocouple converts heat energy to electrical energy.

B. Thermocouple assembly, with protective tube removed, showing junction of element, insulators, and terminal block.

Minneapolis-Honeywell Regulator Co.

In 1834, the French physicist Jean Peltier discovered that when an electric current is made to flow through a thermocouple, heat energy is given off at one end and absorbed at the other.)

Any of a number of combinations of metals may be used for the thermocouple. One such combination consists of platinum and an alloy of platinum and rhodium. Another consists of iron and constantan (an alloy of copper, nickel, manganese, and iron). Each type is suitable for a definite operating range.

To use the thermocouple, it is constructed as a probe (see Figure 19-9B) and the junction end is placed within the device whose temperature is to be measured. This may be an ordinary oven or a metal-melting furnace. The voltage developed across the cool ends is amplified and then fed to some indicating or recording device calibrated to read degrees of temperature. (See Figure 19-10.)

The thermocouple can be used as a sensing device for an electronic relay similar to those described in Chapter 18. The voltage developed across the free ends of the thermocouple is applied to an amplifier and, when the heat and resulting voltage reach a certain predetermined level, sufficient plate current flows to energize the relay.

You are aware, of course, that when a metallic wire is heated, its resistance generally increases in proportion to the heat. Here, then,

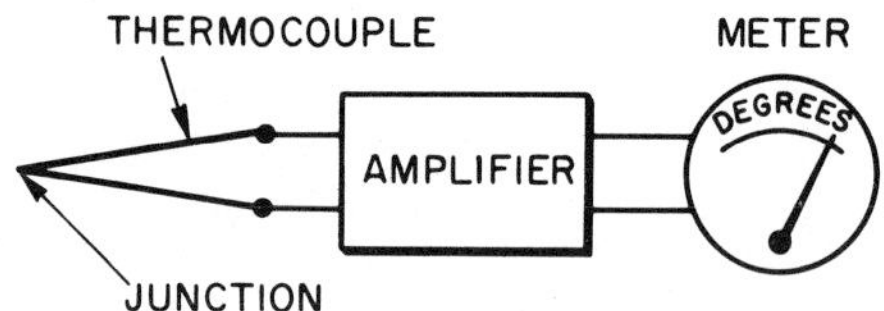

Fig. 19—10. How the thermocouple is used to measure temperature.

is another type of heat-sensing device. Heat a wire and measure its change in resistance. Since this change is proportional to the heat supplied, we then know how much heat has been added.

We can measure the resistance change by measuring the change produced in a steady current flowing through the resistor. Since the resistance change generally is small, the current change, too, will be small. Accordingly, we may first amplify this current before measuring it by means of a meter.

The sensing resistor may be in the form of a coil of fine wire such as copper, nickel, or platinum. (See Figure 19-11.) This coil is enclosed in a metal tube for protection and used as a probe to be inserted into the area where the heat is to be measured. A set of leads running through the tube connects the coil to the rest of the circuit. Another form consists of the wire mounted in paper or plastic which can be cemented to the surface where the heat is to be measured. The precision of these temperature-measuring devices is quite high.

We may measure the resistance of the heat-sensing resistor more accurately by means of a *Wheatstone bridge,* the basic circuit

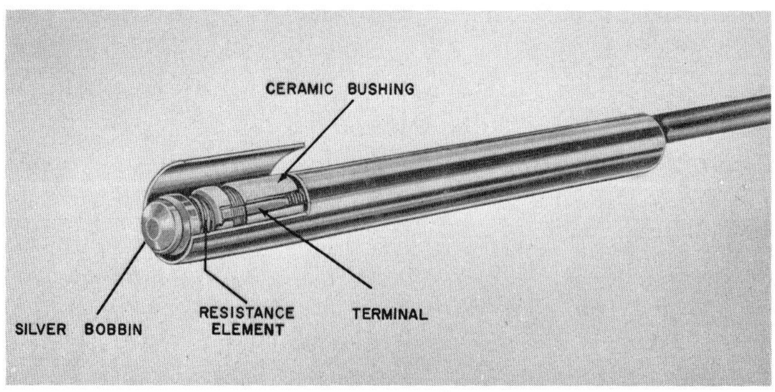

Thomas A. Edison, Inc.

Fig. 19—11. Cutaway view of the resistance-temperature probe.

of which is shown in Figure 19-12. Current flowing from the battery divides at point #1. Part of this current (I_1) flows through resistors R_1 and R_2 to point #3, and back to the battery. The other part of the current (I_2) flows through resistors R_3 and R_4 to point #3 and joins I_1 in flowing back to the battery.

If the resistance of $R_1 + R_2$ is equal to the resistance of $R_3 + R_4$, the current will divide into two equal parts and $I_1 = I_2$. If, further, the resistance of R_1 is equal to that of R_3 (and $R_2 = R_4$), the voltage drop across R_1 will be equal to that across R_3. Accordingly, point #2 will be at the same potential as point #4. There will be no flow of current between these two points, and the zero-center milliammeter connected between them will register zero.

Now, let us suppose that the resistance of R_1 increases, say, as a result of heating. The balance of the bridge is upset, point #2 no longer is at the same potential as point #4, and current flows through the milliammeter. The amount of current flow will depend upon the degree to which the balance of the bridge is upset, that is, upon the resistance increase of R_1.

The currents involved are generally very small since we wish to avoid erroneous results produced by the heating effect of the current

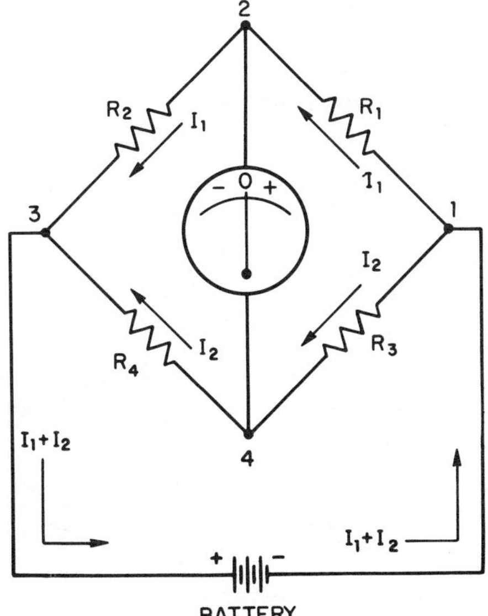

Fig. 19–12.

Basic circuit of the Wheatstone bridge.

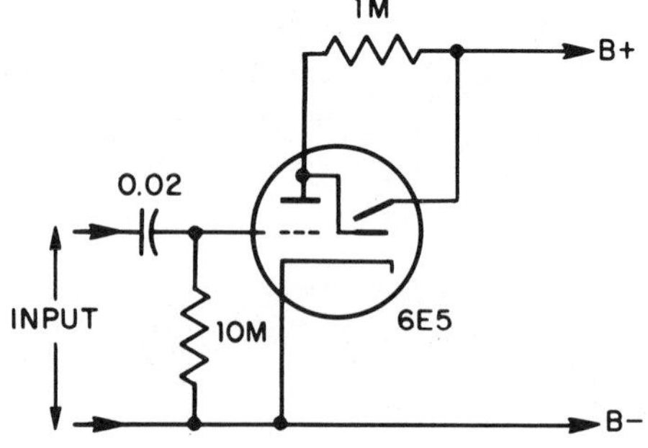

Fig. 19—13.

How the electron-ray tube may be used as a null indicator.

as it flows through the resistors. Accordingly, the current flow between points #2 and #4 may first be amplified and then measured upon a meter whose scale is calibrated in degrees of heat applied to R_1.

There is another way in which the resistance change (and the heat producing it) may be measured. In the bridge illustrated in Figure 19-12, a rheostat is substituted for the fixed resistor R_3. The bridge was balanced and no current flowed through the meter, you will recall, when R_1 was equal to R_3. When R_1 was heated and its resistance was increased, the bridge became unbalanced and current flowed through the meter.

If, now, R_3 is adjusted so that its resistance becomes equal to R_1, the bridge becomes balanced again and the meter reads zero once more. The change in resistance of R_3 can be read on its control dial and this change of resistance may be translated into degrees of heat applied to R_1. The meter, used in this way, is called a *null indicator.*

The electron-ray tube (see Chapter 12, subdivision D) may replace the meter as a null indicator. (See Figure 19-13.) The input terminals are connected in place of the meter of the bridge. With the bridge balanced, no voltage is applied to the null indicator and, as a result, the shadow is at its maximum ("eye wide open"). If the bridge is unbalanced, the potential difference between its arms causes a voltage to appear at the input of the null indicator. As a result, the grid of the indicator tube becomes negative and the shadow is reduced ("eye closes").

We may use an a-c supply instead of the battery for the bridge and employ an a-c meter to show when the bridge is balanced. With this a-c supply, a pair of headphones can be used in place of the meter as a null indicator. When the bridge is unbalanced, alternating current will flow through the headphones and produce a hum. When the bridge is balanced this hum is reduced to a minimum or disappears altogether. When headphones are used it is customary to generate the alternating current by means of an oscillator, hummer, or buzzer that produces an output at about 1,000 cycles per second because the human ear can discern this sound most easily.

If the resistors used in the a-c bridge are wire-wound, there is danger that their inductive reactances may introduce errors. Accordingly, these resistors must be of the *non-inductive* type. Such resistors, then, may be constructed of carbon composition. If they are wire-wound, half the winding is in one direction and half in the other. Hence the inductances balance out.

QUESTIONS

1. a) Explain the basic principle of the electronic voltmeter.
b) Explain the function and operation of an a-c probe used with the electronic voltmeter.

2. a) Explain how the mercury thermometer is used to measure temperature.
b) Explain how the thermometer may be converted into a temperature-control device.

3. a) Explain the structure and operation of the *thermostat.*
b) Explain how the thermostat may be used as a temperature-control device.

4. Explain the structure and operation of the *Klixon switch.*

5. a) Explain the structure and operation of the *thermocouple.*
b) Explain how the thermocouple may be used as a temperature-measuring device;
c) as a temperature-control device.

6. Explain the structure and operation of the *resistance-temperature probe.*

7. Draw the basic circuit of the *Wheatstone bridge* and explain how it is used to measure resistance.

8. a) In the Wheatstone bridge, explain the function of the *null indicator.*
b) Draw and explain the circuit of the electron-ray tube used as a null indicator.

20 Electronic Timers

Timing performs an essential role in industry. Many processes depend upon accurate timing. For example, in X-ray and photographic work it is necessary that the operation cease a definite time after a switch is closed. Also, it frequently is necessary that several operations, each of a definite time duration, take place at definite time intervals between each.

The *timer*, then, is a device which, when activated, measures out a predetermined period of time and, at the end of that period closes or opens a switch, initiating or terminating some process. One type of timer is the clock, either spring-wound or electric-motor, with contact points that are closed or opened after a predetermined lapse of time. The timing intervals may be determined by the positions of the contacts on the face of the clock.

Another type of timer depends upon the heating effect of an electric current. The thermostat, discussed in Chapter 19, subdivision

B, may be used as such a timer. Heat is supplied to the bimetallic strip by a current-carrying coil of resistance wire wound around the strip. After a definite period of time, the strip will bend enough so that the contact points touch. The time interval can be determined by the setting of the adjustment screw.

The circuit of another type of timer depending upon the heating effect of a current is illustrated in Figure 20-1. Current flowing through the diode during the half-cycles when its plate is positive energizes the relay. However, current cannot flow until the switch in the heater circuit is closed. And even then it cannot flow until the heater has made the cathode hot enough to emit electrons in sufficient quantity. Thus the timing period is the interval between the closing of the switch and start of sufficient electronic emission.

The potentiometer R across the heater secondary of the transformer adjusts the voltage applied to the heater and so determines the length of time required to bring the cathode to its proper emission point. Capacitor C across the relay coil charges up during the half-cycles when the plate is positive and there is a flow of current. During the other half-cycles C discharges through the relay coil, thus keeping the relay energized.

For many industrial applications, the clock-type and heater-type timers suffer from certain serious drawbacks. For one, neither type can be adjusted to measure very short time intervals. Some processes require intervals as short as one-hundredth or one-thousandth of a second, or even shorter. Secondly, the heater-type timer cannot be adjusted too accurately.

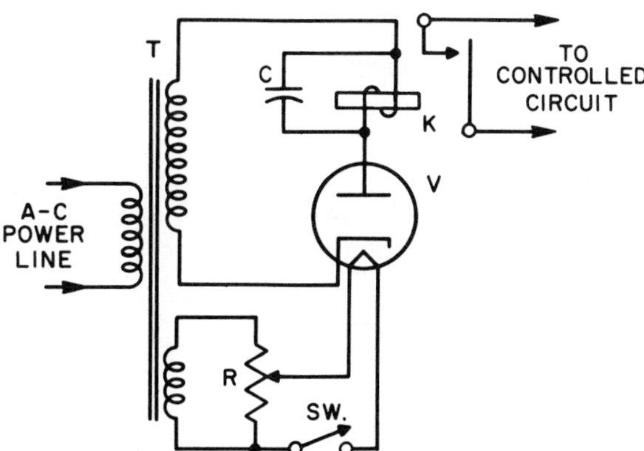

Fig. 20—1. Circuit of timer depending upon the heating effect of a current.

When greater accuracy or extremely short time intervals are desired, another method is used. You will recall that when a capacitor is charged through a resistor it does not reach full charge at once, but requires a period of time depending upon the capacitance of the capacitor and the resistance of the resistor (see Chapter 7, subdivision B). The same delay holds true as a charged capacitor discharges through a resistor.

How this *R-C* timing principle is applied to a timer is illustrated in Figure 20-2. With the switch open, sufficient plate current flows through the tube to energize the relay. When the switch is closed, the negative voltage from the grid battery applied to the grid of the tube biases it to cutoff. Plate current ceases and the relay is deenergized. At the same time C is charged up with the indicated polarity.

Now the switch is opened once again. The grid battery is disconnected from the grid of the tube. However, because of the voltage across C, the grid of the tube still remains negative enough to cut off the flow of plate current. The charge on C starts to leak off through R. When enough of the charge has leaked off, the grid becomes sufficiently less negative so that plate current flows again. As the charge continues to leak off, the plate current rises to the point where it is able to energize the relay.

The time between the opening of the switch and the energizing of the relay depends upon the time required for sufficient charge to leak off C. This, in turn, depends upon the values of C and R. By choosing suitable values for these components, any time lapse may be selected.

In the above circuit the time lapse depends upon the length of time necessary for the discharge of the capacitor. It is possible to

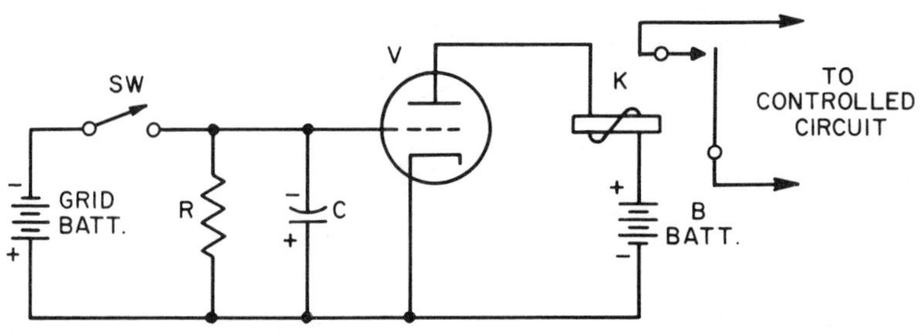

Fig. 20–2. Basic *R*-C timer circuit.

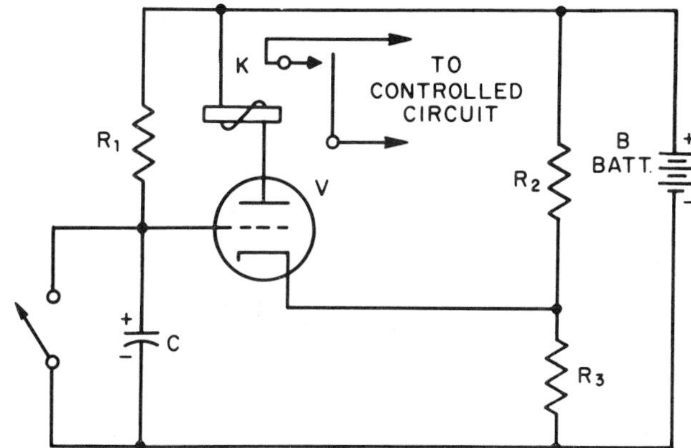

Fig. 20—3. *R-C timer circuit. Time lapse depends upon the length of time it takes C to charge up.*

have the time lapse dependent upon the time required for the capacitor to charge up. Look at the circuit illustrated in Figure 20-3.

With the switch closed, C is shorted out and the grid of the tube is connected to B—. At the same time a positive bias is placed upon the cathode by the voltage drop across R_3. As a result, the tube is driven to cutoff. Accordingly, there is no flow of plate current and the relay is unenergized.

When the switch is opened, C no longer is shorted and starts to charge up through R_1 with the indicated polarity. The voltage across C is applied to the grid of the tube and, when C is sufficiently charged, the grid becomes positive enough to bring the tube out of cutoff. As the charge on C continues to rise, so does the flow of plate current until a point is reached where the relay is energized. The time lapse between the opening of the switch and the energizing of the relay depends upon the values of C and R_1.

So far, we have seen how a relay is energized a short time after a switch is opened. We also are able to energize the relay a short time after a switch is closed. (See Figure 20-4.) With switch Sw_1 at position #1, the grid of the tube is connected to B— through R_1 and the cathode is positive by the voltage drop across R_3. Under this condition the tube is cut off and the relay is unenergized. Capacitor C_1 is shorted out.

Switch Sw_1 now is thrown to position #2. Capacitor C_1 is connected between the grid and cathode of the tube. Because C_1 is uncharged, there is no voltage across it and, therefore, the instantaneous

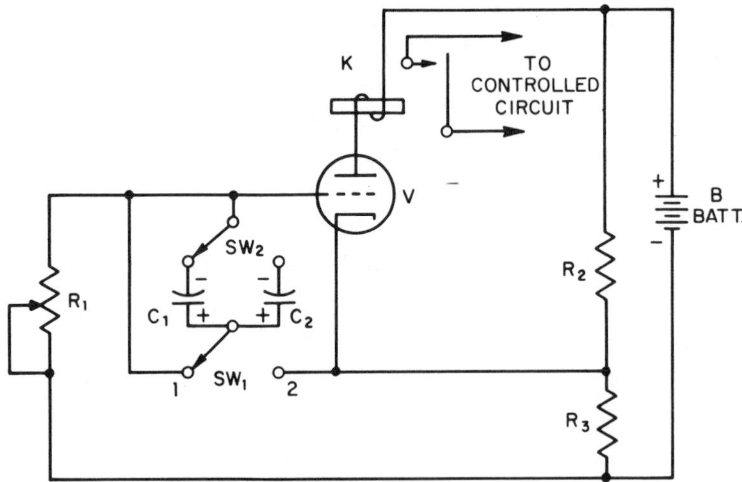

Fig. 20–4. *R-C timer circuit. Timing starts as switch* Sw_1 *is thrown to position #2.*

potential of the grid of the tube becomes that of the cathode. As a result of this zero-bias condition, current flows through the tube and the relay is energized.

Capacitor C_1 now starts to charge up through R_1 with the polarity as indicated. When the charge becomes sufficiently large, the grid becomes negative enough to cut off the tube and the relay is de-energized.

The time interval between throwing the switch to position #2 and the de-energizing of the relay is determined by the values of C_1 and R_1. Different time intervals may be obtained by varying the value of R_1 and substituting C_2 for C_1 by means of switch Sw_2.

The timer may also be a-c operated. A typical circuit is illustrated in Figure 20-5. Let us start with switch Sw open.

Assume a half-cycle when side A of the a-c line is positive and side B is negative. Because its plate is negative, there can be no conduction through the tube, and the relay remains unenergized. But some current does flow from side B through R_3 to the cathode, to the grid, and through R_4 and R_1 to side A. Note in this respect that the grid and cathode form a sort of diode through which current can flow only when the grid is positive with respect to the cathode. This is known as *grid rectification.* As a result of this flow, C_1 is charged with the polarity as indicated.

During the next half-cycle, when side A is negative and side B (and the plate of the tube) is positive, there is no plate current because of the negative bias placed on the grid by the charge on C_1 and the positive bias on the cathode from the positive side of the line through R_3.

Now the switch is closed. Because the cathode is connected to the negative side of the line, the tube would become conductive, were it not for the negative charge placed on the grid by C_1. As C_1 discharges through R_4, its voltage decreases and, after a definite interval depending upon the values of C_1 and R_4, enough current flows through the tube to energize the relay.

During the half-cycles when the plate of the tube is negative there is no plate current, of course. However, during those periods C_2 discharges through the relay coil, thus keeping the relay energized.

The time delay of this circuit may be adjusted by varying potentiometer R_1 which, in turn, varies the voltage charging C_1. If the voltage is lowered, then, when C_1 starts to discharge, this lower voltage is more rapidly reduced to the point at which the tube becomes conductive. This reduces the time delay. If the voltage is increased, the time delay is lengthened.

The circuit of another a-c operated timer, one where closing a switch energizes the relay for a definite period of time and then de-energizes it, is illustrated in Figure 20-6. With switch Sw in position #1, capacitor C_1 is shorted out. Cathode resistor R_2 is large enough so that the tube is cut off. Therefore, during the half-cycle when the

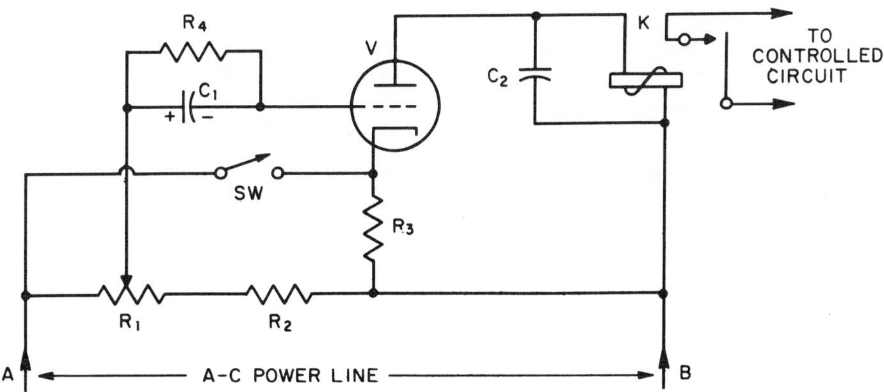

Fig. 20—5. A-c operated *R*-C timer.

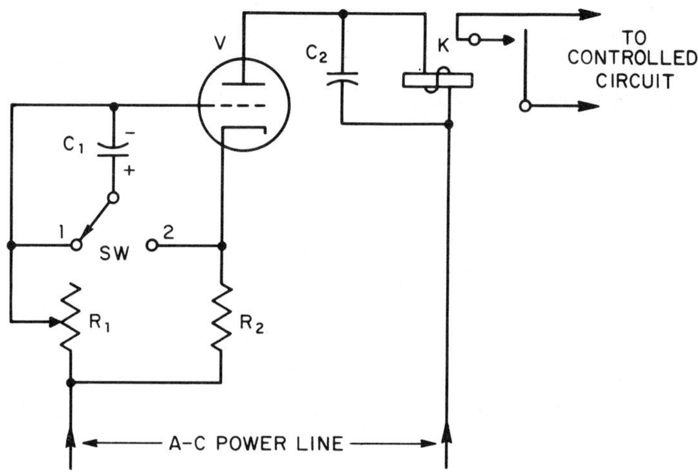

Fig. 20—6. Another a-c operated *R-C* timer.

plate of the tube is positive, no plate current flows because of the bias. During the next half-cycle the plate is negative and, again, there is no flow of plate current. So, as long as the switch is at position #1 the relay is unenergized.

When the switch is thrown to position #2, C_1 is connected between the grid and cathode of the tube. Because it has no charge, there is no voltage across C_1 and the instantaneous potential of the grid becomes that of the cathode. Under this condition of zero-bias,

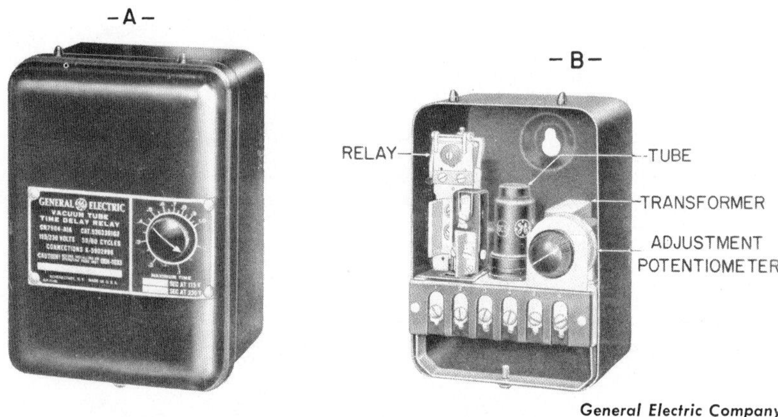

General Electric Company

Fig. 20—7. **A.** Front view of electronic timer.
B. Inside view of electronic timer.

enough plate current flows to energize the relay. (We are considering, of course, the half-cycle when the plate is positive.)

Capacitor C_1 slowly charges, with the indicated polarity, through R_1 which controls the rate of charge. As the charge increases, the grid becomes more negative until a point is reached where the tube is cut off and the relay is de-energized. The length of time the relay remains energized is determined by the setting of R_1.

Certain processes require a series of operations, each of a definite duration, and following in sequence at definite time intervals. For this purpose, two or more timers may be so arranged that as the time delay in one circuit is completed, it initiates the time delay in the following circuit. To do this, a set of contacts in each relay is made to act as the switch for the following circuit. Such an arrangement of timer circuits is known as a *sequence timer.*

QUESTIONS

1. **Explain how a vacuum-tube diode may be employed as an electronic timer.**
2. **Explain the basic principles of the *R-C* circuit employed as a timer.**
3. **Draw and explain the basic circuit of a battery-operated *R-C* timer that starts the timing operation as a switch is opened.**
4. **Draw and explain the basic circuit of a battery-operated *R-C* timer that starts the timing operation as a switch is closed.**
5. **Draw and explain the circuit of an a-c operated *R-C* timer.**
6. **a) What is meant by a sequence timer?**
   b) Explain how it operates.

21 Electronic Control

Control of machinery is an essential part of industry. In the main, it consists of varying the electric current to make a motor run faster or slower, for one example, or to make a weld hotter or cooler, for another. Of course, such control may be performed manually, using rheostats. But this requires the presence of an attendant and, where the currents involved are large, the use of heavy equipment. Further, many of these controls demand split-second timing.

The heart of electronic control consists of apparatus and circuits that will permit more or less current to flow in response to signals. Since these signals may be electrical in nature, the system is very flexible and lends itself to automatic control.

Thus any of the sensing devices previously discussed may generate the signal that initiates the control. For example, a thermocouple inserted in the side of a furnace will generate a voltage proportional to the heat. This voltage may be used as the signal for the control device that regulates the current flowing to the heating elements of the furnace.

In this chapter we will consider a few of the many methods for controlling the flow of current in response to a signal.

A. *Electronic lighting control*

Theatrical productions require lighting systems that may be controlled to meet the demands of the script. Early systems met this need by the insertion of one or more rheostats in series with the lamps. However, since many high-wattage lamps had to be controlled, these rheostats were called upon to pass heavy currents. The power consumed by these rheostats and the heat produced by this wasted power was quite large.

Modern systems frequently employ a *saturable reactor* for this purpose. (See Figure 21-1A.) The reactor, you will notice, resembles a transformer, consisting of two separate and independent windings on an iron core. The a-c, or *load,* winding is divided into two, one half on each outside leg of the core, and connected in series. The d-c, or *control,* winding is composed of many turns wound upon the center leg of the core. Note that the two halves of the a-c winding are wound in opposition so that whatever voltage one half of the a-c winding will induce in the d-c winding will be cancelled out by the induced voltage of the other half.

With no d-c voltage applied and no direct current flowing through the d-c winding, the saturable reactor behaves as an ordinary iron-core inductor. It has a definite inductance and, as such, develops a definite reactance which limits the current flowing through the lamps. If the reactance is large enough, very little current flows through the lamps, hence they barely light.

As a direct current is made to flow through the d-c winding, it generates magnetic lines of force (*magnetic flux*) that flow through the reactor core. Because of the many turns of that winding, even a small direct current is able to bring the core to *saturation,* a point where an increase in current brings practically no increase in magnetic flux. (See Chapter 4, subdivision D, 3.)

As the core becomes saturated, the inductance of the reactor drops. The action is as though the iron core were being removed from the windings. Within certain limits, the drop in inductance is proportional to the increase in direct current flowing through the d-c winding. As the inductance of the reactor is reduced, so is its reactance. More current flows through the lamps which glow more brightly. Hence, by controlling the d-c voltage, which may be accomplished by a relatively small rheostat, the a-c current through the lamps may be controlled.

Note that a small amount of power applied to the control winding controls a large amount of power in the load. Hence the saturable reactor, and its associated components, frequently are called a *magnetic amplifier.* The symbol for the saturable reactor is shown in Figure 21-1B. Its letter symbol is S_x.

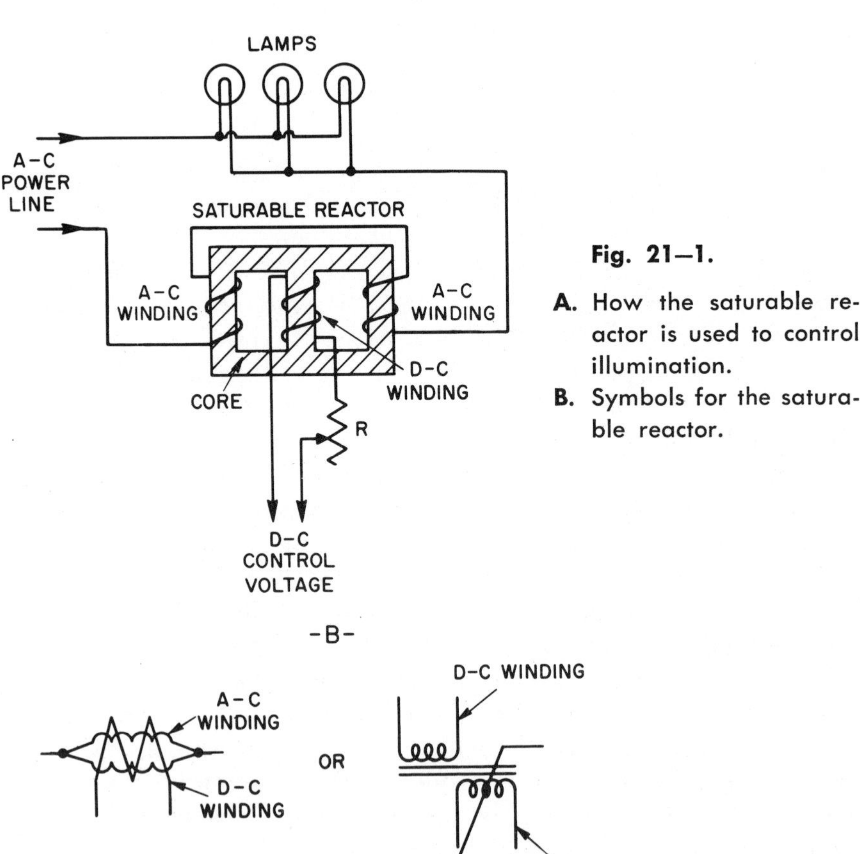

Fig. 21—1.

A. How the saturable reactor is used to control illumination.

B. Symbols for the saturable reactor.

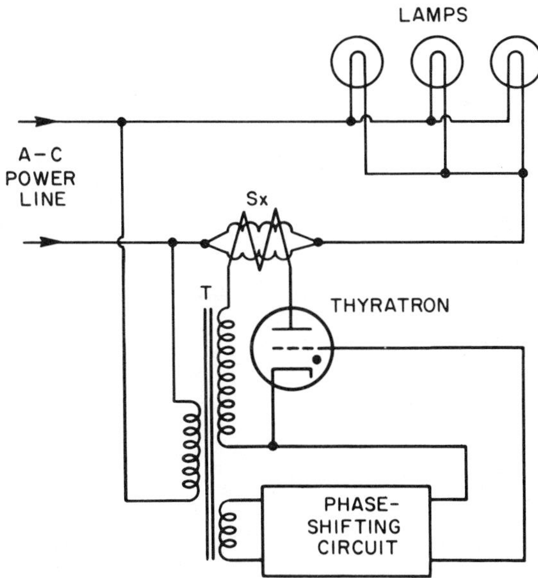

Fig. 21—2. How a thyratron and saturable reactor are used to control lighting.

A controlled thyratron rectifier (previously described in Chapter 13, subdivision G) may be used to supply the d-c control current to the saturable reactor. (See Figure 21-2.) Adjusting the phase shift determines the amount of current flowing through the d-c winding of the reactor and, hence, the degree of saturation of its core. Thus, the brightness of the lamp is controlled.

B. *Electronic welding control*

One of the most frequently employed methods for joining two pieces of metal is *resistance welding.* The two pieces of metal are held together under pressure between two heavy electrodes. Then a heavy current is passed from the electrodes through the metal between them. The contact resistance between the metal pieces is quite low. But because the current is so great, the heating effect (I^2R) is large. As a result, the metal pieces soften at their junction and, under the pressure of the electrodes, fuse together.

The pressures involved may be several thousand pounds or more, usually supplied by compressed air or hydraulic pressure. Small welding machines may employ currents of about 50 to 60 amperes, though at low voltages ranging from about 2 to 10 volts. Large welders may employ currents of thousands, and even hundreds of thousands, of amperes.

The electrodes generally are made of a hard copper alloy strong enough to withstand the high pressure. Frequently, they are hollow to permit the circulation of water through them for cooling purposes.

There are several distinct steps in the process, each of which lasts a definite period of time.

1. *Squeeze time* during which the pressure on the material builds up to the desired point.

2. *Weld time* during which the current flows through the material and the heat is developed at the junction.

3. *Hold time* during which the current is turned off, but the pressure is maintained for the weld to set.

4. *Off time* during which the pressure is relieved and the electrodes are separated so that the welded material may be removed.

Modern welding machines employ electronic timers to control the duration of each of these steps. These timers may be so arranged that each triggers its successor to form a sequence timer.

The various mechanical devices that bring together or separate the electrodes or control the pressure built up between them will not be discussed here. We will however, consider the action that takes place during the weld period.

The basic, and simplified, circuit of the welder is shown in Figure 21-3. In essence, it consists of a step-down transformer (T) controlled by a switch (Sw) in its primary circuit. When the switch is closed, a current flows through the primary winding. Because of the step-down effect, a heavy current flows through the secondary, which consists of a single loop, and through the material to be welded, firmly held between the welding electrodes.

The difficulty with this setup is that, because of the heavy currents involved, the switch must be quite massive. And, because it is so

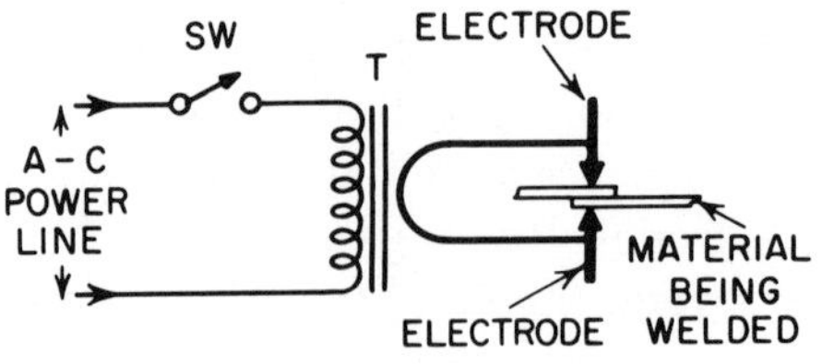

Fig. 21–3. Basic circuit of the resistance welder.

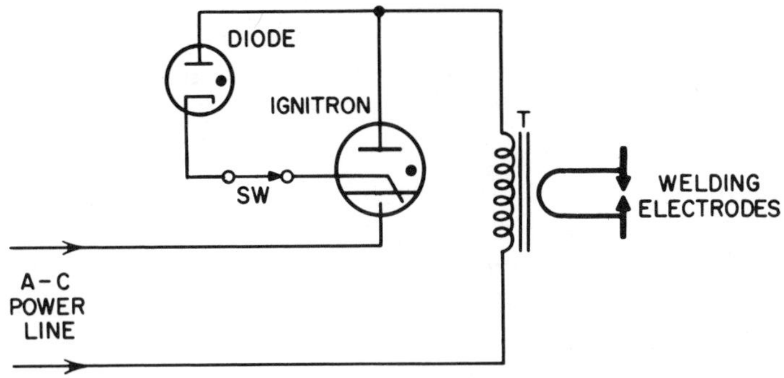

Fig. 21–4. Basic circuit of welder using an ignitron.

heavy, it cannot be manipulated fast enough for the requirements of many welding processes.

The rectifier forms an ideal substitute for the mechanical switch. During the half-cycles when its anode is positive relative to its cathode, current flows through it to the welding transformer. When the polarity is reversed, it is nonconductive. Thus we have a switch without any moving parts and which is capable of extremely fast action.

For small welders, mercury-vapor diodes or thyratrons may be employed. But where the current requirements are high, ignitrons are used. (See Figure 21-4.) In this circuit the ignitron serves as a half-wave rectifier and a pulse of current flows through the transformer each alternate half-cycle when the plate of the tube is positive.

Note, however, that this action can take place only when switch Sw is closed. When it is open, no current flows to the ignitor of the ignitron, hence the tube is nonconductive. Thus the switch can act to initiate and stop the weld. Because it carries only the ignitor current it need not be as heavy as it would have to be were it carrying the entire current, as is the case in the circuit of Figure 21-3. This switch, then, can be the contact section of a relay in an electronic timer. In this way the duration of the weld can be controlled.

For greater efficiency, welders frequently employ two ignitrons in a "back-to-back" circuit as illustrated in Figure 21-5. To initiate the weld, switch Sw, which may be operated by an electronic timer, is closed. If the phase of the a-c supply is such that side A of the a-c line is negative and side B is positive, current flows from side A, through the primary winding of transformer T, and to the cathode of V_1.

Next, the current flows from the cathode to the ignitor, initiating the cathode spot. Then it flows through D_1, switch Sw, through D_4, and to side B of the a-c line. Diode D_2 is a protective device for the ignitor of V_1 and, because of the manner in which it is connected, prevents the current from bypassing the ignitor.

Since the anode of V_1 is connected to side B, it is positive and, once the cathode spot is initiated and emission started, the tube becomes conductive. Therefore, for the remainder of the half-cycle, current flows from side A of the a-c line, through the primary of transformer T, through V_1, and to side B of the a-c line.

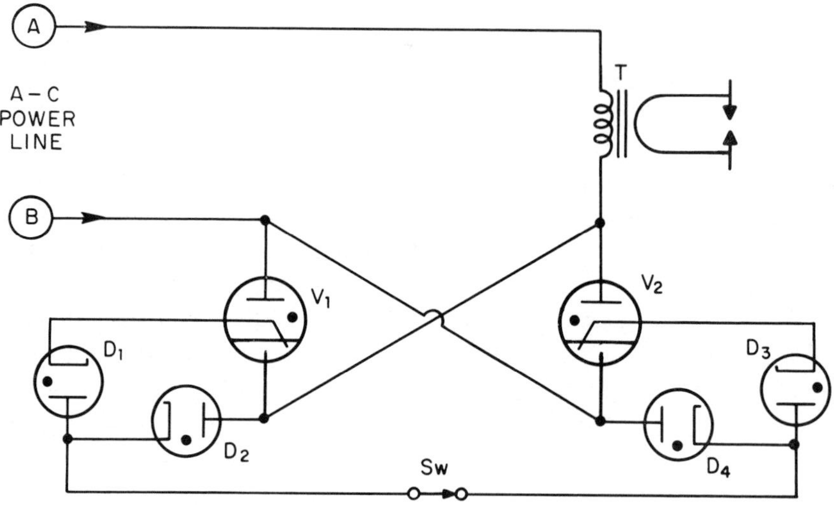

Fig. 21–5. Welder using two ignitrons in a back-to-back circuit.

During the next half-cycle, the polarity is reversed. Current flows from side B of the a-c line, to the cathode of V_2, to its ignitor, through D_3, through switch Sw, through D_2, through the primary of T, and to side A of the line.

Once V_2 becomes conductive, the flow of current is from side B, through V_2, through the primary of T, and to side A of the line. Diode D_4 protects the ignitor of V_2. In practical welders D_1, D_2, D_3, and D_4 are replaced by semiconductor diodes (which will be discussed later in the book).

Note that current flows through the primary of the welding transformer during each half-cycle. But also note that the flow of current is opposite during each alternate half-cycle. Hence there is no rectification.

Neither ignitron can "fire" unless switch Sw is closed. Hence the length of time it is closed, as determined by the electronic timer of which it is a part, determines the duration of the weld time.

So far, we have seen how the duration of the weld time may be controlled. We still need some means for controlling the amount of current flow, which is the other factor determining the heat of the weld. In some welders, this is done by the adjustment of taps on the welding transformer that varies the number of turns of the primary winding.

In our discussion of controlled rectifiers (see Chapter 13, subdivision G) we saw how the thyratron tube may be made to fire at any portion of the cycle through control of its grid. Here, then, is a means for controlling the current output of the ignitron. All we need do is to use the thyratron to initiate the cathode spot of the ignitron.

Thus, by determining the portion of the cycle at which the thyratron fires, we can control the point at which the ignitron fires, and so the portion of the cycle during which current flows through the welding transformer. In this way, we can control the average current flowing through the materials being welded and the heat produced thereby.

A typical "back-to-back" circuit is illustrated in Figure 21-6. Switch Sw, which is operated by an electronic timer, must be closed to initiate the weld. Assume a half-cycle when side A of the a-c line is negative and side B positive.

Ignitor current flows from side A through the load (which is the primary winding of the welding transformer T_1), to the cathode of ignitron V_1, to its ignitor, through diode D_1, to the cathode of thyratron V_4, to its plate, through switch Sw, through diode D_4, and to side B, the other side of the a-c line. Note that this flow of current can take place only if the thyratron V_4 is conductive. This, in turn, depends upon whether its grid (as controlled by the phase-shift circuit) is positive enough to cause the tube to fire.

Once this flow of current takes place, the ignitor of V_1 produces a cathode spot which causes that tube to fire. A heavy current then flows from side A of the a-c line, through the load, through ignitron V_1, and back to the line at side B. During the next half-cycle a similar action takes place, involving ignitron V_2 and thyratron V_3.

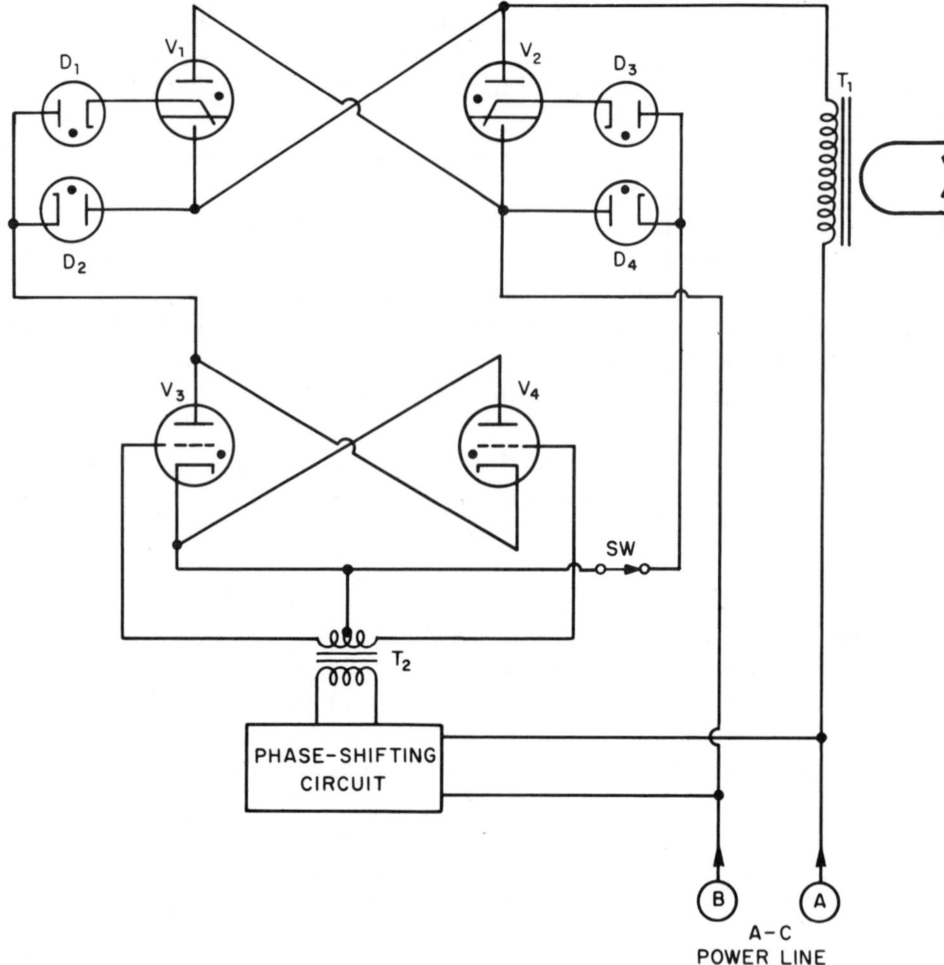

Fig. 21—6. Welder using ignitrons and thyratrons in a back-to-back circuit.

QUESTIONS

1. Explain the basic principles underlying electronic control.
2. Describe and explain the *saturable reactor*.
3. Draw and explain a circuit showing how a saturable reactor may be used to control illumination.
4. Draw and explain a circuit showing how a thyratron may be used to control illumination.
5. Describe the processes involved in resistance welding.
6. Draw and explain the basic circuit of the resistance welder.

7. **Draw and explain the basic circuit of a welder employing an ignitron.**
8. **Draw and explain the circuit of a welder employing two ignitrons in a "back-to-back" circuit.**
9. **Draw and explain the circuit of a welder employing two ignitrons in a "back-to-back" circuit and thyratrons to control the current for the weld.**

22 Miscellaneous Applications

A. *Electronic heating*

1. INDUCTION HEATING

The heating of metals and other conductors by induction is not new, but the application of electronics has expanded the industry tremendously. Basically, it depends upon the heat produced within the conductor in a rapidly changing magnetic field.

The object to be heated is placed within the field of a coil carrying an alternating current. As a result of the current flow, the magnetic field around the coil is built up and collapses in step with the alternations of the current. As the magnetic field sweeps across the object, a voltage is induced that sets up a flow of current within it. This current is called an *eddy current.* Heat in the object is produced by the resistance to the flow of the eddy current.

If the object to be heated is magnetic, there is an additional source of heat. The magnetic field around the coil magnetizes the object, a process which involves the arrangement of its molecules. Each time the magnetic field reverses, which occurs once during each a-c cycle, the molecules are rearranged. As a result of a sort of molecular friction called *hysteresis loss,* this rearrangement produces heat.

Of course, any alternating current flowing through the coil will induce a current in a conductive object within its magnetic field. But the induced voltage, which sets the induced current flowing, is a function of the frequency of the alternating current. The higher the frequency, the greater will be the induced voltage. Hence, high frequencies are employed.

Rotating mechanical generators may be used to produce currents with frequencies up to about 10,000 or 15,000 cycles per second. Such generators are not practical for higher frequencies.

For frequencies between about 50,000 and 200,000 cycles per second, a *spark-gap converter* may be employed. Its basic circuit is illustrated in Figure 22-1. The line voltage is stepped up to a high value by transformer T and is applied to the *L-C* circuit composed of capacitor C and a tapped inductor L. This circuit acts to store up the electrical energy and, at a certain point, discharges through the spark gap in the form of an electric spark. The taps on L are adjusted for optimum results.

This spark is not a one-shot affair. Actually, each discharge produces many rapid back-and-forth surges across the gap in a manner similar to the oscillations of a pendulum set in motion. And, as in the case of the pendulum, the surges are greater at first, rapidly dying away. Then, at the next discharge the process is repeated. Thus the 60-cycle line voltage is converted to a high-frequency alternating voltage across the spark gap. The frequency is determined by the values of C and L.

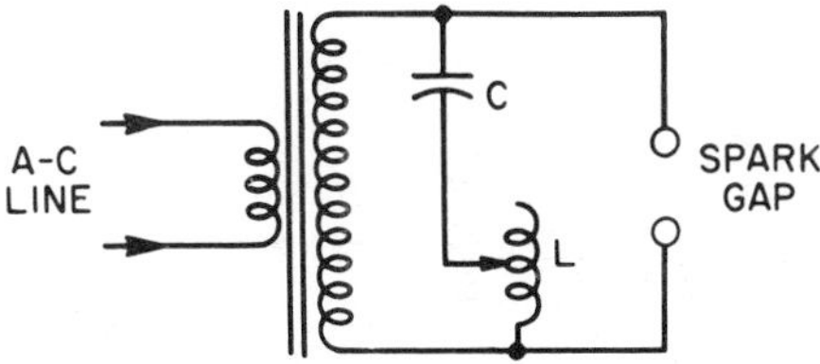

Fig. 22–1. Basic circuit of the spark-gap converter.

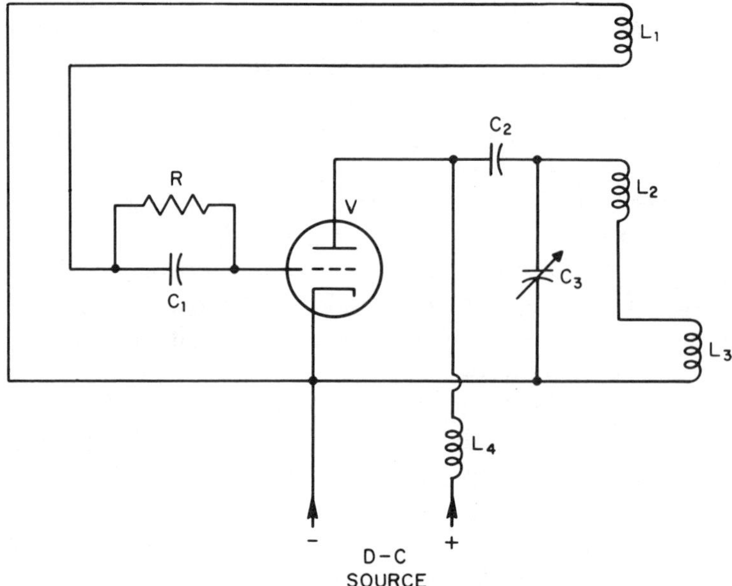

Fig. 22–2. Basic circuit of an oscillator used for induction heating.

We will not discuss further the rotating generator or the spark-gap converter. Instead, we will consider a third, electronic, method for generating the high-frequency currents.

This is the electron-tube oscillator which we studied in Chapter 17, subdivision A. Such oscillators are capable of generating currents and voltages ranging from a few cycles per second to hundreds of thousands, and even millions of cycles per second.

The basic circuit of the oscillator used to generate the high-frequency current needed for induction heating is shown in Figure 22-2. The type illustrated here is the tuned plate–untuned grid regenerative oscillator (although other types, such as the Hartley or Colpitts oscillators may be employed). Because a large power output is required, tube V is a high-power triode such as is employed in broadcast transmission.

Capacitor C_1 and resistor R form the grid leak-capacitor network that develops the bias for the tube. Capacitor C_3 and inductors L_2 and L_3 form the tank circuit that determines the frequency of oscillations. Inductor L_1 is the feedback coil and capacitor C_2 is a blocking capacitor that keeps the d-c plate voltage from the tank.

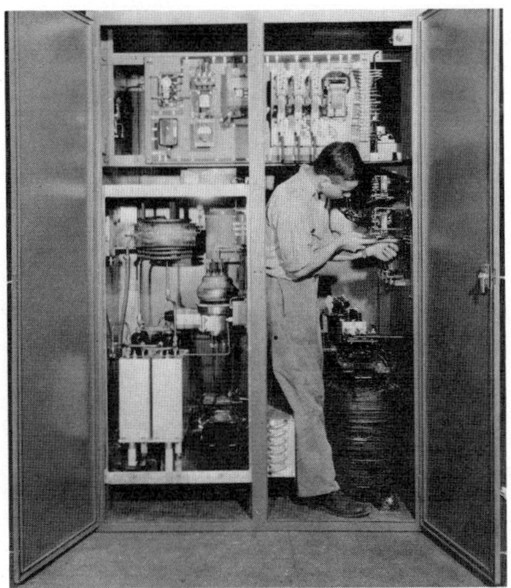

General Electric Company

Fig. 22–3.

Rear view showing interior of induction heating apparatus.

Inductor L_4 is an r-f choke coil that prevents the r-f voltage from entering the d-c source.

Note that the tank inductor is in two parts. Inductor L_3 is the *work coil* that sets up the magnetic field which induces the voltage in the work to be heated. It is important that this coil be shaped to the work in order to concentrate the magnetic field in the object being heated. Frequently, the work coil is formed from copper tubing with water circulating through it for cooling purposes. The frequency of the output voltage may be varied by means of C_3.

For purposes of melting, soldering, brazing, etc., of metals, induction heating performs as well and better than conventional gas and electric heating. Relatively low-frequency currents (up to about 10,000 cycles per second) are employed. Because the heat-producing currents flow inside the work, heating is rapid. Also, because magnetic fields, not heat, are being applied, the air surrounding the work remains cool.

Further, induction heating can perform tasks that are difficult, if not impossible, by other methods. For example, as the glass envelope of the electron tube is being evacuated, it is necessary to heat the electrodes in order that any gas contained in the pores of the metal be driven out. This is easily accomplished by induction heating, which heats the metal electrodes while the glass envelope remains cool.

The advantage of induction heating is expressly marked in the process of surface hardening. An example is the case of a gear wheel whose surface is to be hardened to reduce wear. This is accomplished by heating and quenching in water or oil. The heated steel becomes hard and brittle. It is desirable, however, that the body of the gear wheel remain relatively soft so that it may retain its toughness. This means that a skilled operator must heat the surface just enough so that the body of the wheel remains relatively cool.

In induction heating, the induced currents tend to flow near the surface of the work. Although this tendency is not marked at the lower frequencies, it increases as the frequency of the induced current is increased. By a proper choice of frequency the induced current and, hence, the heat produced by it, may be made to flow through the top few hundredths of an inch of the work. Oscillators used for this process generally operate within a range of about 100,000 to 500,000 cycles per second.

Another use for induction heating is in diathermy where heat is applied to the human body for healing purposes. The high-frequency current flows through a coil surrounding the body or affected member. Since the body is a conductor (though a rather poor one), heat is produced deep within it, though the skin remains cool.

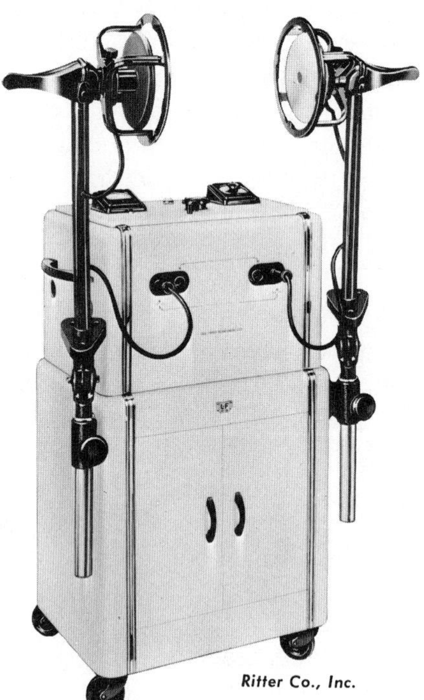

Ritter Co., Inc.

Fig. 22–4.

Diathermy machine.

2. DIELECTRIC HEATING

When the substance to be heated is a nonconductor, or a very poor conductor, another method is employed. This method is called *capacitive,* or *dielectric, heating.* Basically, it consists of placing the substance to be heated between a set of metal plates that are connected to a source of high-frequency voltage. Thus the substance becomes the dielectric of a capacitor consisting of itself and the metal plates.

As you have learned, when an alternating voltage is applied to the plates of a capacitor the orbits of the electrons around the nuclei of the dielectric are distorted, first in one direction and then in the other, in step with the alternations of the voltage. (See Chapter 6, subdivision C.) If the frequency of the applied voltage is high, these changes in orbit, too, have a high frequency. As a result of a sort of internal friction, considerable heat is produced in the dielectric. As in the case of induction heating, an oscillator is employed to generate the high-frequency voltage.

A typical method for coupling the load circuit, containing the work being heated, to the tank circuit of the oscillator is illustrated in Figure 22-5. Capacitor C_1 and inductor L_1 form the tank circuit that is tuned by C_1. Inductor L_2, capacitor C_2, and the capacitor formed by the work between two metal plates constitute the load circuit. This circuit is tuned to the frequency of the tank circuit by C_2.

Inductors L_1 and L_2 form an air-core transformer whereby energy is fed from the tank to the load circuit. Note that the coupling between L_1 and L_2 is variable, hence the energy transferred may be varied.

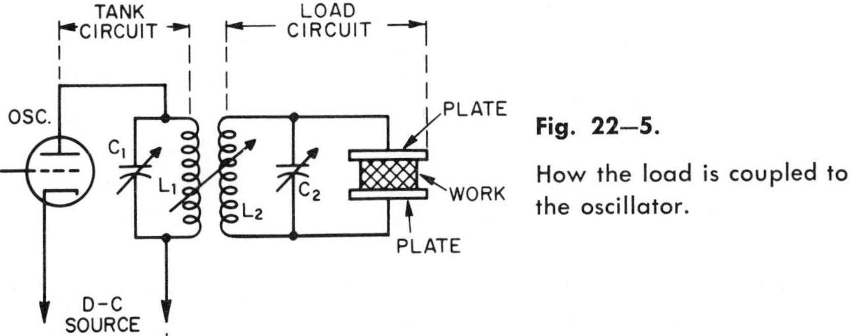

Fig. 22–5. How the load is coupled to the oscillator.

The frequencies used for dielectric heating generally are considerably higher than those employed for induction heating. Frequencies up to about 40,000,000 cycles per second are fairly common. For some applications the frequencies are even higher.

There are two characteristics of dielectric heating that are quite advantageous. Unlike customary heating where the heat travels from the outside toward the center of the object being heated, the heat is generated throughout the entire object. Hence, heating is more rapid and uniform. Also, since the dielectric properties of different substances vary, the heating effect upon these substances, too, will vary.

These qualities are shown to good advantage in the manufacture of plywood which is made by gluing several sheets of wood together. Previously, it was made by placing a layer of glue between the sheets of wood and inserting the sandwich thus formed between steam-heated plates of a press in order to melt and set the glue. Not only did it take days for the heat of the steam to penetrate the wood, but even so, the glue would tend to harden unevenly. For this reason, plywood generally was limited to a thickness of about one inch.

Today, the wood-glue sandwich, acting as the dielectric of a capacitor, is placed between the two metal plates of the press. The plates are connected to the high-power source of high-frequency alternating voltage and a job that used to take days is performed in hours. Because of the difference in their dielectric properties, the glue is quickly melted whereas the wood remains comparatively cool. And because the heat is uniformly distributed through the layers of glue, it melts and hardens evenly. As a result, plywood today can be manufactured with much greater thickness and in a great variety of shapes.

Another widespread use of dielectric heating is in the "electronic sewing" of flexible plastic materials. Such plastics (vinyl, nylon, Orlon, saran, etc.) become soft when heated. Where two pieces are to be joined, the ends are overlapped and placed between the two metal plates that press them together. As the high-frequency voltage is applied to the plates, the ends of the plastic quickly soften and fuse together. When cool, the plastic returns to its normal state and the seam is made.

Dielectric heating is also used for the heat-sterilization of foods that are packed in non-metallic containers. By employing suitable frequencies, the food may be heated while the container remains

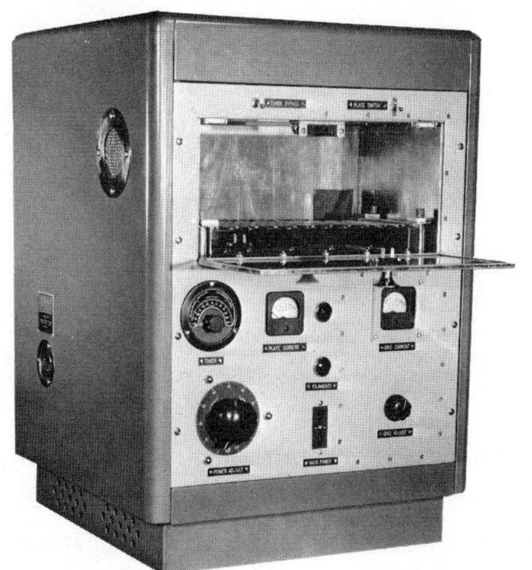

Fig. 22–6.

Electronic drying oven using dielectric heating.

Erdo Engineering Corp.

comparatively cool. Frequencies as high as 150,000,000 cycles per second are employed for this purpose. Because the food is raised to its proper temperature very quickly, its flavor and nutritional values are retained.

We have listed here but a few of the rapidly-growing list of applications for high-frequency heating.

B. *Ultrasonics*

Sound is produced by pressure waves traveling through a medium, such as air. However, such waves are not confined to air. They will travel as well and, in many instances, better, through liquid and solid mediums.

Sound waves are set up by the vibration of some body which transforms its mechanical energy into sound energy. This body may be set vibrating in any of a large number of ways. It may be plucked, as the string of a guitar. It may be struck with a hammer, as in the xylophone. In the human voice, the vocal chords are set in vibration by puffs of air blown over them.

The human ear can respond to sound frequencies ranging from approximately 30 to 15,000 vibrations per second. Some other animals are known to be able to hear sounds of higher frequencies. The Galton whistle, used for signaling to dogs, produces a sound of high frequency which is inaudible to human ears, but is clearly heard by dogs.

Sound with frequencies higher than human ability to hear, is called *ultrasonic*. Its use in industry is widespread and growing rapidly. It is employed in cleaning, drilling, soldering, homogenizing, emulsifying, therapeutic treatment, and lots more.

One of the advantages of ultrasonic sound is that it is inaudible to human ears. Where, as in many industrial applications, high power is required, audible sound would produce an unbearable noise. Also, because of its high frequency, ultrasonic sound can be transmitted in very brief pulses of energy and in narrow beams, a requirement in certain applications.

We may produce ultrasonic sound by mechanical means. For example, one type of remote-control tuning of television receivers employs ultrasonic sound produced by causing a metal bar to vibrate at its natural frequency. Ultrasonic sound may also be produced electronically. For this we need a generator to provide alternating current at the ultrasonic frequencies (currents of approximately 20,000 to 40,000 cycles per second generally are employed for industrial applications) and a *transducer* that changes the electrical energy to sound energy. The generator may be any of the oscillators described in Chapter 17, subdivision A. Where high-power sound is required, the electronic method generally is employed.

Two types of transducers are in common use. One, the *magnetostrictive transducer*, consists of a stack of thin nickel, or nickel alloy, strips bound together and surrounded by a coil through which the alternating current flows. As the current flows through the coil, the stack expands and contracts in step with the alternations of the current, setting up mechanical vibrations. (This type of transducer

Fig. 22–7.

Open view of magnetostrictive transducer.

Acoustica Associates, Inc.

operates on the principle that magnetic substances will change in length under the influence of a magnetic field.)

The other type is the *piezoelectric transducer*, consisting of slabs of certain materials such as quartz, barium titanate, or various other ceramics. As alternating electrical energy is impressed across these slabs, they vibrate at their natural frequencies (see Chapter 17, subdivision A, 4).

One of the earliest uses of ultrasonic sound was in the sonar systems whereby vessels are able to communicate with each other through underwater signals and the positions of submerged objects can be located. The transducer is connected to a diaphragm set in the side or bottom of the ship and pulses of ultrasonic sound of relatively low power are sent through the water. The receiving vessel picks up these pulses on a microphone mounted in its side and converts them to an audible or visible signal.

For depth-finding purposes, the pulses are directed downward and, striking any solid object, are reflected back to a microphone. Knowing the speed at which the sound wave travels through the water (approximately 4,700 feet per second as contrasted with about 1,100 feet per second in air) and the time interval between the transmitted and received pulses, the depth of the object producing the echo may easily be determined.

When high-power ultrasonic sound is directed through a liquid, millions of microscopic bubbles are formed and collapse thousands of times per second. This process is called *cavitation* and results in tremendous localized pressures. One of the chief uses of cavitation is in the cleaning of small or intricate objects where the ordinary cleaning process is long and tedious, or even well-nigh impossible.

The object to be cleaned is immersed in a tank containing water or a detergent solution and the ultrasonic sound is sent through the liquid. Cavitation produces a scrubbing action which literally blasts all grease and dirt from the object to be cleaned in a matter of seconds.

Cavitation is also used in the homogenization and emulsification of liquids that normally are not miscible. Here the action reduces the liquids to extremely small particles which then can mix.

Intricate patterns can be cut or carved into metals, glass, or ceramics by ultrasonic sound. The cutting head consists of a metal rod attached to the transducer. An abrasive slurry, such as carborundum and water, is applied at the point where the cutting head

Fig. 22–8.

Ultrasonic cleaning of radio equipment.

Westinghouse Electric Corp.

touches the material to be cut. As a result of cavitation, the slurry cuts into the material at high speed. Since it is the abrasive that does the cutting, the metal rod may actually be softer than the material that is cut. The same principle is applied in the so-called "painless dental drilling."

The resistance-welding process, you have learned, consists of joining two pieces of metal by passing a high current through their junction, thereby producing a temperature high enough to cause them to soften and fuse. Ultrasonic power can be used to weld metals without the high temperatures. The high pressures produced by the vibrations literally pound the molecules of the metals together, thus forming the weld. Because the heat created is low, fine work can be so welded.

The soldering process consists of joining metals by cleaning and heating them to the point where they can melt solder which forms a bond that holds them together. However, as the metals are heated, they form an oxide film on their surfaces that prevents the solder from taking hold. Accordingly, a flux must be applied that prevents or dissolves away the oxide coating. Some metals, such as aluminum, form oxide films so rapidly that the ordinary flux is ineffective.

Aluminum, however, can be soldered by means of ultrasonic power which is applied to the work surface of the metal through the molten solder. The action of cavitation removes the oxide film from the metal. Hence no flux is needed.

There are many other uses for ultrasonics. To measure the thickness of metal sheets, for example, an ultrasonic pulse is applied to one surface and the time required for the pulse to traverse the metal and be reflected back from the opposite surface is measured. Knowing the rate at which the pulse travels through the metal (approximately 15,000 feet per second), the thickness can be determined.

The same principle can be applied to detect flaws in a thick beam. An ultrasonic pulse is applied to one surface. When the pulse strikes the flaw it will be reflected back from that region and the time it takes for the echo to return reveals its location. If the beam is flawless, the pulse will not be reflected until it reaches the opposite surface.

Another use for ultrasonics is as a liquid-level sensor. A piezoelectric transducer in the shape of a small probe is connected to an oscillator. The probe then is installed into the side of the tank at a certain predetermined level. When liquid surrounds the probe, it produces a loading or damping effect that prevents the oscillator from oscillating. But when the liquid level falls sufficiently so that the probe is surrounded by air, the load is removed and the oscillator starts oscillating. As a result, a relay in the oscillator circuit is energized and this relay, in turn, actuates some indicating device which calls attention to the fact that the liquid level has dropped below that of the probe.

Still another use for ultrasonics is in the field of medical therapy. Sufferers from arthritis have obtained some relief when ultrasonic energy is applied to the afflicted areas. There also is some evidence that it can reactivate paralyzed muscles and stimulate the glandular production of hormones.

These are but a few of the many applications of ultrasonics. New uses are being found almost daily.

C. X-rays

In 1895 Wilhelm Konrad Roentgen, a German physicist, accidentally discovered that when a material is bombarded with high-speed electrons a mysterious invisible ray is emitted. These rays he called *X-rays*. We now know that when the bombarding electrons have enough energy to penetrate the atoms of the bombarded material and disturb the electrons of their inner orbits, X-rays, an electromagnetic wave somewhat similar to light rays, are given off.

X-rays resemble visible light in a number of respects. Both are electromagnetic waves and travel at the same speed, approximately 186,000 miles per second. And both will affect a photographic plate.

There are some differences, however. The frequencies of X-rays are higher, that is, their wavelengths are shorter. Whereas the wavelengths of visible light lie in a range of about 4,000 to 8,000 angstrom units (see Chapter 12, subdivision B), the wavelengths of X-rays lie in a range of about 0.1 to 100 angstrom units. Also, X-rays are able to penetrate materials that are opaque to light rays.

It is this latter property that makes X-rays so useful in medicine. The denser the material, the less the X-rays can penetrate it. Ordinary flesh can be penetrated easily. Bones, which are denser, are more opaque. Hence, if the human body is exposed to X-rays, the fleshy areas are more easily penetrated and produce a black area on a photographic plate at the other side of the body. The denser bones retard the rays and thus they appear white on the photographic plate when it is developed. In this way the structure of the bones can be studied and a bone fracture, for example, will be clearly revealed.

Tuberculosis causes calcified spots to appear in the lung tissue. Since these calcified spots are more opaque to X-rays than the lung tissue, an X-ray photograph of the lungs can be used to detect the presence of the disease.

The taking of X-ray pictures is called *radiography*. X-rays also have the property of causing certain chemicals, called *phosphors*, to *fluoresce*, that is, to glow and give off visible light. Hence, if a screen is coated with these phosphors and placed behind the body, the penetrating X-rays will produce a picture in light upon the screen whereby the denser portions appear darker than the less dense portions. This process is called *fluoroscopy*. It is used by doctors to observe the functioning of internal organs. For example, to study the action of the digestive tract, the patient is given a barium solution to drink. Because barium is quite dense, the solution is opaque to X-rays. Then, by exposing the patient to the X-rays and observing the picture produced on a fluorescent screen, the progress of the barium solution through the digestive tract may be seen and any irregularities detected.

Prolonged exposure to X-rays can destroy human tissue. In this way certain skin diseases may be treated. The rays are also used in a sort of surgery where deep-lying tumors or other growths are destroyed without the use of a scalpel. The rays are focused to a needle-like beam which penetrates the body to the growth.

The X-ray tube is a diode, consisting of a filament that emits electrons when heated and an anode, or target, usually of tungsten,

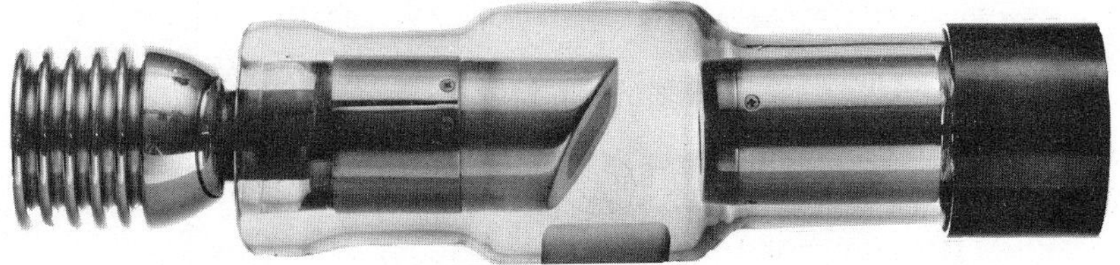

General Electric Company

Fig. 22–9. X-ray tube.

that attracts these electrons. (See Figure 22-9.) The whole is enclosed in an evacuated glass envelope. When the electrons strike the slanted face of the target, X-rays are produced that shoot off in a beam through the side of the envelope.

To give the electrons the necessary high speeds (approaching that of light), very high filament–anode voltages are required. The higher the anode voltage, the greater will be the speed of the electrons. And the greater the speed of the bombarding electrons, the higher will be the frequency of the X-rays produced. Rays whose frequencies are relatively low, are called *soft rays* and their penetrating power, too, is relatively small. Where the frequencies are high, the rays are called *hard rays* and their penetrating power is great.

For ordinary medical purposes soft rays, produced by potentials of about 20,000 to 60,000 volts, are employed. For deep-seated surgery, hard rays are used. These rays are produced with 1,000,000, or more, volts on the anode. The heat produced as the electrons bombard the target is quite high and tungsten generally is employed because it can stand this heat. The fins attached to the anode of the tube shown in Figure 22-9 are for the purpose of radiating away this heat into the surrounding air. In certain high-power installations the anode is water-cooled.

A typical X-ray circuit is illustrated in Figure 22-10. Transformer T_1 is an autotransformer that varies the voltage applied to the primary of the step-up transformer T_2 (through the adjustment of the taps on T_1 by its associated switch Sw). Hence the voltage output of T_2 can be varied and, thus, the frequency and penetrating power of the resulting X-rays. Transformer T_3 is a step-down transformer that reduces the line voltage to that required by the filament.

Resistor R varies the temperature of the filament and, so, the number of electrons it emits. This, in turn, varies the quantity or intensity of the X-rays produced.

Note that the X-ray tube is self-rectifying. That is, it is operative only on those half-cycles when the anode is positive. This generally does not interfere with the operation of the equipment and the necessity for high-voltage rectifiers is thus eliminated.

There are many industrial uses for X-rays. For example, flaws and air pockets that can weaken it structurally may exist deep inside of a heavy metal casting. The casting may be checked easily and quickly by means of X-rays. The flaws and air pockets are less dense to the rays than the surrounding metal and their presence will be revealed on a photographic plate or fluorescent screen. For this purpose, hard X-rays are used because of their greater penetrating power.

In Figure 22-11 you see how X-rays are used to check the weld joining large metal pipes. The entire apparatus is portable. The X-ray tube is contained in a protective housing and is placed inside the pipe. The photographic film is attached to the outside of the pipe at the seam. The developed film readily reveals any defects, such as air pockets, that may be present in the weld.

In another application, metallic and other impurities in food contained in cardboard cartons can be detected. The cartons are conveyed on a moving belt in front of an X-ray tube. Because the impurities are denser than the surrounding food, their presence can readily be detected on a fluorescent screen at the other side of the carton.

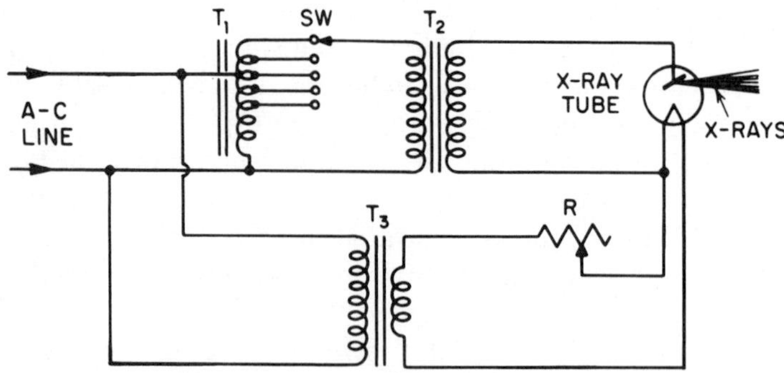

Fig. 22—10. Typical X-ray circuit.

General Electric Company

Fig. 22–11.

How X-rays are used to check the seams in welded pipes.

X-rays may be used in a thickness gage to measure the thickness of materials. As the rays pass through the material, a certain amount of the X-rays is absorbed. The thicker the material, the more it will absorb. Hence, by knowing the original intensity of the X-rays and measuring their intensity after they pass through the material, we can calculate the thickness of the latter.

To measure the intensity of the X-rays, we may permit them to fall upon a fluorescent screen. The more intense the rays, the brighter the screen will glow. The brightness of glow can be measured by means of a phototube and a meter that measures its output. Hence the intensity of the X-rays may be determined.

Another means for measuring the intensity of the X-rays is to permit them to strike a crystal of cadmium sulfide. This crystal has the property of changing its resistance when struck by X-rays. The more intense the X-rays, the lower will be the resistance of the crystal. Thus, by measuring the changes in a current flowing through the crystal, the intensity of the X-rays striking it can be determined. (The cadmium-sulfide crystal will be discussed further, later in the book.)

In another application, X-rays are used to determine whether opaque containers are properly filled. As the containers pass by on a moving belt, an X-ray beam is directed at the side of each container on a level with the top of the material each is supposed to contain. Should a container be underfilled, the air in the space will absorb less of the rays than would the material in a properly-filled container. As a result, a cadmium-sulfide crystal at the other side of the container would indicate a less-than-normal resistance. The resulting greater current then can be used to operate a relay that removes the defective container from the belt.

X-rays are exceedingly dangerous. Prolonged exposures may cause burns and cancer. Further, they readily affect the genes of the reproductive cells of the body and may produce sterility. Great care must be taken when working with them. To avoid scattering of the rays, all but a small window in the tube is shielded with lead. This lead, being very dense, absorbs practically all the rays, permitting very few to get through. In addition, the operator must avoid getting in the direct path of the rays. Lead-shielded screens usually are provided behind which the operator stands while operating the apparatus by remote control. Where a fluorescent screen is employed, a mirror generally is set up so that the operator may observe the screen while standing to one side, and so out of the direct path of the rays.

D. *Radiation counters*

When the nucleus of an atom is split, as in the atomic reactor, it emits particles and rays called *radiations*. Some atoms, such as those of radium, emit radiations spontaneously. So do some of the man-made radioactive materials. An examination of these radiations reveals that they consist, for the most part, of two types of particles, called *alpha* and *beta,* and a ray called *gamma.*

The alpha particle was found to be the nucleus of the helium atom, consisting of two protons and two neutrons, and hence, charged positively. The beta particles are electrons carrying negative charges. The gamma ray is an electromagnetic wave somewhat similar to X-rays.

The gamma rays travel at the speed of light. Hence, they are extremely penetrating. Such a ray may pass through seven or eight inches of lead.

The beta particles travel at nearly the speed of light. Because of their greater mass, however, they are not nearly as penetrating as the gamma rays. A beta particle will be stopped by several yards of air.

Alpha particles travel at speeds ranging from about one-tenth to one-hundredth the velocity of light. Because of their lower speed and greater mass, these particles are the least penetrating. An alpha particle will be stopped by a single sheet of paper or several inches of air.

These radiations are extremely dangerous to man, especially the gamma rays because of their penetrating ability. But they also have been put to use. For example, radioactive chemicals are used in medicine to trace the course of drugs through the body. In agriculture, they have been used to measure the degree to which a fertilizer is absorbed by a plant and to map its course. These applications depend upon the fact that the radioactive particle can be located by suitable devices which detect and measure its radiations.

The development of atomic power has made available a large supply of radioactive byproducts that are finding increasing employment in industry. For example, the thickness of various materials can be measured by placing some of the radioactive substance below the sheet being gaged and a radiation counter above it. As the radiations pass through the sheet, some are absorbed. The thicker the sheet, the greater is the absorption. Hence, by measuring the amount of radiation that gets through, we can determine the thickness of the sheet.

In another application, the efficiency of a lubricating oil in a gasoline engine may be determined. Some of the radioactive material is incorporated into the pistons of the engine. As these pistons wear, slight amounts of the radioactive material, too, are worn away and find their way into the crankcase. The less efficient the lubricating oil, the greater is the wear, and the more radioactive material in the crankcase. Thus, by measuring the radioactivity of the crankcase material we can determine the lubricating ability of the oil. These are but a few of a large, and ever-growing number of applications.

One of the most frequently employed radiation detectors is the *Geiger-Müller tube.* (See Figure 22-12B.) This tube is a diode. Its cathode is a long metal cylinder and the anode is a fine wire running through the center of the cylinder. Both are mounted in a thin-walled glass envelope. The air is evacuated from the envelope and a very small amount of gas, such as argon, is added.

If the anode is made positive and the cathode negative, and the potential difference is increased gradually, a point will be reached where the gas will ionize. Current will flow through the tube as the negative electrons move toward the positive anode and the positive ions go toward the negative cathode.

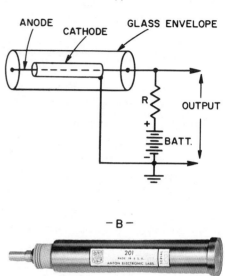

-A-

ANODE
CATHODE
GLASS ENVELOPE
R
OUTPUT
BATT.

Fig. 22–12.

A. Circuit of the Geiger-Müller tube.

B. Geiger-Müller tube using a mica end window for detecting alpha, beta, and gamma rays.

-B-

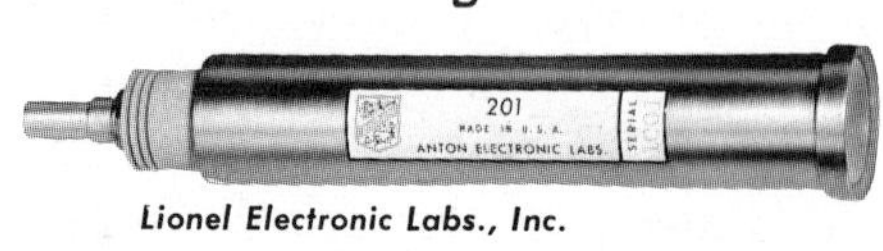

Lionel Electronic Labs., Inc.

Suppose, however, that the voltage across the electrodes is stopped short just below the ionizing point of the gas in the envelope. Obviously, no current will flow through the tube. If, at this point, some radiation should succeed in penetrating the glass envelope, it would collide with some of the gas atoms, causing them to ionize. The ions and electrons so produced would rush to their respective electrodes and, in their progress, would strike against and ionize other gas atoms. The action would be as if a trigger were pulled and complete ionization would take place almost immediately. A pulse of current would flow through the tube and through R, which is in its plate circuit. The resulting voltage drop across R would be the output voltage.

Once ionization is established, we would expect the current to continue flowing through the tube, just as it does in the gas diodes previously discussed. Note, however, that resistor R (which has a high resistance) is in series with the anode of the Geiger-Müller tube and the battery that is the voltage source. As long as the gas is not ionized (and no current flows through the tube), there is, of course, no voltage drop across this resistor. Hence the voltage on the anode of the tube is the same as that of the source.

However, as the gas ionizes and current flows through the tube (and, hence, through R), a large voltage drop takes place across the resistor. This voltage drop is sufficient to reduce the voltage on the tube below the ionizing point of the gas. The gas is de-ionized and current ceases flowing through the tube—until the radiations again penetrate the tube envelope.

In this way, a series of alpha or beta particles, or bursts of gamma rays may cause a series of current pulses to flow through the tube and its plate circuit. The output pulses from the tube can be amplified and registered on some indicating device as flashes of light or clicks in a loudspeaker. By counting the number of flashes or clicks, we may tell how many particles entered the Geiger-Müller tube in a given period of time and so have an indication of the intensity of radiation. In some indicators the pulses are stored up and the cumulative result shown on a meter. The combination of tube, amplifier, and indicating device is known as a *Geiger-Müller counter*.

In addition to its other applications in industry, the Geiger-Müller counter is used extensively to check the radiation intensities of materials to see if they are safe to handle and of areas to determine if they are safe to enter. The counter also is used as a prospecting tool to locate uranium ore (which is radioactive).

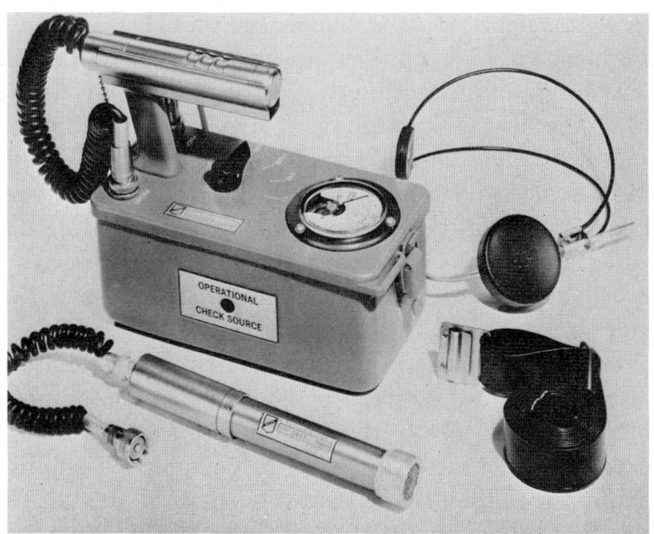

Lionel Electronic Labs., Inc.

Fig. 22–13. Radiation detector using a side-wall tube for beta and gamma radiations and an end-window tube for alpha radiations.

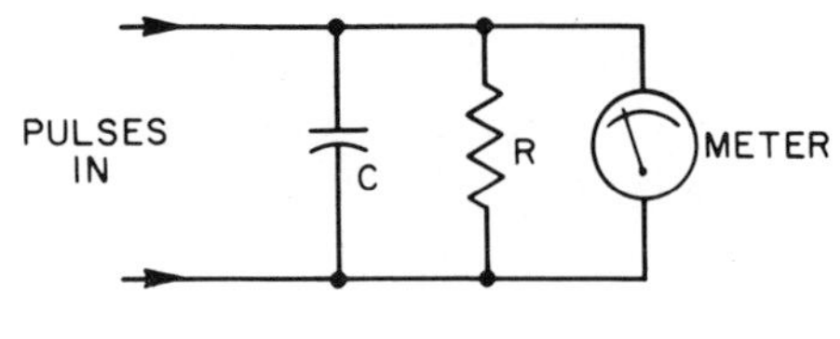

Fig. 22—14.

How a meter is used to indicate the frequency of the pulses.

There are a number of different types of Geiger-Müller tubes, each designed for a specific purpose. Tubes to be used for detecting particles with low penetrating power must be made of extremely thin glass or else have special windows made of aluminum foil, mica, or some other substance that can be penetrated by these particles. Of course, this is not necessary for high-penetrating gamma rays.

To convert the frequency of the incoming pulses into a reading upon a meter, the output pulses are stored in a capacitor that is connected across the meter. Thus the meter measures the voltage of the charge on the capacitor. (See Figure 22-14.)

During the interval between pulses, capacitor C partially discharges through resistor R. If the frequency is low, that is, the pulses are relatively far apart, most of the charge on C leaks off through R. If the frequency is high (pulses close together), C has very little time to discharge and, accordingly, retains most of its charge. You can see that the average charge on C depends upon the frequency with which it receives the pulses. This charge is indicated by the meter whose scale is calibrated to read the frequency of the pulses.

This holds true if all the pulses from the Geiger-Müller tube are of the same amplitude. Unfortunately, this is not always so. Accordingly, we require some device that will equalize all the pulses before they are stored in the capacitor.

The circuit of the pulse equalizer is illustrated in Figure 22-15. This is, essentially, a two-stage resistance-coupled amplifier with certain modifications. (Note that it closely resembles the circuit of the triggered multivibrator discussed in Chapter 17, subdivision B, 3.) Normally, since the grid of V_1 is slightly positive with respect to its cathode, that tube is conducting. Because of this current flow, the voltage drop across R_2 places a bias upon V_2 sufficient to cut it off.

The input pulse from the Geiger-Müller tube is negative-going and, when applied to the grid of V_1 decreases its plate current and increases the voltage at its plate. Since this plate voltage is applied to the grid of V_2 through C_2, V_2 is brought out of cutoff and starts to conduct. As plate current flows in V_2, the corresponding decrease in its plate voltage is coupled through C_3 to the grid of V_1, driving it still more negative. This, in turn, increases the plate voltage of V_1 and makes the grid of V_2 still more positive. This process continues until V_1 is cut off and V_2 is conducting heavily.

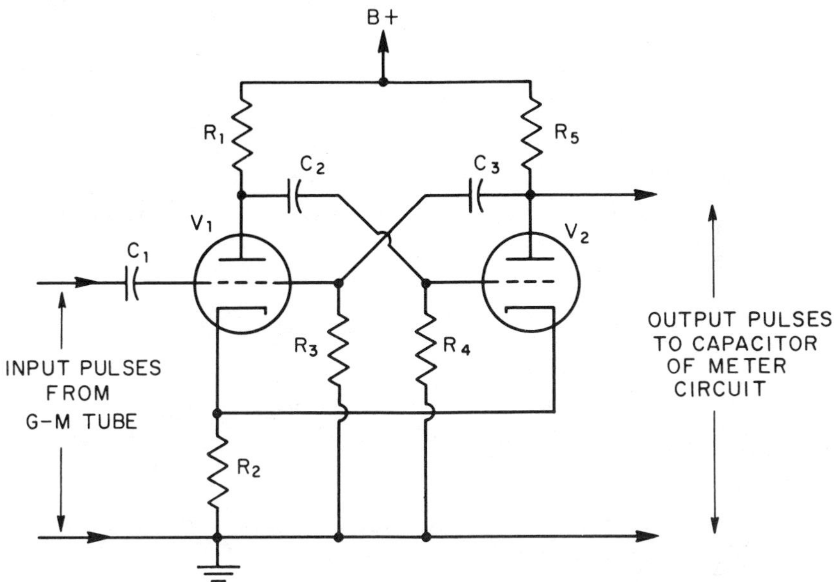

Fig. 22—15. Circuit of the pulse equalizer.

The negative bias driving V_1 to cutoff is maintained by the charge on C_3, which leaks off gradually through R_3. When this charge is reduced to the point where V_1 comes out of cutoff, the entire process is reversed. As V_1 conducts, its plate voltage drops, making the grid of V_2 less positive. This reduces the plate current and increases the plate voltage of V_2. As a result, the grid of V_1 becomes more positive and this, in turn, increases its plate current and reduces its plate voltage. The process continues until the circuit once again attains its original state with V_1 conducting, and V_2 cut off. It now is ready for the next pulse from the Geiger-Müller tube.

Although the explanation of how this circuit works is lengthy, the action is quite rapid. Hence there is one output pulse produced for each incoming pulse. The amplitude of the output pulse is determined by the values of the components surrounding V_2. Thus all the pulses fed to the capacitor of the meter circuit are alike and their frequency is determined by the frequency of the pulses coming from the Geiger-Müller tube.

A *scintillation counter,* using the photomultiplier tube described in Chapter 12, subdivision C, may be used instead of the Geiger-Müller counter. Certain crystals, such as sodium iodide, produce a brief flash of light each time the crystal is struck by a particle or gamma ray from a radioactive substance. These flashes are "piped" to the window of the tube by means of a lucite rod. Thus each particle or ray produces a pulse of plate current at the output of the tube. The balance of the circuit is similar to that previously described —pulse equalizer and meter circuit. The advantage of the scintillation counter over the Geiger-Müller counter is the greater sensitivity of the former.

QUESTIONS

1. a) **Explain the principle upon which *induction heating* operates.**
   b) **What are the advantages for induction heating over conventional methods?**
   c) **Give an example where induction heating is employed.**
2. a) **Explain the principle upon which *dielectric heating* operates.**
   b) **What are the advantages for dielectric heating over conventional methods?**
   c) **Give an example where dielectric heating is employed.**
3. **Explain what is meant by *ultrasonic sound.***
4. **Describe and explain the *magnetostrictive transducer.***
5. **Explain what is meant by *cavitation.***
6. **Explain how ultrasonic sound is used to locate submerged objects.**
7. **Explain how ultrasonic sound is used for cleaning purposes.**
8. **Explain how ultrasonic sound is used for soldering aluminum.**
9. **Explain how ultrasonic sound may be used for detecting flaws in metal castings.**

10. Describe and explain the theory of the *X-ray tube.*
11. Explain the difference between *soft* and *hard* X-rays.
12. Explain how X-rays are used to produce photographs of the organs of the human body.
13. Explain how X-rays may be used to detect flaws in metal castings.
14. Explain how the cadmium-sulfide crystal may be used to measure the intensity of X-rays.
15. Explain how X-rays may be used as a thickness gage to measure the thickness of materials.
16. Describe and explain the theory of the *Geiger-Müller tube.*
17. Draw and explain the basic Geiger-Müller circuits.
18. a) Describe and explain the *Geiger-Müller counter.*
    b) Explain how it is used to measure the intensity of radiations.
19. Explain how the Geiger-Müller counter may be used to measure the efficiency of a lubricating oil.
20. a) Describe and explain the *scintillation counter.*
    b) Explain how it is used to measure the intensity of radiation.

Section 5

The Electron Tube in Communication and Entertainment

Electronics, in the modern sense of the word, had its birth in the field of communication. The diode and triode were invented during the search for a better radio receiver. As a matter of fact, for a considerable period of time "electronics" and "radio" were practically synonymous. It was during the World War II period that electronics made its greatest advances in the field of industry.

In this section we shall consider the role of the electron tube in radio broadcasting, television broadcasting, and radar. Of necessity, the treatment of each subject cannot be exhaustive; space does not permit it. Nor can we delve deeply into the circuitry involved. Entire volumes have been written on each topic.

Instead, we shall draw upon what we have already learned and see how it is applied. New material shall be introduced as needed.

23 Radio Broadcasting

Communication is the act of sending information from one place to another. Thus it entails two factors. One is the information or intelligence we wish to transmit. The other is the carrier that transports the intelligence. At the radio transmitter the carrier—the radio wave—is generated and the intelligence—sound in the form of speech or music—is "mounted on its back." At the radio receiver the carrier and its burden are received. Then the two are separated and only the intelligence is retained.

A. *The radio transmitter*

1. THE CARRIER CURRENT

The electric current can serve as the carrier. Indeed, it is employed as such in the telegraph and telephone systems. Its chief drawback lies in the fact that it requires a conductor—a wire—

stretching between the transmitter and the receiver. Hence it cannot be used for communication with an airplane in flight or a ship at sea.

As a current flows through a conductor, the latter is surrounded by a magnetic field that extends into space. Hence the magnetic field may be used as a carrier were it not for the fact the field falls off rapidly the further it extends.

In 1887, Heinrich Rudolph Hertz, a German scientist, discovered that when an alternating current flows through a conductor it radiates into space an electromagetic wave which we now call the *radio wave.* This wave can travel great distances at the speed of light (186,000 miles per second) and its field does not fall off as rapidly as does the magnetic field. Here, then, is our carrier which can be used without the necessity for connecting wires.

As the alternating current flows through a conductor known as an *antenna,* the radio waves are radiated into space. Although alternating currents of all frequencies send out these radiations, only if the frequencies are high—upward of about 15,000 cycles per second—can these radio waves be sent out and received readily.

Each transmitter is assigned a definite operating frequency by the Federal Communications Commission. Electron-tube oscillators are employed to generate the alternating current. Any of the *L-C* oscillators previously discussed (see Chapter 17, subdivision A) may be used, though the crystal oscillator, because of the frequency stability of its output, generally is preferred.

The desired frequency is obtained by choosing suitable values of inductance and capacitance for the oscillating circuits and choosing a crystal with the proper natural frequency. Where the final frequency is to be a whole-number multiple of the oscillator frequency, frequency multipliers as described in Chapter 16, subdivision A, 1, are employed. R-f amplifiers are used to bring the output of the oscillator up to the power level suitable for feeding the antenna. This power level may be hundreds, sometimes even thousands of kilowatts.

The waveform of the *carrier current* is sinusoidal and of a constant frequency and amplitude. (See Figure 23-1.) The waveform of the radio wave is similar to that of the antenna current producing it. If, as we shall see, the waveform of the carrier current is modified by the addition of the intelligence we wish it to carry, the waveform of the radio wave produced by it will carry similar modifications.

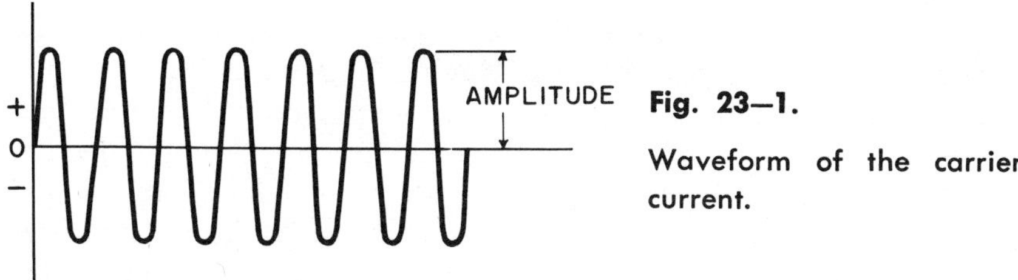

Fig. 23—1. Waveform of the carrier current.

2. THE SOUND CURRENT

The intelligence we wish to convey is sound. Any elastic substance, be it the human vocal chords or the string of a violin, will vibrate if it is struck or plucked. The number of vibrations per second depends upon the size, shape, and material of the thing vibrating. We call the number of vibrations per second the *frequency.*

Let us see what happens as a violin string vibrates. Look at Figure 23-2. As it comes forward (Figure 23-2A), the string crowds the molecules of air in front of it, producing a compression, or *condensation.* This condensation travels away from the string, passing its energy from molecule to molecule of the air in its path.

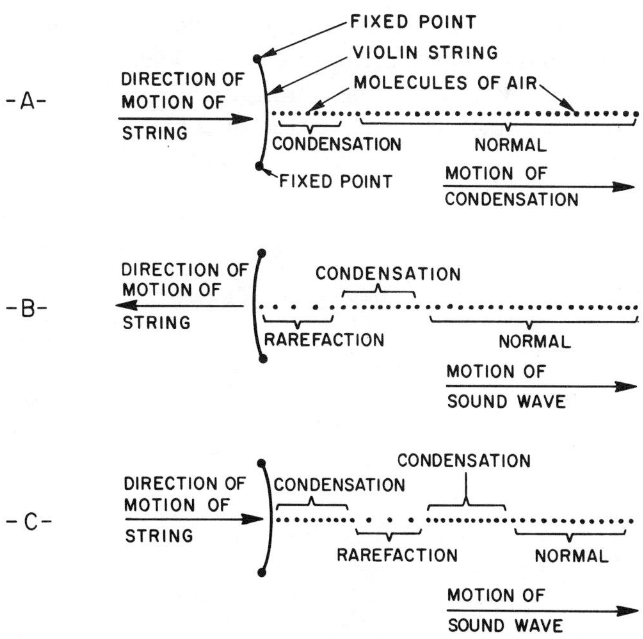

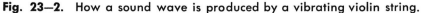

Fig. 23—2. How a sound wave is produced by a vibrating violin string.

As the string moves back (Figure 23-2B), it leaves a space into which the air molecules can rush. Having more room, the molecules are spaced further apart. We say that a *rarefaction* has been produced. The rarefaction follows the condensation through the air. As the string continues to vibrate, alternate condensations and rarefactions are produced in the air next to it. These alternate condensations and rarefactions travel through the air and are called *sound waves* (Figure 23-2C).

A condensation and its adjacent rarefaction constitute a cycle. The number of cycles per second is the frequency of the sound wave. If the frequency lies between approximately 100 and 15,000 cycles per second, the sound can be heard by human ears.

The *pitch* of the sound is determined by its frequency. Sounds of low frequency are called *low-pitched;* sounds of high frequency are called *high-pitched.* The *loudness* of the sound depends upon the force, or intensity, with which the object producing the sound vibrates. Thus, if the violin string of our illustration is plucked gently, it will set up feeble condensations and rarefactions in the air. On the other hand, should the string be plucked vigorously, the condensations and rarefactions will be stronger.

The next step is to transform the sound wave into an electric current that fluctuates in step with the condensations and rarefactions. This is the function of the *microphone*. There are many different types of microphones, but the simplest is the *carbon-button* type illustrated in Figure 23-3. This consists of a large number of small

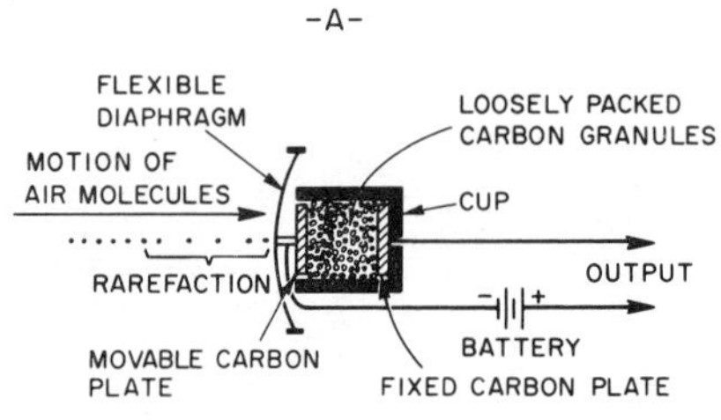

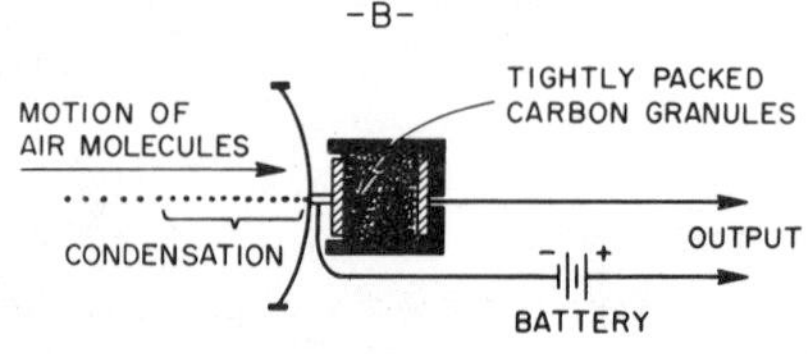

Fig. 23–3. The carbon-button microphone.

A. Rarefaction.

B. Condensation.

carbon granules loosely packed in a cup between two carbon plates. The back plate is held firmly in place, but the front one can move back and forth. This movable carbon plate is fastened to a thin, flexible diaphragm that is held firmly at its rim.

Suppose, as in Figure 23-3A, that a rarefaction of the sound wave approaches this diaphragm. Because of the low pressure produced by this rarefaction, the diaphragm bulges out, carrying with it the movable carbon plate. As a result, the carbon granules are more loosely packed than before. When a condensation approaches the flexible diaphragm (Figure 23-3B) the latter is pushed in and the carbon granules become more tightly packed.

The electrical resistance between the two carbon plates through the carbon granules depends upon how tightly packed the granules are. If they are loosely packed, the resistance is greater. If they are tightly packed, the resistance is smaller. Thus the granules form a variable resistor whose resistance depends upon whether a rarefaction or a condensation of the sound wave approaches the diaphragm and their strengths.

Now, suppose that we apply a steady direct current across the two carbon plates. The current will remain steady as long as the carbon granules are not disturbed. But as a rarefaction approaches the diaphragm, the resistance of the granules increases. As a result, the current drops. As a condensation approaches the diaphragm, the resistance decreases and the current rises.

You can see this relationship in the graphs of Figure 23-4. Note that the closer the air molecules are crowded together, the lower is the resistance of the microphone, and the larger the current. The further the molecules are spread apart, the greater is the resistance of the microphone, and the smaller the current. Thus we have an electric current whose variations reflect the variations of the sound wave.

This is the *sound current* that contains the intelligence we wish to transmit. Note that the variations are at audio frequency. Note, too, that the waveform here is sinusoidal. This would be so if the sound were of a single frequency. In nature, however, we rarely find such a single-frequency sound. Most generally, sound contains waves of a number of frequencies produced simultaneously. The result, then, is a sound current whose waveform is quite complex. In our discussion here, however, we will consider only the simple sinusoidal waveform for the sake of simplicity.

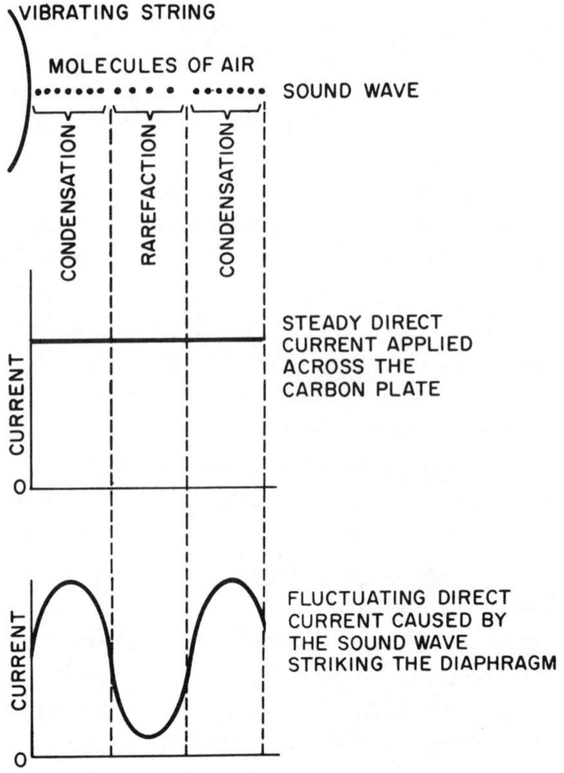

Fig. 23–4. Graph showing how the sound wave striking the diaphragm of the microphone converts a steady direct current into a fluctuating direct current.

The sound current from the microphone is fed to the control grid of an a-f amplifier by means of a transformer (T). (See Figure 23-5.) Frequently, one or more additional stages of a-f amplification are required before the current reaches a level suitable for mixing with the carrier current. The final a-f stage generally is a push-pull power stage operating in Class B. Resistor R is a potentiometer that functions as a *volume control* by varying the voltage placed on the control grid of the first a-f stage and thus controls the final output.

3. AMPLITUDE MODULATION

The primary purpose of radio transmission is to convey intelligence from the sender to the receiver. Thus, merely to generate and radiate the carrier wave is meaningless, as meaningless as sending a letter that contains a blank sheet of paper. Something must be done to the carrier wave if it is to convey a message.

The carrier current, and the radio wave produced by it, have two inherent characteristics which lend themselves to our purpose. These are the *amplitude* and *frequency.* Varying either of these characteristics according to a prearranged plan will permit us to "write" on the "blank sheet" of the carrier wave.

This process of varying either the amplitude or frequency of the carrier current is called *modulation.* If the amplitude of the current is varied, we call the process *amplitude modulation* (AM). If the frequency is varied, it is called *frequency modulation* (FM). The ordinary radio broadcast stations, lying in a band of frequencies ranging from 535 to 1,605 kc, employ amplitude modulation. FM broadcast stations, occupying a band of frequencies ranging from 88.1 to 107.9 mc, are frequency-modulated.

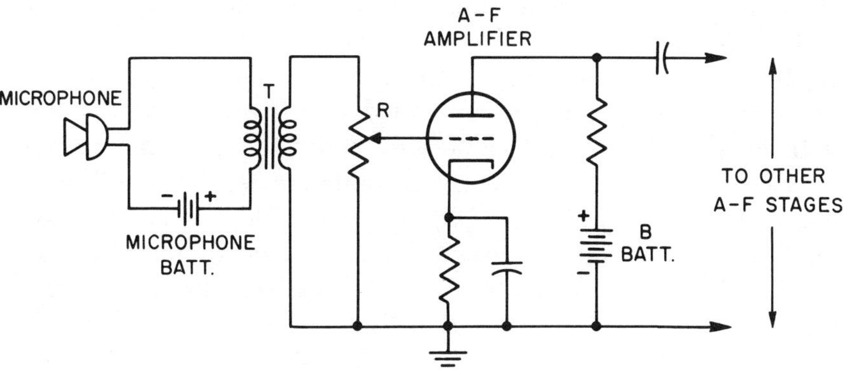

Fig. 23—5. How the microphone is connected to the a-f amplifier.

We will consider first amplitude modulation. The carrier current, generated by the oscillator and amplified by the r-f amplifiers, is a radio-frequency current of constant frequency and amplitude. See Figure 23-6A. The sound current, produced as the sound waves strike the microphone and amplified by the a-f amplifiers, is an audio-frequency current whose waveform varies in step with the variations of the sound waves (refer to Figure 23-4). The a-c component of this current is shown in Figure 23-6B. (Note that we are still using the simple sine wave for purposes of illustration.)

When the two currents are combined, the result is a *modulated carrier current* whose waveform is shown in Figure 23-6C. Note that it still is an r-f current whose frequency is the same as the carrier current. Its amplitude, however, varies in accordance with the variations of the sound current, as shown by the dotted line (called the *envelope*) connecting its peaks. Note, too, that the negative portion of the modulated carrier current is the mirror-image of the positive portion.

There is more. When the carrier current is combined with the sound current the result is not only the modulated carrier current, but two more currents. One is a current whose frequency is equal to the sum of the frequency of the carrier current and the frequency of the sound current. The other is a current whose frequency is equal to the frequency of the carrier current minus the frequency of the sound current.

Let us assume that the frequency of the carrier current is 1,000 kc. If we were to modulate this carrier by an audio-frequency current whose frequency is 1 kc, we would get two new frequencies, 1,000 kc plus 1 kc, or 1,001 kc, and 1,000 kc minus 1 kc, or 999 kc. These two new frequencies are called *side frequencies.*

If, instead of using a 1-kc note to modulate our carrier current, we were to use audio-frequency currents created by sound waves whose frequencies vary up to, say 5 kc, we would get, not two single side frequencies, but two whole bands of side frequencies. One band would lie between 1,000 kc and 1,005 kc, whereas the other would lie between 1,000 kc and 995 kc. We call these bands, *sidebands.*

Thus, a modulated carrier occupies a band of radio frequencies, or *channel,* rather than a single frequency as in the case of the unmodulated carrier. The channel width is twice the highest modulation frequency. In the example given here, the channel width would be twice 5 kc, or 10 kc. As the modulated carrier current and its sidebands flow through the antenna, a radio wave of similar frequencies and waveform is radiated.

Modulation may be applied to any stage of the carrier current. However, it is not desirable to apply the modulation to the oscillator stage since this may interfere with the frequency stability of its output. Accordingly, modulation is applied to subsequent stages, frequently the final r-f stage.

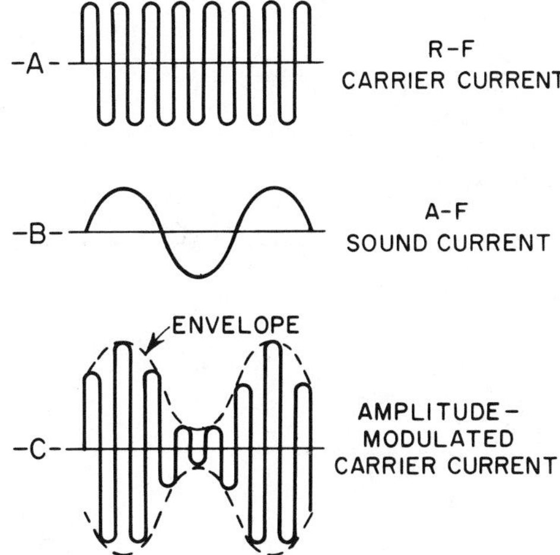

Fig. 23—6.

Graphs showing how the amplitude-modulated carrier current is produced.

A typical modulation circuit is illustrated in Figure 23-7. The final a-f amplifier stage, now called the *modulator stage,* consists of a stage of a-f push-pull amplification. The only difference between the modulator stage and an ordinary stage of a-f amplification lies in the fact that whereas the a-f amplifier feeds into an output transformer, which, in turn couples the loudspeaker to the amplifier, the modulator stage feeds into a *modulation transformer* that couples the a-f variations to the plate circuit of the r-f amplifier.

Note that the plate voltage of the r-f amplifier is applied through the secondary of the modulation transformer. The a-f voltages in this secondary vary this d-c plate voltage, and the result is that a varying direct voltage (varying in step with the a-f variations) is applied to the plate of the r-f amplifier. These variations cause the current fed to the antenna to vary with the a-f variations, and hence a modulated wave is radiated.

The r-f choke coil (RFC) between the secondary of the modulation transformer and the plate tank of the r-f amplifier is used to offer a high impedance to the r-f currents, and thus keep them from flowing down into the plate-voltage supply. Note that separate plate-voltage supplies are indicated for the modulator and the r-f amplifier. This practice is followed where powerful transmitters are employed. For low or medium power, the same plate-voltage supply may be used for both.

This method of coupling the a-f variations to the carrier is called *transformer coupling.* There are other coupling methods. Also, modulation may be applied to the grid circuit or the cathode circuit

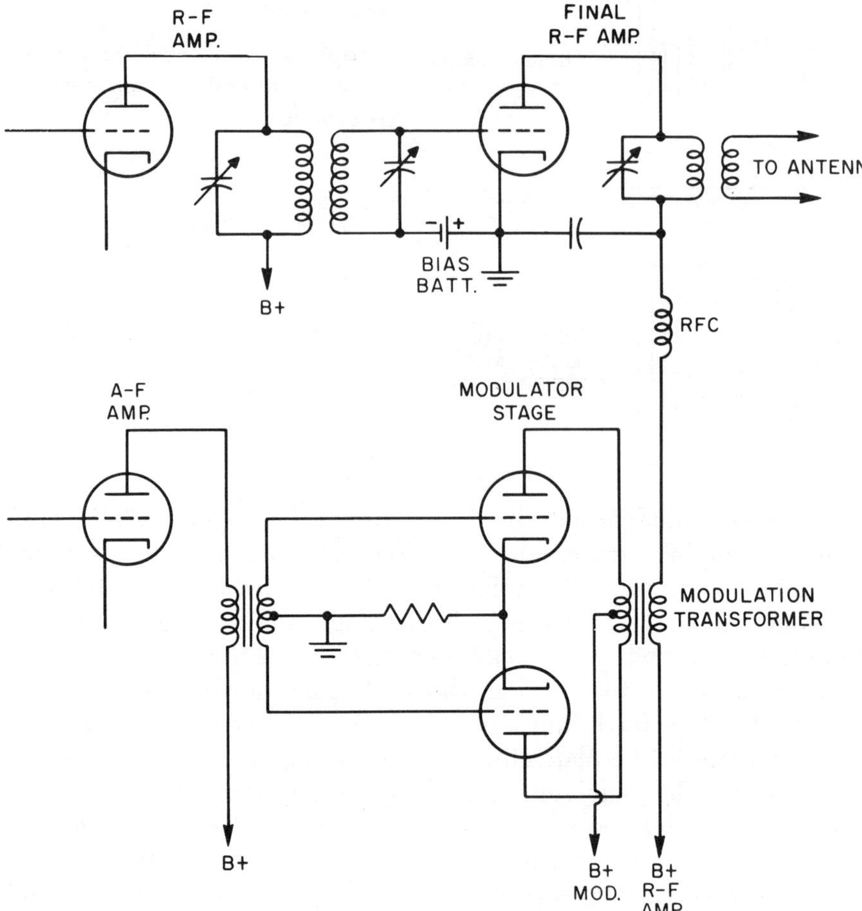

Fig. 23—7. Plate modulation using a modulation transformer.

of the r-f stage instead of to the plate circuit as illustrated. However, since this is not an exhaustive study of the transmitter, such alternate methods will not be considered here.

The block diagram of the transmitter is shown in Figure 23-8.

4. FREQUENCY MODULATION

As stated earlier in this chapter, intelligence may be conveyed by varying the frequency of the carrier current in step with the a-f variations caused by the sound wave. This method of radio communication is called *frequency modulation.* Although it is beyond

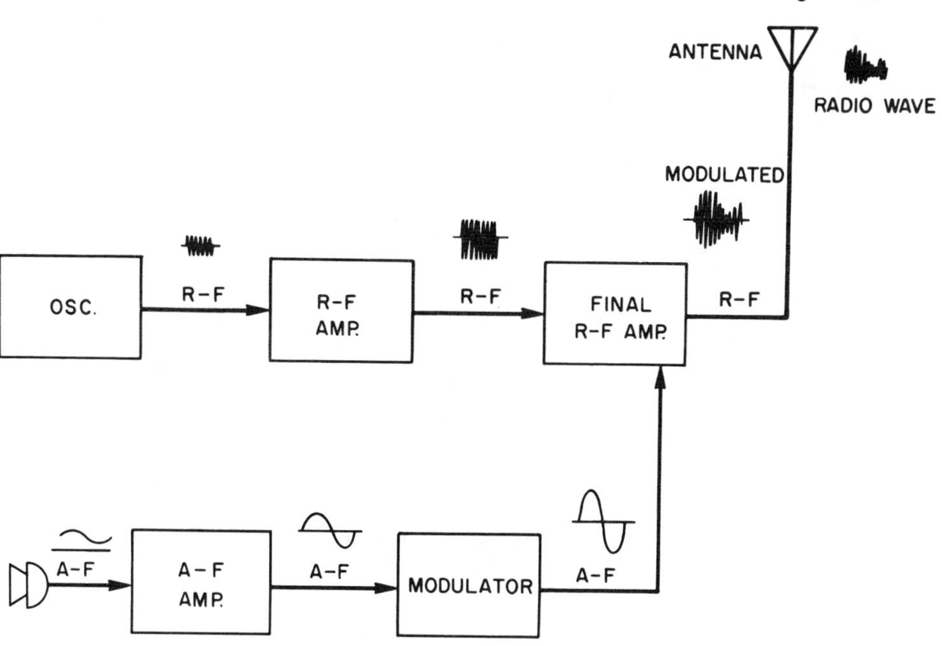

Fig. 23—8. Block diagram of the radio transmitter.

the scope of this book to go into the procedure in detail, we shall attempt to outline the basic principles.

How the carrier current is frequency-modulated is illustrated graphically in Figure 23-9. In Figure 23-9A you see the waveform of the unmodulated carrier current. The waveform of the sound current appears in Figure 23-9B. The condensations or compressions of the sound wave produce the positive half-cycles, the rarefactions the negative half-cycles. (Note we are still using the simple sine wave for purposes of illustration.)

When the two are combined, the result is as appears in Figure 23-9C. The amplitude of the carrier is not affected. But a condensation of the sound wave reduces the frequency of the carrier and a rarefaction increases it. Hence the frequency of the carrier current varies in step with the variations of the sound current.

The greater the amplitude of the sound current (the louder the sound striking the microphone), the greater will be the frequency variation, or *deviation*, of the FM carrier. Under present regulations, it has been decided that when sounds of maximum intensity or loudness strike the microphone, the frequency of the carrier current shall vary from 75 kc above to 75 kc below the normal or

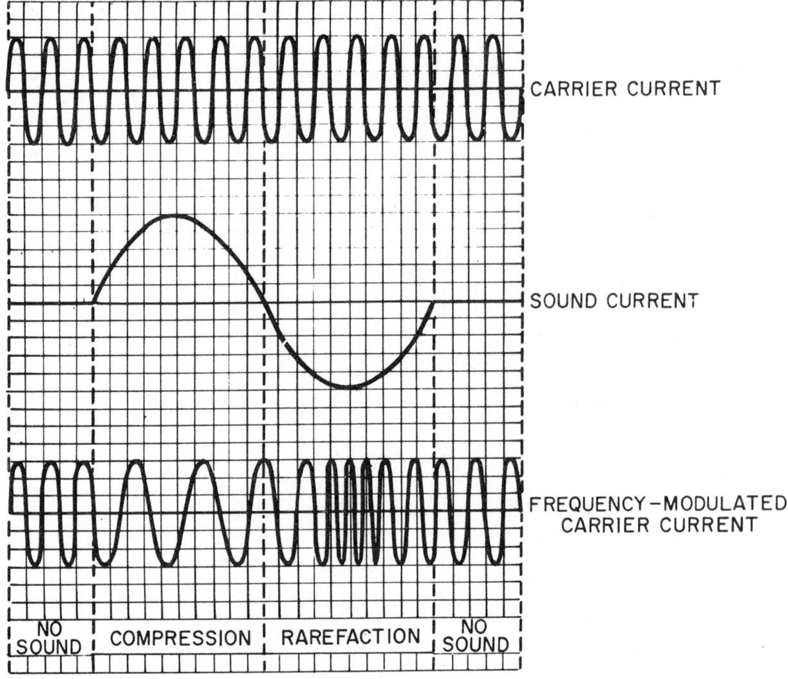

Fig. 23—9. Graph showing how the sound current frequency-modulates the carrier current.

central carrier frequency. Sounds of lesser intensity produce proportionally smaller frequency deviations. The number of such frequency deviations per second is determined by the frequency (pitch) of the sound current.

Another important difference between amplitude and frequency modulation lies in the matter of sidebands. In the AM wave, the width of the sideband is equal to the highest audio frequency (about 15 kc). Thus, the bandwidth of the transmitted signal would have a maximum width of about 30 kc. In practice today, most AM broadcast stations operate on a bandwidth of 10 kc, and thus the range of the audio signal that is heard in the speaker is limited to about 5 kc.

In the case of the FM wave, there is no such limit since the frequency of the sound signal varies only the number of frequency deviations per second of the carrier current. Modern FM transmitters operate on a bandwidth of 150 kc, thus permitting a frequency deviation of 75 kc to each side of the carrier frequency. To get the necessary wide channels and prevent one station from interfering with its neighbors, extremely high frequencies are employed. Hence the FM broadcast band lies between 88.1 and 107.9 mc, a spread of 19,800 kc.

An idea of how frequency modulation may be produced is illustrated in Figure 23-10. You will recognize this diagram as the regenerative oscillator (see Chapter 17, subdivision A). Capacitor C is a *capacitor microphone* which, together with L_1, tunes the grid tank circuit. This microphone is a small, air-spaced capacitor containing one fixed plate (the heavy bar) and one movable. As the sound wave strikes the movable plate it moves closer to or further away from the fixed plate, depending upon whether a condensation or rarefaction is striking it.

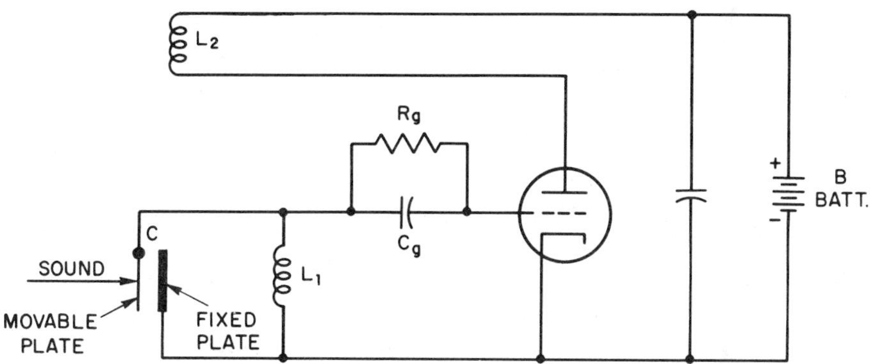

Fig. 23—10. Theoretical circuit showing how sound can frequency-modulate the oscillator output.

As a condensation causes the movable plate to move closer to the fixed plate, the greater will be the capacitance and hence the lower the frequency of the oscillator. A rarefaction will cause it to move further away, the capacitance will be reduced, and the frequency of the oscillator will increase. Thus, the frequency of the sound wave determines the number of frequency deviations per second.

The louder the sound, the greater will be the intensity of the condensations and rarefactions. Hence the louder the sound, the greater will be the variation in capacitance, and the greater will be the frequency deviation of the oscillator.

In practice, this method of frequency modulation is not employed and the illustration is used merely to show the principle involved. It is beyond the scope of this book to discuss the actual circuits employed.

B. *The radio receiver*

1. THE AM RECEIVER

As the radio wave broadcast by the transmitter sweeps across the receiver's antenna, it generates in it an alternating current whose waveform is a miniature replica of the modulated carrier current and its sidebands. The first function of the receiver is to separate the desired signal from the signals broadcast by other transmitters. This is done by feeding the current to the *L-C* tank circuit of an r-f amplifier tuned to the frequency of the desired signal. (See Figure 23-11. The symbol Y indicates the antenna.)

Since the desired signal is in resonance with the tank circuit, it will add its energy and so the oscillations will be built up. If the signal is of some other frequency it will be out of resonance with the tank circuit and the oscillations will die down quickly.

We may draw an analogy with a man pushing a swing. If he applies the pushes (no matter how feeble) in step with the oscillations of the swing, these oscillations will build up and soon the swing will

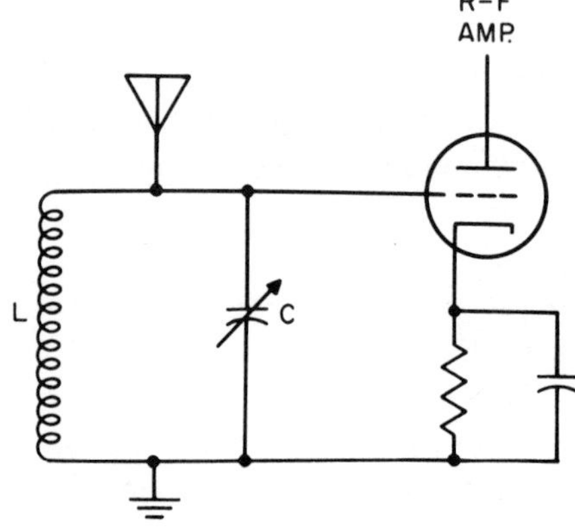

Fig. 23—11.

How the antenna signal is applied to the r-f amplifier.

be oscillating vigorously. Should he apply his pushes out of step with the oscillations of the swing, it will soon come to a halt.

The next step is to build up the feeble signal. This is accomplished by several stages of r-f amplification. Early receivers employed two or three stages of tuned r-f amplification. Not only was the signal amplified, but, as each stage introduced a tuned circuit, the ability to reject unwanted signals (the *selectivity*) was increased (see Chapter 7, subdivision C, 1).

In the early stage of radio development, the tuned r-f amplifier served admirably. But as radio developed, greater sensitivity was required to receive weak and distant stations. And, because there now were a larger number of radio stations on the air, there was a need for greater selectivity to prevent them from interfering with each other. The tuned r-f amplifier could meet this demand by adding more stages.

The increased number of stages, in turn, introduced a number of difficulties. Since each stage of r-f amplification added at least one more tuned circuit, the difficulty of keeping the tuned circuits in line—that is, all tuning to the same frequency at the same time—became much greater. Secondly, since the r-f amplifier must be able to amplify signals from broadcasting stations whose frequencies range from 535 to 1,605 kc, the tuning must cover a wide band of frequencies. To cover this wide band, the efficiency of the r-f transformer is sacrificed. A transformer designed to operate only at a single frequency (or, at most, at a small percentage of the frequency band) may be made considerably more efficient than a type designed to cover a very wide range.

Thus the gain per stage is relatively low. This requires many stages and introduces further complications. For instance, more tubes are required, thus increasing the amount of tube noise.

If a signal of only one frequency (or rather, a narrow band of frequencies) were to be received, the problem of tuning all the circuits to that frequency would be easy. And, since the tuned circuits would be fixed rather than continuously variable, more efficient r-f transformers could be designed and the gain per stage would be greater.

In essence, this is how the modern *superheterodyne* receiver operates. The desired signal is received and all others are rejected. Then the frequency of the signal is converted to a certain predetermined frequency (the *intermediate frequency*). The r-f amplifiers (now called *intermediate-frequency amplifiers*) are tuned to that predetermined frequency, and the signal is amplified at that new frequency.

The process of changing a signal of one frequency to another frequency is called *frequency conversion.* The mathematics of frequency conversion is quite complicated and it is not our purpose to explain it here. Sufficient to say, if two voltages of different frequencies, say frequency A and frequency B, are applied to the input to a special tube used for that purpose, both voltages produce a plate current which has a complicated waveform. This plate current contains components of four chief frequencies: A, B, A — B, and A + B. The process of mixing the two voltages of different frequencies in this manner is called *heterodyning,* and the resultant current so produced is called the *beat-frequency current.*

(A somewhat similar effect takes place when two sounds of slightly different frequencies are mixed. The result is a sort of rise and fall in sound which is called the *beat note.* The frequency of this beat note is the difference between the frequencies of the original sounds producing it.)

In addition to producing a beat-frequency current, if one of the voltages used in mixing is modulated, this modulation is passed on to the beat-frequency current.

In the superheterodyne receiver, one of the mixing voltages is the modulated r-f signal coming from the antenna (see Figure 23-12). The other is an unmodulated voltage produced by an oscillator, called the *local oscillator,* within the receiver. A Hartley oscillator is shown here, though any of the *L-C* oscillators previously discussed may be employed.

The r-f signal is fed directly to the grid of the *mixer* tube. The voltage of the local oscillator is fed to the same grid by means of a small coupling capacitor (C_3). The output of the mixer stage is the beat-frequency current which carries the modulation of the original r-f signal. The beat-frequency current also is called the *intermediate-frequency* (I-F) *current* and is coupled to the intermediate-frequency amplifier by means of the intermediate-frequency transformer (T).

You can readily see that by ganging the tuning controls (C_1 and C_2) of the circuits that tune in the incoming signal and the local oscillator, we can arrange to have, at all times, a constant frequency difference between the incoming signal and the oscillator output.

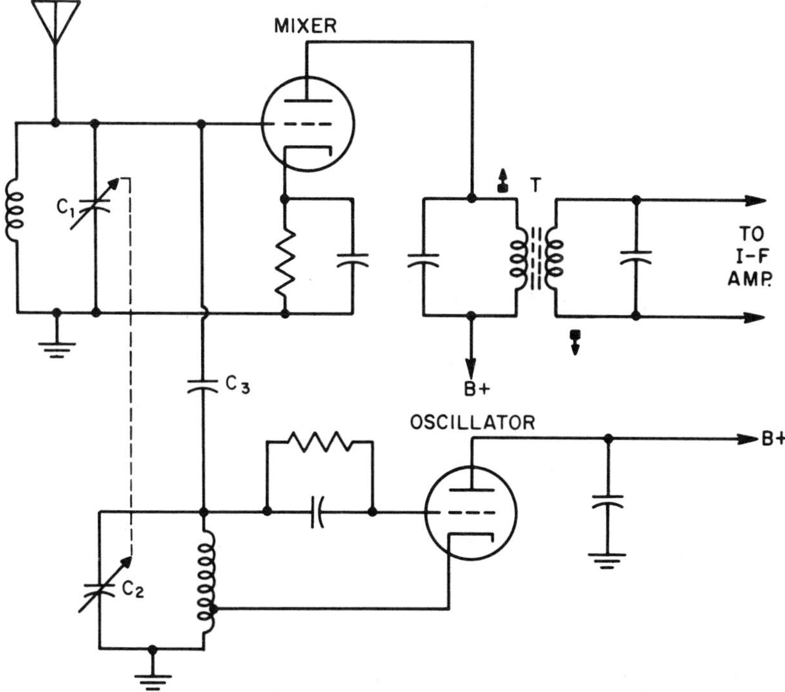

Fig. 23—12. Frequency-conversion circuit using separate oscillator and mixer tubes.

(The dotted line connecting C_1 and C_2 indicates that the two capacitors are ganged.) Thus, the beat frequency, or intermediate frequency, remains constant regardless of the frequency of the incoming signal.

We have stated that heterodyning occurs when the r-f signal voltage is mixed with the oscillator output voltage. The frequency of the signal is determined by the broadcasting station. Since the intermediate frequency can be obtained by mixing the signal with a voltage that is either higher or lower in frequency than the r-f signal frequency, we can tune our local oscillator above or below the signal frequency. For receivers in the AM broadcast band, the oscillator usually is tuned to a frequency that is higher than that of the r-f signal.

The intermediate frequency of receivers in the AM broadcast band generally is 455 kc. Thus, if the r-f signal has a frequency of 1,000 kc, for example, the oscillator is tuned to 1,455 kc. If the receiver is tuned to a signal whose frequency is 1,500 kc, the oscillator would be tuned to 1,955 kc. In both instances the intermediate-frequency current has a frequency of 455 kc.

There are a number of different types of frequency-conversion circuits. One frequently employed variation is the circuit shown in Figure 23-13. Here a single pentagrid converter tube (see Chapter 11, subdivision D) is used as both the local oscillator and mixer. The r-f signal is fed to grid #3 of the tube. The local oscillator is a tuned-grid type (although, as previously stated, any of the *L-C* oscillators described in Chapter 17, subdivision A, may be employed).

The oscillator is tuned by C_1 which is ganged to the control that tunes in the r-f signal. L_2 is the feedback coil. R is the grid leak. The voltage of the oscillator is fed to grid #1 of the tube through C. Mixing takes place in the electron stream from cathode to plate.

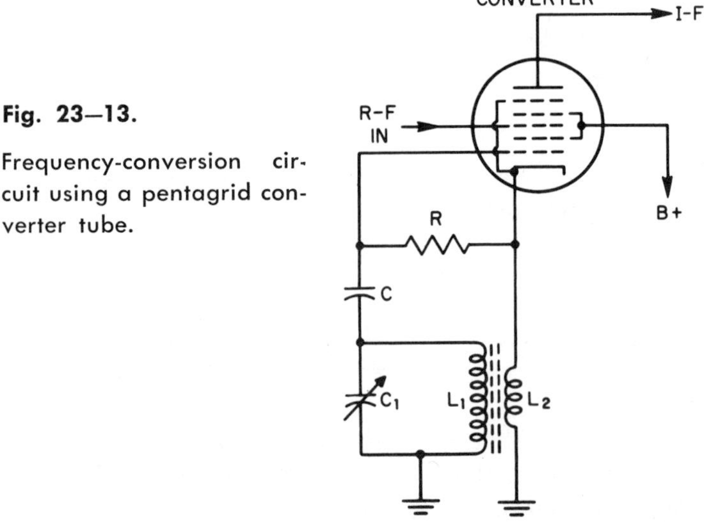

Fig. 23—13.

Frequency-conversion circuit using a pentagrid converter tube.

After the output from the frequency-conversion stage has been amplified by two or three stages of i-f amplification, the intelligence (the a-f sound signal) must be removed from the r-f carrier. This process is called *demodulation* or *detection* and the stage where this is accomplished is called the *detector stage*. There are several different types of detectors, but the one most frequently employed is the *diode detector* whose circuit is shown in Figure 23-14A.

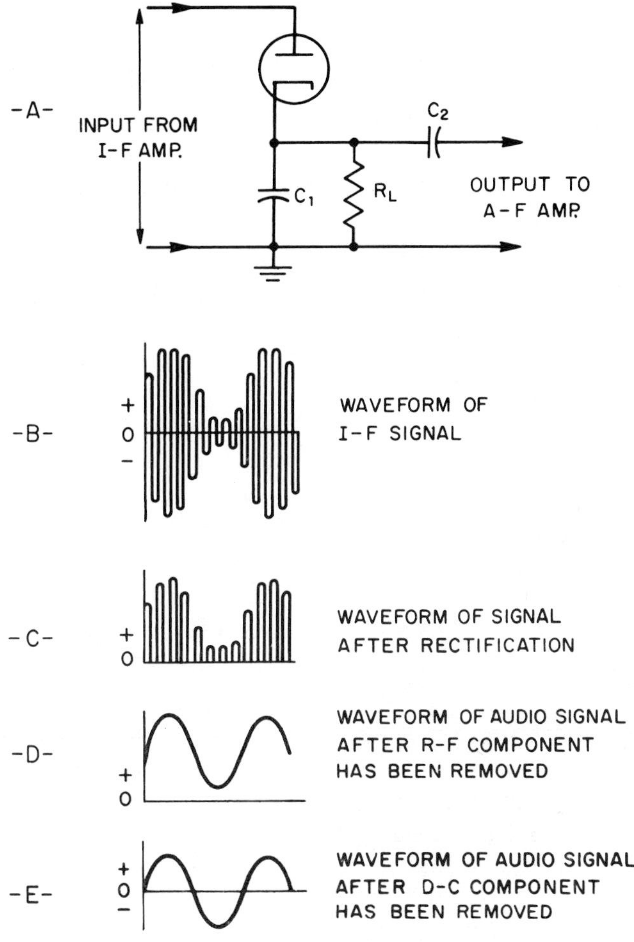

Fig. 23—14. **A.** Circuit of the diode detector. **B, C, D, E.** Graphs showing how the diode detector demodulates the i-f signal.

Examine Figure 23-14B where the waveform of the modulated i-f signal is illustrated. This signal contains an r-f component (the carrier) and an a-f component (the sound). In addition, you will note that the sound component also has a d-c component. Since both halves of each cycle are mirror-images of each other, the positive and negative half-cycles would cancel out, leaving a net result of zero.

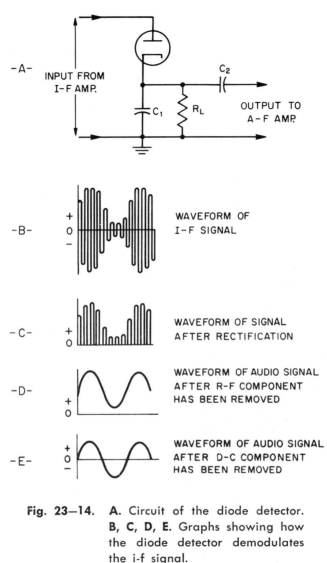

Hence the first step in detection consists of rectification which removes the negative half-cycles. See Figure 23-14C. This is accomplished by the diode.

The next step is to remove the r-f component. This is accomplished by capacitor C_1 whose impedance is practically zero at the radio frequencies but is large as compared to R_L at the audio frequencies. As far as the d-c component is concerned, C_1 offers an infinite impedance. As a result, the r-f component is bypassed to ground through C_1, but the a-f and d-c components appear across R_L (see Figure 23-14D).

The final step consists of removing the d-c component. This is accomplished by C_2 which blocks out the d-c component, but passes the a-f component to the a-f amplifiers (see Figure 23-14E).

The a-f signal next is amplified by several stages of a-f amplification and all that remains is to convert the a-f signal into sound waves that are identical with the sound waves striking the microphone at the transmitter. This is the function of the *loudspeaker.*

The most frequently employed loudspeaker is the *permanent-magnet dynamic* type illustrated in Figure 23-15. This speaker operates on the principle of interacting magnetic fields. A strong, steady magnetic field is set up by means of a permanent magnet of *alnico* or other similar alloy. Suspended in this field and free to move (within certain limits) is a small, light coil, consisting of a few turns of wire on a thin Bakelite ring or shell, called the *voice coil.* The coil rides back and forth on a cylindrical *core* or *pole.* Through this coil flow the a-f currents from the final a-f amplifier. As a result, a magnetic field that fluctuates in step with the variations of the a-f currents is set up around the voice coil.

The fluctuating magnetic field reacts with the steady field and, as a result, the voice coil is made to move back and forth in step with the variations of the a-f currents. Attached to this voice coil is the apex of a large paper cone whose rim is fixed. As the voice coil moves, this cone vibrates, setting the air in motion and producing sound waves. Since the motion of the cone is determined by the variations of the a-f currents, the sound waves produced are identical with those striking the microphone at the transmitter.

Since the voice coil consists of a few turns, its impedance is low. Hence an *output transformer* is required to match its impedance to the plate circuit of the final a-f stage. The symbol for the permanent-magnet dynamic speaker is P.M. .

Fig. 23—15.

Cutaway view of permanent-magnet dynamic loudspeaker.

The block diagram of the AM receiver is shown in Figure 23-16. A stage of tuned r-f amplification sometimes is used before the frequency-conversion stage. This increases the sensitivity and selectivity of the receiver somewhat. It also serves an additional function. The local oscillator is, in reality, a small transmitter. Should any of its current appear in the antenna, a spurious radio wave would be broadcast that would interfere with neighboring receivers. The stage of r-f amplification blocks out these currents. Where a stage of tuned r-f amplification is used, its tuning control is ganged to those used to tune the mixer and local-oscillator stages and all three stages are tuned simultaneously.

The antenna may consist of an elevated wire about 100 feet long. However, the superheterodyne receiver is so sensitive that it generally employs a small self-contained wire-wound loop antenna.

2. THE FM RECEIVER

The FM receiver resembles the AM type. There are some differences, however, arising from differences in the transmitted signals and the frequency ranges in which they operate. The AM signal consists of the amplitude-modulated carrier and its sidebands, constituting a band of frequencies usually 10 kc wide. For broadcast purposes this signal lies in a band ranging from 535 to 1,605 kc. The AM receiver is of the superheterodyne type and the intermediate frequency most frequently employed is 455 kc.

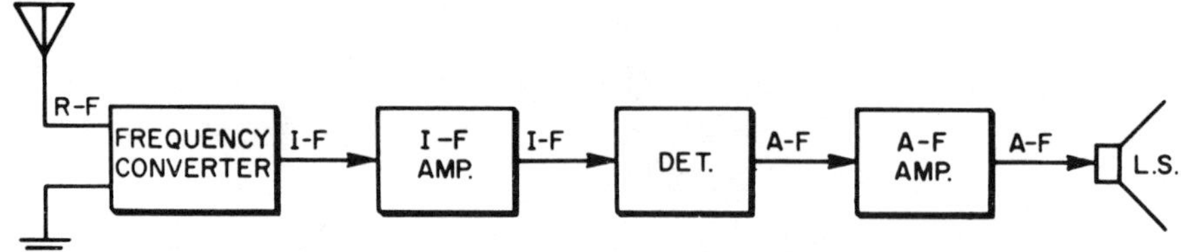

Fig. 23—16. Block diagram of the radio receiver.

The FM signal consists of the frequency-modulated carrier which occupies a band of frequencies 150 kc wide. For broadcast purposes this signal lies in a band ranging from 88.1 to 107.9 mc. The FM receiver is also of the superheterodyne type, but the intermediate frequency most frequently employed is 10.7 mc.

The frequency-converter stage of the FM receiver generally employs separate mixer and local oscillator tubes. This is done to reduce interference that may be caused if both stages were combined so as to employ a single tube. Unlike the AM receiver, the local oscillator usually is tuned to a frequency *below* the signal frequency.

Generally, two or three stages of i-f amplification are employed. In common with all the other tuned circuits of the receiver, the i-f transformers must be able to pass a band of frequencies 150 kc wide. This broadband characteristic generally is obtained by overcoupling the primary and secondary windings of the transformer (see Chapter 7, subdivision D).

The demodulator of the FM receiver must be, essentially, a frequency-sensitive rather than an amplitude-sensitive device. In some types of demodulation, any amplitude variations that may be present in the signal are removed by the demodulator stage itself. In other types the amplitude variations must be removed before the signal reaches that stage.

Amplitude variations present in the signal may be eliminated by the *limiter* stage, which is the final i-f amplifier stage modified so that it provides an output of constant amplitude regardless of the amplitude variations in the signal input. In essence, the limiter consists of a sharp-cutoff pentode operating with low plate and screen-grid voltages. Thus, the tube overloads easily on the positive half-cycles of the signal and quickly reaches cutoff on the negative half-cycles. A graphic representation of the action of the limiter is shown in Figure 23-17.

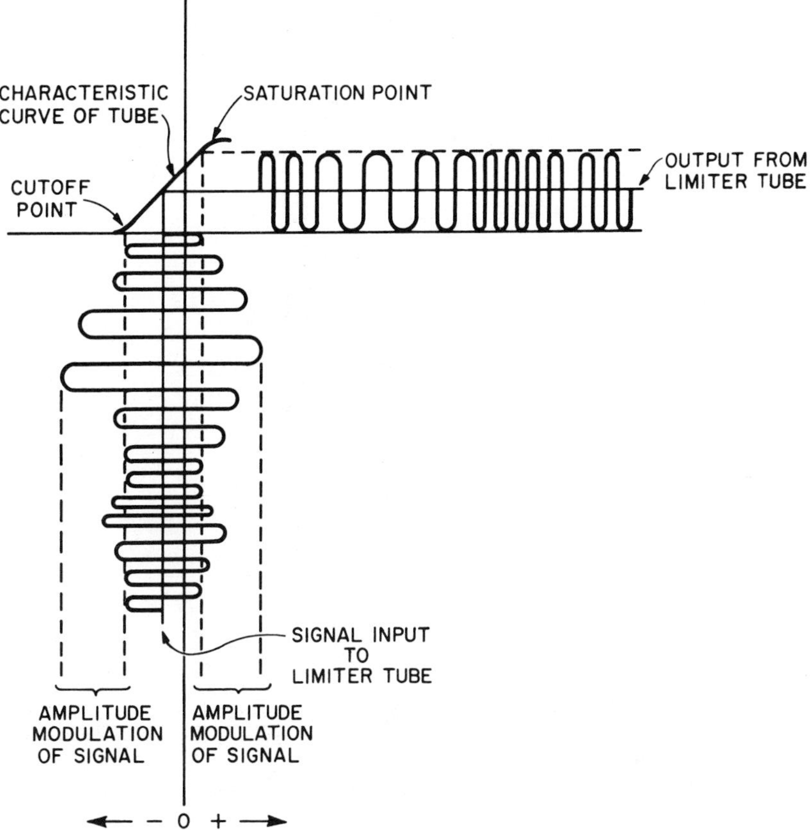

Fig. 23—17. Graph showing action of the limiter stage.

A typical limiter circuit appears in Figure 23-18. Note that the cathode of the limiter tube is connected to ground and that normally there is no bias on the grid. On the positive half-cycles, the tube is quickly driven to saturation. Any further increase in the amplitude of the positive half-cycles does not increase the output of the tube. Thus, all the positive half-cycles produce an output of constant amplitude.

As the grid is driven positive, current flows in the grid circuit. A voltage drop appears across the grid leak R_g which places a charge on the grid capacitor C_g. This charge acts to place a negative bias on the grid of the tube so that, as the negative half-cycles of the signal reach the grid, the tube is quickly cut off. Again, any increase

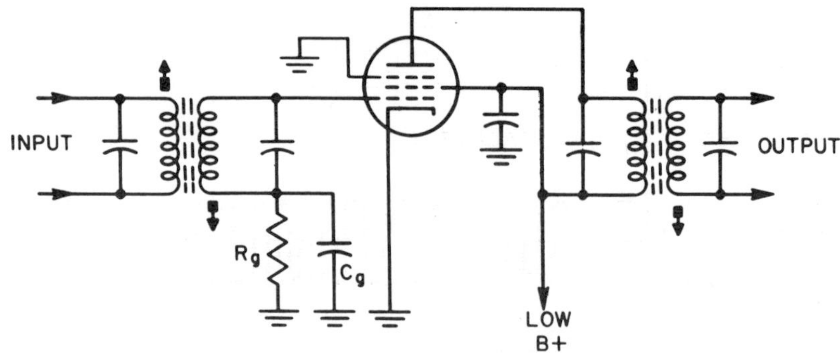

Fig. 23—18. Circuit of the limiter stage.

in the amplitude of the negative half-cycles of the signal does not affect the output of the tube. In this way, all the negative half-cycles produce an output of constant amplitude.

Since the i-f signal carries the intelligence in the form of frequency variations, the function of the FM demodulator is to produce an output that is determined by the frequency of the signal. There are several different types of such demodulators. Of course, all late circuitry uses semiconductor diodes which are close to ideal because of their minute size and excellent high-frequency characteristics.

The circuit of one of these is illustrated in Figure 23-19. It is called a *triple-tuned discriminator* circuit. Coils L_1, L_2, and L_3 form the primary and two secondary windings of a *discriminator transformer*. The primary (L_1) is in the plate circuit of the limiter stage and is tuned by C_1 to the intermediate frequency (10.7 mc or 10,700 kc).

Secondary L_2 is tuned to 75 kc *above* the intermediate frequency (10,775 kc) by means of C_2 and forms tuning circuit #1. Secondary L_3 and C_3 form tuning circuit #2, which is tuned to 75 kc *below* the intermediate frequency (10,625 kc).

Diode D_1 acts as a rectifier for the voltage developed in tuning circuit #1, and its output is developed across its load resistor R_1. Diode D_2 performs a similar function for tuning circuit #2, and its output is developed across its load resistor R_2. Since R_1 is equal to R_2, we thus have a balanced circuit.

Suppose an unmodulated i-f signal comes along. The output from the limiter stage then will be at the intermediate frequency (10,700 kc). Since the frequency-mismatch between the plate tank (L_1 - C_1) and tuning circuit #1 is the same as the mismatch between it and

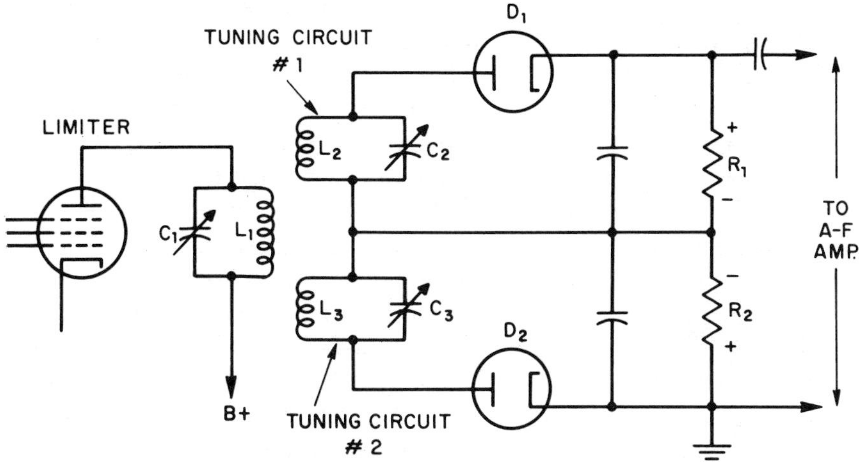

Fig. 23—19. Circuit of the triple-tuned discriminator.

tuning circuit #2, equal voltages will be induced in L_2 and L_3, and the voltage drops across R_1 and R_2 will be equal. Since the polarities of these voltage drops are in opposition to each other, the voltage drops will cancel each other, and no signal will be passed to the a-f amplifier.

But if a modulated i-f signal comes along which raises the frequency of the output from the limiter stage, the mismatch between the plate tank and tuning circuit #1 will be less than with tuning circuit #2. Hence a greater voltage will be induced in L_2 and this, in turn, will produce a greater voltage drop across R_1. Since the two voltage drops no longer cancel out, a signal voltage will be passed to the a-f amplifier. The higher the frequency of the i-f signal, the greater will be the voltage sent to the a-f amplifier.

Similarly, if the frequency of the i-f signal falls below the intermediate frequency, a voltage of opposite polarity will be passed to the a-f amplifier. Thus, the output of the discriminator stage is determined by the frequency of the i-f signal. Since the frequency variations are determined by the sound modulation, the sound signal appears at the output of the discriminator stage to be amplified by the a-f amplifier.

If the i-f signal should contain amplitude as well as frequency modulation, such amplitude modulation would create a voltage variation across R_1 and R_2 that would interfere with the audio output of the discriminator stage. You can see now why we need the limiter stage to remove any amplitude modulation that may be present.

Some of the more expensive receivers use two limiter stages to make doubly sure such modulation is removed.

Another type of FM demodulator that is frequently employed is the *ratio detector*, whose basic circuit is illustrated in Figure 23-20. Coils L_1, L_2, and L_3 form the primary, secondary, and tertiary windings, respectively, of a transformer. The primary L_1 is part of the plate-tank circuit of the final i-f amplifier. The secondary L_2 is part of the resonant circuit in the input to the ratio detector. The tertiary winding L_3 is closely coupled to the bottom of the primary winding. Both resonant circuits are tuned to the intermediate frequency.

Note that L_2, diode D_1, R_1, R_2 and diode D_2 form a series circuit. If an unmodulated i-f signal is impressed across L_2, current will flow through this circuit as indicated by the arrows and, as a result, voltage drops will appear across R_1 and R_2, with the indicated polarities. Since R_1 is equal to R_2, the voltage drops across them, too, will be equal.

The output voltage is taken from across R_2. Since the i-f signal is unmodulated, this output voltage is a steady direct voltage that is blocked out by C_2. Hence there is no signal fed to the a-f amplifier.

Should the average amplitude of the i-f signal increase, the voltage across R_1 and R_2 in series would increase too. Should the average amplitude of the signal decrease, the voltage would decrease. However, should any amplitude modulation appear in the signal, such amplitude changes, which are merely momentary changes, would be absorbed and ironed out by capacitor C_1, which is quite large.

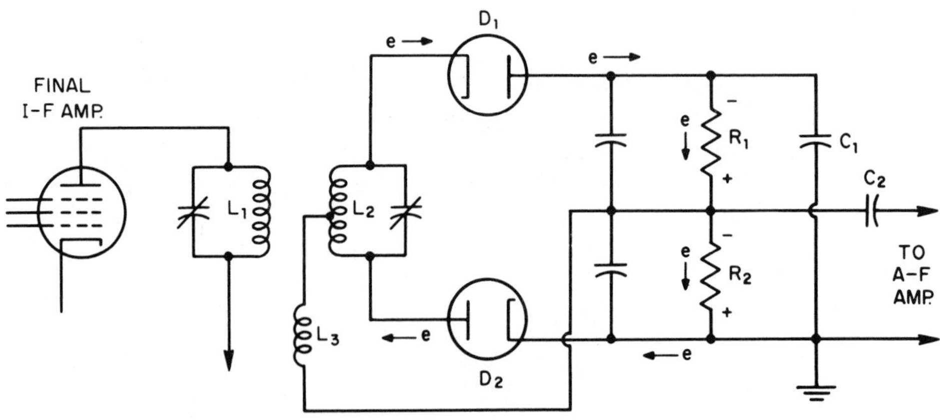

Fig. 23–20. Circuit of the ratio detector.

Thus, you see, we need no limiter stage to remove amplitude modulation from the signal. The effect of C_1, then, is to keep a constant voltage across R_1 and R_2 in series, a voltage that will vary only with the average signal strength, not with amplitude modulation. This voltage, as we have seen, is divided equally between R_1 and R_2.

Now what happens when the signal is frequency-modulated? Note that L_3, which is inductively coupled to L_1, has one of its ends connected to the center-tap of L_2 and the other end to the junction of R_1 and R_2. Thus current flowing through L_3 has two parallel paths. One is through L_3, R_2, D_2, the bottom half of L_2, and back to L_3. The other is through L_3, the top half of L_2, D_1, R_1, and back to L_3.

To show what happens requires a rather complicated mathematical process, but suffice it to say that as a modulated i-f signal is applied, the currents flowing in each of the two paths become unequal. Hence different voltage drops will appear across R_1 and R_2. These differences will vary as the frequency variations of the modulated signal.

Thus you see that, while the *total* voltage across R_1 and R_2 in series remains constant (its magnitude depending upon the average signal strength), frequency modulation of the signal changes the *ratio* in which the voltage is divided between R_1 and R_2. Hence the name, ratio detector. Thus the variations in voltage across R_2 will reflect the sound variations which frequency-modulated the carrier current at the transmitter.

The a-f section of the FM receiver is similar to that employed for an AM type. However, the AM signal, because of its 10-kc bandwidth, can be modulated by sound waves whose frequencies are limited to 5 kc. The FM signal has no such limitation. Hence the full audio range of about 15 kc may be employed. Accordingly, the a-f amplifier and loudspeaker used in FM receivers generally are of the "high-fidelity" type to take advantage of the greater audio range.

The FM receiver has another advantage. Extraneous noise, called *static,* has a tendency to affect the amplitude of the radio signal. Since the FM demodulator is not sensitive to amplitude changes, its output is a static-free signal.

The disadvantage of the FM system, a defect it shares with all other high-frequency systems, is the fact that its broadcast range is considerably smaller than that of the AM system which operates at lower frequencies.

QUESTIONS

1. a) Explain how the radio wave is produced.
   b) Describe the waveform of the *carrier current.*
2. a) Explain how the *sound current* is produced.
   b) Describe the effect of *pitch* and *loudness* of the sound upon the waveform of the sound current.
3. a) Explain what is meant by *amplitude modulation.*
   b) Describe the waveform of the *amplitude-modulated carrier current.*
   c) Explain the production of *sidebands.*
4. a) Draw and explain the block diagram of the radio transmitter.
   b) Compare the waveforms of the modulated carrier current and the radio wave.
5. a) Explain what is meant by *frequency modulation.*
   b) Describe the waveform of the *frequency-modulated carrier current.*
   c) Explain the effect of the *pitch* and *loudness* of the sound upon this waveform.
6. a) What is the function of the *antenna* of the radio receiver?
   b) Compare the waveforms of the radio wave and the current set flowing in the antenna system.
7. a) What is the function of the r-f amplifiers in the receiver?
   b) Explain how the desired signal is tuned in and all others kept out.
8. a) Explain the process of *frequency conversion* in the *superheterodyne* receiver.
   b) What are the advantages of the superheterodyne receiver over one employing ordinary r-f amplification?
9. Draw and explain the circuit of a frequency-conversion stage employing a *pentagrid converter tube.*
10. Explain the process of *demodulation.*
11. Draw and explain the circuit of a diode employed as a demodulator.
12. a) What is the function of the a-f amplifiers in the receiver?
    b) Describe and explain the operation of a *permanent-magnet dynamic loudspeaker.*
13. Draw and explain the block diagram of the superheterodyne receiver.
14. Explain the action of the *triple-tuned discriminator* of the FM receiver.
15. a) Explain the advantages of the FM receiver over the AM receiver.
    b) Explain the advantage of the AM receiver over the FM receiver.

24 Television Broadcasting

Before considering the television system, let us examine the process by which we see. When the human eye views a scene, it sees in terms of rays of light coming from the scene to the eye. These rays vary in intensity, the most intense rays coming from the brightest or whitest portion of the scene, the least intense coming from the darkest or blackest portions. Rays of light from portions of the scene lying between the brightest and the darkest (that is, the grays) will have intermediate intensities. Thus the eye sees the scene in terms of rays of light of varying intensities. (For the sake of simplicity we will consider scenes in black and white only. Colors add further complications.)

These rays of light are focused by the eye's lens to form an image on a screen, called the *retina,* at the back of the eye. This image is made up of light areas, dark areas, and areas of intermediate shades of gray, in the same arrangement as found in the original scene.

If we examine the retina closely we find that it is not a continuous sheet. Rather, it is made up of millions of tiny specks or elements, called *rods* and *cones*. Since each rod and cone has received a small portion of the image, you can see that the image has been broken up into millions of tiny elements. Each rod and cone is sensitive to light, generating an electric charge that is proportional to the intensity of the light striking it. Hence each rod and cone generates an electric charge that is proportional to the intensity of the light forming its portion of the image on the retina. The light portions of the image produce greater electric charges on the rods and cones affected; the darker portions produce lesser electric charges.

Thus the image formed by light rays of varying intensities on the retina is changed to an image formed by electrical charges of corresponding intensities. Each element—that is, each rod or cone—transmits its electrical charge to the brain through the optic nerve. The brain reassembles the millions of electrical charges to reproduce the image that appeared on the retina.

Here, then, is a clue for our television system. Project an image of the scene to be viewed onto a screen composed of a great many tiny, photosensitive elements. Each such element then acquires an electrical charge proportional to the intensity of the light striking it, just as did the rods and cones of the human eye. Thus, the image formed by light rays of varying intensities on the screen is changed to an image formed by electrical charges of corresponding intensities.

In the human eye each rod and cone has its private line to the brain. Thus, the optic nerve is made up of millions of such lines and the brain receives the electrical messages from all of them simultaneously. In the television eye it would be impractical to run a separate line to each of the photosensitive elements. Hence we are unable to send a simultaneous electrical image. Instead, we pass from element to element in some orderly sequence, picking up the electrical charge from each as we pass, and in this way we convert this image into a varying electric current. This current may be sent through a wire to the receiver or may be used to modulate a radio-frequency carrier current to produce the radio wave that is broadcast.

At the receiving end of the television system we can convert the varying current back into the original picture or scene in the following manner. A thin beam of electrons is focused on a phosphorescent screen, producing a pin-point of light where it strikes. The more intense this electron beam, the brighter will be the spot of light produced. The less intense the electron beam, the dimmer will be the spot of light.

The electron beam is moved from spot to spot on the phosphorescent screen at exactly the same time and in the same sequence as we passed from element to element of the screen at the transmitting station. At the same time, the varying electric current is used to vary the intensity of the electron beam. Thus, a white spot of the original scene will produce a bright spot on the phosphorescent screen. A black spot of the scene will leave its corresponding spot on the screen dark. Spots of intermediate shades will reproduce spots of intermediate brightness. Hence a replica of the original scene at the transmitter is painted in light on the phosphorescent screen at the receiver.

A. *The television transmitter*

1. THE ICONOSCOPE

In the human eye, the picture is broken up into tiny spots as its image is focused on a screen (the retina) composed of millions of light-sensitive elements (rods and cones). In television we use an electronic "eye" to do the same thing. One such device is the *iconoscope,* whose structure is shown in simplified form in Figure 24-1A.

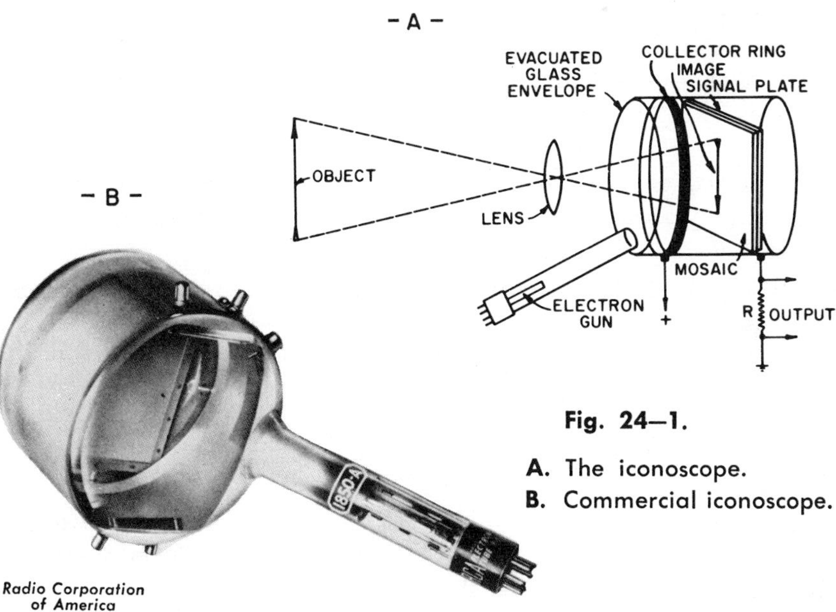

Fig. 24—1.

A. The iconoscope.

B. Commercial iconoscope.

(Modern television transmitters generally use electronic "eyes" that are much more sophisticated than the early iconoscope. However, because the iconoscope performs essentially the same function as these others and its operation is easier to understand, we will confine our discussion only to this device.)

The screen of the iconoscope is a sheet of mica that is mounted within a glass envelope from which the air has been evacuated. The front surface of the mica sheet is covered with millions of tiny light-sensitive metallic globules, each acting as a photoelectric cell and each insulated from the others. This surface is called the *mosaic.* The back of the mica sheet is coated with a metallic plate, called the *signal plate.* (See Figure 24-2.) You can see that each metallic globule forms a tiny capacitor with the signal plate, the mica sheet acting as the dielectric and the signal plate being common to all the capacitors.

As the image is focused on the mosaic, each light-sensitive globule receives its portion of light from the scene. As light strikes a globule, electrons are knocked off. The number of electrons emitted by each globule is proportional to the intensity of the light striking it. The more intense the light, the more electrons are emitted. Thus bright spots of the scene, reflecting more intense light to the globules, cause more electrons to fly off. Darker spots of the scene cause fewer electrons to be emitted.

As each globule loses an electron (which has a negative charge) it acquires a positive charge. Thus, bright spots of the scene cause their respective globules to acquire greater positive charges. The globules corresponding to the darker spots of the scene acquire lesser positive charges. Because the globules are insulated from each other, the electrical charges remain fixed and do not flow from one globule to another. In this way light from the scene forms a sort of electrostatic image on the mosaic.

Turn back to Figure 24-1. You will notice that the signal plate is connected to ground through resistor R. As a globule is struck by a light ray, loses electrons, and acquires a positive charge, electrons are drawn up from the ground, through resistor R, and on to a spot on the signal plate opposite the positively-charged globule where they are held in place by the attraction between opposite electrical charges. The more intense the light striking the globule, the greater is its positive charge, and the greater the number of electrons on the signal plate opposite it. Thus, you see, light from the scene forms an electron image on the signal plate.

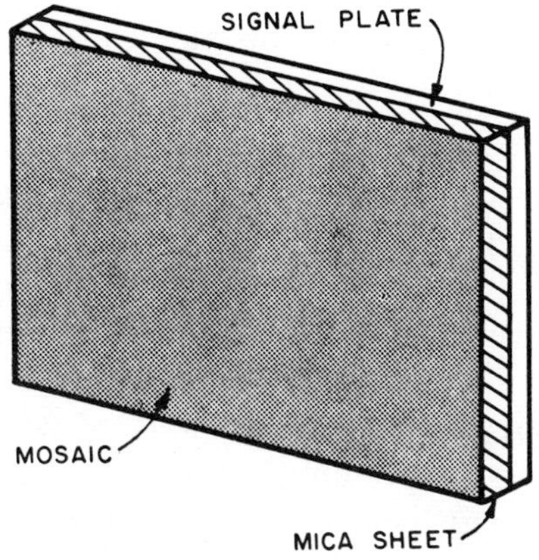

Fig. 24–2.

The mosaic and signal plate.

Now what would happen if the globule were to lose its positive charge? The electrons on the signal plate opposite it, no longer attracted, would flow back through resistor R to ground. This flow constitutes an electric current, and consequently a voltage is developed across the resistor. The strength of this voltage depends upon the number of electrons flowing through the resistor. This, in turn, depends upon the intensity of the light striking the globule. So you see that each spot of the scene can cause a voltage to be developed across the resistor, and the strength of this voltage will be proportional to the brightness of that spot.

To neutralize the positive charges on the globules, all we need do is to shoot a thin beam of electrons at them and so replace the electrons that have been knocked off by the light rays. The electron beam is generated by an *electron gun* similar to the type employed in the cathode-ray tube (see Chapter 12, subdivision E). A deflection system, identical to the electromagnetic type used by the cathode-ray tube, moves the electron beam from globule to globule.

As we move from globule to globule, neutralizing the positive charges on each in turn, a varying voltage, varying in proportion to the charges, is developed across the resistor. And since these charges vary in proportion to the light and dark spots of the scene, the voltage across the resistor, too, will vary in proportion to these light and dark spots. Thus the scene is converted into a varying voltage. If we amplify the varying voltage across resistor R and use it to modulate a carrier current, we have in effect, our television transmitter.

As the light rays strike the light-sensitive globules of the mosaic, electrons are emitted. These electrons are attracted to the positively-charged *collector ring* in the envelope in front of the mosaic and thus are drawn out of the envelope and so out of the way.

2. THE CAMERA SIGNAL

The movement of the electron beam from globule to globule of the mosaic is called *scanning* and the resultant voltage output from the iconoscope is known as the *camera signal.* It is necessary that the scanning be done in a regular, logical order, rather than in a haphazard, erratic manner.

There are a number of possible logical methods of scanning. However, the system employed in present-day television has the electron beam start at the top left-hand corner of the mosaic and move horizontally to the right, dropping slightly as it moves, until it reaches the extreme right of the mosaic. Next, the beam is moved back rapidly to the left-hand side of the mosaic. Then, once again, it moves horizontally to the right-hand side. This is continued until the entire mosaic is scanned by a series of left-to-right sweeps. (See Figure 24-3.) The effect is as if we cut the mosaic up into a series of horizontal strips, or lines, and then scanned each such line in turn, just as your eyes scan the lines of this page.

One complete scan from top to bottom is called a *frame*. After the frame is completed, the electron beam is returned to the top left-hand corner (under the influence of the vertical deflection system of the iconoscope) and the entire process started again.

As each line is scanned, the globules contained in that line are neutralized in sequence as the electron beam passes over them. Thus

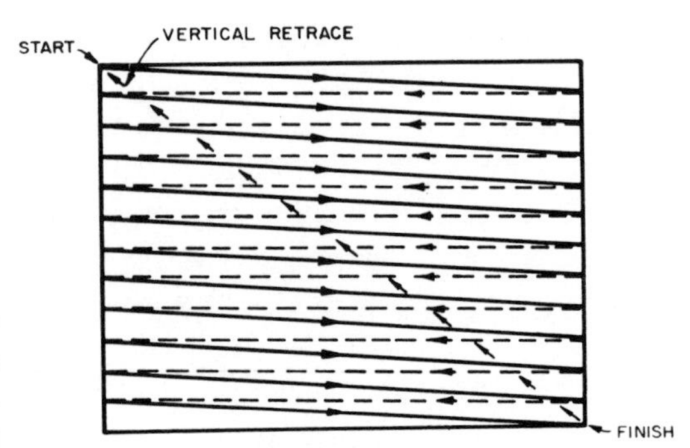

Fig. 24—3.

Path followed by the electron beam as it sweeps across the mosaic, resulting from the simultaneous action of the horizontal and vertical deflection systems.

a voltage is set up across resistor R (Figure 24-1A). This voltage will vary in proportion to the variations in charges on the globules, and these charges will vary in proportion to the intensity of light the globules receive from the scene.

Look at Figure 24-4A. Here we represent an image focused on the mosaic. Suppose a line on the mosaic is being scanned as shown in the figure. As the beam encounters globules illuminated at maximum intensity, the resulting voltages set up will be at the *100%,* or *white,* level. Where the image is darkest, the voltages will be at the 25%, or *black,* level. At intermediate shades, the voltages are at corresponding intermediate levels. (The region between zero and 25% is called the *blacker-than-black region,* and, as we shall see later, is reserved for additional information.)

The waveform of the voltage produced as the line under consideration is scanned as shown in Figure 24-4B. As you see, each line carries its own intelligence in the form of a varying voltage. If each line is scanned in turn, the entire picture is converted into a varying voltage (the camera signal).

In motion pictures the illusion of motion is obtained by flashing on the screen about 30 stationary pictures per second, each picture being slightly different in progression from its predecessor. A peculiarity of the eye, called *persistence of vision,* causes these pictures to blend, forming an illusion of smooth, continuous motion.

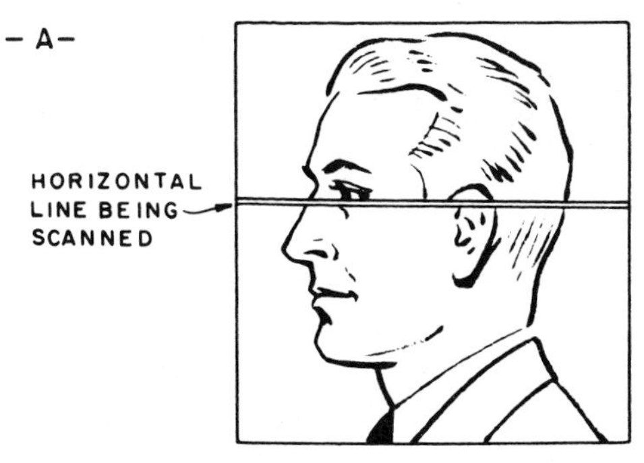

Fig. 24–4.

A. A horizontal line of the picture is scanned.

B. The resulting camera signal.

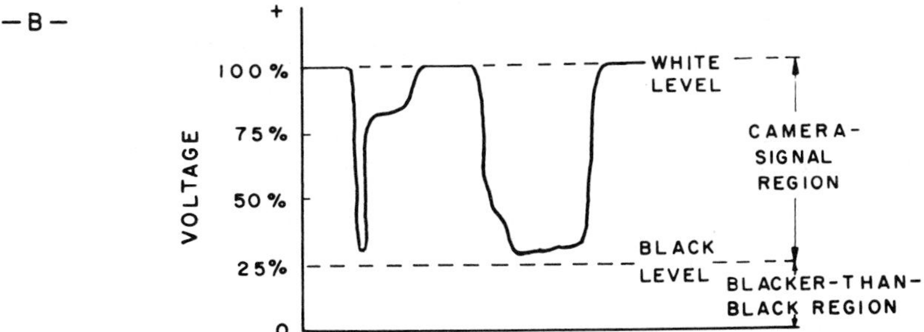

Thus, if we are televising a moving scene we must scan at least 30 frames per second. Since, under present-day standards, each frame contains 525 horizontal lines, 15,750 (525 × 30) horizontal lines must be scanned per second. (Some foreign TV systems use more lines per frame.) The frequency of the sweep oscillator used for the horizontal deflection system of the iconoscope, accordingly, is 15,750 cycles per second.

Since we are to scan 30 frames per second, the frequency of the sweep oscillator used for the vertical deflection system of the iconoscope should be 30 cycles per second. However, it was found to be advantageous to scan each frame twice, each scan covering alternate lines. Hence, the vertical sweep oscillator has a frequency of 60 cycles per second.

3. THE BLANKING AND SYNC PULSES

In the television receiver, the picture is reproduced on the screen of a cathode-ray tube as its electron beam sweeps its screen exactly in step with the sweep of the electron beam of the iconoscope over the surface of the mosaic. The time it takes the beam to scan a single horizontal line corresponds to the *rise*, or *sweep, time* of the horizontal sweep oscillator. The time it takes the beam to come back to start another scan is the *retrace*, or *flyback, time* (see Chapter 12, subdivision E). In modern practice, the horizontal sweep time is 53.5 microseconds (a microsecond is 1/1,000,000 of a second) and the retrace time is 10 microseconds. The time it takes the beam to move from top to bottom is the sweep time of the vertical sweep oscillator (15,416.6 microseconds) and the time required for its return is the vertical retrace time (1,250 microseconds).

Note that during both retrace periods the beam of the iconoscope sweeps across globules of the mosaic, thus producing a voltage output. If this voltage were permitted to affect the beam of the cathode-ray tube, it would interfere with the picture produced on its screen.

Accordingly, a *horizontal blanking pulse* is applied to the signal at the end of each horizontal sweep which drops the camera signal to the black level. (See Figure 24-5.) The duration (width) of this pulse corresponds to the horizontal retrace time. As a result, the electron beam at the cathode-ray tube is cut off and its screen goes black. Since the time is so short (10 microseconds) the eye cannot detect the black interval.

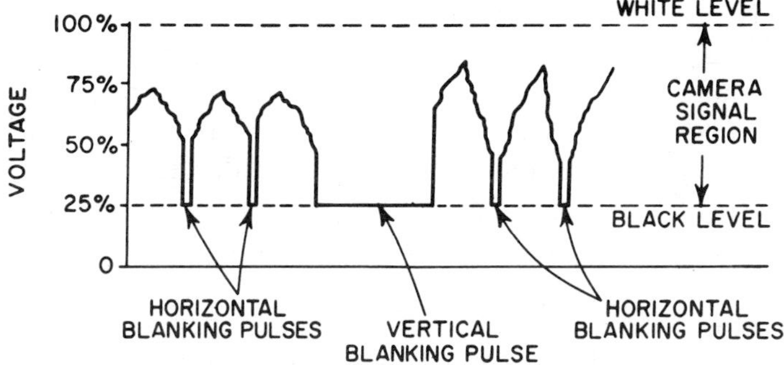

Fig. 24—5. The television picture signal, showing the camera signal, the horizontal blanking pulses, and the vertical blanking pulse.

Similarly, a *vertical blanking pulse* is applied to the signal at the end of each vertical sweep. Although the duration of the vertical blanking pulse is longer (1,250 microseconds), it still is too short to be detected by the eye.

As previously stated, the electron beam of the cathode-ray tube must sweep its screen in exact step (*synchronization*) with the electron beam that sweeps the mosaic of the iconoscope. To obtain this sweep, both the transmitter and receiver employ horizontal and vertical sweep oscillators that oscillate at frequencies of 15,750 and 60 cycles per second, respectively. At the transmitter, where more expensive equipment can be afforded, these oscillators are kept exactly on frequency. The oscillators of the receiver, on the other hand, tend to drift away somewhat from their rated frequencies. This would destroy the synchronization and, hence, the picture.

In our discussion of oscillators (see Chapter 17, subdivision B, 3), we saw that it is possible to lock the frequency of the oscillator in synchronization with a series of pulses. Hence the transmitter generates and transmits such a synchronizing (*sync*) pulse just before the start of each horizontal sweep. This pulse triggers the horizontal sweep oscillator of the receiver and thus keeps it in synchronization with the transmitter's horizontal sweep oscillator.

The *horizontal sync pulses* that lock the horizontal sweep oscillator of the receiver in synchronization with the transmitter are pulses of about 5-microsecond duration that are added to the horizontal blanking pulses. (See Figure 24-6.) These sync pulses extended into the blacker-than-black region and thus cannot affect the picture at the receiver. (Besides, they occur during the period when the electron beam is cut off by the horizontal blanking pulse.)

In a similar manner, the vertical sweep oscillator of the receiver, too, is locked in synchronization with that of the transmitter by a *vertical sync pulse* of about 190-microsecond duration that is added to the vertical blanking pulse. At the receiver, these sync pulses are removed from the signal and sent on to trigger their respective oscillators.

The composite picture, or *video,* signal, then, consists of the camera signal plus the various blanking and sync pulses. This signal is used to modulate the amplitude of the carrier current.

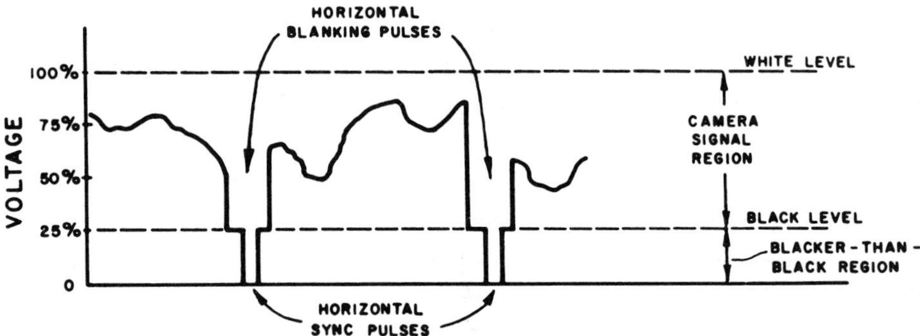

Fig. 24—6. The television picture signal, showing the camera signal, the horizontal blanking pulses and the horizontal sync pulses.

4. THE SOUND SIGNAL

The television sound system employs a frequency-modulated signal. Hence it resembles the FM radio system previously discussed. However, in the FM radio system the signal is permitted a maximum frequency variation of 75 kc above and below the center carrier frequency. Thus the band of frequencies is 150 kc wide. In the television system the maximum frequency variation permissible is 25 kc above and below the center carrier frequency. The frequency band, then, is 50 kc wide.

5. THE TELEVISION SIGNAL

It has been calculated that the video signal employed in present-day television encompasses a frequency range of 4 mc. Thus, if this signal is used to modulate the carrier current, the current and its two sidebands would spread 8 mc. This 8-mc spread would occupy too great a portion of the entire frequency spectrum assigned to the television band (from 54-216 mc for the VHF band and from 470 to 890 mc for the UHF band).

Accordingly, some method must be found to reduce the bandwidth of the television signal. From our discussion of amplitude modulation in Chapter 23, subdivision A, 3, you may have noted that each side-band carries the full modulation information. Hence, if we were to eliminate one sideband, the bandwidth of the television signal would be cut in half. It has been found impractical to suppress one of the sidebands completely, but it is possible to suppress most of one sideband. Such transmission is called *vestigial sideband transmission.*

Figure 24-7 illustrates graphically the picture, or video, portion of the standard television channel used in this country. Note that the sideband below the picture carrier frequency has been amputated, leaving only a vestigial band 1.25 mc wide. The lower 0.5 mc of this vestigial band, where the signal strength decreases rapidly, is a *guard band* to prevent interference by signals from the television channel immediately below.

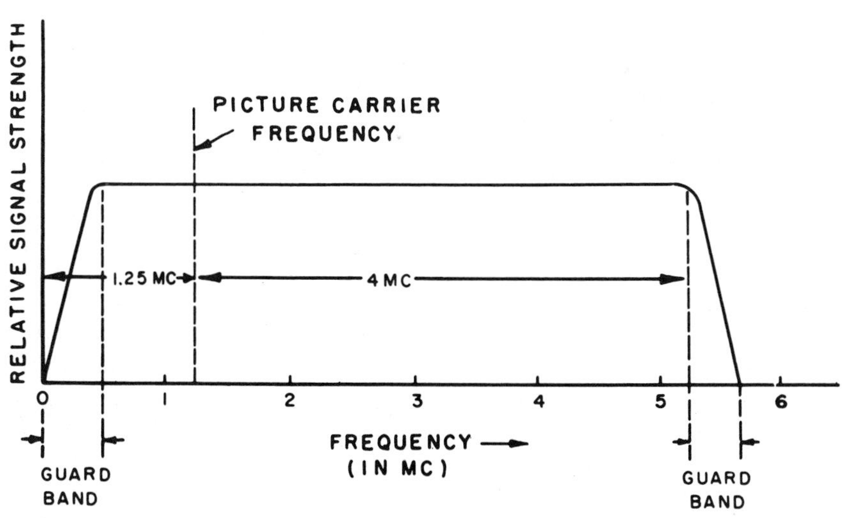

Fig. 24–7. Graphic representation of the picture portion of the standard television channel.

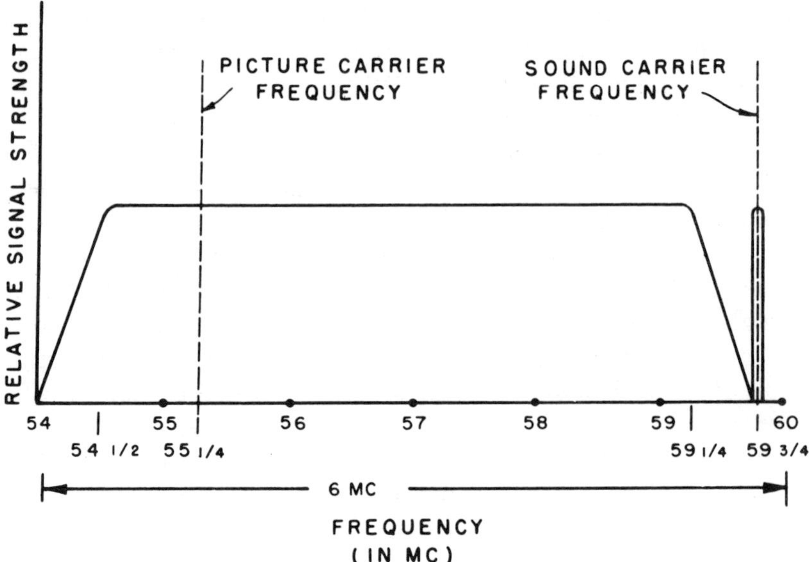

Fig. 24—8. Graphic representation of television Channel 2.

The picture signal occupies a band 4 mc wide and is located in the upper sideband. A 0.5-mc guard band at the upper end of the band prevents interference from the sound signals and from the television channel immediately above.

The sound signal occupies a band 50 kc wide, immediately following the picture band. Another guard band, 0.25 mc wide, separates the sound signals from the television channel immediately above. Thus the channel for the picture and sound signals (plus their guard bands) is 6 mc wide.

The graphic representation of a typical television channel (Channel 2) is shown in Figure 24-8. This channel is 6 mc wide, extending from 54 to 60 mc. The picture carrier frequency is at 55¼ mc. The vestigial sideband extends from 54 to 55¼ mc. The lower guard band is from 54 to 54½ mc. The picture-signal information is contained in a 4-mc spread from 55¼ to 59¼ mc. Then comes another guard band, from 59¼ to 59¾ mc. At 59¾ mc we have the 50-kc sound band with the sound carrier frequency located at 59¾ mc. Finally, we have another 0.25-mc guard band extending to 60 mc.

About 1945 the Federal Communications Commission authorized 12 television channels in the Very-High-Frequency (VHF) range

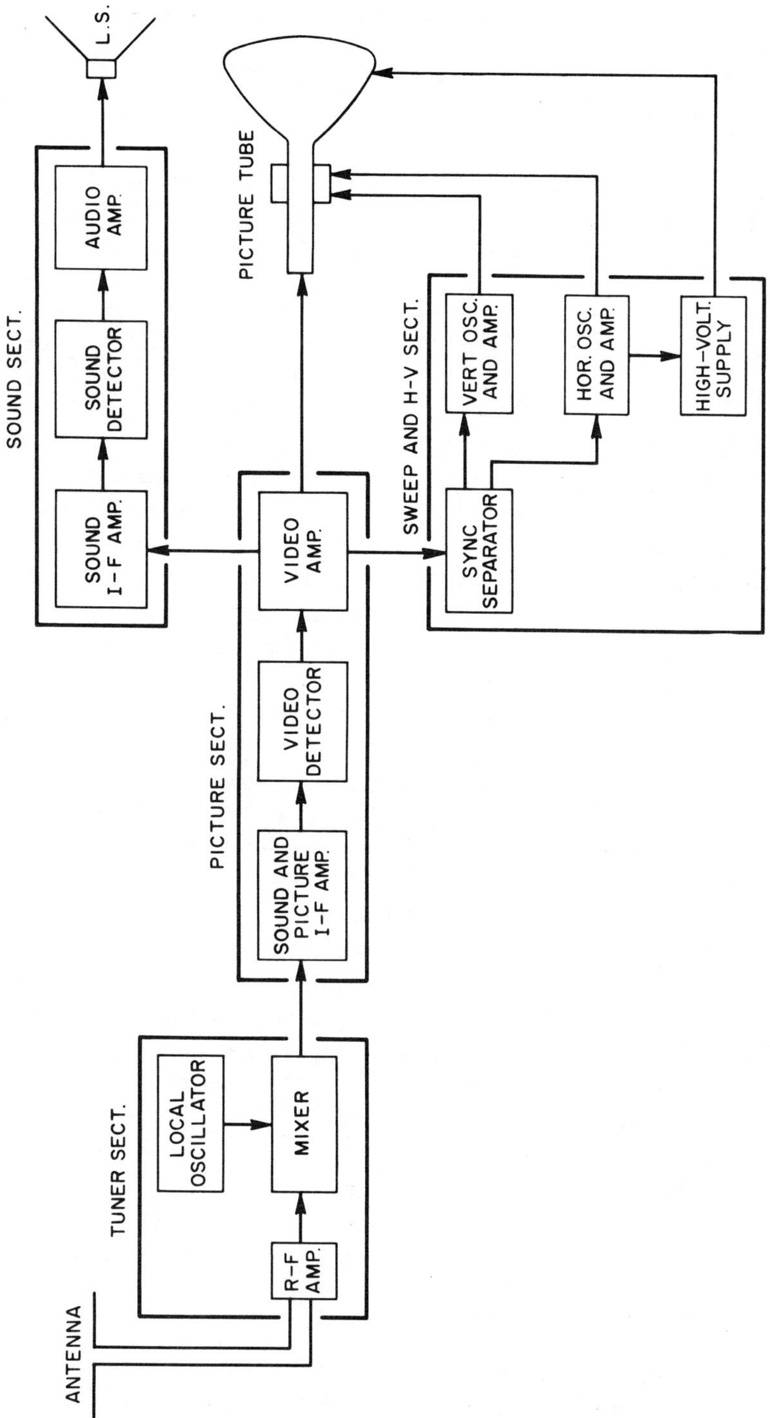

Fig. 24–9. Block diagram of the television receiver.

for commercial broadcast. These channels are each 6 mc wide and their frequency allotment is as follows:

Channel	Frequency Range	Channel	Frequency Range
2	54-60 mc	8	180-186 mc
3	60-66 mc	9	186-192 mc
4	66-72 mc	10	192-198 mc
5	76-82 mc	11	198-204 mc
6	82-88 mc	12	204-210 mc
7	174-180 mc	13	210-216 mc

Channel 1, with a frequency range of 44 to 50 mc, was originally allotted, but it was never used and has been discontinued.

Later, the FCC allocated 70 more television channels located in the Ultra-High-Frequency (UHF) range. These channels, too, are 6 mc wide; they lie between 470 and 890 mc. Since more than one station can use the same channel without interference, provided the stations are at least 150 miles apart, these 82 channels can accommodate thousands of stations.

B. *The television receiver*

Of necessity, our discussion of the television receiver must be presented here in simplified form. A block diagram of the receiver is shown in Figure 24-9. Modern receivers have a number of refinements which have been omitted in our diagram, though the basic sections do appear. Each section has a specific function. Exclusive of the antenna, the loudspeaker, and the picture tube, the television receiver may be considered as consisting of the following sections:

1. The tuner section.
2. The picture section.
3. The sound section.
4. The sweep and high-voltage section.

In addition, there is a low-voltage power supply that supplies the heater, plate, and screen-grid voltages to the tubes of each section.

1. THE TUNER SECTION

The input from the antenna to the tuner section is the television signal which contains three components: the video signal, the sound signal, and the blanking and sync pulses. The function of the

Standard Coil Products, Inc.

Fig. 24—10.

The television tuner showing how bandswitching of the inductors is accomplished.

tuner is to select the signal from the desired television station and reject all others, to amplify this signal, and to convert it to a lower-frequency i-f signal.

Basically, the tuner resembles its counterpart in the ordinary superheterodyne receiver. It contains an r-f amplifier stage, a local-oscillator stage, and a mixer stage. However, because of the high frequencies involved and the large range of frequencies to be covered, special precautions must be taken. The tuner must be constructed for extreme stability. Also, instead of using one inductor for each tuned circuit, a bandswitch is used to switch in a separate inductor for each channel. (See Figure 24-10.) In this way we can use components of optimum values for each channel instead of compromising upon values to cover them all. The bandswitch is calibrated in channel numbers.

The bandwidth of each television channel, you will recall, is 6 mc wide. Accordingly, the r-f amplifier and other portions of the tuner must be able to pass and amplify this 6-mc band of frequencies.

Frequency conversion is accomplished by means of a local oscillator and a mixer. The local oscillator operates on a frequency that is higher than that of the television signal. Since, as you can see from Figure 24-8, the picture carrier frequency and sound carrier frequency are 4.5 mc apart, the output from the tuner contains two intermediate frequencies, also 4.5 mc apart. One is the *picture intermediate frequency* usually at 45.75 mc and the other is the *sound intermediate frequency* at 41.25 mc.

2. THE PICTURE SECTION

The input to the picture section consists of the picture i-f signal containing the video information and the blanking and sync pulses, and the sound i-f signal containing the sound information. The picture i-f signal is amplitude-modulated and the sound i-f signal is frequency-modulated.

The first step is to pass both i-f signals through two or three stages of amplification. Since the video signal has a bandwidth of 4 mc, the i-f amplifiers must be able to pass and amplify such a wide band. One method for obtaining the proper bandwidth employs double-tuned, overcoupled i-f transformers (see Chapter 7, subdivision D, 2). As a result of this overcoupling, the gain of each stage is reduced, but the bandwidth is increased.

Another method employs transformers which are *stagger-tuned.* (See Figure 24-11.) Each stage is tuned to a peak frequency that is somewhat different, though all are within the overall bandpass desired. Note that the combined result of all these stages is a broadband amplifier. Both methods may be employed.

The output of the sound-and-picture i-f amplifier then goes to the video detector. This may be a diode tube or a crystal diode (which will be described later in the book). One function of this stage is to remove the picture i-f component and retain the video signal which contains the camera signal and the various blanking and sync pulses.

It has a second function. It mixes the picture and sound i-f signals to produce, in addition to the video signal, a beat signal called the *intercarrier signal.* Since the intercarrier signal is produced by the picture i-f and sound i-f signals, which are 4.5 mc apart, the central frequency of the intercarrier signal is 4.5 mc. Also, since it is produced by the amplitude-modulated picture signal and the frequency-modulated sound signal, the intercarrier signal carries both types of modulation.

Thus the input to the video amplifier is the demodulated video signal and the intercarrier signal. Both signals are amplified and separated in this stage. The intercarrier signal is passed to the sound section. The video signal is applied to the input of the cathode-ray picture tube.

The camera-signal portion of the video signal varies the intensity of the electron beam and this produces brighter or darker spots of light on the screen. It is these spots of light that produce the picture. At the proper time, the blanking pulses drive the picture tube to cutoff, thus blanking out the screen for the duration of each pulse (which, as you know, corresponds to the retrace time of the sweep oscillators). The sync pulses, which occur during the blanking periods, have no effect upon the picture.

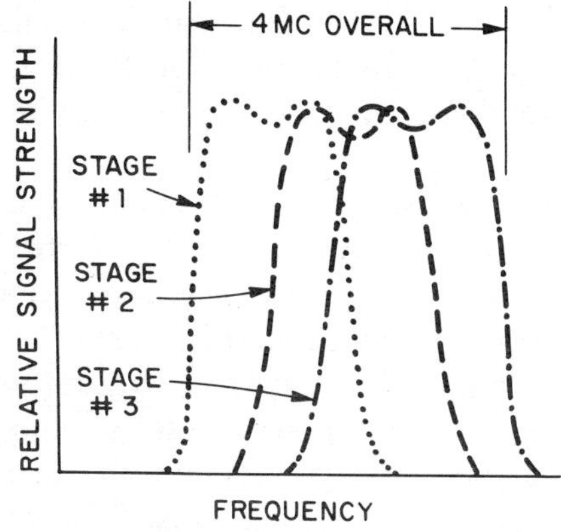

Fig. 24–11.

Graph showing how three stages are stagger-tuned to produce an overall wideband effect.

3. THE SWEEP AND HIGH-VOLTAGE SECTION

The video signal is also fed to the sync separator stage of the sweep and high-voltage section. Here the sync pulses are removed from the composite video signal and the horizontal pulses are separated from the vertical pulses to be fed to their respective sweep oscillators.

The sync pulses are removed from the video signal by means of a *clipper*, whose circuit is illustrated in Figure 24-12. The composite video signal, with the sync pulses negative-going, is applied to the cathode of the clipper diode through capacitor C. As the negative portion of the signal is applied to the cathode of the diode, the tube becomes conductive. Current flows through it, charging capacitor C as indicated. This charge acts as a bias which makes the cathode positive, and the tube becomes nonconductive. Before the tube can become conductive again, the positive bias must be overcome by the negative portion of the signal.

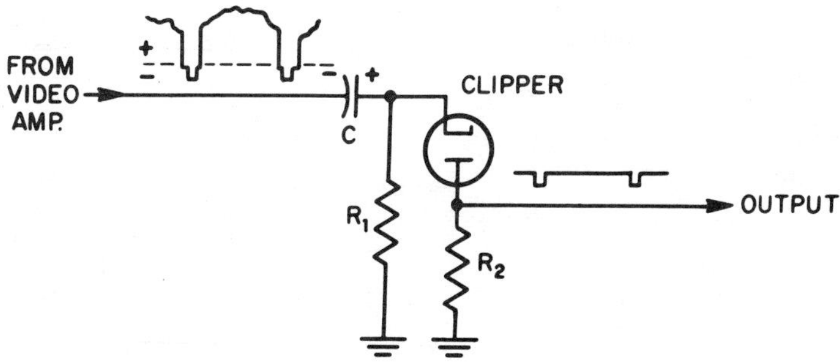

Fig. 24–12. Circuit of the diode clipper.

By the use of proper values of C and R_1, this bias can be established so that only the sync pulses of the signal are sufficiently negative to overcome it and make the diode conductive. Hence, only the sync pulses appear at its output.

The output of the clipper stage consists of the horizontal sync pulses, at a frequency of 15,750 cycles per second, and the vertical sync pulses, at a frequency of 60 cycles per second. The separation of the horizontal and vertical pulses is accomplished by taking advantage of the difference in their frequencies. Part of the output from the clipper stage is fed to a *high-pass filter* circuit (see Chapter 7, subdivision E) illustrated in Figure 24-13A. Its output, then, consists of only the higher-frequency horizontal sync pulses that are fed to trigger the horizontal sweep oscillator.

Another portion of the clipper-stage output is fed to a *low-pass filter* circuit (Figure 24-13B). Its output consists of only the lower-frequency vertical sync pulses that are fed to trigger the vertical sweep oscillator.

Note that the sync pulses are negative-going. In our discussion of oscillators we indicated that the oscillators are triggered by positive-going pulses. The phase-inversion may be accomplished by passing the pulses through a stage of amplification where, as you know, a phase-inversion takes place.

Multivibrators or blocking oscillators (see Chapter 17, subdivision B, 3 and 4) generally are employed as the horizontal and vertical sweep oscillators. These oscillators are locked in synchronization

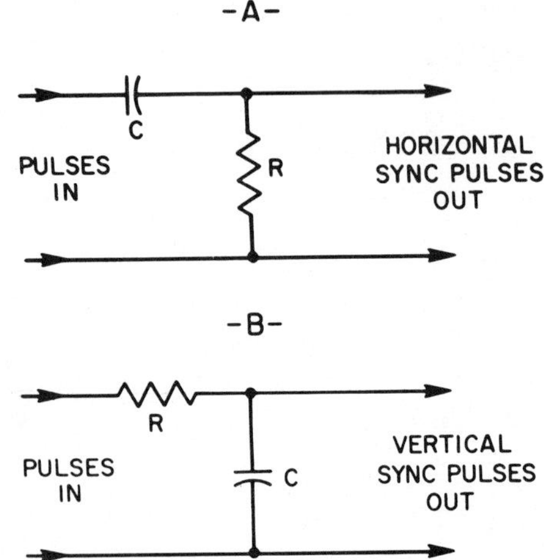

Fig. 24—13.

A. High-pass filter circuit.
B. Low-pass filter circuit.

with their counterparts at the transmitter by the horizontal and vertical sync pulses. The sawtooth-voltage outputs from the oscillators are applied to their respective portions of the deflection system of the picture tube. This deflection system generally is of the electromagnetic type.

The output voltage of the horizontal sweep oscillator has a sawtooth waveform at a frequency of 15,750 cycles per second. The duration of the sweep, you will recall, is 53.5 microseconds. The retrace time is 10 microseconds. Thus, if this sweep voltage is applied to an inductor, a very high voltage will be built up by self-induction across the coil during the extremely short retrace period (see Chapter 6, subdivision B, 1).

A portion of the horizontal sweep oscillator voltage is fed to the primary of a step-up transformer, called a *flyback transformer.* The high voltage thus produced is stepped up by transformer action to thousands of volts. This voltage then is rectified and filtered, and is applied to the high-voltage terminal of the cathode-ray picture tube.

4. THE SOUND SECTION

The intercarrier signal is obtained from the video amplifier. Its central frequency is 4.5 mc and it is both amplitude-modulated (picture information) and frequency-modulated (sound information). The signal first is applied to the sound i-f amplifier which also acts as a limiter stage to remove the amplitude modulation. It then is demodulated by an FM demodulator and amplified by an a-f amplifier, just as in the radio receiver.

QUESTIONS

1. a) Describe the *iconoscope.*
b) Explain how the image focused on its mosaic is converted to a varying voltage.
2. a) Explain what is meant by *scanning.*
b) Explain what is meant by a *frame.*
3. a) Explain the function of the *blanking pulses.*
b) Explain the function of the *sync pulses.*
4. What are the components of the *video signal?*
5. Draw and explain the graphic representation of the standard television signal.
6. Explain the functions of the *tuner section* of the TV receiver and its component stages.

7. **Explain the functions of the *picture section* of the TV receiver and its component stages.**
8. **Explain the functions of the *sweep and high-voltage section* of the TV receiver and its component stages.**
9. **a) Explain how the sync pulses are removed from the video signal.**

   b) Explain how the horizontal and vertical sync pulses are separated from each other.
10. **Explain the functions of the *sound section* of the TV receiver and its component stages.**

25 Radar

Radar—RAdio Detection And Ranging—is one of the outstanding electronic developments in recent years. Essentially, it is a device for detecting objects at a considerable distance and for determining the object's direction and distance between it and the radar station. Developed for war purposes, radar was quickly adapted to peaceful needs, especially in the field of navigation.

A consideration of the actual circuits employed in radar equipment is beyond the scope of this book. Instead, the general basic principles will be presented.

Assume that an airplane is flying high above the earth on a dark night. A searchlight station on the ground sends out a narrow beam of light. When this beam strikes the airplane, light is reflected back from the surface of the plane to the eyes of an observer stationed near the searchlight. The plane is seen, or *detected*.

With radar, a narrow radio beam is used instead of the light beam. This invisible beam, striking the plane, is reflected back to a radio receiver located near the transmitter and thus the plane is detected.

Thus far, the radio beam acts as the light beam. However, whereas clouds or fog render the light beam inoperative, the radio beam easily penetrates these obstacles. Further, the light beam is visible; the radio beam is invisible and the plane may be detected even though its occupants may not be aware of the fact.

It is not enough to detect the plane. We must know how far away it is, and how high up, and its bearing (that is, its compass position in relation to the observer). The searchlight permits only an approximation to the answers to these questions since it affords no accurate information concerning the distance of the plane from the observer.

With radar equipment, however, we are able to measure the time it takes the radio beam to travel from the transmitter to the plane and back again to the receiver. Knowing the speed at which the radio beam travels (approximately 186,000 miles per second), it is relatively easy to calculate the distance between the plane and the observer at the radar station.

Because of the enormous speed of the radio beam, the time intervals are very small—in the order of *microseconds* (a microsecond, you will recall, is one-millionth of a second). Thus, during one microsecond (abbreviated μs) the radio beam will travel approximately 328 yards.

To measure these extremely small time intervals we may employ the cathode-ray tube. If we apply a sawtooth sweep voltage to the horizontal deflection system, the electron beam will sweep horizontally from left to right across the screen of the tube during the rise, or sweep, time of the voltage, and back from right to left during the flyback, or reverse, time. This constitutes a single cycle and will be repeated for all the succeeding cycles.

Assume we have a cathode-ray tube and sweep voltage where the electron beam sweeps the screen at the rate of one inch per 100 microseconds. The eye, of course, cannot follow the moving spot of light at this speed. But since the phosphors of the screen continue to glow for a short period after the electron beam has passed, the eye will see a horizontal line of light.

If the radio beam and the moving spot of light on the screen of the cathode-ray tube are started simultaneously, the beam will have traveled 32,800 yards in the time it takes the spot to travel one inch across the screen (100 microseconds). Should the radio beam strike an object at this point and be reflected back, the spot will

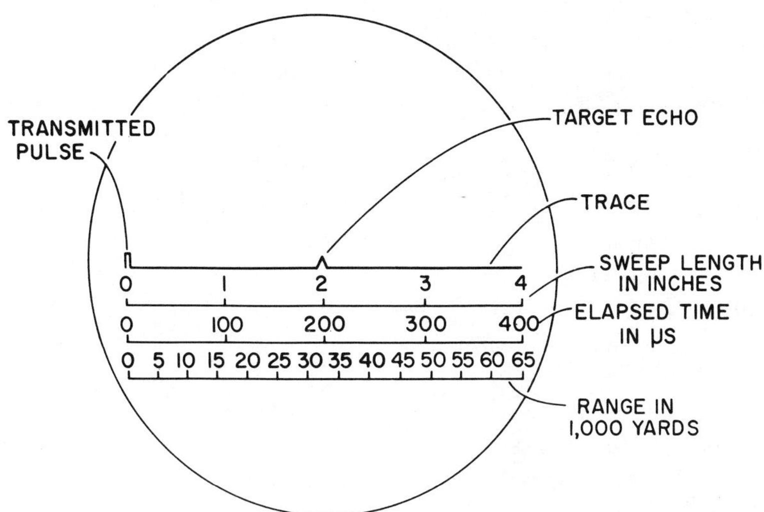

Fig. 25—1. Trace on the face of cathode-ray tube, showing pips.

travel another inch by the time the beam returns to its source. Thus, if we notice that the spot has traveled two inches across the screen in the time it takes the radio beam to reach the object and return, we know that the object is 32,800 yards (approximately 18½ miles) away.

The radio beam is sent out as a short *pulse*, or *burst*, of energy rather than as a continuous wave. The duration of this pulse usually is only about one microsecond. Part of this pulse is sent to the vertical deflection system of the cathode-ray tube, and its effect is to produce a sharp vertical bump, or *pip*, on the trace. When the reflected pulse is received by the radar receiver, it, too, goes to the vertical deflection system of the tube. Thus a second pip appears on the trace (see Figure 25-1).

In the 200 microseconds required for the radio pulse to reach its object and be reflected back to the receiver, the spot of light will have traveled two inches across the screen of the tube. Thus the two pips will appear two inches apart. By means of a scale printed on the face of the cathode-ray tube, we can translate the distance between the two pips of the trace into distance between the radar station and its target.

The radio pulses may be sent out several hundred times per second. Each pulse is also used to trigger the horizontal sweep oscillator of the cathode-ray tube. Thus, since the sweep circuit of the tube is synchronized to start with each transmitted pulse, all the traces will coincide, producing the effect of a single trace.

Also, since the duration of each pulse is extremely brief and the time between pulses relatively great, the average power consumed is small. Thus, small tubes and other components may be employed, even though the power of each pulse is large.

The radio beam must be narrow so that it may be directed toward a particular spot, just as is the light beam of the searchlight. This requires special antenna arrays. To make the equipment portable and allow the beam to be rotated easily, the antenna array must be quite small. This necessitates the use of very high frequencies—thousands of megacycles. Because of these high frequencies, radar operates on line-of-sight transmission, similar to light beams.

For peacetime use, radar equipment may be mounted on ships, airplanes, or other vehicles to detect obstacles that normally would not be seen because of darkness or fog.

Fig. 25–2.

Radar antenna, mounted in the nose of an airplane.

Radio Corporation of America Educational Services

A. *IFF*

It is not enough to detect a target by radar. Before opening fire on it, one should know whether the target is friendly or hostile. To establish this identification, auxiliary apparatus has been developed for use with radar. This apparatus is known as *Identification–Friend* or *Foe* (IFF).

Essentially, it consists of an automatic receiver and transmitter set which is carried by all friendly craft. When the target is detected by radar, a special coded signal, at a frequency other than the radar

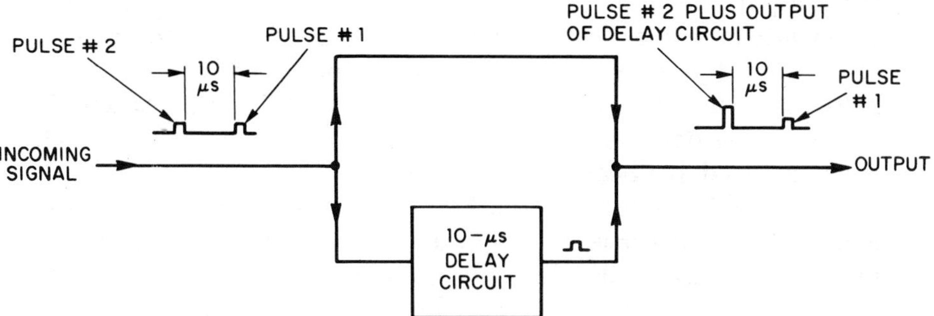

Fig. 25—3. Circuit showing how the coded signal is converted to a pulse large enough to trigger the automatic transmitter.

frequency, is transmitted. This challenging signal, when it is received by the receiver aboard the friendly craft, causes the transmitter to send out automatically a coded reply which is received by the receiver of the challenging station. If there is no reply, or if the reply is not in the code previously agreed upon, the craft is assumed to be hostile.

There are a number of methods that may be used to establish this code. One method is to employ a signal that consists of several pulses separated by definite time intervals. At the receiver, the signal is made to pass through a network containing circuits that introduce certain time delays. If these time delays correspond to the time intervals between the pulses of the transmitted signal, all the pulses will coincide, producing a large pulse that is able to trigger the automatic transmitter. If the time delays of the receiver do not correspond to the time intervals between the pulses, the automatic transmitter will not be triggered and there will be no reply.

As you know, if a voltage is applied to an *R-C* circuit, a certain time must elapse before the capacitor is charged up (see Chapter 7, subdivision B). Here, then, is a method for obtaining the time delay we need. Assume that the coded signal consists of two pulses 10 microseconds apart. One portion of the received signal is sent through a circuit that has no time delay. Another portion is sent through an *R-C* network that requires 10 microseconds for the capacitor to charge up sufficiently to produce a voltage great enough to overcome a fixed opposing voltage. (See Figure 25-3.)

Neither of the two pulses flowing through the undelayed circuit is sufficient to trigger the transmitter. But 10 microseconds after the first pulse enters the R-C network the capacitor has charged up sufficiently to send out its own pulse which coincides with the arrival of the second pulse of the signal. The two pulses add, producing a pulse large enough to trigger the transmitter. (This explanation of how the time delay is produced, though it demonstrates the principles involved, has been greatly simplified.)

The cathode-ray tube is used here, too. If the reply to the challenge is the proper one, a pip appears on the screen. If there is no pip, the craft is hostile. In some installations, the reply signal is superimposed on the radar screen through suitable circuits. In this way it changes the echoing pip produced by the target and thus identifies it as friendly.

In Figure 25-4 there appears a radar screen which contains a trace showing two pips, and hence two targets, at different distances from the radar station. The nearest target (#1) is friendly, as shown by the downward pip appearing beneath the original pip. This downward pip is produced by the answer to the IFF challenge. The furthest target (#2) contains no downward pip and thus is assumed to be hostile.

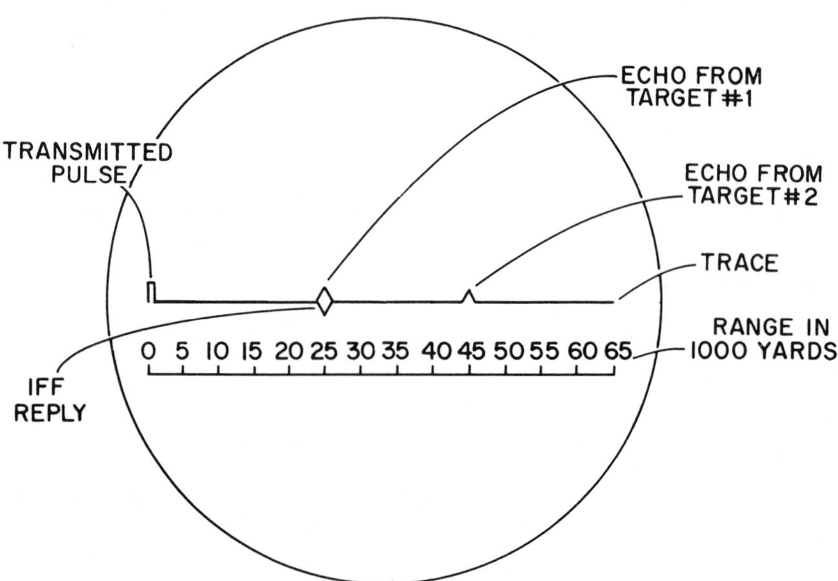

Fig. 25—4. Trace, showing the two targets. One, the nearest, is friendly, as indicated by the IFF reply.

For peacetime use, the automatic receiver and transmitter may be placed at definite, known points on the ground to serve as beacons for aircraft. As an airplane flies overhead, it might send out a challenging signal. The nearest beacon then would reply. The reply of each beacon would be coded differently for purposes of identification. Since the pilot would know the locations of these beacons from his maps, he thus would know his position.

B. *PPI radar*

More wonderful than the ordinary radar is the *PPI* radar (derived from Plan Position Indicator). With this instrument a radio beam is sent out and brings back a picture of the area surrounding the station.

Assume that a ship containing this apparatus is located near a shore (Figure 25-5A). The radar antennas are located at the top of a mast and are rotated so that they point at, or *scan*, the horizon. Below deck, and connected to the antennas, is the radar equipment.

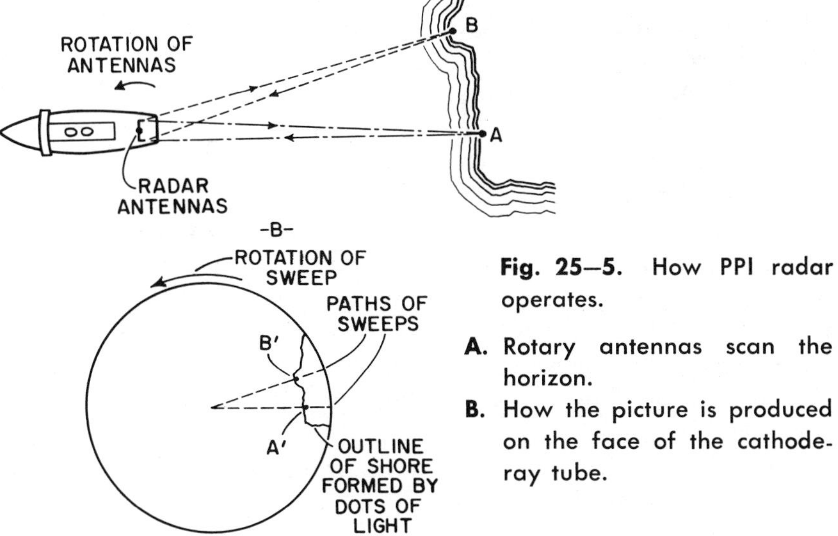

Fig. 25—5. How PPI radar operates.

A. Rotary antennas scan the horizon.

B. How the picture is produced on the face of the cathode-ray tube.

Fig. 25–6.

Radar screen, as seen by the pilot of an airplane.

Radio Corporation of America Educational Services

The cathode-ray tube employed is of a special type whereby the electron beam sweeps from the center of the tube's face, along a radius to the periphery, and back to the center (Figure 25-5B). In addition, the electron beam can be rotated around this center. The rotation of this electron beam is synchronized with the rotation of the radar antennas.

As the radar beam is transmitted from the antennas to a point on the horizon, the electron beam of the tube starts from the center toward a corresponding point on the periphery of the tube's face. The trace produced is just too faint to be seen. If the radar beam strikes a reflecting object (A), the echo causes the trace to brighten and a bright spot appears at a corresponding point on the trace (A^1). Both the antennas and electron beam of the tube move to the next angular position. The echo from point B produces a corresponding bright spot on the trace (B^1).

This continues until a whole series of bright spots, corresponding to the outline of the shore, has appeared on the screen of the tube. Since the screen of the cathode-ray tube is of the high-persistence type, the bright spots will remain for some time after the sweep has moved on to other angular positions. The result is a picture of the area surrounding the ship, whose position is indicated by the center of the screen.

The PPI radar has done much to improve the safety of the commercial airlines by showing the pilot a picture of the weather conditions around him. In Figure 25-6 is such a picture as seen on his radar screen by the pilot of an airplane. The plane's position is in the center of concentric circles, each of which marks a distance of five miles. The plane is heading toward the top of the picture. It is approaching the tail of a major storm which extends nearly 15 miles to the left. (The radar beam is able to produce echoes from the storm clouds.) Straight ahead and ten miles away are two imposing thunderheads. This information enables the pilot to decide whether to fly into the storm-free area at his right and return to course through the gap between thunderheads seen at the 15-mile circle, or to attempt to weave through non-turbulent areas of the storm ahead.

The PPI radar also aids the men in the control tower of a busy airfield. All nearby airplanes are indicated as spots of light on a radar screen and so they know when the way is clear for an airplane to land or take off.

QUESTIONS

1. **Explain how, by means of *radar*, the position and distance of an airplane may be determined by an observer.**
2. **Explain how, by means of *IFF*, a radar operator may determine whether the plane is friendly or hostile.**
3. **Explain how, by means of *PPI radar*, an observer may obtain a picture of the surrounding area.**

Section 6

Semiconductor Devices

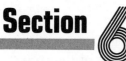

Perhaps the greatest innovation in the field of electronics in recent years has been the development of semiconductor devices to replace and supplement the electron tube. Semiconductors have been known for many years. As early as the beginning of this century semiconductors were used as rectifiers and demodulators. But when, in 1948, W. H. Brattain and J. Bardeen of the Bell Telephone Laboratories announced the development of the transistor, a tremendous impetus was given to the study of semiconductors. This, in turn, gave rise to a better understanding and the development of many new semiconductor devices.

26 Theory of Semiconductors

Semiconductors are a class of elements, such as germanium and silicon, and compounds, such as copper oxide and cadmium sulfide, whose electrical properties lie in an area between conductors and insulators, hence their name. As an example, at ordinary room temperature a cubic centimeter of pure copper (a conductor) will offer a resistance to current flow of about 0.000,0017 ohm. A cubic centimeter of slate (an insulator) has a resistance of about 100,000,000 ohms. A cubic centimeter of germanium (a semiconductor) has a resistance of about 60 ohms.

As you have learned, we consider the atom as consisting of a nucleus, which contains the protons and neutrons, surrounded by orbiting electrons located in one or more concentric shells. The positive charges of the protons are neutralized by the negative charges of the electrons. In the neutral atom there are as many electrons in orbit as there are protons in the nucleus.

Because of the attraction between positive and negative charges, the electrons are held in place around the nucleus. The electrons in the shell closest to the nucleus are most tightly held. The further the electrons are from the nucleus, the less firmly they are held. Hence the electrons in the outer shell are held the least firmly.

The electrons in the outer shell of an atom of a conductor are held very loosely. In fact, at ordinary room temperatures, some of these outer electrons may leave their shells and wander aimlessly from atom to atom. Such electrons are called *free electrons.* In the presence of an electric field (voltage) these free electrons will be attracted to the positive side of the field. The electric field may also cause some of the other outer-shell electrons to pull away from their atoms and join the stream to the positive side. This flow of negative charges is called an electric current and the electrons that carry these charges are called *carriers.*

The electrons in the outer shell of an atom of an insulator are held firmly to their nucleus. Hence there are very few free electrons and an electric field will produce a very small current flow. This accounts for the high resistance of insulators.

In nature, the atom seldom exists as an individual. Most frequently, it is found in combination with other types of atoms, forming compounds, or in combination with other atoms, both similar and different, forming crystals. In the crystal, the atoms are arranged in definite geometric patterns, called *lattices.*

Let us consider the atom of the semiconductor *germanium.* Its theoretical structure is illustrated in Figure 26-1. At the center is the nucleus containing 32 protons. Around this nucleus orbit 32 electrons in four concentric shells. The shell nearest the nucleus contains two electrons. The next shell contains 8 electrons and the third shell has 18 electrons. The outermost shell has four electrons. These outermost electrons are known as the *valence electrons.* Because the 32 negative charges of the electrons neutralize the 32 positive charges of the protons, the atom is neutral.

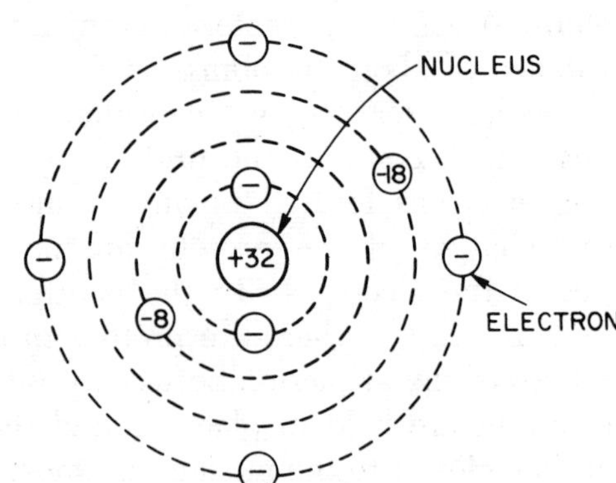

Fig. 26—1.

Theoretical structure of the germanium atom.

When the germanium atom joins with other similar atoms to form a crystal, it always assumes a definite pattern, illustrated in Figure 26-2. Each germanium atom (Ⓖe) is a relatively great distance from its neighbor. The atoms are held in place as one valence electron of one atom pairs up with a valence electron of a neighboring atom to form a *covalent bond* (shown in the illustration by the parallel lines linking the atoms). Note that each germanium atom is linked to four others, thus accounting for its four valence electrons. (The crystal actually is a three-dimensional structure, but in Figure 26-2 we show a theoretical two-dimensional view of it for the sake of simplicity.)

The net electrical charge of the pure germanium crystal is zero. Although the valence electrons of one atom pair up with those of its neighbors to form covalent bonds, the positive charges of its nucleus still neutralize the negative charges of its electrons.

If energy, in the form of an electric field, heat, or light is added to the atoms, some of the outer electrons forming the covalent bonds may be broken loose to become free electrons, leaving gaps, or *holes*, behind. At ordinary room temperatures the crystal has enough heat energy to account for the presence of a number of free electrons and holes.

As you know, when a neutral atom loses an electron, it acquires a positive charge. Since the atom acquires a hole at the same time it acquires its positive charge, we may consider the hole as bearing a positive charge. Note, however, the crystal still is neutral. For every free electron with a negative charge there is a hole with a positive charge.

Just as electrons are carriers of negative electrical charges, holes may be carriers of positive charges. Consider the pure germanium crystal as shown in Figure 26-2. If an electric field is applied to it, the attraction of the positive side of the field may be great enough to cause a valence electron of a nearby atom to leave its covalent bond and become a free electron. The departure of the electron leaves a hole and the atom, having lost a negative charge, becomes a positive ion.

The attraction of the positive ion pulls a valence electron from the covalent bond of an atom further back. As this electron falls into the hole left by the previous electron, the positive ion becomes a neutral atom once again. But now the second atom has lost an electron, obtained a hole, and has become a positive ion. This process is repeated with an atom still further back, and so on.

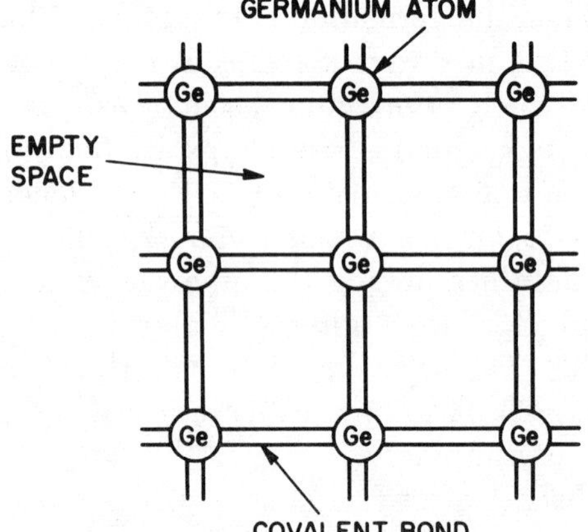

Fig. 26–2.

The structure of the germanium crystal.

The result, then, is the progressive movement of the hole and its accompanying positive charge toward the negative side of the field. Just as electrons are called *negative* carriers of electricity, holes are known as *positive* carriers. These carriers move in opposite directions, the negative carriers to the positive side of the field and the positive carriers to the negative side.

If an impurity, such as an atom of arsenic or antimony, is introduced into the germanium crystal, the result illustrated in Figure 26-3 is produced. The impurity atom (arsenic, shown as (As)) replaces one of the germanium atoms in the crystal. It has five valence

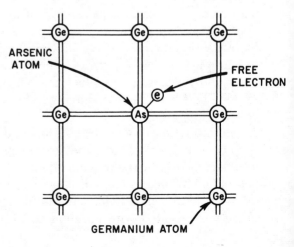

Fig. 26–3.

The structure of the N-type crystal.

electrons in its outer shell, four of which pair up with the valence electrons of its four germanium neighbors to form covalent bonds. The fifth valence electron is unpaired and remains rather loosely attached to its parent nucleus.

The result of adding such an impurity to the semiconductor crystal, then, is to add a number of excess electrons. This type of impurity is known as a *donor* impurity. The crystal with its impurity is called an *N*- (for *negative*) type since the electrons carry negative charges.

Thus, in the N-type crystal at room temperatures we have both negative (electrons) and positive (holes) carriers of current. But because of the presence of the donor impurity, there are far more excess electrons than holes. In this type crystal the electrons are considered as the *majority* carriers and the holes as the *minority* carriers.

The germanium crystal may also be altered by the addition of an impurity in the form of an atom having only three valence electrons in its outer shell. Examples of such atoms are indium, gallium, boron, etc. (See Figure 26-4.)

The impurity atom (indium, shown as (In)) replaces one of the germanium atoms in the crystal. Its three valence electrons form covalent bonds with three of its germanium neighbors. With the fourth, it can form only a *single-valence bond,* leaving a hole for the missing electron.

The result of adding such an impurity to the semiconductor crystal, then, is to add a number of holes. This type of impurity is known as an *acceptor* impurity. The crystal with its impurity is called a *P*- (for *positive*) type since the holes carry positive charges.

At room temperatures this crystal has both negative and positive carriers. But because of the presence of the acceptor impurity, there are far more holes than free electrons. In the P-type crystal the holes are considered as the *majority* carriers and the electrons as the *minority* carriers.

Note that the semiconductor crystal still has no net electrical charge. Where heat energy has broken covalent bonds, the negative charge of each freed electron is balanced by the positive charge of the hole it left behind. For every excess electron introduced by a donor atom there is a balancing positive charge in its nucleus. And for every hole introduced by an acceptor atom there is one less positive charge in its nucleus.

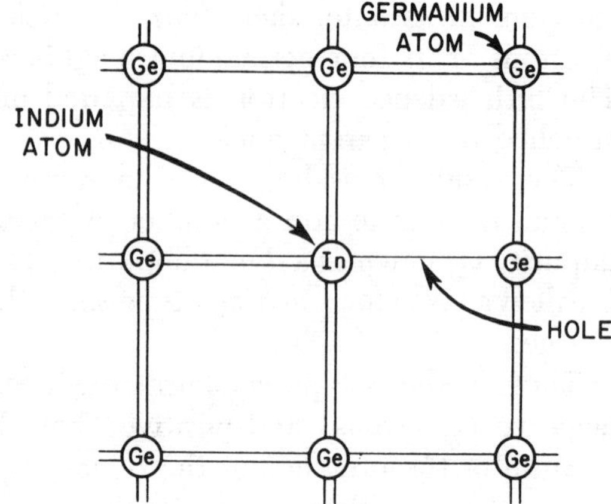

Fig. 26–4.

The structure of the P-type crystal.

In Figure 26-5A, an N-type crystal is shown joined to a P-type. The minus (−) signs indicate free or excess electrons. The plus (+) signs indicate holes. Some of the free electrons from the N-crystal move into the P-crystal and some of the holes move from the P-crystal to the N-crystal. The crystals are no longer neutral. The P-crystal now has more electrons than before and the N-crystal has fewer. Hence the P-crystal has a net negative charge and the N-crystal has a net positive charge.

(In practice, two separate crystals are not used. Instead, an *N-region* and a *P-region* are created side by side in the same crystal. Hence the N-region has a net positive charge and the P-region a net negative charge.)

Because of these opposite charges, an electric field (that is, a voltage) exists at the junction of the two regions. This is a permanent feature of a P-N junction and the magnitude of the voltage depends upon the amount of impurities in both regions. This voltage is known as the *potential barrier* of the P-N junction. When the two regions were first joined, the flow of electrons from the N-region to the P-region and holes from the P-region to the N-region stopped when the resulting potential barrier became high enough to prevent further flow and a condition of equilibrium was established.

If an external voltage is applied to the crystal as indicated in Figure 26-5B, the free electrons in the N-region tend to move toward the junction, as do the holes in the P-region. Also, the positive

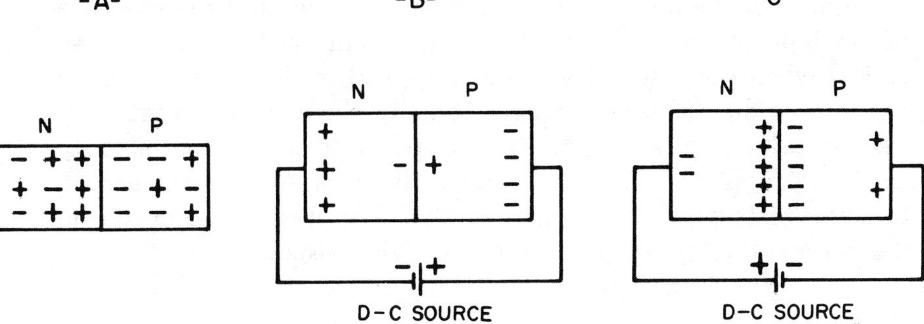

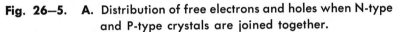

Fig. 26–5. **A.** Distribution of free electrons and holes when N-type and P-type crystals are joined together.
B. Distribution of free electrons and holes when a voltage is applied in the forward direction.
C. Distribution of free electrons and holes when a voltage is applied in the reverse direction.

charge of the N-region is reduced by the negative voltage of the external source and the negative charge of the P-region is reduced by the positive voltage of the source.

As a result of the reduction in the charges of the regions, the electric field at the junction and, hence, the potential barrier, is lowered. Accordingly, additional electrons from the N-region are able to move into the P-region and more holes from the P-region can move into the N-region. The net result, then, is a flow of electrons through the crystal from the N-region to the P-region, the external voltage source supplying additional electrons to the N-region and removing them from the P-region.

Thus the crystal is conductive. When an external voltage is applied in this manner, we say it is in the *forward* direction.

If an external voltage is applied to the crystal as indicated in Figure 26-5C, the free electrons in the N-region tend to move away from the junction, as do the holes in the P-region. The absence of majority carriers of both regions from the vicinity of the junction creates a *depletion zone* there.

Also, the positive charge of the N-region is increased by the positive voltage of the source and the negative charge of the P-region is increased by the negative voltage of the source. This results in an increase of the electric field at the junction and, hence, the potential barrier is raised.

Because of the increase in the potential barrier and the presence of the depletion zone, we would expect no current flow through the crystal when the external voltage is applied in this manner, which we say is in the *reverse* direction. Actually, the presence of minority carriers, which move in a direction opposite to that of the majority carriers, causes a small current, called the *back*, or *reverse, current*, to flow. Thus when an external voltage is applied to the crystal in the reverse direction it encounters a high resistance.

QUESTIONS

1. a) Explain what is meant by a *semiconductor*.
b) Give two examples that are widely used in electronics.

2. a) Explain, in terms of electrical charges, what is meant by a *neutral atom*.
b) What is the effect upon a neutral atom as an electron is added?
c) What is the effect upon a neutral atom as an electron is subtracted?

3. a) Explain what is meant by a *free electron*.
b) Explain the difference between conductors and insulators in terms of free electrons.
c) Explain the behavior of free electrons in the presence of an electric field.

4. a) Draw and explain the theoretical structure of the *germanium atom*.
b) What is meant by the *valence electrons?*

5. a) Describe the *lattice structure* of a pure germanium crystal.
b) What is meant by *covalent bonds?*

6. a) In the germanium crystal, what is meant by a *hole?*
b) Explain the movement of holes through the crystal under the influence of an electric field.

7. a) Explain the formation and structure of an *N-type crystal*.
b) Which, in this type crystal are the majority carriers and which the minority carriers?

8. a) Explain the formation and structure of a *P-type crystal*.
b) Which, in this type crystal are the majority carriers and which the minority carriers?

9. a) In a crystal containing an N-region and P-region, explain the normal distribution of majority and minority carriers in each region.
b) Explain the formation of the *potential barrier* at the junction of the regions.

10. a) Explain what happens in a P-N junction crystal as an external voltage is applied in the *forward direction.*
b) What happens to the resistance of the crystal?

11. a) Explain what happens in a P-N junction crystal as an external voltage is applied in the *reverse direction?*
b) What happens to the resistance of the crystal?

12. Explain the presence of a *back current* when an external voltage is applied to a P-N junction crystal in the reverse direction.

27 Semiconductor Diodes

A. *Crystal diodes*

Semiconductor diodes are widely employed in communications and industry. Those primarily intended for use at radio frequencies as, for example, detectors in radio receivers, are called *crystal diodes*. Such diodes generally operate at low voltages and very low currents (in the order of milliamperes).

Those primarily intended for use at power-line frequencies as, for example, rectifiers in the power-supply circuits, are called *power diodes*. Such diodes generally operate at power-line voltages and currents up to 200 amperes, or more.

The crystal diodes may be divided into two general categories, *point-contact* and *junction* diodes.

1. POINT-CONTACT DIODES

In the early 1900's it was found that a semiconductor crystal, *galena* (lead sulfide), could act as a rectifier and, as such, was used in radio receivers for detection purposes. The galena crystal forms one electrode of a sort of diode. The other electrode consists of a

A crystal diode of this type using a catwhisker as one electrode, is called a *point-contact diode.* Its symbol is —▶|—. The arrow points in the direction *opposite* to the flow of electrons through the diode. Thus, in an analogy to the electron-tube diode, the bar of the symbol corresponds to the cathode and the arrowhead to the plate or anode.

Because of the difficulty of locating a sensitive spot on the crystal and other faults, the galena diode gave way to the electron-tube diode. However, with a better understanding of semiconductors and new methods of manufacture, the improved point-contact diode made a comeback.

One electrode of the improved point-contact diode consists of a tiny semiconductor crystal, generally of N-type germanium or silicon. (Both types have four valence electrons in their outer shells.) The other electrode is a fine catwhisker wire. The crystal represents the cathode, the catwhisker the anode. The whole is enclosed in a ceramic or glass tube, and the entire diode may be only a fraction of an inch in length. (See Figure 27-1.)

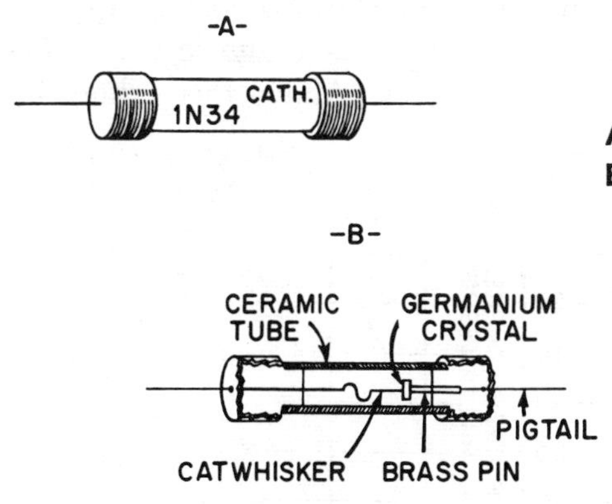

Fig. 27—1.

A. Point-contact diode.
B. Cross-sectional view, showing how the diode is constructed.

How the point-contact diode works, is not too well understood. It is thought that in some manner a sort of P-type region is established in the crystal at the spot where it is touched by the catwhisker. (See Figure 27-2.) Since the crystal is N-type, we thus have a P-N junction, complete with a potential barrier between the two regions. The brass plate, which is used merely for making contact with the crystal, has no effect upon it and thus forms no regions where it touches.

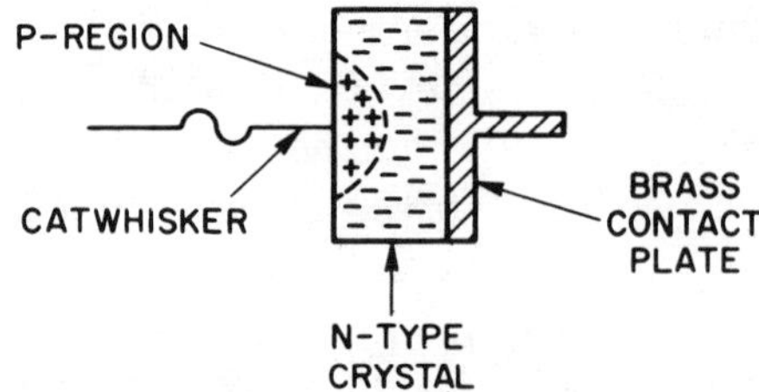

Fig. 27—2. How the point-contact diode works.

The diode, then, behaves as the P-N junction previously described. With an external voltage applied in the forward direction (negative to the crystal cathode and positive to the catwhisker anode), the diode exhibits a low resistance to current flow. With the applied voltage in the reverse direction (polarity reversed), the resistance is very high and only a small back current may flow.

The characteristic curve of a typical point-contact diode is shown in Figure 27-3. (Note that, whereas the forward current is shown in milliamperes, the reverse current appears in units of microamperes.) The forward current increases rapidly as the forward voltage applied to the diode is increased. The maximum safe average forward current for this diode is about 50 milliamperes. Above this value, the

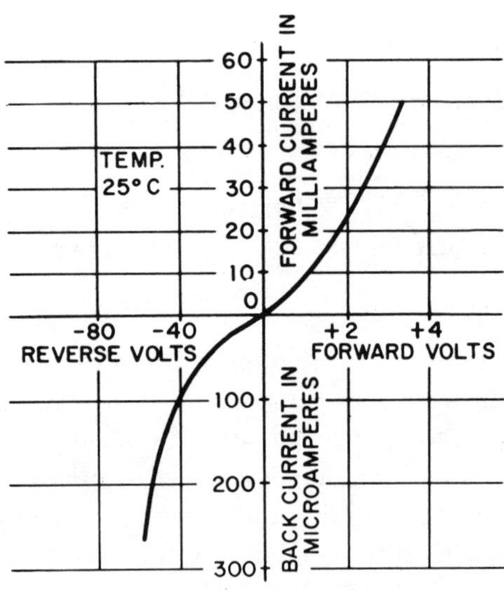

Fig. 27—3. Characteristic curve of a typical point-contact diode.

current may generate heat sufficient to destroy the rectifying action of the crystal.

With the voltage applied in the reverse direction, the back current rises slowly until, at about 20 volts, a back current of about 25 microamperes is attained. From there on the back current rises sharply with each small increase in reverse voltage. At 60 volts the back current is about 275 microamperes. This is the peak inverse-voltage rating for this diode. If the reverse voltage is increased beyond this point, the flow of back current may become great enough to ruin the crystal.

The characteristics of the crystal diode vary greatly with changes in temperature. Note that the characteristic curve of Figure 27-3 is taken at 25° Centigrade.

The point-contact diode is quite inefficient. Also, it can handle only very small currents since the entire current must pass through a point.

Since the diode consists of two electrodes close to each other, it may be considered as a sort of capacitor. One of the factors that determines its capacitance is the area of its plates. Because of the fine wire of its catwhisker, this area is small for the point-contact diode.

2. JUNCTION DIODES

The two-region crystal illustrated in Figure 26-5 may act as a diode, too, each region being considered as one electrode. Such a diode is called a *junction diode.*

The junction diode is more efficient than its point-contact counterpart. Also, since its junction has a larger area than that of the end of a catwhisker, it can handle greater currents without overheating.

The chief disadvantage of the junction diode is its higher shunt capacitance due to the greater area of its junction. Hence it is not as suitable for high-frequency applications as the point-contact type. However, modern manufacturing technique has reduced the shunt capacitance of the junction diode to a point where its approaches that of its point-contact counterpart. Accordingly, it rapidly is replacing the latter, even in some radio-frequency applications.

The characteristics of the junction-type crystal diode resembles that of the point-contact type. Its maximum forward current and peak inverse voltage, though generally somewhat greater than that for its point-contact counterpart, are quite small. (Note, however, that junction-type power diodes, which will be discussed later, are able to handle much greater currents and voltages.) The junction diode generally is enclosed in glass, ceramic, or plastic and is quite small, a quarter-inch or so in length. Its symbol is similar to that of the point-contact diode.

Both types of crystal diodes are quite sensitive to temperature. Care must be taken that neither the *ambient temperature* (the temperature of the surrounding air) nor the temperature due to the heating effect of the current rise much beyond the recommended value.

B. *Power diodes*

1. METALLIC RECTIFIERS

One of the major uses for the diode is as a rectifier where considerable currents are involved. The crystal diodes we have discussed, obviously, are not suitable for this purpose since they can handle only small currents.

A diode that can pass currents rated in amperes (instead of milliamperes) is called a *power diode*. In the early 1920's it was found that copper oxide, a semiconductor, could serve this purpose. A thin copper-oxide film is deposited upon one surface of a copper base plate. A lead plate is pressed against the copper-oxide film for contact purposes. (See Figure 27-4.) The entire sandwich is called a *copper-oxide rectifier cell*. The center hole is for mounting.

The theory of operations is somewhat similar to that of the junction diode. The copper oxide is considered to be a P-type semiconductor, that is, it has a deficiency of electrons or, what is the same thing, an excess of holes. The copper base plate, a good conductor, has a great many free electrons. Thus it may act somewhat as an N-type material. A potential barrier is created at the junction between the copper-oxide film and the copper base plate.

When voltage is applied to the rectifier cell in the forward direction (negative to the copper and positive to the copper oxide) electrons from the copper and holes from the copper oxide move toward the junction and the potential barrier is reduced. As a result, the electrons may move easily into and through the copper oxide and so we have a considerable current flow.

When voltage is applied in the reverse direction (positive to the copper and negative to the copper oxide) electrons from the copper and holes from the copper oxide move away from the junction, thus creating a depletion zone here. The potential barrier is raised and only a small reverse current may flow through the cell.

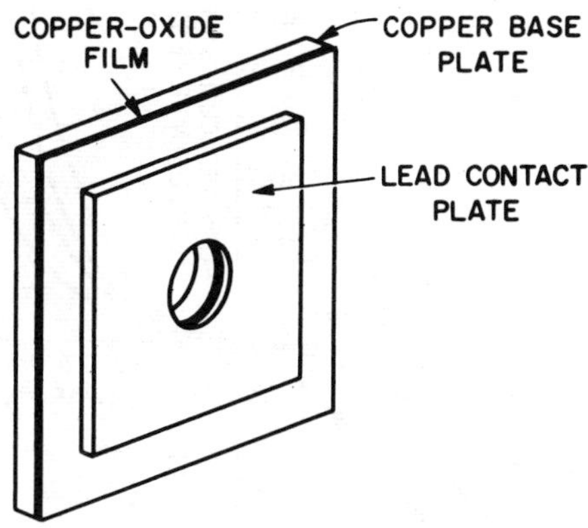

Fig. 27—4. The copper-oxide rectifier cell.

In this diode the copper corresponds to the cathode and the copper oxide to the anode. Its symbol is the same as for the crystal diode. This type of rectifier sometimes is called a *metallic*, or *dry-disk, rectifier.*

The characteristic curves for a typical copper-oxide rectifier taken at various operating temperatures is shown in Figure 27-5. As in the crystal diode, current through the cell increases rapidly as the forward voltage is increased. The maximum safe current in the forward direction depends upon the junction area; the greater the area, the larger this maximum current can be. In general, if the copper-oxide cell is self-cooled, a current of about 0.16 ampere per square inch of junction is considered a safe maximum. If the cell is cooled by an air draft, it may safely pass about twice this amount. Should the maximum be exceeded, the heat so created may ruin the rectifier.

With the voltage applied in the reverse direction, the reverse current increases slowly up to a point and rapidly beyond that. The peak inverse-voltage rating for this cell generally is in the neighborhood of 8 to 10 volts.

The cell is quite susceptible to changes in temperature. An increase in temperature causes an increase in both the forward and reverse currents. Accordingly, the rectifier must be operated within specified temperature limits.

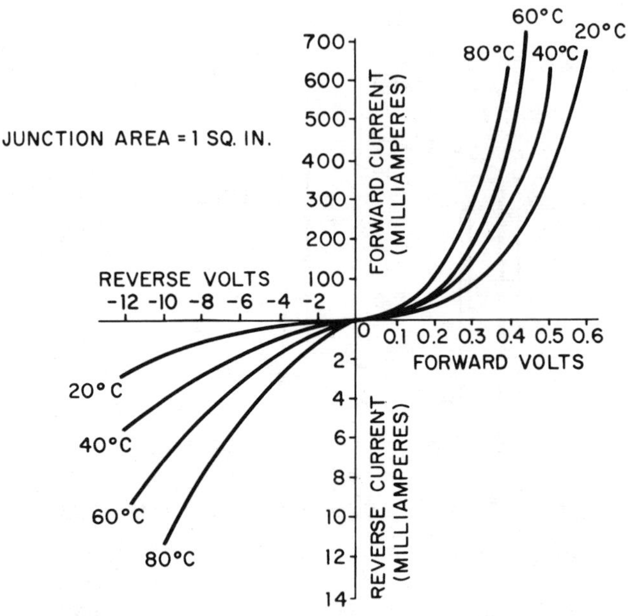

Fig. 27—5. Characteristic curves for a typical copper-oxide cell, taken at various temperatures.

To increase the maximum forward current a cell may safely pass, we either may increase its area or connect two or more cells in parallel. To increase the peak inverse-voltage rating of a rectifier, we may connect two or more cells in series. Such groups of cells are called *stacks*.

As indicated, for a greater forward current the stack consists of two or more cells in parallel. For a greater peak inverse-voltage rating it consists of two or more cells in series. Where both a greater forward current and a greater peak inverse-voltage rating are required, the cells may be connected in a series-parallel arrangement. An insulated bolt through the holes in the centers of the cells holds the stack together. (See Figure 27-6.)

General Electric Company

Fig. 27–6.

Copper-oxide rectifier stack.

Thus, rectifiers can be constructed to pass currents ranging from a few milliamperes to many thousands of amperes. Also, the peak inverse-voltage rating may range from a few volts to many thousands. The fins of the stack are used to radiate away heat and thus keep the rectifier from overheating. In some operations fans are used for further cooling.

About ten years after the copper-oxide rectifier was developed, the *selenium rectifier,* which closely resembles it, came into general use. A plate of aluminum or steel is coated with selenium, properly processed, and then sprayed with a metallic alloy generally consisting of tin, cadmium, and bismuth. The aluminum or steel plate acts as the cathode, the selenium coating as the anode. The alloy coating serves as a contact plate.

The characteristics of the selenium cell resemble those of the copper-oxide cell. The chief advantage of the selenium cell over the copper-oxide type is the higher peak inverse-voltage rating of the former. Selenium cells generally have a peak inverse-voltage rating of about 26 volts, though special types have been made with even higher ratings. Thus, for a given voltage, fewer selenium cells are needed per stack. (See Figure 27-7.)

Both types of rectifiers have an extremely long life if operated within their rated limits. Although they are being displaced by germanium and silicon power rectifiers, they still enjoy wide use. The copper-oxide type generally is employed in applications requiring low voltages as, for example, in the charging of storage batteries.

Fig. 27—7.

Selenium rectifier.

P. R. Mallory & Co. Inc.

The selenium type generally is used at power-line voltages as, for example, rectifiers in the power supply for a radio receiver.

2. JUNCTION-TYPE POWER RECTIFIERS

The junction-type crystal diodes, as we have seen, are low-power devices and, hence, not suitable for rectification where appreciable currents are involved. However, as our scientists obtained a better understanding of semiconductors and our engineers developed improved manufacturing techniques, better rectifiers were produced. The germanium junction diode, for example, was developed to the point where it could pass safely about 300 amperes per square inch and the peak inverse-voltage rating was raised to about 300 volts per junction.

Hence the junction-type power rectifier came into being. The P-N germanium diode generally is encased in metal for protection and to help radiate away the heat produced. As the metallic rectifier, it may be fitted with fins for better cooling and arranged in stacks for applications at higher voltages and currents.

The chief disadvantage of the germanium junction diode is that it is extremely sensitive to heat. Operating precautions generally state that its temperature must not be permitted to rise above 65° to 75° Centigrade. Beyond that the currents may increase to the point where the diode will be ruined.

Chiefly because of its better temperature characteristics, silicon quickly replaced germanium in the power diode. The silicon junction diode is able to function safely at temperatures of 200° Centigrade, and higher. Also, it can handle larger currents and withstand a higher inverse voltage than can its germanium counterpart.

The silicon power diode generally is encased in metal for protection and heat dissipation. (See Figure 27-8.) Those with higher

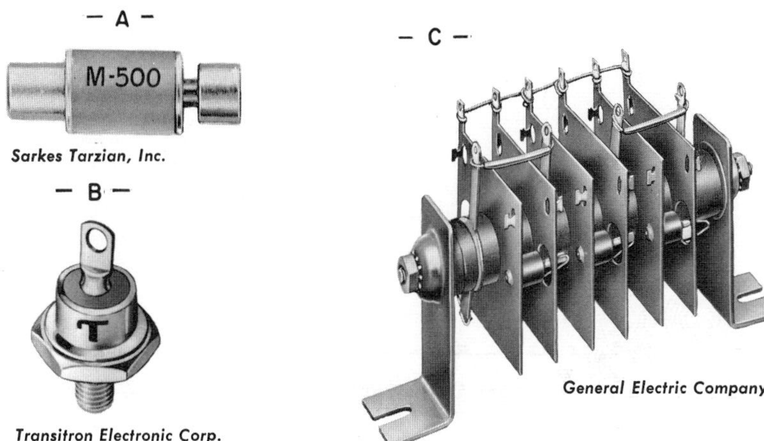

Fig. 27—8. A. Clip-mounted silicon rectifier.
B. Stud-mounting silicon rectifier.
C. Stacked silicon rectifiers.

current ratings usually are equipped with a stud whereby they may be bolted securely to a heavy metal plate, called a *heat sink.* Because of its mass, the sink is able to absorb a great deal of heat from the diode.

The diode may be furnished with fins for greater heat dissipation. Also, it may be stacked for applications at higher voltages and currents.

QUESTIONS

1. **Describe and explain the theory of the modern *point-contact crystal diode.***
2. **Describe and explain the theory of the *junction diode.***
3. **Explain the advantages and disadvantages of the point-contact and junction diodes.**
4. **Describe and explain the theory of the *copper-oxide metallic rectifier cell.***
5. **Describe and explain the theory of the *selenium metallic rectifier cell.***
6. **Describe and explain the theory of the *junction power diode.***
7. **Explain the advantages and disadvantages of the junction power diode and the electron-tube diode.**

Transistors

A. Theory of operation

The *junction transistor* consists of a small semiconductor crystal, generally of germanium or silicon, containing three regions. The center region (which is thinner than the others) may be of the N- or P-type. If, as illustrated in Figure 28-1A, the center region is of the P-type, the two outer regions are of the N-type. Such a crystal is called an *N-P-N transistor*. If the center region is of the N-type (as in Figure 28-1B), the two outer regions are of the P-type. This is called a *P-N-P transistor*.

The center region is called the *base*. The left-hand region is called the *emitter* and the right-hand one the *collector*. Between the emitter and base we have an *emitter-base junction*. Between the collector and base we have a *collector-base junction*.

The symbols for both type transistors are also shown in Figure 28-1. The horizontal line marked B represents the base region. The slanted lines represent the emitter and collector regions. The emitter is identified by an arrowhead on its line. If the arrowhead points out (Figure 28-1A), it indicates that the emitter region, and hence the collector region, too, is of the N-type. If the arrowhead

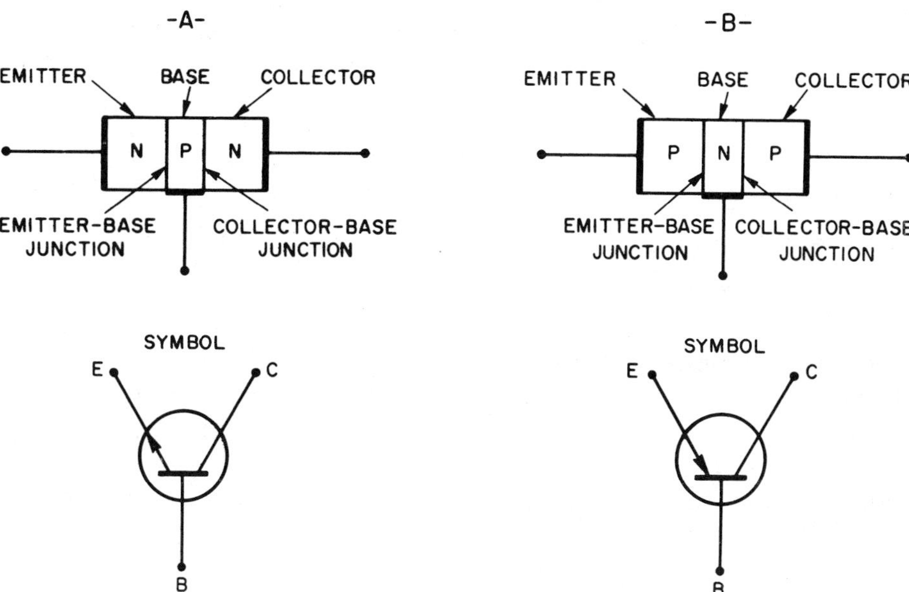

Fig. 28—1. **A.** N-P-N transistor.
B. P-N-P transistor.

points in (Figure 28-1B), the emitter and collector regions are of the P-type. The collector line has no arrowhead.

The basic N-P-N transistor circuit is illustrated in Figure 28-2. Battery B_1 biases the emitter-base junction in the *forward* direction. Battery B_2, because its positive terminal is connected to the N-type collector region, biases the collector-base junction in the *reverse* direction.

Since it is biased in the forward direction, the potential barrier of the emitter-base junction is low. And because it is biased in the reverse direction, the potential barrier of the collector-base junction is high.

Because of the low potential barrier, electrons flow readily from the emitter to the base. At the same time, holes from the base flow into the emitter. The base region is made very narrow and the percentage of impurities is kept lower than that of the emitter region. Thus there normally are fewer holes in the base than electrons in the emitter. As a result of the flow there is a concentration of electrons in the base region. Because of the mutual repulsion between these electrons, and under the influence of the attraction of

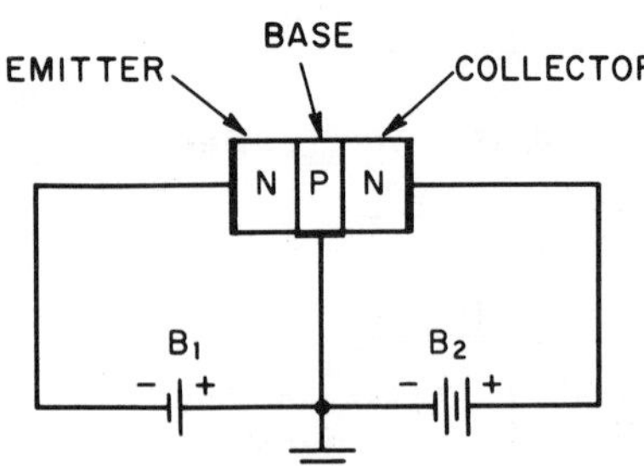

Fig. 28–2.

Basic N-P-N transistor circuit.

the positive terminal of battery B_2, nearly all of the electrons coming from the emitter diffuse across the base into the collector region.

As you have learned, when a P-N junction is formed, the N-region obtains a positive charge (Chapter 26). Hence the collector region normally has a positive charge. As the electrons from the emitter diffuse through the base region into the collector region, they reduce its positive charge, thus lowering the potential barrier of the collector-base junction. Hence many more electrons can flow across this barrier, through the collector region, and to the positive terminal of battery B_2. Thus the current flow between the emitter and base produces a current flow out of the collector.

In the P-N-P transistor the polarities of the biasing batteries are reversed. See Figure 28-3. As in the N-P-N type, the emitter-base junction is biased in the forward direction and the collector-base junction in the reverse direction. Holes flow from the emitter into the base region and diffuse into the collector region. As a result, electrons from the negative terminal of battery B_2 flow into the collector region.

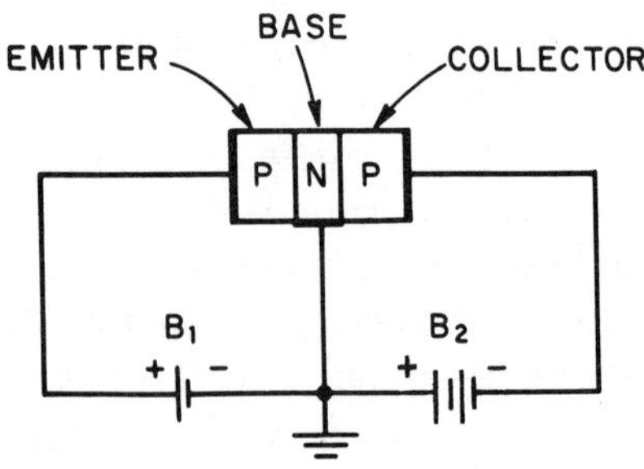

Fig. 28–3. Basic P-N-P transistor circuit.

The basic N-P-N transistor amplifier circuit is illustrated in Figure 28-4. The input signal is applied across R_1 and the output signal is taken from across R_2. Battery B_1 biases the emitter-base junction in the forward direction. Battery B_2 biases the collector-base junction in the reverse direction. Because the base is common to the input and output circuits, this configuration is called a *common-*, or *grounded-, base* circuit.

The biasing batteries, supplying a steady voltage, produce a steady current flow in the transistor. Since we are interested in the changes produced by the incoming signal, the current flow produced by the batteries may be disregarded in our discussion at this point.

Assume that the input signal is such as to increase the emitter-to-base voltage (from that of the bias) in the forward direction. More electrons are made to move from the emitter to the base. For example, let us suppose that an additional 100 microamperes flow in this manner.

At the same time, the signal voltage causes holes to move from the base to the emitter. Since there are less holes available to move from the base to the emitter than electrons from the emitter to base, this current flow will be less than the electron flow in the opposite direction. Assume that the current flow due to the movement of holes from the base to the emitter is 10 microamperes.

Since the base region is very narrow and because of the mutual repulsion between electrons, practically all the electrons entering the base region from the emitter drift across the collector-base junction and flow through the collector region, drawn by the attraction of the positive terminal of battery B_2. Hence 100 microamperes of current flow out of the collector region. This is the output current.

Thus the current flow between the emitter and base causes a current flow through the collector and a voltage drop across R_2. If the voltage across the emitter and base is increased by the incoming signal across R_1, the current flow and, hence, the voltage drop across R_2 is increased.

If the signal voltage is reduced, the current, too, is reduced, and so is the voltage drop across R_2. You see, then, that a signal voltage applied to the input produces a corresponding output voltage. The action is the same as in the electron-tube amplifier.

When considering the input current, we must add the 10 microamperes carried by the holes moving from the base to the emitter to the 100 microamperes carried by the electrons from the emitter. Thus the input current is 110 microamperes.

The current amplification (which for this type of circuit is designated by the Greek letter α, pronounced *alpha*) is found by dividing the output current by the input current, or 100/110, which is less than 1. This is always the case when a junction transistor is operated in a common-base configuration. Practical transistors may have an α that is between about 0.80 and 0.99.

Note, however, that the input impedance of the transistor is fairly low (usually about 100 ohms), thanks to the fact that the emitter-base junction is biased in the forward direction. The output impedance is very high (about 1,000,000 ohms) because of the reverse bias of the collector-base junction. Because of this high output impedance of the transistor, we may employ a high-resistance load resistor (R_2), say, about 50,000 ohms.

Since $P = I^2 \times R$, the power input is $(0.00011 \text{ ampere})^2 \times 100$ ohms, or 0.0012 milliwatt. The power output is $(0.0001 \text{ ampere})^2 \times 50{,}000$ ohms, or 0.5 milliwatt, providing a power gain of over 400. Also, since $E = I \times R$, the voltage input is 0.00011 ampere $\times$ 100 ohms, or 0.011 volt. The voltage output is 0.0001 ampere $\times$ 50,000 ohms, or 5 volts. The voltage gain for our amplifier thus is over 450.

The transistor itself, of course, does not generate any power. The gain is provided by the collector battery (battery B_2). The feature of the transistor is that a small input power may produce a large output power, or that a small varying input voltage may cause a large output voltage to vary in like degree.

The ratio between the collector (output) current and the base current is designated by the Greek letter β (pronounced *beta*). Thus, for our amplifier, β is equal to 100/10, or 10. Transistors are constructed that have a β of 100 or more.

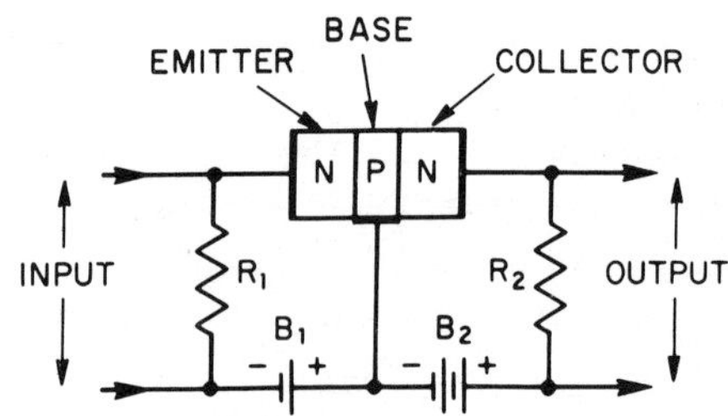

Fig. 28–4. Basic N-P-N transistor amplifier circuit.

If a P-N-P transistor were to be substituted for the N-P-N transistor, the amplifier would act in the same way. The polarities of the biasing batteries would be reversed, of course, and the flow of current would be in terms of holes, rather than electrons.

The amount of power that can be dissipated by the collector is limited by the possibility of damage due to overheating. In the case of transistors used for handling small signals, this limit usually is about 50 to 150 milliwatts. The maximum collector current generally is between 10 and 20 milliamperes and the maximum voltage supplied by the biasing battery is between 20 and 40 volts.

From the above, you can see that the maximum collector current and maximum collector voltage may not *both* be applied at the same time. If, for example, the maximum voltage (40 volts) is applied and the maximum current (20 milliamperes) flows, the power ($E \times I$) would be 800 milliwatts, well above the maximum power that can be dissipated by the collector.

Because of improved manufacturing techniques and by the use of fins and heat sinks to carry away excess heat, *power transistors* have been made with a maximum collector dissipation of 60, or more, watts. Such transistors frequently employ silicon rather than germanium because the former can operate satisfactorily at higher temperatures.

B. *Transistor types*

The earliest transistors were of the *point-contact* type which consists of a tiny slab of N-type germanium crystal whose upper surface is touched at two points very close together by the tips of two fine wire catwhiskers. (See Figure 28-5A.) The entire unit is cased in metal or sealed in plastic.

The explanation of how this type of transistor operates is somewhat complicated, but in effect, it acts as though small P-regions are created in the crystal around the points where it is touched by the catwhiskers. (See Figure 28-5B.) The entire unit, then, behaves as a P-N-P transistor.

The point-contact transistor suffers in comparison with the junction-type previously described, chiefly because of its low power-handling capacity. This is because of the relatively high heat generated at the contact point between the crystal and the collector catwhisker through which point the entire output current must flow.

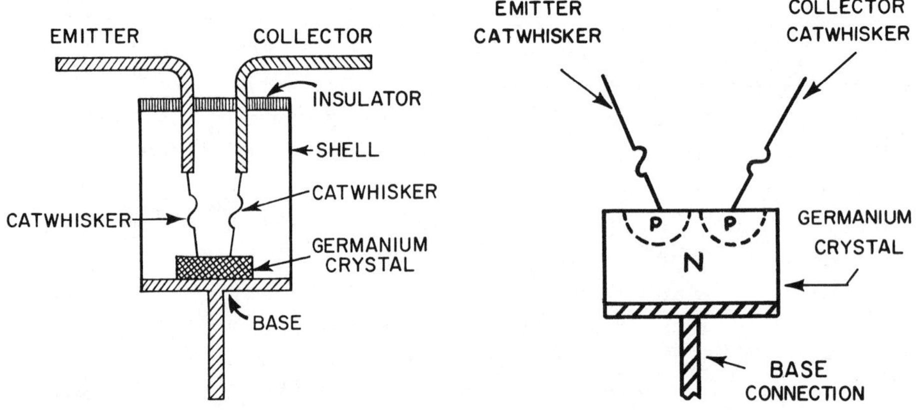

Fig. 28—5. **A.** Cross-sectional view of the point-contact transistor. **B.** Point-contact transistor showing regions.

The chief advantage of the point-contact transistor is the fact that it can function at rather high frequencies. You will recall that, as the carriers from the emitter region reach the base region, they diffuse across into the collector region. The time it takes them to cross into the collector region is important. If this interval is significant in relation to the time required for one cycle of the incoming signal, the amplifying action of the transistor will be destroyed. Hence, the narrower the base region is, the less the time required for the carriers to diffuse across, and the higher the frequency rating of the transistor.

The point-contact transistor may have a very narrow base region since the two catwhisker tips can be placed very close together. However, as new techniques resulted in the manufacture of junction transistors having very narrow base regions and, hence, able to function at higher frequencies, this advantage for the point-contact transistor was overcome. As a result, very few of this type transistor are made today.

There are several methods for producing the junction transistor. In the *grown-junction* method, the semiconductor material and its impurities are melted and a single crystal grown from the melt. By carefully controlling the rate of crystal growth, either an N-region or P-region can be produced. Thus the entire N-P-N or P-N-P crystal results. Leads are attached to the various regions and the whole encased in plastic and placed in a metal case for protection and heat dissipation.

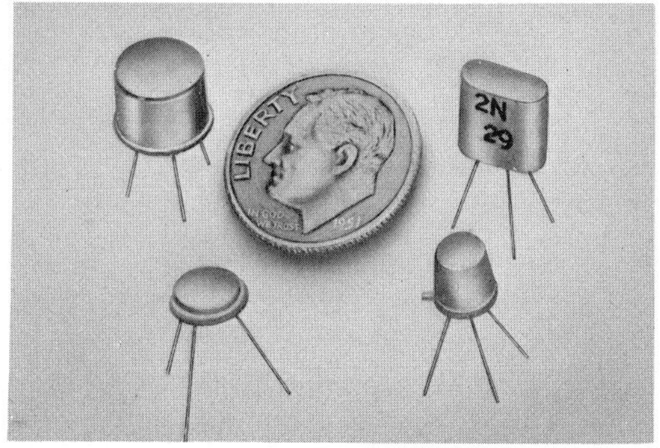

Sylvania Electric Products, Inc.

Fig. 28—6. Various types of small-signal transistors.

In the *alloy-junction* method, dots of indium (acceptor impurity) are attached to each surface of a small slab of N-type semiconductor material. As the whole is carefully heated, the indium alloys with the semiconductor and diffuses into it, forming a P-region. (See Figure 28-7A.) The result is a P-N-P transistor.

The collector dot is about twice the diameter of the emitter dot. Because the extent of the P-region can be controlled, the base region between them can be made very narrow. Hence the frequency response of the alloy-junction transistor is quite good. Transistors having a frequency range in the order of megacycles per second can be made in this way.

Leads are attached to the various regions and the whole is mounted within a metal case. A glass bottom holds the leads in place and seals the case. (See Figure 28-7B.) Such transistors can be made very small, some are only about a quarter of an inch, or less, in height.

N-P-N transistors may also be constructed in this manner. Here the semiconductor slab is of the P-type and the dots are made of donor material, such as arsenic.

In the *surface-barrier* method, some of the material of a slab of N-type semiconductor about 0.003 inch thick is etched away to form a pit in either side. Indium is plated on the surfaces of these pits.

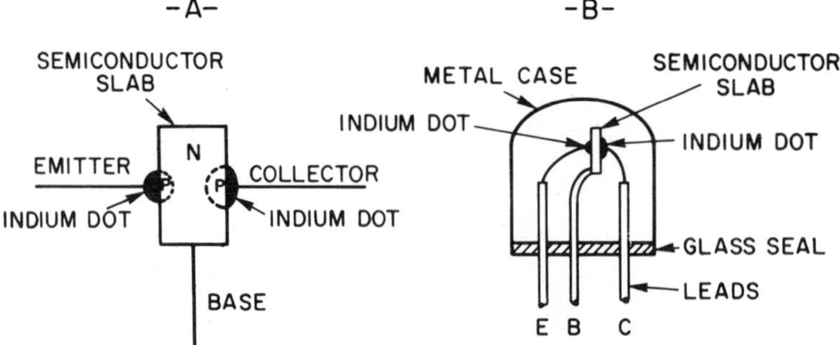

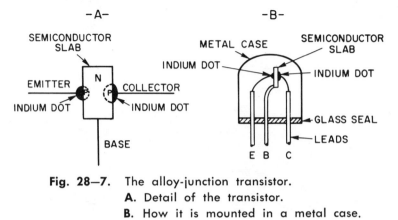

Fig. 28—7. The alloy-junction transistor.
A. Detail of the transistor.
B. How it is mounted in a metal case.

See Figure 28-8. As a result, chemical bonding between the indium and the semiconductor form a P-type zone at the surface of the semiconductor. This zone does not extend very deeply and the potential barrier is practically at the surface. Hence we have a P-N-P transistor.

Because the etching process can be controlled closely, the width of the base region between the two P-zones can be made extremely narrow, frequently only a few thousandths of a millimeter. As a result, transistors having a high frequency range (up to about 75 megacycles per second) can be made.

Power transistors, which are heavier and larger than the small-signal type, generally are made by the alloy-junction method. To prevent overheating due to the heavier currents involved, they may have cooling fins and provisions are made for intimate connection to a heat sink.

A newer method of construction is shown in Figure 28-9 in the mesa and planar epitaxial transistors. They are made possible by

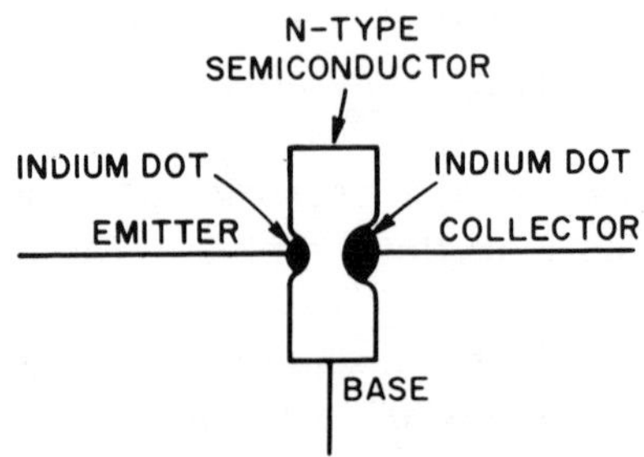

Fig. 28—8. The surface-barrier transistor.

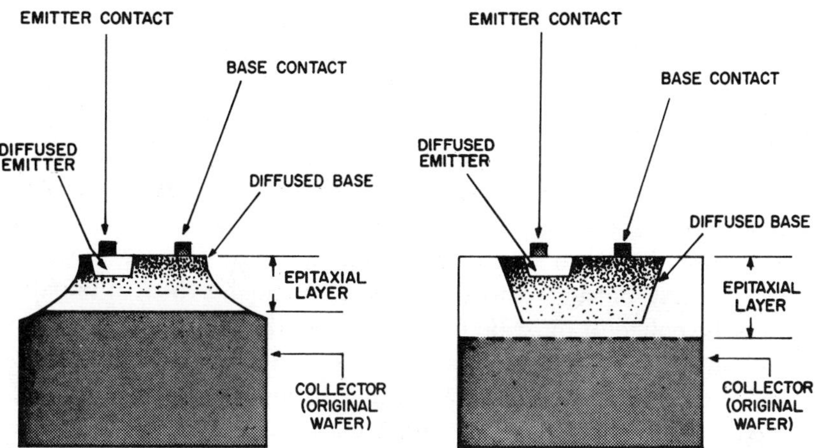

Fig. 28-9. A. Double-diffused epitaxial mesa transistor. **B.** Double-diffused epitaxial planar transistor.

improved diffusion masking materials and photolithographic techniques to create transistors where the base and emitter junctions are diffused into a collector wafer and buried under a protective layer.

The construction of a typical power transistor is illustrated in Figure 28-10. The thin semiconductor wafer is strengthened by a

Fig. 28—10. Cross-sectional view showing the construction of a typical power transistor.

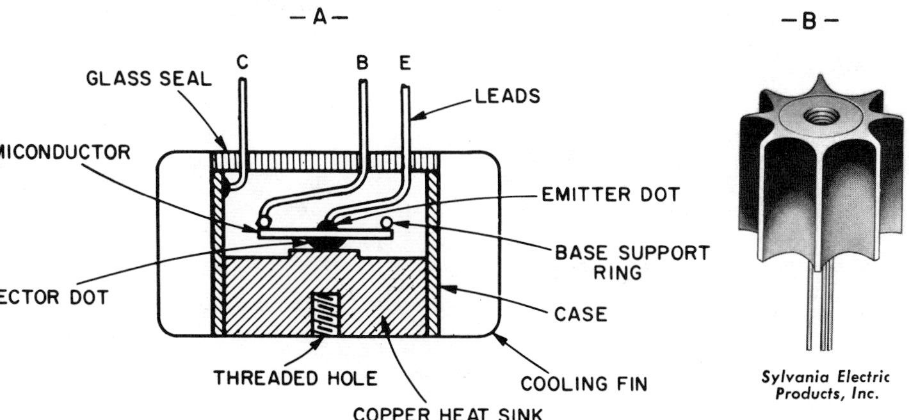

metal ring fastened to it. The emitter and collector dots are attached and fused into the sides of the wafer. The transistor is mounted so that its collector dot presses against a copper block that acts as a heat sink. The whole is enclosed in a metal case.

The base lead is attached to the metal ring, the emitter lead to the emitter dot, and the collector lead to the metal case which makes electrical contact with the heat sink and so to the collector dot. A glass base holds the leads in place and seals the case. The metal fins attached to the case are for cooling purposes.

The threaded hole in the heat sink is there so that the entire transistor may be bolted to the chassis for further heat dissipation. Where the collector must be insulated from the chassis, an anodized-aluminum or mica washer can be used for insulation.

The power transistor may take other forms. (See Figure 28-11.) The collector dot may be in contact with a heavy metal flange or a metal stud used for fastening the transistor or the chassis. If, as previously described, the collector must be insulated from the chassis, an insulating washer may be used. Input signals, being applied to the forward-biased base-emitter junction at the transistor, are required to supply power because input resistance of the junction is low.

A later type of transistor called the field-effect transistor overcomes this occasionally objectionable characteristic. In this type of transistor, the output circuit is connected between 2 contacts imbedded in a channel of N-type material. These two contacts are called the source and drain. (See Figure 28-12.) The conductivity of the channel is controlled by a bias applied to the gate which attracts or repels carriers. The gate is either a reverse-biased P-N junction (Figure 28-12A) or a metal-gate electrode separated from

-A-

Sylvania Electric Products, Inc.

-B-

Minneapolis-Honeywell Regulator Co.

Fig. 28—11. Various types of power transistors.

A. Flange-type.
B. Stud-type.

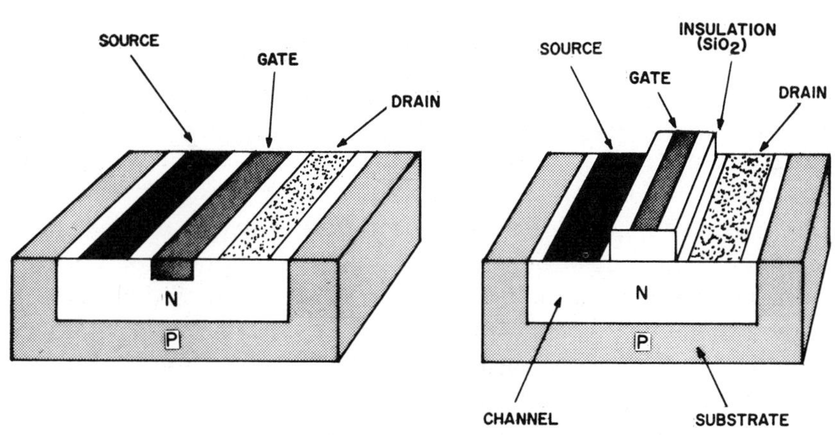

Fig. 28—12. **A.** PN junction field-effect transistor.
B. MOS field-effect transistor.

the semiconductor material by an insulator. (See Figure 28-12B.) In general, there are two types of F. E. T.'s—the enhancement type where the gate must be forward biased to produce active carriers and permit conduction through the channel and the depletion type where there are normally charge carriers in the channel, and reverse biasing of the gate reduces the active carriers and therefore the channel conductivity. (See Figure 28-13 for the schematic symbol for the MOSFET.)

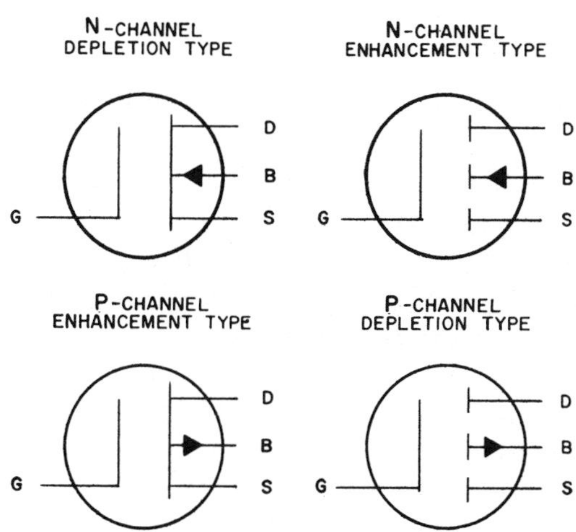

Fig. 28-13.

Schematic symbols for MOS field-effect transistors (G = gate, D = drain, B = active bulk, S = source).

C. *Comparison between transistor and electron tube*

It might be well, at this point, to compare the transistor with its electron-tube counterpart, noting similarities and differences. The emitter of the transistor corresponds to the cathode of the tube since both supply the carriers that eventually reach the output circuit. The cathode supplies the carriers (electrons), generally as a result of thermionic emission. The emitter becomes a supplier of carriers (electrons or holes, as the case may be) as a result of the forward bias across the emitter-base junction.

The base of the transistor corresponds to the control grid of the tube. We may consider the transistor and electron tube as controlled resistors. By varying the controls, we vary the resistances and, hence, the current flowing through them (the output current). The grid-to-cathode voltage determines, to a large extent, the internal resistance of the electron tube. The base current regulates, to a great degree, the internal resistance of the transistor.

The collector of the transistor corresponds to the plate of the tube. Both are the recipients of the carriers which constitute the output current. The plate receives the electrons from the cathode, the collector receives the electrons or holes, as the case may be, from the emitter.

The electron tube is a voltage-controlled device. Input voltage, generally impressed across the grid and cathode, controls the output current. The transistor is a current-controlled device. Input current, generally flowing across the emitter-base junction controls the output current, except in the case of the field-effect transistor, which is also a voltage-controlled device.

The transistor has a number of advantages over the electron tube. It is smaller; some are only a few thousandths of an inch long. It is more rugged; its internal structure is simple since there are no filament grid wires, or plate assembly.

The transistor consumes less power. It has no filament, hence requires no filament power. Also, it operates at lower voltages than does the electron tube. And because it consumes less power it operates at lower temperatures. This all results in a longer life for the transistor, if it is operated properly.

Another important difference between transistors and tubes is the possibility of making use of *complementary* transistors. That is, an N-P-N transistor may be constructed with characteristics similar to a

P-N-P transistor except that in normal operation the polarities of all voltages and currents will be reversed. This permits the use of many circuits which would not be possible with tubes (since a tube cannot be operated with a negative plate voltage).

In addition, the transistor has a lower intrinsic cost than its electron-tube counterpart since it has fewer and simpler parts.

Should the maximum power rating of the tube be exceeded by an instantaneous peak value, the tube generally will not be damaged, provided the major portion of the cycle is below the maximum. A transistor in this position would probably be ruined. Also, if the applied voltages to the tube were accidentally reversed, the tube would not function but would not be otherwise damaged. When the voltages are corrected, the tube would function normally. If reversed voltages were applied to the transistor, on the other hand, it might be ruined.

QUESTIONS

1. a) **Describe and explain the theory of the *N-P-N transistor.***
   b) **Explain the theory of its operation.**
2. a) **Describe and explain the theory of the *P-N-P transistor.***
   b) **Explain the theory of its operation.**
3. **Draw and explain the basic biasing circuit of the N-P-N transistor.**
4. **Draw and explain the basic biasing circuit of the P-N-P transistor.**
5. **Draw and explain the basic circuit of the N-P-N amplifier.**
6. **Draw and explain the basic circuit of the P-N-P amplifier.**
7. **In the transistor amplifier, explain what is meant by**
   a) ***alpha;***
   b) ***beta.***
8. a) **Describe and explain the structure of the *point-contact transistor.***
   b) **What are its advantages and disadvantages compared to a junction-type transistor?**
9. **Describe and explain the *grown-junction transistor.***
10. **Describe and explain the *alloy-junction transistor.***
11. **Describe and explain the *surface-barrier transistor.***
12. **Compare the transistor with its electron-tube counterpart.**

29 Semiconductor Circuits

With very few exceptions, semiconductor diodes and transistors are able to replace their electron-tube counterparts. Thus we have all-transistor radio receivers. We even have all-transistor television receivers (except for the picture tube and high-voltage rectifier). In this chapter we will discuss semiconductor versions of some of the electron-tube circuits we have encountered. Note that there are many variations of these circuits. Of necessity, not all of them can be presented here. Instead, we shall examine a number of typical circuits.

A. *Power supply circuits*

As we have seen, the semiconductor diode can function as a rectifier. Thus the crystal diode is used extensively as a detector in AM, FM, and television receivers. Its low-current capacity, however, precludes its use as a rectifier in power supplies where larger currents must be passed. Instead, stacked dry-disk rectifiers or junction-type power rectifiers are employed.

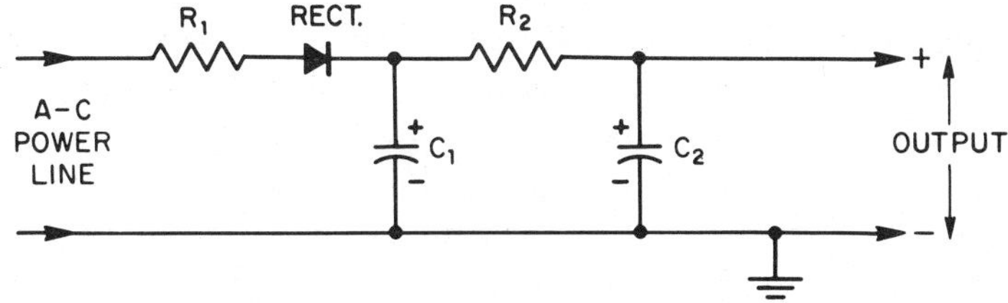

Fig. 29—1. Half-wave power supply.

The circuit of a typical half-wave power supply operating from the a-c power lines is illustrated in Figure 29-1. The rectified line voltage is filtered by the filter circuit consisting of R_2, C_1, and C_2. Resistor R_1 is a *surge resistor* used to protect the rectifier from any sudden rises in the line current. This is a resistor of small value. Normally, there is a small voltage drop across it. But should the current increase, so will the voltage drop across the resistor. As a result, the voltage applied to the rectifier is reduced to a safe level. In some installations R_1 is arranged to burn out on large overloads, thus serving the additional function of a fuse.

As you have already learned, the negative side of the power supply generally is grounded to the chassis. One side of the a-c line, too, is grounded to an external ground. All is well as long as the grounded side of the line coincides with the power-supply ground. But if the ungrounded ("hot") side of the a-c line is connected to the grounded side of the power-supply line (as, for example, the plug is reversed at the a-c outlet), the chassis becomes "hot." Then, if a person should touch the chassis while in contact with an external ground (such as a water pipe), a dangerous shock may result.

Fig. 29—2.

Decoupling network used with transformerless power supply.

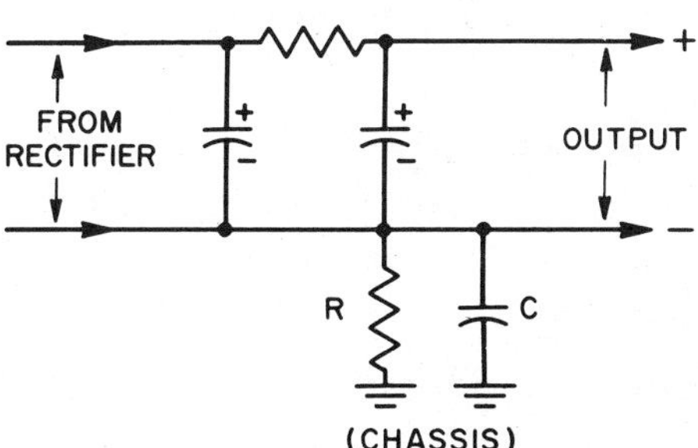

To get around this danger, the negative side of the power supply is connected to the chassis through a *decoupling network*. (Figure 29-2.) R is a resistor of large value between the chassis and the negative side of the power supply. Thus, even if this negative side were connected to the "hot" side of the a-c line, the danger of shock is eliminated. Capacitor C furnishes a path from the negative side of the power supply to the chassis only as far as the a-f and r-f signals are concerned. For currents at the line frequency, C offers a very high impedance.

(Note that the power supply illustrated in Figure 29-1 can function from a d-c power line, if its connection to the line be with the correct polarity, that is, with the anode of the diode to the positive side of the line, as well as from an a-c power line. Hence it sometimes is called an *AC-DC power supply.*)

The output voltage of this power supply is approximately equal to the line voltage. Where a larger output voltage is desired, the line voltage may be stepped up by means of a transformer before being applied to the rectifier. (This can be done, of course, only if the power supply is operating from an a-c line.) Note that now a decoupling network is not needed since the power line no longer is connected directly to the power supply.

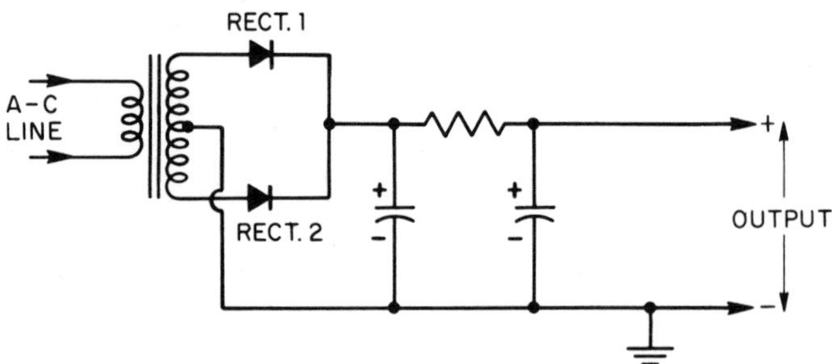

Fig. 29—3. Full-wave power supply.

The full-wave rectifier circuit is illustrated in Figure 29-3. Except for the fact that semiconductor diodes are used instead of electron tubes, the circuit is the same as the electron-tube circuit described in Chapter 13.

The full-wave bridge rectifier circuit is shown in Figure 29-4. This circuit, except that semiconductor diodes are employed, is also

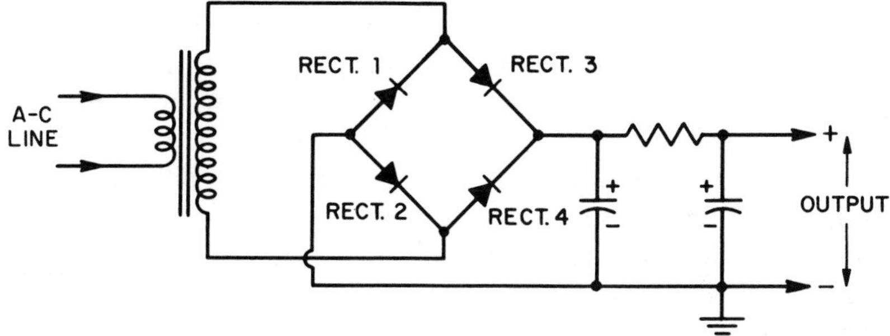

Fig. 29—4. Bridge-rectifier power supply.

similar to its electron-tube counterpart. The chief advantage of the semiconductor circuit lies in the fact that these diodes have no heaters, thus eliminating the need for complicated heater circuits.

The electron-tube voltage-doubler circuit discussed in Chapter 13, too, has its semiconductor counterpart. (See Figure 29-5.) Resistor R is a surge resistor. Note that the negative side of the power supply is connected directly to the a-c line. Thus a decoupling network is required if the possibility of shock is to be avoided.

In our discussion of the electron-tube diode used as a rectifier in the power supply, we saw how a glow-discharge tube may be employed to regulate the voltage output (see Chapter 13, subdivision E). Semiconductor diodes can perform a similar function.

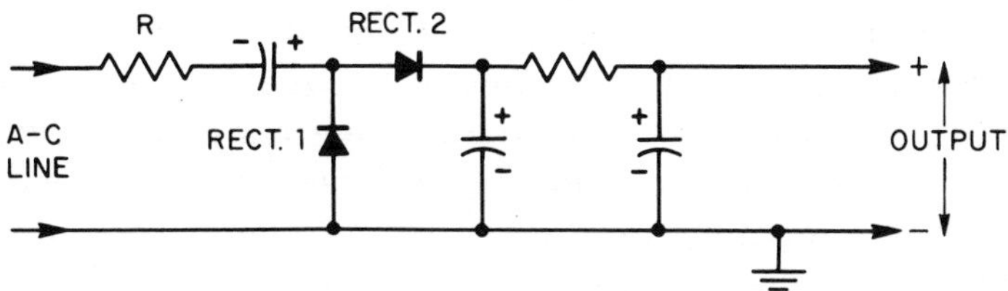

Fig. 29—5. Voltage-doubler circuit.

A typical current-voltage characteristic curve for a special junction-type silicon rectifier is shown in Figure 29-6. Note that the curve shows three regions. Between points #2 and #3 is the *saturation* region in which the current through the diode is very small and practically constant for a wide range of voltages. At point #1 we enter the *forward breakdown* region where the current increases very rapidly with slight increases in forward voltage. At point #3 we enter the *reverse breakdown,* or *zener,* region where the reverse current increases very rapidly with slight increases in reverse voltage.

The saturation current consists of a two-way flow across the junction of minority carriers—electrons from the P-region to the N-region and holes from the N-region to the P-region. Hence it is quite small and practically independent of the applied voltage.

The breakdown currents are of two general types. The *avalanche* breakdown is the result of collisions between the carriers and the atoms of the semiconductor. As the applied voltage is increased, so is the acceleration of the carriers. At a certain critical value of voltage, the acceleration is so great that the force of impact is large enough to break some of the covalent bonds, thus producing more electrons and holes. This process is cumulative and results in a sharp increase in current flow.

The *zener* breakdown results from the formation of additional carriers due to the breaking of covalent bonds when the applied voltage reaches a point where the electric field produced by it becomes large enough. Unlike the avalanche breakdown, the effect of the zener breakdown is not cumulative.

Neither of these breakdowns is, by itself, damaging to the diode, provided that they are kept within safe bounds. It is the power and resulting heat developed by the high current flow under breakdown conditions that may permanently damage the diode. Thus the diode may cycle into and out of the breakdown condition as long as the current flow, and the resulting heat, do not become too large.

Note that in the saturation region (between points #2 and #3) the current is fairly constant for relatively large changes in voltage. Hence this region may be considered as a constant-current region. In the zener region (beyond point #3), a large increase in current is the result of a small increase in reverse voltage. Hence it can be classified as a constant-voltage region where a relatively large change in current can occur with a very small change in voltage.

Diodes that exhibit these properties are called *zener diodes.* By regulating the amount of impurities in the N- and P-regions introduced at the time of manufacture, the voltage at when the zener effect occurs may be determined. Such diodes may be made with the zener point at any desired level from about 2 to 300 volts. They may be made with safe power dissipation from a fraction of a watt up to about 50 watts.

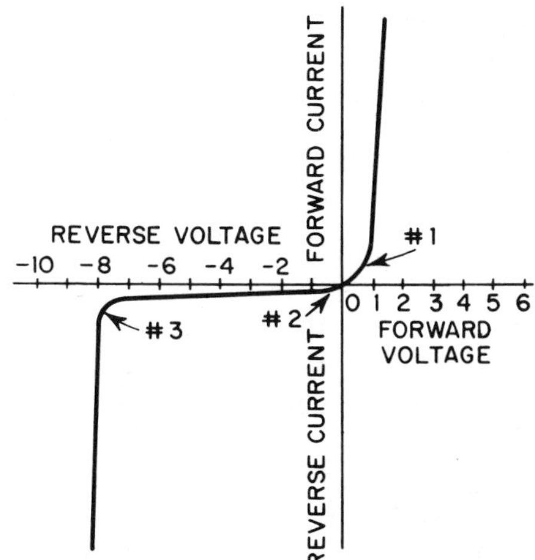

Fig. 29—6.

Current-voltage characteristic curve for a zener diode.

In appearance the zener diode is similar to the ordinary junction diode. To differentiate between the two in electrical diagrams, the symbol for the zener diode is shown as [zener symbol] or [zener symbol] B .

It is the constant-voltage property of the zener diode that enables it to be used for voltage regulation in a manner similar to the glow-discharge tube. There are some differences, however. The glow-discharge tube is able to regulate voltages within a small limited range (generally between 75 and 150 volts). The zener diode has a much greater range (from about 2 to 300 volts). Where greater voltages are desired, the zener diodes may be connected in series. Also, the glow-discharge tube can handle currents up to about 40 milliamperes. The zener diode has a safe power dissipation of up to about 50 watts.

A voltage-regulating circuit employing a zener diode is illustrated in Figure 29-7. Note that the diode is connected with its cathode to the positive side of the line and its anode to the negative, that is, in the reverse direction. Resistor R is a current-limiting resistor whose resistance is such as to limit the current flowing through the diode to the maximum power it can safely handle.

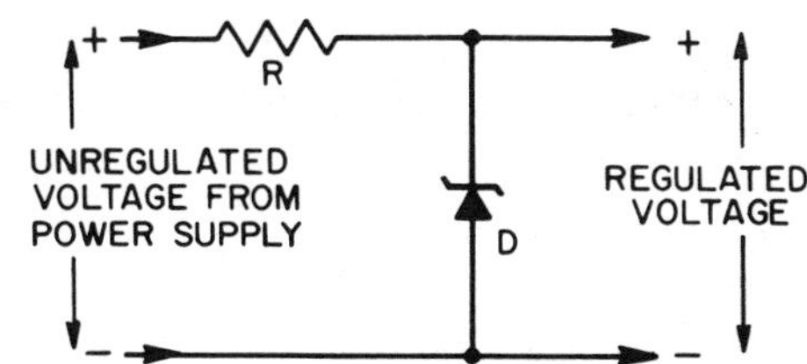

Fig. 29—7.

How the zener diode is used as a voltage regulator.

Semiconductor diodes are also used as rectifiers in many other power circuits. For example, dry-disk copper-oxide rectifiers are employed instead of electron-tube diodes in the back-to-back ignitron circuit discussed in Chapter 21. In this way complicated heater circuits are eliminated.

B. *Amplifier circuits*

When considering the transistor as an amplifier, there are a number of comparisons with electron tubes that must be kept in mind. The electron tube is a *voltage-operated* device, that is, the input to the tube is a voltage (generally applied between the grid and cathode) that causes a corresponding flow of current in the plate circuit. The transistor, on the other hand, is a *current-operated* device; the input is a current that causes a corresponding flow of output current.

Unlike the electron tube, the input and output circuits of a transistor can never be independent of each other. The output impedance of a transistor depends on the input impedance, and vice versa. Even the impedance of the signal source must be considered in determining the gain of a transistor amplifier.

The input impedance of transistor amplifiers generally is much lower than that of equivalent electron-tube amplifiers. Depending on the circuit used, a transistor amplifier may have input impedances of about 60 to several hundred thousand ohms. The input impedance of an electron-tube amplifier may be many megohms. Because of the lower impedance, the values of resistors employed in transistor circuits are lower and the values of capacitors are higher than their equivalent components used in electron-tube amplifiers.

Nevertheless, there are many points of similarity. The emitter of the transistor is analogous to the cathode of the electron tube, the base to the grid, and the collector to the plate. Just as for the tube, there are three general configurations for supplying the input and obtaining the output from a transistor. For the tube, the most common configuration is the *common-*, or *grounded-*, *cathode* circuit shown in Figure 29-8A. The signal is applied to the grid and cathode (which is common to the input and output circuits). Output is obtained from the load resistor (R_L) between the plate and cathode.

Fig. 29—8.

A. Electron-tube common-cathode configuration.

B. Transistor common - emitter configuration.

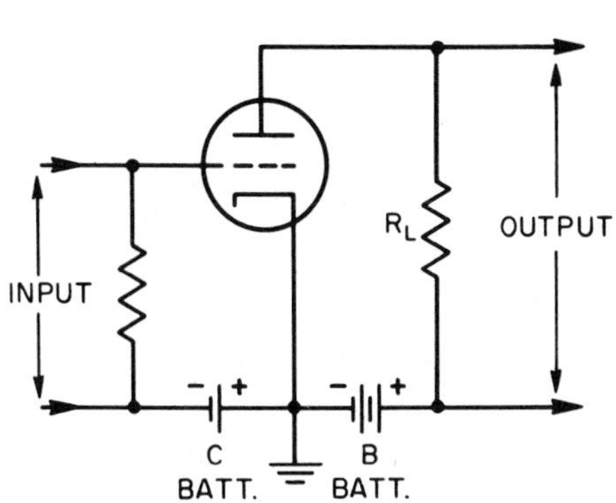

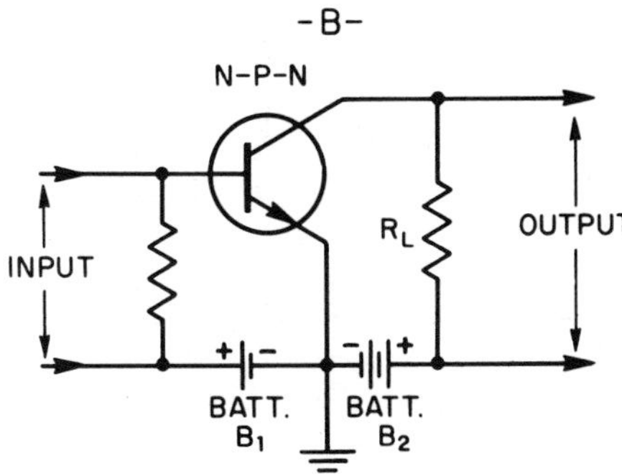

The analogous transistor circuit is the *common-*, or *grounded-*, *emitter* circuit shown in Figure 29-8B. Input is between the base and emitter; output is between the collector and emitter. As in its electron-tube counterpart, the signal undergoes a phase reversal. Battery B_1 biases the emitter-base junction in the forward direction. Battery B_2 biases the collector-base junction in the reverse direction. Both input and output impedances are moderate in value. This circuit provides the greatest power gain.

A second configuration is the *common-*, or *grounded-*, *grid* circuit shown in Figure 29-9A. The signal again is applied between the grid and cathode, though this time it is the grid that is common to both input and output circuits. The output is obtained from the load resistor (R_L) between the plate and grid.

The analogous transistor circuit is the *common-*, or *grounded-*, *base* circuit shown in Figure 29-9B. Input is between the emitter and base; output is between the collector and base. There is no

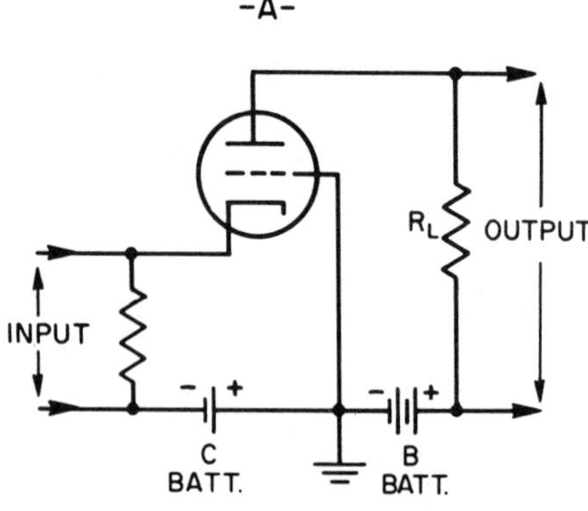

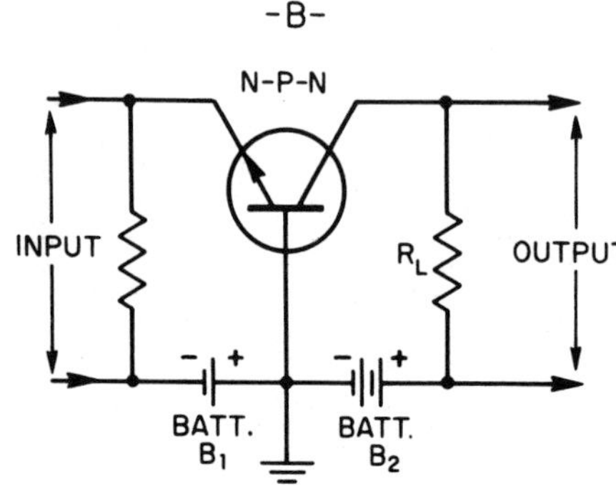

Fig. 29—9.

A. Electron-tube common-grid configuration.

B. Transistor common-base configuration.

phase reversal of the signal. The input impedance is low; the output impedance is high. This circuit provides the greatest voltage gain.

The third configuration is the *common-*, or *grounded-*, *plate* circuit (also known as the *cathode-follower* circuit) shown in Figure 29-10A. The signal is applied between the grid and plate (which is common to the input and output circuits). The output is obtained from the load resistor (R_L) between the plate and cathode.

The analogous transistor circuit is the *common-*, or *grounded-*, *collector* circuit shown in Figure 29-10B. Input is between the base and collector; output is between the collector and emitter. There is no phase reversal of the signal. The input impedance is high; the output impedance is low. This circuit provides the greatest current gain.

Note that in Figures 29-8B, 29-9B, and 29-10B, N-P-N transistors were illustrated. If P-N-P transistors are to be substituted, the circuits are the same, except the polarities of the biasing batteries must be reversed.

Fig. 29—10.

A. Electron-tube common-plate (cathode-follower) configuration.

B. Transistor common-collector configuration.

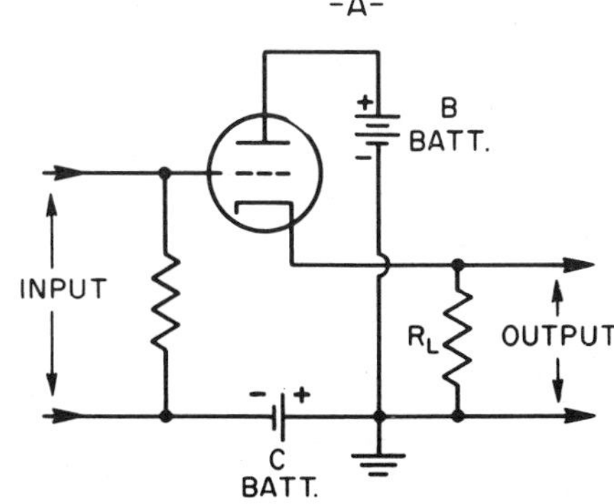

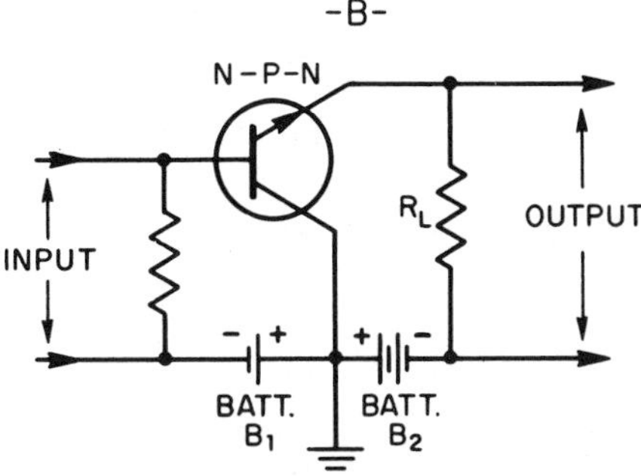

To operate properly, the base-emitter junction must be biased in a forward direction and the base-collector junction in a reverse direction. The relative voltage relationships existing at the elements of the transistor is shown in Figure 29-11.

In Figure 29-11A we have the relative voltage relationships for an N-P-N transistor. This does not mean that the base and collector must be positive and the emitter negative. What it does mean is that the base must be more positive (or less negative) than the emitter and that the collector must be more positive (or less negative) than the base.

The following table of typical voltages (with respect to ground) on the elements of N-P-N transistors used as amplifiers will illustrate the point:

Example	Base Voltage	Emitter Voltage	Collector Voltage
#1	+0.07 v	0 v	+ 7.4 v
#2	+7.1 v	+6.8 v	+18.5 v
#3	—2.8 v	—3.2 v	+ 4.0 v

The differences in voltages in these examples arise from the use of transistors with different characteristics, different circuit components, batteries with different voltages, and whether the positive or negative side of the battery is grounded.

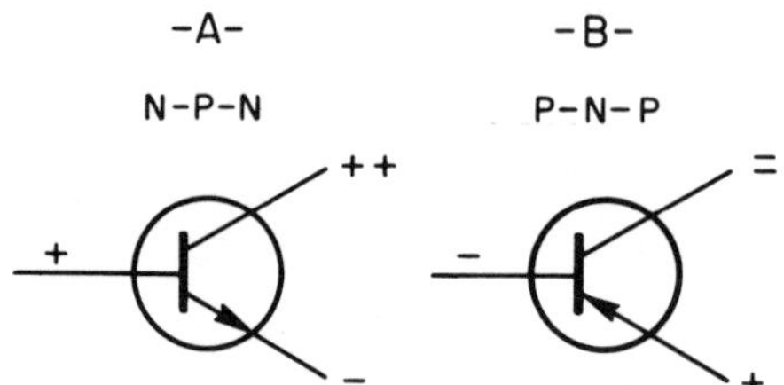

Fig. 29—11. **A.** Relative voltage relationships for an N-P-N transistor.
B. Relative voltage relationships for a P-N-P transistor.

The relative voltage relationships for a P-N-P transistor is shown in Figure 29-11B. The following table further illustrates the point:

Example	Base Voltage	Emitter Voltage	Collector Voltage
#1	+ 4.25 v	+ 4.7 v	+1.7 v
#2	+11.0 v	+11.2 v	0 v
#3	— 3.2 v	— 3.0 v	—6.0 v
#4	— 0.2 v	0 v	—3.0 v

Note: Due to the low base-to-emitter resistance, the voltage difference between the base and emitter is small. On the other hand, because the base-to-collector resistance is large, the voltage difference between the base and collector, too, is large.

1. AUDIO-FREQUENCY AMPLIFIERS

(See Chapter 15.)

A typical circuit of a stage of a-f amplification is illustrated in Figure 29-12. Here the N-P-N transistor is connected in a common-emitter configuration, which is most frequently employed. Typical component values are indicated. Because the transistor is a low-impedance device, values of resistors are low and those of capacitors high, as compared to those employed in high-impedance tube circuits. (If a P-N-P transistor were employed, the polarity of the battery would need be reversed. Note that the polarities of electrolytic capacitors C_3 and C_4, too, would have to be reversed.)

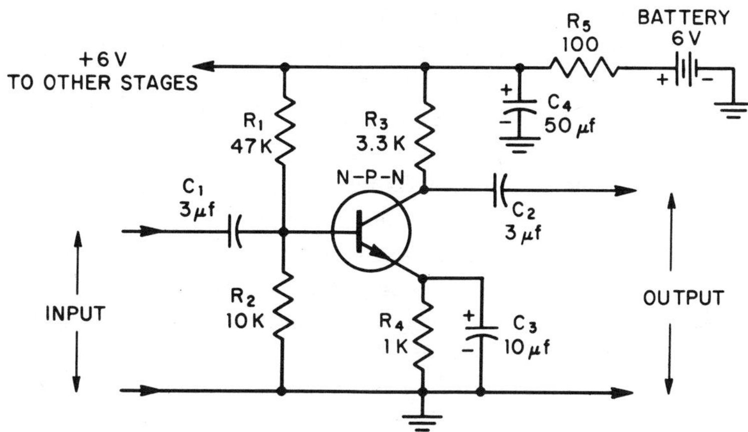

Fig. 29—12. Typical circuit of a stage of transistor a-f amplification.

Forward base-emitter bias is established and held firm by the battery through a voltage divider composed of R_1 and R_2 whereby the base is made positive relative to the emitter. Reverse base-collector bias is established by the battery through R_3 whereby the collector is made positive relative to the base.

The chief function of R_4 in the emitter circuit is to stabilize the circuit and protect the transistor against a runaway condition that may occur if the temperature increases. Such a rise in temperature might cause the collector current to increase. Since the input and output circuits of the transistor are interdependent, the rise in collector current produces a rise in base current which, in turn, produces a further rise in collector current. This action is cumulative and may end by damaging the transistor.

However, with R_4 in the emitter lead, a rise in collector and base current causes an increased voltage drop across R_4 in a direction opposing the input bias voltage. This reduces the base current and the chain reaction is stopped. Thus the resistor acts to stabilize the circuit and, hence, it is called the *stabilizer resistor.* It also acts in a similar manner to counteract variations produced as the transistor is replaced by one having slightly different characteristics. Capacitor C_3 serves R_4 in a manner similar to the action of a cathode-resistor bypass capacitor in electron-tube circuits.

Since all the stages of the receiver are connected to the battery, there is danger of feedback from one stage to another through this common path. Accordingly, a decoupling filter (C_4 and R_5) may be employed. Resistor R_5 opposes the flow of the a-c signals to the battery and capacitor C_4 offers them a low-impedance path to ground.

Since the transistor is a current-operated device, care must be taken so as to properly match the output impedance of a stage of amplification to the input impedance of the following one. Where a common-emitter stage (whose output impedance is moderate) is to be coupled to another common-emitter stage (whose input impedance, too, is moderate), ordinary *R-C* or direct coupling may be employed. But if, for example, a common-emitter stage is to be coupled to a common-base stage (whose input impedance is low), an impedance-matching coupling transformer generally is employed.

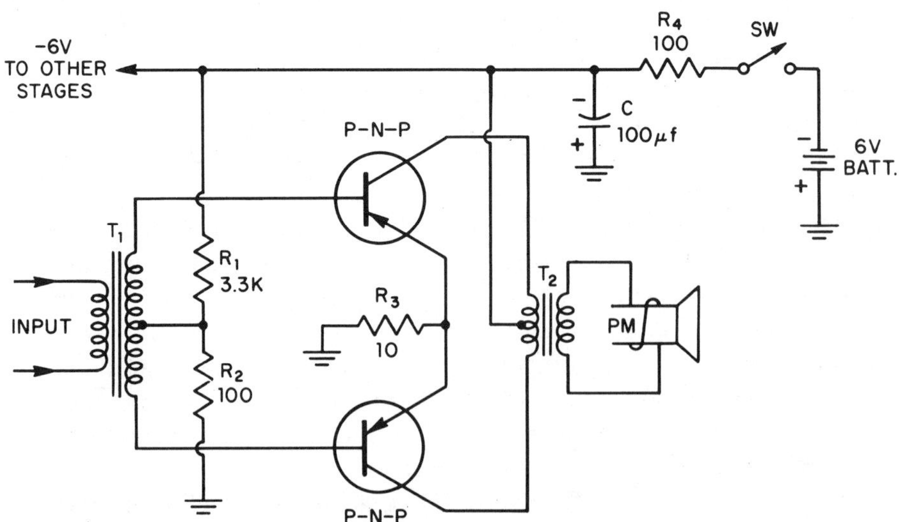

Fig. 29—13. Typical circuit of a push-pull power output stage.

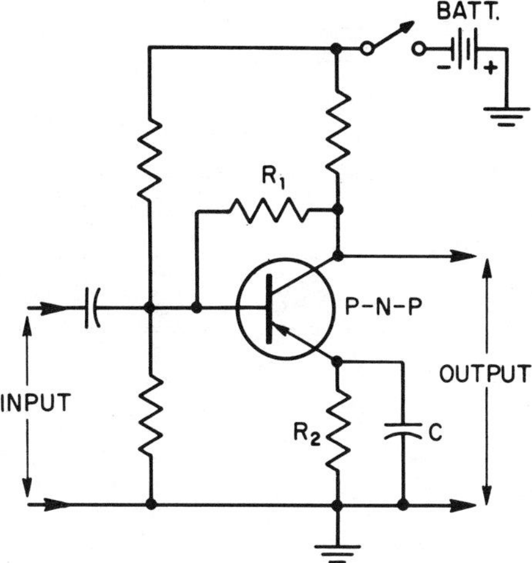

Fig. 29—14.

Inverse feedback circuit.

A typical circuit of a push-pull power output stage, showing typical values for the components, is illustrated in Figure 29-13. The two matched P-N-P transistors of the stage are connected in a common-emitter configuration. This stage generally is operated in Class B.

In addition to supplying the 180-degree phase difference in signals required by the push-pull circuit (see Chapter 15, subdivision C, 1), the input transformer (T_1) matches the output impedance of the previous stage to the relatively low input impedance of this one. Resistors R_1 and R_2 form the voltage divider that furnishes forward bias to the base-emitter junctions of the transistors. Resistor R_3 is the stabilizing resistor common to both emitters. Note that R_3 has no bypass capacitor since this is a push-pull circuit. Capacitor C and resistor R_4 form the decoupling filter and T_2 is an output transformer that matches the impedance of the collector circuit to the low impedance of the voice coil of the speaker.

Inverse feedback (see Chapter 15, subdivision C, 3) can be applied to the transistor amplifier. See Figure 29-14. Since, in the common-emitter amplifier, the output voltage is out of phase with the input voltage, voltage feedback is accomplished by feeding back a portion of the voltage on the collector to the base of the transistor through R_1. Current feedback may be accomplished by omitting capacitor C from across the stabilizing resistor R_2. (Note that the effect is the same as the omission of the cathode-resistor bypass capacitor in the electron-tube amplifier.)

2. RADIO-FREQUENCY AMPLIFIERS
(See Chapter 16, subdivision A.)

The circuit of a typical transistorized tuned r-f amplifier with typical component values is shown in Figure 29-15. Transformers T_1 and T_2 are the tuned input and output transformers, respectively. Note that the collector of the transistor is connected to a tap on the primary winding of T_2. This tap is chosen at a point where the impedance of T_2 will match that of the collector.

Forward emitter bias is furnished by the battery through the voltage divider formed by R_1 and R_2. Reverse collector bias is applied by the battery through the primary of T_2. R_3 is the stabilizing resistor and C_2 is its bypass capacitor.

Because of the relatively high internal capacitance of the transistor, feedback from the output to input circuits must be neutralized, just as in an r-f amplifier employing an electron-tube triode. The secondary winding of T_2 is wound so that the voltage at point X

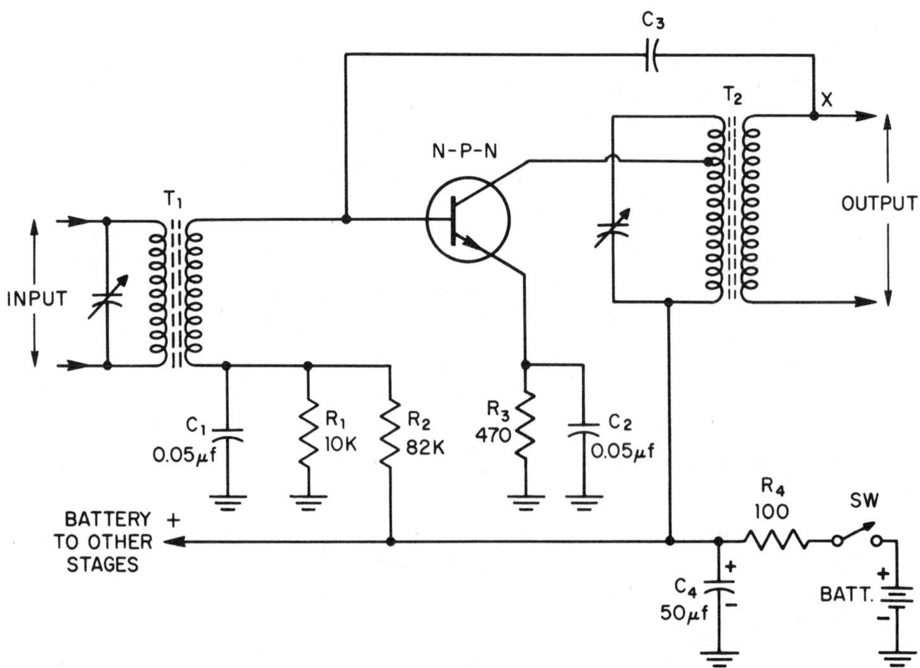

Fig. 29—15. Circuit of a typical transistorized tuned radio-frequency amplifier.

is 180 degrees out of phase with the voltage appearing at the base of the transistor. This out-of-phase voltage then is fed from point X to the base through neutralizing capacitor C_3 whose value is equal to the internal capacitance of the transistor (generally a few micro-microfarads).

(Modern technique has permitted the manufacture of transistors with extremely low internal capacitance. Where such transistors are employed, neutralization is not necessary.)

The circuit of the intermediate-frequency amplifier resembles that of the tuned r-f amplifier except that the transformers are tuned to a fixed frequency. Hence fixed capacitors are employed instead of variable capacitors. These transformers usually are supplied with movable powdered-iron cores whereby the inductances may be adjusted for the purpose of aligning the tuned circuits.

3. VIDEO-FREQUENCY AMPLIFIERS

(See Chapter 16, subdivision B.)

The circuit of a typical video amplifier is shown in Figure 29-16. The component values, too, are typical. Note the large coupling capacitors required by the low frequencies and impedances. Coil L is a peaking coil which helps compensate for the drop in amplification at high frequencies. Sometimes an additional peaking coil is inserted between the collector of the transistor and the output coupling capacitor for the same purpose.

R_2 and C_2 are the stabilizer resistor and its bypass capacitor. R_1 and C_1 form an *equalizing network* that reduces the low-frequency response of the amplifier relative to the high-frequency response, thus producing a more even overall response.

The presence of R_1 in the emitter circuit would tend to introduce inverse, or *degenerative*, feedback that would reduce the signal strength in a manner similar to the action of the stabilizer resistor R_2. The presence of capacitor C_1 tends to bypass the signal current around R_1, thus eliminating the degenerative feedback. But note the low value of C_1. Thus it offers a high impedance to low-frequency signals, and a low impedance to high-frequency signals. Hence, the high-frequency portion of the signal is bypassed, but the low-frequency portion is not. As a result, the low-frequency currents are reduced whereas the high-frequency currents are barely affected.

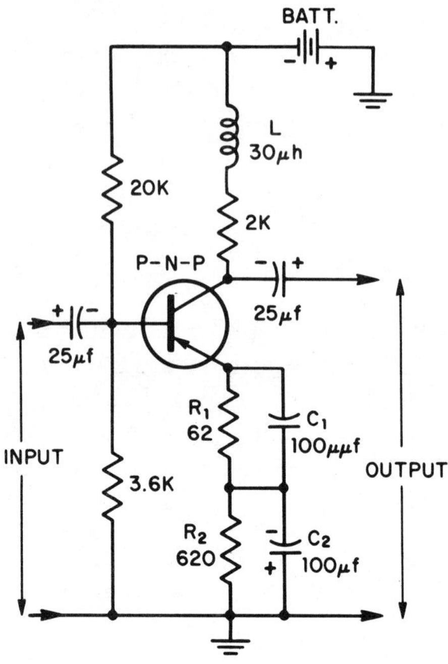

Fig. 29—16.

Circuit of a typical transistorized video amplifier.

4. DIRECT-COUPLED AMPLIFIERS

(See Chapter 16, subdivision C.)

A typical direct-coupled amplifier circuit is illustrated in Figure 29-17. Since, in the common-emitter configuration employed here the input and output impedances are both moderate, the output of Tr_1 may be coupled directly to the input of Tr_2.

The problem of stability is aggravated in this type of amplifier because, not only will temperature variations affect the collector current of Tr_1, but Tr_2 will amplify such changes and will also add its own temperature variations. (Note that where a-c amplifiers are concerned, the coupling capacitor separates the d-c conditions of each stage. Thus the cumulative effect is avoided.)

Because of the inherent greater instability of the direct-coupled amplifier, in addition to the customary emitter stabilizers (R_3 and R_5 for Tr_1 and Tr_2, respectively), inverse feedback is also employed. Thus a rise in collector current will produce a corresponding fall in base current. The inverse feedback is produced by R_7 for the first stage and R_8 for the second.

Note that the collector bias for Tr_1 and base bias for Tr_2 are supplied by the battery through R_4. In the P-N-P transistor the voltage on the collector must be more negative than the voltage on the base. This requires that the battery voltage must be larger than normal for the circuit illustrated here.

As its electron-tube counterpart, the direct-coupled amplifier can operate on both d-c and a-c signals. Because it tends to be unstable, more than two stages rarely are employed.

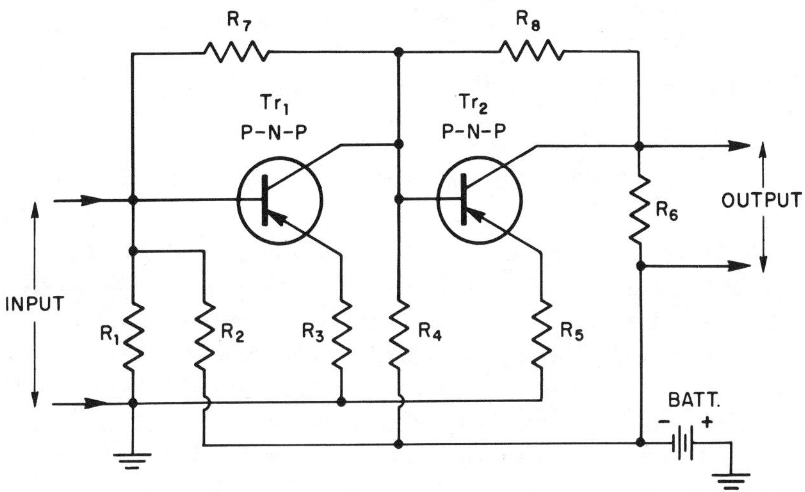

Fig. 29—17. Circuit of a typical transistorized direct-coupled amplifier.

C. *Oscillator circuits*

1. *L-C* OSCILLATORS

(See Chapter 17, subdivision A.)

The transistor also may replace the electron tube in oscillator circuits. In Figure 29-18 is shown the transistorized version of the regenerative oscillator. The transistor is used in a common-base configuration. L_2 and C_2 form the tuned tank in the collector circuit. Feedback is accomplished by means of feedback coil L_1 to the base of the transistor. As in the electron-tube oscillator, L_1 must be so coupled that the feedback is regenerative, that is, it will add to the oscillations in the tank circuit. L_3 is the output coil which is inductively coupled to L_2.

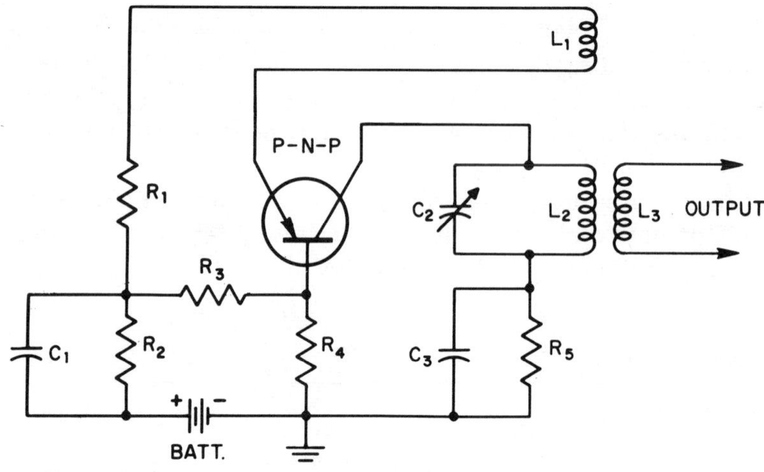

Fig. 29—18. Transistorized regeneratve oscillator.

Proper base-emitter bias is established by R_1 and R_3. R_4 is the stabilizer resistor. The bottom of the tank circuit is connected to ground, as far as the r-f current is concerned, through C_3. Collector bias is established through R_5 which, together with C_3, form a decoupling network that keeps the r-f voltage out of the battery. A similar decoupling network for the same purpose is formed by C_1 and R_2. The frequency of oscillation is determined by the values of L_2 and C_2.

The circuit of a Hartley oscillator employing a transistor in a common-emitter configuration is shown in Figure 29-19. L_2 and C_1,

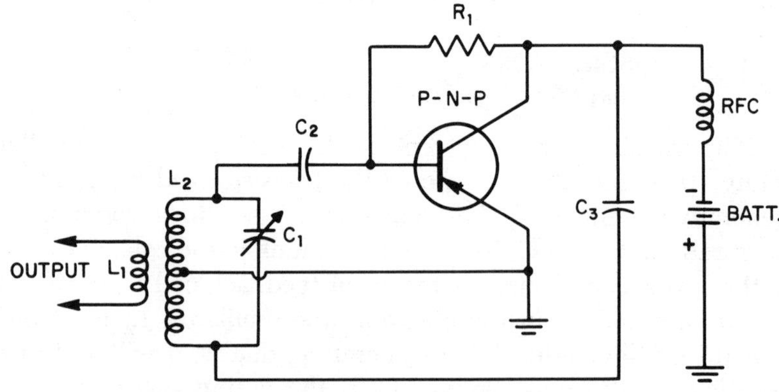

Fig. 29—19. Transistorized Hartley oscillator.

whose values determine the frequency of oscillation, form the tank circuit. L_1 is the output coil that is inductively coupled to L_2. Since, for proper bias in a P-N-P transistor the base must be negative (relative to the emitter) and the collector still more negative, resistor R_1 establishes the proper voltage relationship. Blocking capacitor C_2 keeps the positive voltage from the battery off the base.

Voltage is applied in the shunt-feed method. The r-f current is kept from entering the battery by the r-f choke coil (RFC) and is shunted around the battery through C_3.

The circuit of the transistorized Colpitts oscillator is shown in Figure 29-20. The Colpitts oscillator is similar to the Hartley type, except that the tap to the emitter is taken from between the junction of the two tuning capacitors C_1 and C_2. The frequency of oscillations is determined by the values of L_2, C_1 and C_2. In this circuit C_1 and C_2 are shown as fixed. L_2 is variable and its inductance may be varied by moving its powdered-iron core in or out. Class C self-bias is obtained by the action of C_g and R_g.

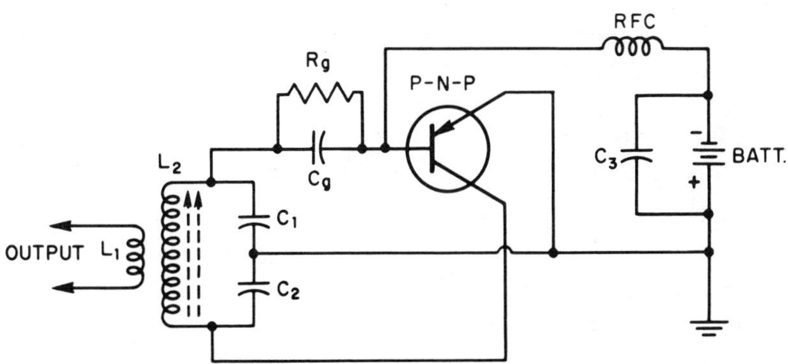

Fig. 29—20. Transistorized Colpitts oscillator.

The circuit of the transistorized crystal oscillator is shown in Figure 29-21. The collector tank circuit, consisting of L_1 and C_1, is tuned in resonance to the frequency of the crystal. A positive charge is applied to the emitter through RFC. A smaller positive charge is placed on the base through R_1. A negative charge is placed on the collector through L_1. This, you will recognize, is the proper voltage relationship for a P-N-P transistor.

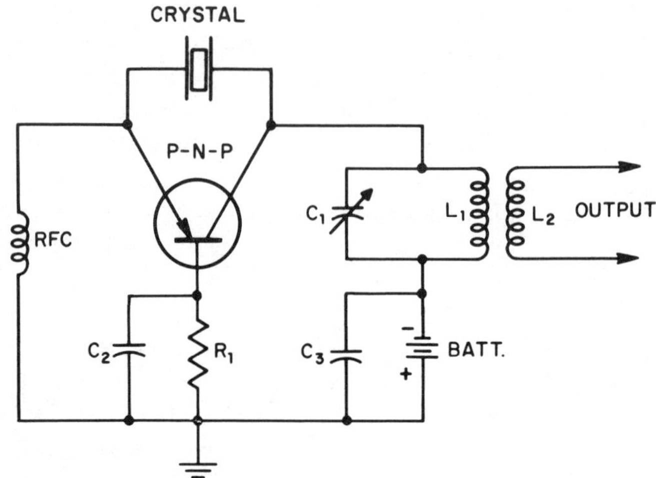

Fig. 29—21. Transistorized crystal oscillator.

Since the circuit employs a common-base configuration, there is no phase reversal between the voltage at the emitter and that at the collector. Proper feedback thus is accomplished by feeding some of the collector energy to the emitter by means of the crystal which acts as a tuned circuit.

2. *R*-C OSCILLATORS

(See Chapter 17, subdivision B.)

Semiconductor devices also can replace the electron tube in *R*-*C* oscillator circuits. As an example, compare the circuit of a simple sawtooth oscillator employing a zener diode (Figure 29-22) with the neon-lamp oscillator discussed in Chapter 17, subdivision B, 1.

As in the neon-lamp oscillator, capacitor C is charged up relatively slowly from the direct-voltage supply through resistor R. Because diode D is connected in its reverse direction (cathode positive, anode negative), it exhibits a very high resistance and so has little effect on the voltage across C.

When the voltage across C reaches the zener point of the diode, a breakdown occurs. The resistance of the diode drops suddenly to a very low value, discharging C. Then the entire cycle is repeated.

The output voltage, taken from across C, is a sawtooth voltage with a relatively long rise time corresponding to the charging period

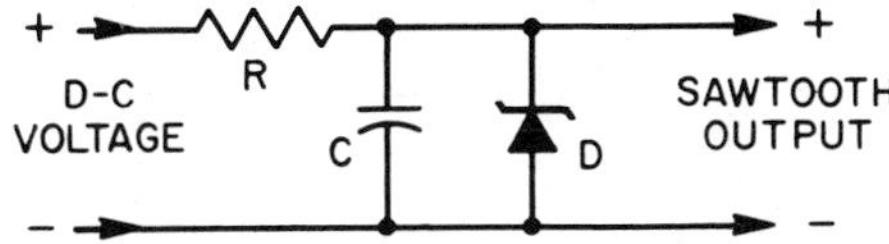

Fig. 29–22. Simple sawtooth oscillator using a zener diode.

of C and a short flyback, or retrace, time corresponding to the discharge period. The frequency is determined by the values of R and C. Resistor R also serves as a limiting resistor to protect the diode from overloads.

The circuit of the transistorized phase-shift oscillator is shown in Figure 29-23. Since the common-emitter configuration provides a 180-degree voltage phase shift between the base and collector, the feedback network of three *C-R* sections produces a 180-degree phase shift so that collector voltage may be fed back in phase to the base and so sustain the oscillations. Each *C-R* section produces a 60-degree phase shift.

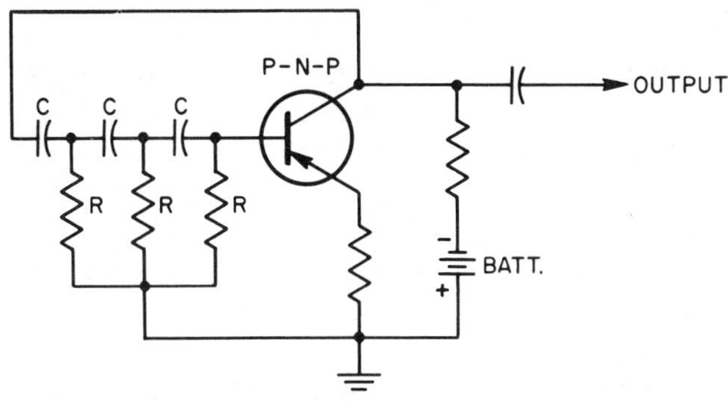

Fig. 29–23. Transistorized phase-shift oscillator.

A transistorized version of the multivibrator is illustrated in Figure 29-24. Bias for transistor Tr_1 is established through a network consisting of R_1 and R_2, and for Tr_2 through R_3 and R_4.

Let us start with Tr_1 conducting more heavily than Tr_2. The increase in Tr_1 collector current increases the voltage drop across R_2, causing the collector potential to become less negative. Because the collector of Tr_1 is coupled to the base of Tr_2 through C_1, the potential of that base, too, becomes less negative.

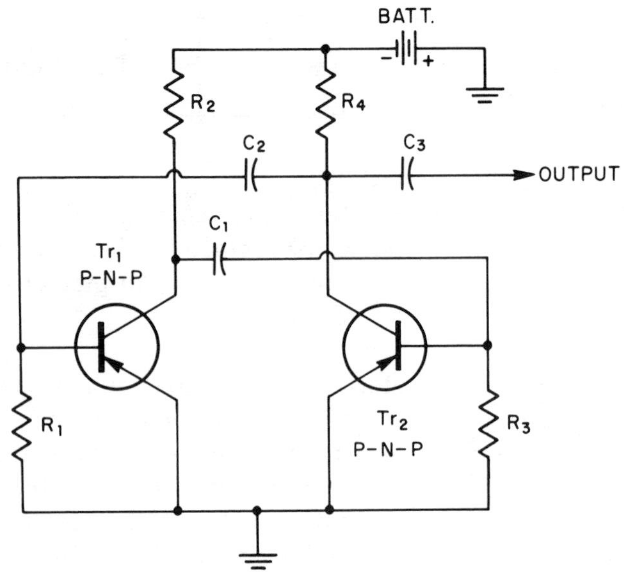

Fig. 29—24. Transistorized multivibrator.

As a result, the forward base bias of Tr_2 is reduced, reducing its collector current. Hence the voltage drop across R_4 is reduced, making the collector potential of Tr_2 more negative. Because the collector of Tr_2 is coupled to the base of Tr_1 through C_2, the potential of that base, too, becomes more negative. As a result, the forward base bias of Tr_1 is increased and the transistor conducts more heavily. This process continues until Tr_1 is conducting at maximum and Tr_2 is cut off by the charge on C_1.

As this charge leaks off to ground through R_3, the base of Tr_2 becomes negative enough so that Tr_2 starts to conduct. The increase of its collector current produces a larger voltage drop across R_4, making its collector potential less negative. Now the base of Tr_1 begins to become less negative and the cycle ends as Tr_2 is conducting at maximum and Tr_1 is cut off by the charge on C_2. When this charge leaks off through R_1, a new cycle starts.

The free-running frequency of the multivibrator is determined essentially by the values of R_1, R_3, C_1, and C_2. Triggering may be accomplished by negative pulses fed to the base of either transistor.

The circuit of a transistorized blocking oscillator is shown in Figure 29-25. The transistor is connected in a common-emitter

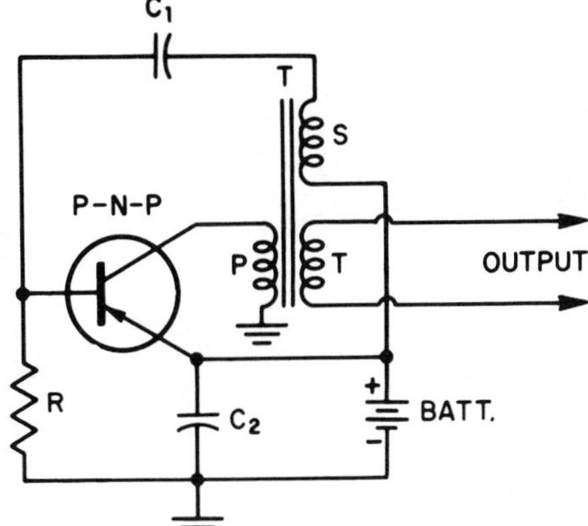

Fig. 29–25.

Transistorized blocking oscillator.

configuration. Forward base-emitter bias is obtained from the battery through R. Reverse bias for the collector is obtained through the primary winding of transformer T.

As the transistor becomes conductive, the increasing flow of collector current through the primary winding of T induces a voltage in the secondary winding. This voltage charges C_1 and places a positive charge on the base of the transistor, driving it to cutoff. As the charge on C_1 gradually leaks off through R, the transistor again becomes conductive, thus completing the cycle.

The frequency of this oscillator is determined essentially by the values of C_1 and R. Output is obtained by means of a tertiary winding on T. The oscillator may be triggered by pulses fed in to the base of the transistor.

3. FREQUENCY CONVERTERS

(See Chapter 23, subdivision B, 1.)

The transistor, too, may be employed as a frequency converter in the radio receiver. A typical circuit, showing typical values of components, is illustrated in Figure 29-26. Here separate transistors are used for the oscillator and mixer stages.

The r-f signal is fed to the base of the mixer transistor. Also feeding this base through C_4 is the output from the oscillator whose tank circuit consists of L_3 and C_1. Capacitor C_5 furnishes the signal-return path to ground. The tank voltage is fed to the base of the oscillator transistor through L_4 and coupling capacitor C_6. The amplified voltage at the collector is fed back to the tank by means of a tap on L_3. (Note that L_3 and L_4 are connected and inductively coupled, forming, in effect, a single tapped coil.)

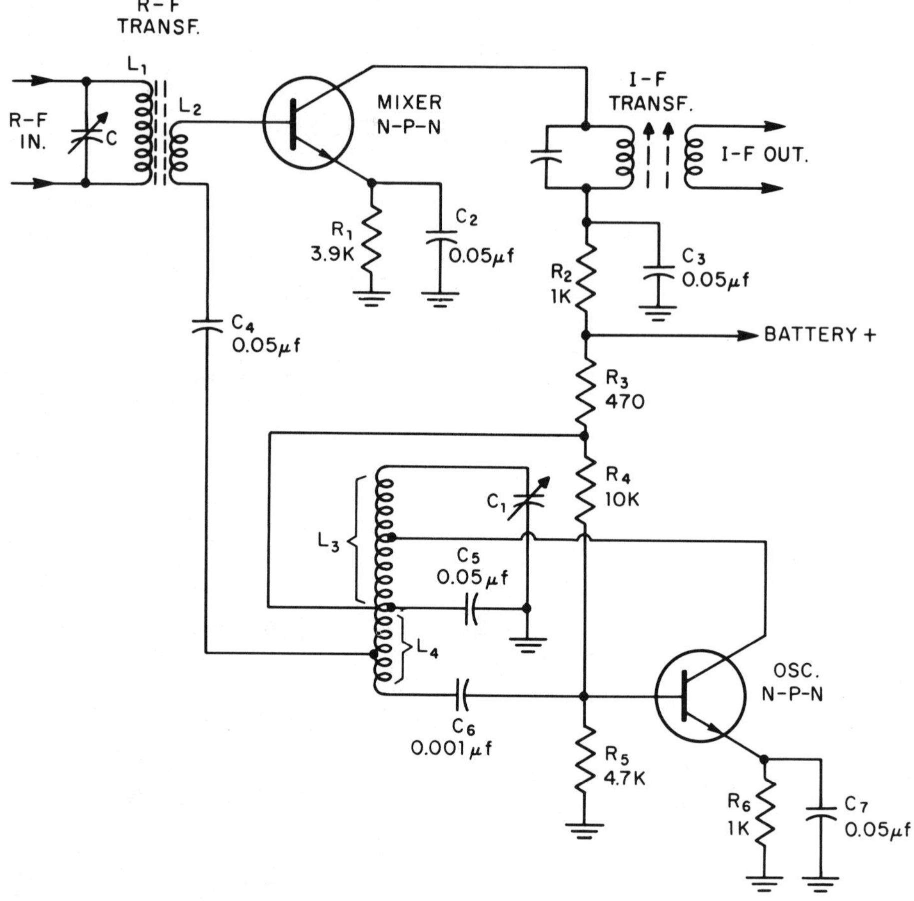

Fig. 29—26. Frequency converter, using separate transistors for the oscillator and mixer stages.

Resistor R_3 is a voltage-dropping resistor. Collector bias for the oscillator is applied through R_3 and a tap on L_3. Emitter bias is applied by means of the voltage divider consisting of R_4 and R_5. Resistor R_6 is the emitter stabilizer and C_7 is its bypass capacitor.

The incoming r-f signal, after mixing with the oscillator voltage, emerges as the i-f signal that is fed to the i-f transformer. Resistor R_1 is the mixer emitter stabilizer and C_2 is its bypass capacitor. Resistor R_2 and capacitor C_3 form a decoupling filter to keep the signal from entering the battery.

A single transistor may be used to perform the functions of both oscillator and mixer. Such a circuit is shown in Figure 29-27.

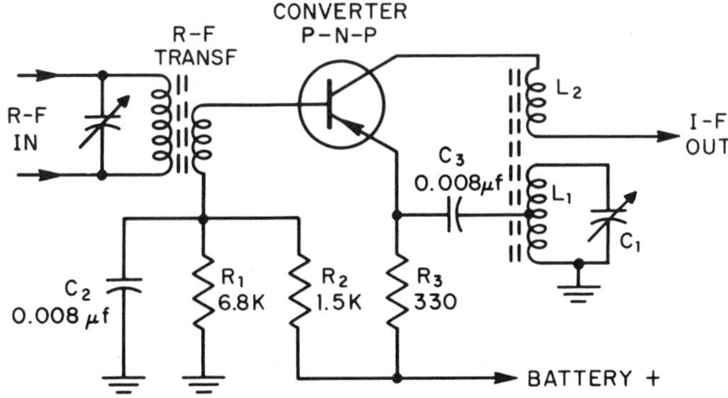

Fig. 29—27. Frequency converter using a single transistor.

The oscillator is a regenerative type. The tank circuit consists of L_1 and C_1. The oscillator voltage is fed to the emitter through coupling capacitor C_3. Part of the amplified voltage is fed back to the tank from the collector through L_2 which is inductively coupled to L_1.

Meanwhile, the r-f signal is fed to the base of the transistor by means of the r-f transformer. Mixing takes place and the resulting i-f signal is fed from the collector through L_2 to the i-f amplifiers. Resistors R_1 and R_2 form the voltage divider that applies bias to the base. Resistor R_3 is the stabilizer resistor and C_2 furnishes the signal-return path to ground.

QUESTIONS

1. a) **Draw and explain the circuit of the half-wave power supply using a semiconductor diode.**
   b) **Explain the function of the *surge resistor.***
2. **Draw and explain the circuit of the full-wave power supply using semiconductor diodes.**
3. **Draw and explain the circuit of the series voltage doubler using semiconductor diodes.**
4. a) **Describe and explain the theory of operation of the *zener diode.* Explain what is meant by**
   b) ***avalanche breakdown;***
   c) ***zener breakdown.***
5. **Draw and explain a voltage-regulating circuit employing a zener diode.**
6. **Explain the differences between amplifiers using transistors and electron tubes.**

7. Draw and explain the basic circuit of a transistor amplifier using a *common-emitter configuration* and that of its electron-tube counterpart.
8. Draw and explain the basic circuit of a transistor amplifier using a *common-base configuration* and that of its electron-tube counterpart.
9. Draw and explain the basic circuit of a transistor amplifier using a *common-collector configuration* and that of its electron-tube counterpart.
10. Explain the relative voltage relationships at the electrodes of
    a) an N-P-N transistor;
    b) a P-N-P transistor.
11. a) Draw and explain the circuit of an a-f amplifier using an N-P-N transistor in a common-emitter configuration.
    b) Explain how the various biases are obtained.
    c) Explain the function of the *stabilizer resistor*.
12. Draw and explain the circuit of a stage of push-pull a-f amplification using P-N-P transistors in common-emitter configurations.
13. Draw and explain the circuit of an r-f amplifier using an N-P-N transistor in a common-emitter configuration.
14. Draw and explain the circuit of a regenerative oscillator using a P-N-P transistor in a common-base configuration.
15. Draw and explain the circuit of a Hartley oscillator using a P-N-P transistor in a common-emitter configuration.
16. Draw and explain the circuit of a sawtooth oscillator using a zener diode.
17. Draw and explain the circuit of the multivibrator using P-N-P transistors in common-emitter configurations.
18. Draw and explain the circuit of the blocking oscillator using a P-N-P transistor in a common-emitter configuration.
19. Draw and explain the circuit of the frequency converter using a P-N-P transistor in a common-emitter configuration.

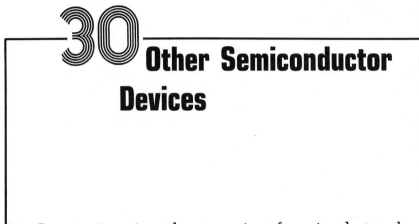

30 Other Semiconductor Devices

Investigations into the properties of semiconductors have opened a new world of research. Not only have semiconductor replacements for electron tubes been developed, but many entirely new applications have been found. And this is only the beginning. There is promise of much more to come.

In this chapter we will discuss a few of these other applications.

A. *Integrated Circuits (Microelectronics)*

As transistor technology has improved, transistor size has been reduced drastically to the point where thousands of uncased transistors can be held in a thimble. This led the way to integrated circuitry wherein a complete circuit with many transistors diodes, resistors, and possibly capacitors are mounted on a tiny chip called a substrate. These components can be deposited by various techniques. In one technique called thin-film, the elements of the circuit are

Radio Corporation of America

Fig. 30–1.
Mask used in the manufacture of integrated circuits. The image shown is reduced in size to make a photomask containing hundreds of circuit patterns, each less than five-thousandths of a square inch in area.

deposited in their layers by an evaporation technique. Using a series of masks for the various materials, complete circuits are built up. (See Figure 30-1.) There are various hybrid techniques wherein some of the active elements are included in the form of tiny chips with the rest of the circuit deposited as above or in the form of pastes that are silk-screened onto the substrate.

1. OPERATIONAL AMPLIFIERS

One of the most popular circuits available in integrated form is the operational amplifier. (A typical circuit is shown in Figure 30-2.) This is a multi-stage high-gain amplifier that can be used for a multitude of purposes.

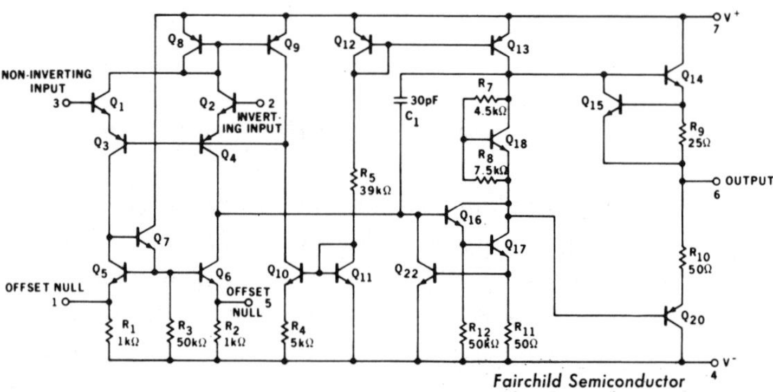

Fairchild Semiconductor

Fig. 30–2. High operational amplifier integrated circuit.

Fig. 30–3. **A.** TO-99 transistor case.
B. Flatpack (dual in-line) case.

Fairchild Semiconductor

The most frequent case configurations are the typical transistor case (TO-5 style) and the flat pack. (See Figure 30-3.) The symbol for the operational amplifier is with the input at the left and the output at the right.

B. *Double-base junction diode*

The ordinary junction diode consists of a crystal containing N- and P-regions (Figure 30-4A). When voltage is applied in the forward direction (negative to the N-region and positive to the P-region) the resistance of the diode is very low. When the voltage is applied in the reverse direction, the resistance is very high.

An interesting variation is the *double-base junction diode* illustrated in Figure 30-4B. Here a P-type pellet is alloyed into an N-type base to form a P-N junction diode. Leads are taken from both ends of the base. If a current should flow through B_1 and B_2, a voltage drop will appear along the entire length of the base region, owing to the distributed resistance of the path. If a voltage is applied to the P-region at the center of the diode and this voltage is equal to that existing at the center of the base, one portion of the P-N junction will be forward-biased and the other reversed-biased.

Thus the forward-biased portion of the base will act as an emitter and the reversed-biased portion will act as a collector. As a result, we have a sort of transistor. However, there is no gain to be had and, accordingly, this type of device is not used as an amplifier. It may be employed, nevertheless, in switching circuits.

A switch is a device that exhibits a very high resistance when it is open and a very low resistance when it is closed. The electron-tube diode is such a switch. When the anode is positive, relative to its cathode, the tube is conductive, that is, its resistance is low. When the polarities are reversed, the resistance is high. The semiconductor diode, too, acts in this manner. The advantage of electronic switches over the mechanical type is that the former have no moving and wearing parts and are easily actuated by electrical inputs. And because they have no moving parts, they can operate at higher frequencies.

The electron tube and ordinary semiconductor diode can operate well at relatively low and medium frequencies. At high frequencies the electron tube is limited by the time it takes the electrons to overcome the space charge. In the semiconductor diode the limiting

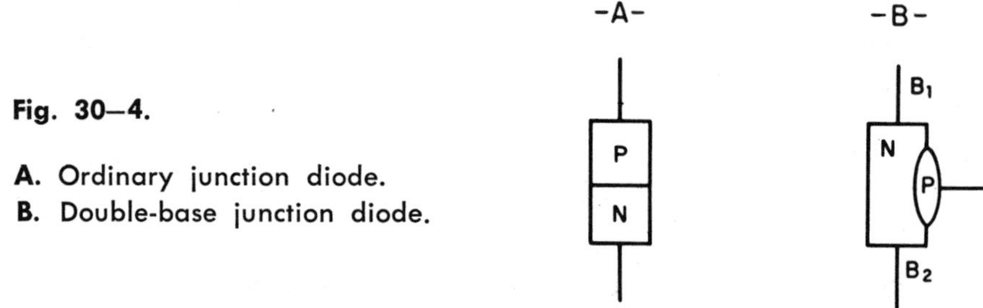

Fig. 30–4.

A. Ordinary junction diode.
B. Double-base junction diode.

factor is the time it takes for the charge stored in the device to disappear as a result of the recombination and neutralization of the minority carriers before it can be "closed" again. The double-base diode can operate at much higher frequencies since it is "closed" when the voltage of the incoming pulse reaches the potential required to enable the diode to act in the forward direction and is "opened" when it falls to the point where the diode acts in the reverse direction.

C. *Three-junction devices*

The ordinary transistor is a three-region device that contains two P-N junctions. By adding another region, it is possible to construct a three-junction device. See Figure 30-5. Here an N-region (N_2) has been added to the collector end of a P-N-P transistor (whose regions are designated as P_1, N_1, and P_2). This N_2-region now becomes the collector.

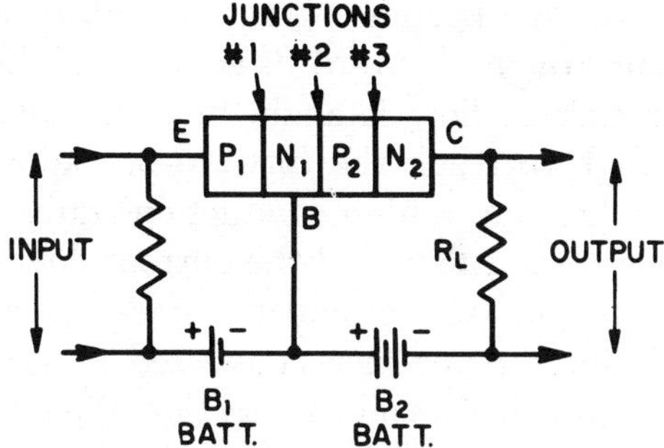

Fig. 30—5.

Basic amplifier circuit employing a P-N-P-N transistor.

Note that both junctions #1 and #3 are biased in the forward direction. As the input signal becomes more positive, more holes flow from the emitter (P_1), through the base (N_1), and into the P_2-region. Since junction #3 is biased in the forward direction, electrons can flow easily from the collector (N_2) to the P_2-region. As a result, the potential barrier is reduced and a greater output current may flow from this device, which is known as a "hook" transistor.

You will recall that α, which is the ratio between output current and input current, always is less than unity for a transistor connected in a common-base configuration. The chief advantage for the hook transistor is that it may produce an α which is greater than unity. Yet, because such a transistor is more difficult to manufacture and because it has an inherent higher noise level, it seldom is employed.

However, the three-junction device may be used as a *controlled rectifier* whose action is similar to that of the thyratron tube. See Figure 30-6. Disregard the *gate* connection for the time being. With the rectifier connected in the reverse direction (anode negative, cathode positive) the current flow through it is blocked until

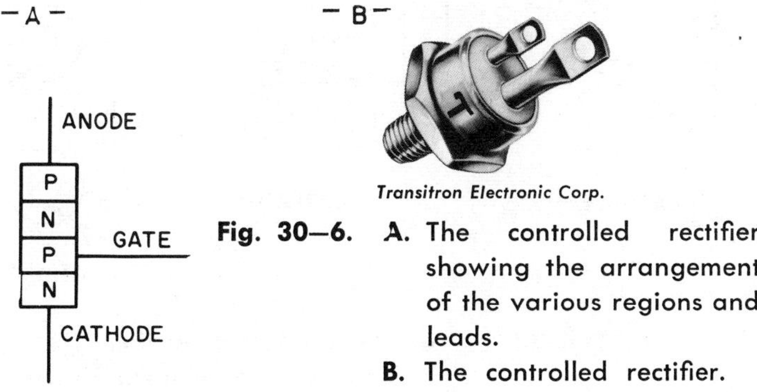

Fig. 30—6. **A.** The controlled rectifier showing the arrangement of the various regions and leads.
B. The controlled rectifier.

the voltage becomes great enough to reach the reverse-breakdown point. In the forward direction (anode positive, cathode negative) the current flow is again blocked until the voltage becomes great enough to reach the forward-breakdown point. Then the rectifier switches into a high-conduction state, its internal resistance falls to a very low value, and the current flow through it is limited only by the supply voltage and the impedance of the external circuit.

With the rectifier connected in the forward direction and the anode–cathode voltage less than the breakdown voltage, the rectifier can be switched to its high-conduction state by means of a small pulse applied between the gate and cathode. Once in this state, the rectifier conducts continuously, even after the gate signal is removed, until the current is interrupted or diverted by some external means for a short period of time (about 20 microseconds). This permits the rectifier to regain its forward-blocking capacity.

We may understand the operation of the gate if we look at Figure 30-7. Here we see a P-N-P-N device (Figure 30-7A) and its two-transistor equivalent (Figures 30-7B and C). Note the equivalent consists of a P-N-P transistor and an N-P-N type, internally connected so that the base and collector of one is connected, respectively, to the collector and base of the other. As a result, there is positive feedback from one transistor to the other.

Assume that the positive voltage applied to the anode of the rectifier is below the value required for forward breakdown and

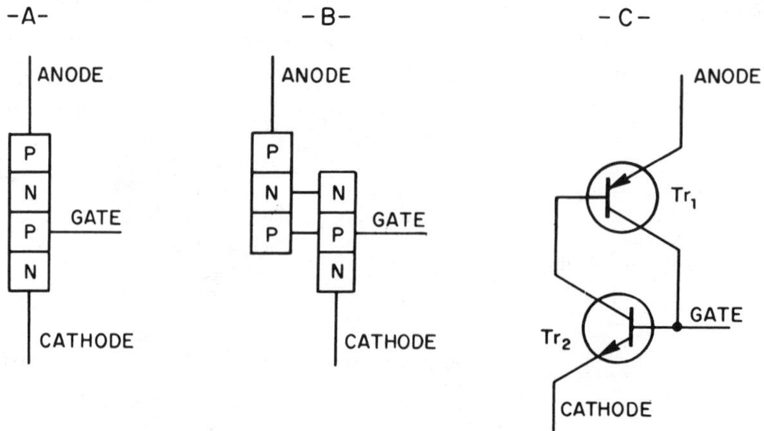

Fig. 30–7. **A.** The controlled rectifier.
B. and **C.** Its equivalent circuits.

there is no voltage on the gate. The resistance of the rectifier is high and only a very small leakage current flows through the device. If the positive voltage on the anode is made sufficiently large, forward breakdown occurs, the resistance of the rectifier is sharply reduced, and a relatively heavy current flows through it.

If, instead of increasing the positive voltage on the anode, a much smaller positive voltage is applied to the gate, the same result is accomplished. The injection of holes to the base of Tr_2 (Figure 30-7C) causes an increased flow of holes from its collector to the base of Tr_1 and holes from the collector of Tr_1 flow to the base of Tr_2. Once again, the resistance of the entire rectifier is sharply reduced and a heavy current flows through it. Because of the positive-feedback relationship between Tr_1 and Tr_2, this low-resistance, heavy-current condition will persist, even if the positive voltage on the gate is removed.

You see now why the controlled rectifier is compared to a thyratron tube. In electrical circuits the symbol for the controlled rectifier is

To stop the flow of current through the rectifier, the anode voltage must be removed. Another method is to reduce the current flowing in the feedback loop between Tr_1 and Tr_2 to a value too low to maintain the low-resistance condition of the rectifier. This may be done by placing a small negative voltage on the gate. Thus the device now acts as a switch. When a positive pulse is applied to the gate it goes into its "on" (low-resistance) condition. When a negative pulse is applied to the gate it goes into its "off" (high-resistance) condition. The symbol for a three-junction device used as a switch is

D. *Thermistors*

The resistance of most substances increases with increases in temperature. The resistance of many semiconductors, on the other hand, decreases with increases in temperature. The *thermistor* is such a device. It is a P-type semiconductor composed of the oxides of certain metals such as nickel, manganese, iron, etc. As such, a majority of its electrons are locked in covalent bonds with their neighbors. As heat energy is applied, a number of these bonds are broken, releasing free electrons. As a result, the resistance of the thermistor is reduced.

The temperature–resistance relationship for a typical thermistor is illustrated graphically in Figure 30-8A. Note how sharply the resistance drops with increases in temperature.

In Figure 30-8B you see the current–voltage graph of this thermistor. Note that between zero and point #1 of this graph the current increases with increases in voltage. Since $E = I \times R$, we call such a relationship *positive resistance.* But beyond point #1 the current increases with decreases in voltage. Hence this relationship is called *negative resistance.*

As the current through the thermistor rises, the voltage drop across it rises rapidly, exhibiting positive resistance. But the negative-resistance region is quickly reached. After that an increase in current produces a decreasing voltage drop.

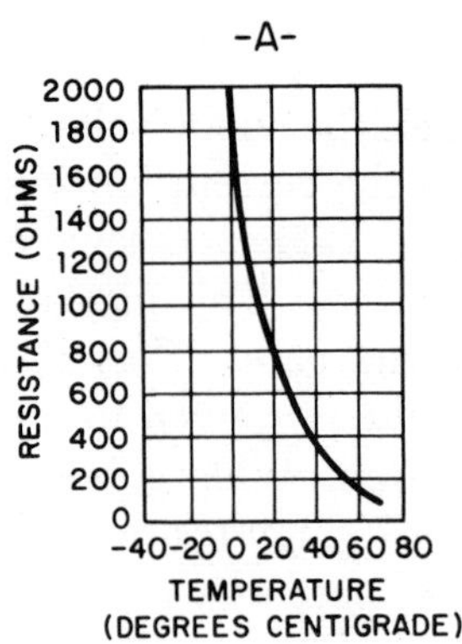

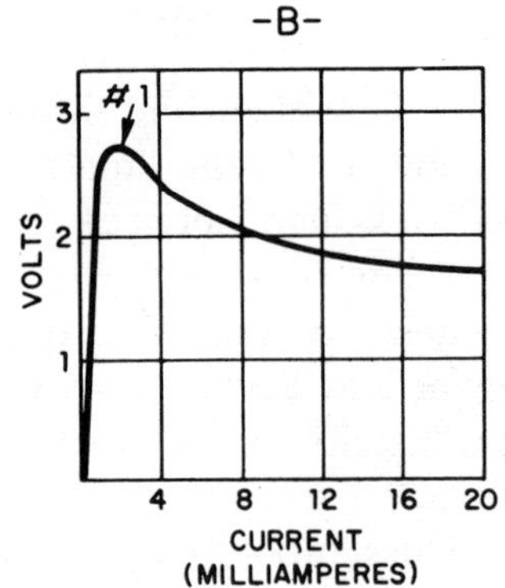

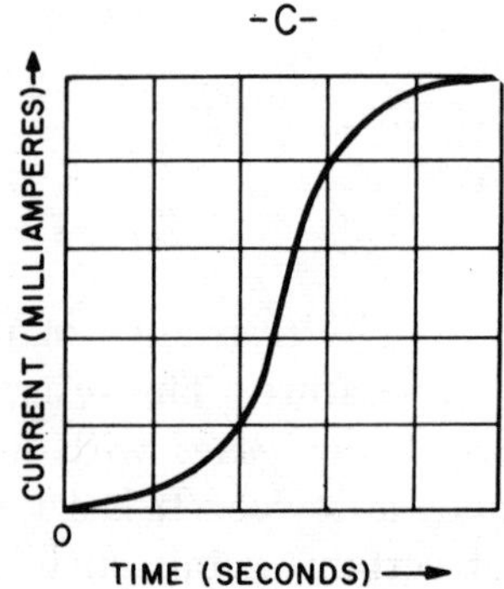

Fig. 30–8. Typical characteristic curves for the thermistor.

A. Temperature-resistance curve
B. Current-voltage curve.
C. Time-delay curve.

Another characteristic of the thermistor is the fact that the current does not instantly reach its full value (as indicated on the graph of Figure 30-8B) as a given voltage is applied. Instead, there is a thermal lag that holds back the current for a certain period of time. This time-delay characteristic is shown in Figure 30-8C.

Thermistors can be manufactured to display these characteristics in varying degrees. On the whole, however, they follow the pattern shown by the graphs of Figure 30-8. They can be made in the form of disks, rods, and beads, and can be extremely small in size. The symbol for a thermistor is .

Fenwal Electronics, Inc.

Fig. 30–9. Thermistors come in various sizes and shapes.

There are many applications to which the thermistor can be put. For example, it can be used as a resistance thermometer to measure temperature by change in resistance. The change in current flowing through the thermistor can be registered on a meter calibrated in degrees. Because they are so small, they can be inserted in areas that normally are not accessible. Also, since they are temperature-sensitive, they may be used as sensing devices in various types of control circuits (see Section IV).

Since thermistors are not rectifiers, they will function with either direct or alternating current. Care must be taken that the currents employed are minute to reduce the extraneous heating effect of the current flowing through them.

The thermistor also may be applied as a temperature-compensating resistor to offset the resistance rise of some other device. Thus it may be connected in series or parallel with this device whose resistance rises as its temperature increases. The same rise in temperature reduces the resistance of the thermistor. Hence the overall resistance is kept constant.

The thermal-lag characteristic of the thermistor also finds many applications where a time delay is required. For example, suppose we wish a relay to become activated a certain length of time after a switch is closed. The circuit for such a device is illustrated in Figure 30-10.

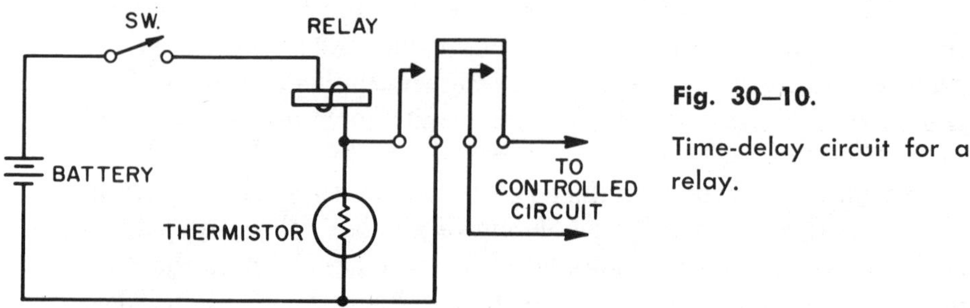

Fig. 30–10.

Time-delay circuit for a relay.

When the switch is closed, the resistance of the thermistor in series with the relay coil is large enough to prevent the flow of sufficient current to activate the relay. But as current flows through the thermistor it heats up. After a definite period of time (depending upon the characteristics of the thermistor) the resistance drops and the relay is activated by the increased flow of current. In order that the thermistor should be ready to repeat its performance, a second set of relay contacts are closed at the same time, shorting out the thermistor and so permitting it to cool.

The negative-resistance characteristics of the thermistor present some interesting possibilities. For example, we know that an *L-C* oscillatory circuit would continue oscillating indefinitely were it not for the presence of resistance which robs the circuit of its energy and thus stops the oscillations. In the oscillator we feed back to the circuit additional electrical energy to compensate for this energy loss.

Suppose, now, we were to insert a thermistor in series with the coil and capacitor of the oscillatory circuit. As current flows through that circuit, the resistance of the coil rises due to the heating effect of the current. The same heating effect, on the other hand, causes the resistance of the thermistor to fall, thus balancing out the resistance of the coil. The result, then, is an oscillator that requires no feedback energy from an outside source.

These are but a few of the many applications of this versatile device.

E. *The tunnel diode*

If we examine the characteristic curve of the conventional semiconductor diode (see Figure 27-4) we see that in the forward direction the current increases as the voltage is raised. Thus it exhibits *positive resistance.* (The diode also exhibits positive resistance in the reverse direction.)

In 1958, Leo Esaki, a Japanese scientist, discovered that if he heavily doped a semiconductor with impurities, he could make a junction diode that would exhibit negative resistance in certain regions in the forward direction. Such a diode is called a *tunnel diode.*

The characteristic curve of a typical tunnel diode is shown in Figure 30-11.Note that as voltage is applied in the forward direction the forward current rises quickly, as in a conventional diode. As the voltage reaches point #1, however, the current decreases and continues to decrease as the voltage is increased until point #2 is reached. The region between points #1 and #2 is the negative-resistance region. Beyond point #2, increasing the voltage results in an increase in current. Hence the diode exhibits positive resistance again.

In the conventional junction diode, the potential barrier of the junction retards the flow of charge carriers from one region to the other until the potential reaches a level high enough to overcome the barrier. Although this level is low (approximately 0.6 volts for silicon diodes and even less for germanium diodes) it does slow up conduction and is especially troublesome in high-frequency applications.

In the tunnel diode, however, it was found that the charge carriers seemed to "tunnel" their way through the potential barrier without the necessity for waiting for the potential to be raised high enough to get them over the barrier. Hence the name. This "tunneling" action takes place at nearly the speed of light and so makes this diode particularly suitable for high-frequency applications.

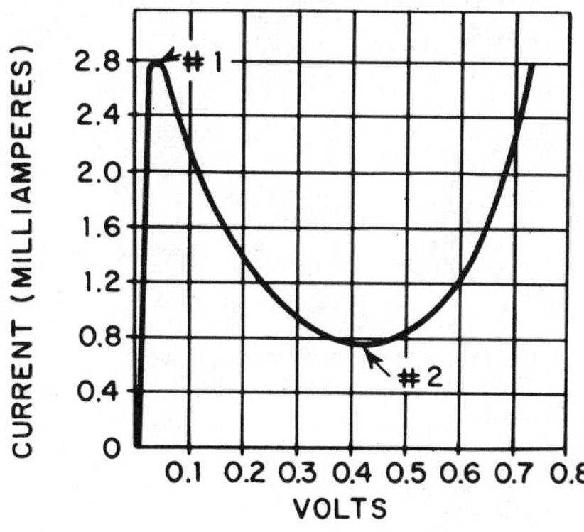

Fig. 30—11.

Typical characteristic curve of the tunnel diode.

A small forward bias will cause electrons in the N-region to tunnel across the junction into the P-region even though the electrons do not have the energy to surmount the potential barrier. As the forward bias is increased, the flow of electrons through the junction is increased (between zero and point #1 of Figure 30-11). This accounts for the positive resistance of the diode.

At point #1 the forward bias has reached a point where the energy of the free electrons of the N-region is greater than the energy of the valence electrons in the P-region. The tunneling current diminishes and continues to diminish as the forward bias is increased (between points #1 and #2). This accounts for the negative resistance of the diode.

As the forward bias is increased further (beyond point #2) the free electrons and holes obtain enough energy to flow over the potential barrier of the junction in a manner similar to that of a conventional diode. Hence the tunnel diode exhibits positive resistance in this region.

Various types of semiconductors are employed for tunnel diodes such as germanium, silicon, gallium arsenide, etc. In appearance the tunnel diode resembles the ordinary junction diode and some types have been made about as small as a match head. Its symbol is

Anode Cathode or Anode Cathode.

The negative-resistance property of the tunnel diode lends itself to a number of applications. For example, we have learned that we may apply negative resistance to an oscillating circuit in order to balance out the positive resistance, and thus keep the oscillations going. Such a circuit employing a tunnel diode is illustrated in Figure 30-12.

The oscillating tank circuit consists of L_1 and the capacitance supplied by the tunnel diode. The diode is biased to its negative-resistance region by means of the battery and potentiometer R. The

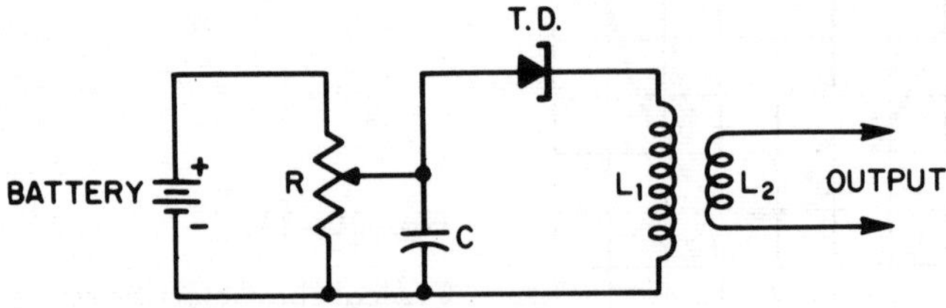

Fig. 30—12. Circuit of the tunnel diode as a sinusoidal oscillator.

positive resistance of the tank circuit is balanced out by the negative resistance of the tunnel diode and thus the oscillations are sustained. L_2 is an output coil that is inductively coupled to L_1.

The tunnel diode may also be employed as a relaxation oscillator, as shown by the circuit illustrated in Figure 30-13. The diode acts as a sort of capacitor that is charged up by the battery. Hence the voltage across it approaches that of the battery. As this charge leaks off through R, the voltage across the diode drops. Then it starts charging up again. The result is that the voltage across the tunnel diode rises and falls at a frequency determined by the value of its capacitance and the resistance of R.

Since the characteristics of the diode depend upon the voltage across it, these characteristics change in step with the rise and fall of the voltage. As these characteristics change, inductor L absorbs and releases energy, thus producing a sawtooth voltage at the output.

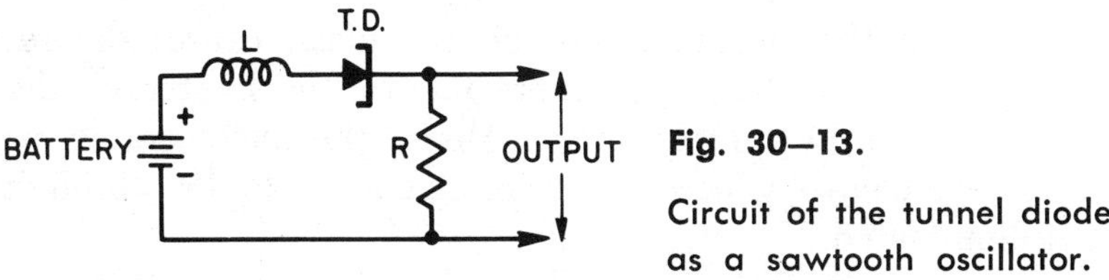

Fig. 30—13. Circuit of the tunnel diode as a sawtooth oscillator.

F. *Photosensitive devices*

1. PHOTOCONDUCTIVE DEVICES

Certain substances possess the property of exhibiting drastically reduced resistances in the presence of light. Such substances are called *photoconductive*. The symbol of a cell constructed of photoconductive material is .

The semiconductor selenium is such a substance. Normally, the electrons of its atoms are firmly held by their nuclei. Hence there are relatively few free electrons and, as a result, the resistance is very high. When light strikes the semiconductor, some of the electrons absorb sufficient energy to break away and become free electrons. If an electric field is applied, these electrons will flow to the positive terminal as new electrons to take their places enter from the negative terminal. Thus the resistance drops to a low level.

The structure of a typical selenium cell is illustrated in Figure 30-14. Two metal grids, which are to serve as the terminals, are cemented to a sheet of glass or some nonconductive transparent plastic. A thin layer of selenium is deposited between the grids. The whole then is covered by another sheet of glass or plastic.

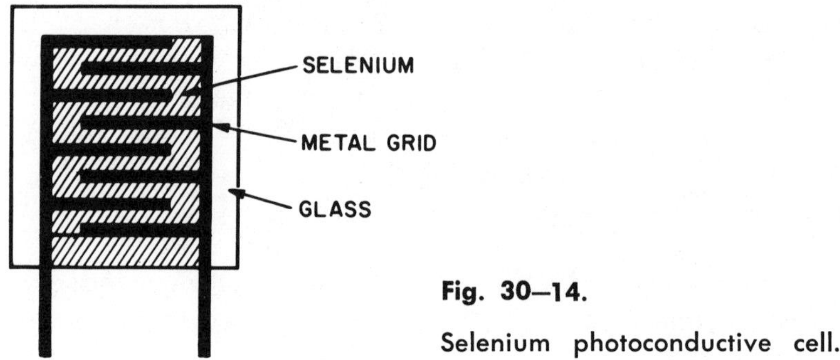

Fig. 30–14.

Selenium photoconductive cell.

Light, striking the selenium through the glass, causes the resistance of the cell to drop. The more intense the light rays, the greater will be the drop in resistance. Hence the photoconductive cell can act as a control which permits more current to flow through it as it is illuminated.

In recent years, the selenium cell has largely been replaced by the crystal of another semiconductor, cadmium sulfide (CdS). As in the case of selenium, the electrons of the crystal are firmly held by their nuclei under normal conditions. Hence the resistance is very high (hundreds of megohms). Light striking the crystal imparts sufficient energy to some of the electrons as to enable them to break their bonds and so become free electrons. As a result, the resistance drops to a few hundred ohms.

The number of freed electrons, and hence the current flow through the crystal under the influence of an electric field, depends upon the intensity of the light (among other factors). Thus, the more intense the light, the greater will be the current flow. (After the light is removed, the electrons return to their normal states, surrendering the energy they have absorbed in the form of a reddish glow.)

The cadmium-sulfide crystal is much more sensitive than selenium. Because of its high sensitivity, it can be made into extremely small cells. For example, one model consists of a crystal contained in a housing ¼ inch in diameter and ⅜ inch long. And this includes a built-in lens for focusing the light upon the crystal.

Fig. 30–15.

Various types of cadmium-sulfide cells.

Sylvania Electric Products Co.
Radio Corporation of America

The cadmium-sulfide cell thus performs somewhat as does the phototube and can be used as a light-sensing device in similar circuits (see Chapter 18). However, its action is more sluggish than that of the phototube and, accordingly, it cannot be used in fast-acting circuits.

On the other hand, it possesses a number of advantages over the phototube. It is smaller, more rugged, and cheaper. Also, it is more sensitive and in some applications may operate a relay directly without the need for amplification.

Such a circuit is shown in Figure 30-16. The rectifier converts the line current to direct current. R and C form a filter that changes the pulsating direct current coming from the rectifier to a steady direct current. The cadmium-sulfide cell is in series with the coil of the relay and, when the cell is darkened, its resistance is so high that not enough current may flow through it to activate the relay. But when the cell is illuminated, its resistance drops drastically and sufficient current flows through to operate the relay.

Where amplification is required, an a-c operated circuit such as the one illustrated in Figure 30-17 may be employed. Resistor R_1 is

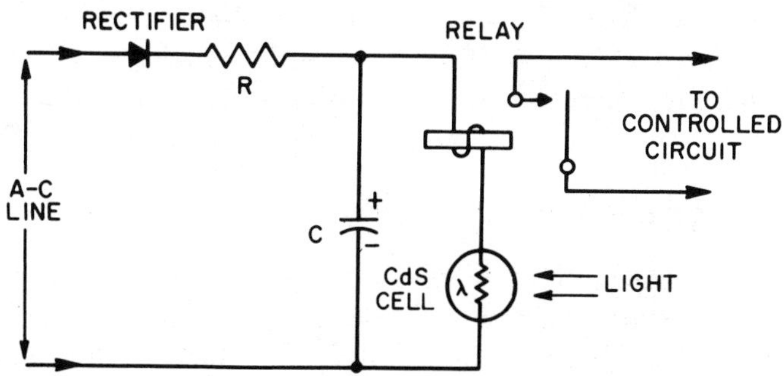

Fig. 30–16. Photosensitive relay circuit using a cadmium-sulfide cell.

a limiting resistor which prevents too great a current from flowing through the cell. The cell and R_2 form a voltage divider across the line. With no light illuminating the cell, its resistance is so great that the voltage drop across R_2 is insufficient to fire the thyratron. Hence the relay is not activated.

But upon illumination, the resistance of the cell drops, a greater current flows through it, and the voltage drop across R_2 becomes large enough to fire the thyratron, thus activating the relay. When the illumination is removed, the resistance of the cell rises again and the relay is de-activated. Capacitor C_1 is used to prevent relay chatter. So sensitive is the cadmium-sulfide cell that this circuit may be operated at ordinary room illumination, thus eliminating the need for a special light source.

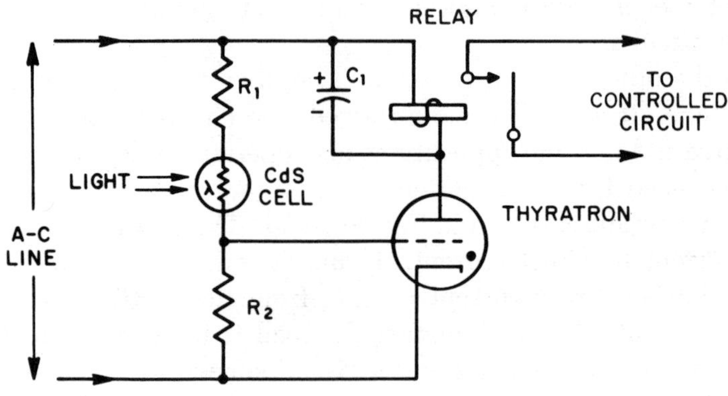

Fig. 30—17. A-c operated photosensitive relay circuit using a cadmium-sulfide cell.

The cadmium-sulfide cell has the additional advantage that it is also sensitive to X-rays, gamma rays, and other high-energy radiations such as electrons, protons, and alpha particles. Thus it can be used in applications where a light-phototube combination cannot function.

2. PHOTOVOLTAIC DEVICES

In the photoconductive cell, light energy, striking a semiconductor, reduces its resistance drastically. In the *photovoltaic,* or *photronic, cell* the light energy, striking a semiconductor, generates a voltage. Such a cell is illustrated in Figure 30-18.

It consists of an iron disk, one side of which is covered with a thin layer of selenium. This layer then is sprayed with a transparent film of a conducting metallic lacquer and contact is made to the lacquer coating by means of a metal ring. The iron disk forms one electrode of the cell and the selenium layer, through the lacquer and metal ring, the other. (Note the similarity to the iron–selenium rectifier described in Chapter 27, subdivision B, 1.) The whole is enclosed in a case of glass and plastic.

Light, passing through the transparent layers and striking the selenium, causes some of its electrons to break away. These high-energy electrons have sufficient energy to get over the potential barrier to the iron, thus giving the latter a negative charge. The selenium, having lost electrons, acquires a positive charge. If an external load is connected to the two, current will flow.

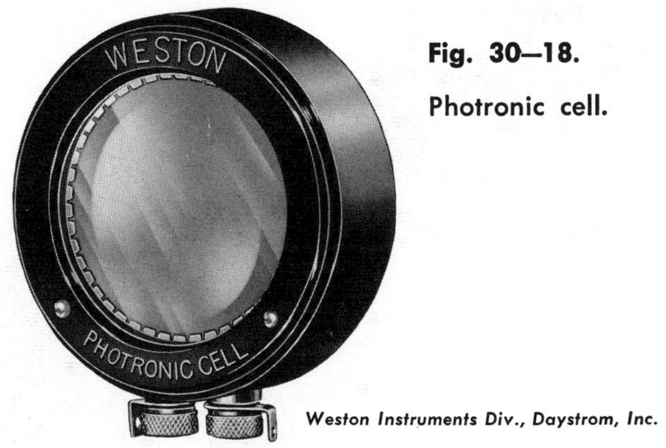

Fig. 30—18.

Photronic cell.

Weston Instruments Div., Daystrom, Inc.

This load may be a microammeter that measures the amount of current. Since the current depends upon the intensity of the light striking the cell (among other factors), we thus have a means for measuring the intensity of the light. This is the principle upon which the photographer's exposure meter (see Figure 30-19) operates.

The photronic cell can be used to operate an extremely sensitive relay (such as described in Chapter 19). Or else the output of the cell may first be fed to an amplifier before being applied to a more power-consuming device.

Another type of photovoltaic cell consists of a P-N junction diode made from a single silicon crystal. If light strikes the P-region of the crystal, sufficient energy is supplied to break a number of the covalent bonds, thus releasing free electrons and holes. Normally, the potential barrier at the junction would keep the free electrons from migrating into the N-region. However, the energy furnished by the light enables some of the electrons to surmount the barrier, thus making the N-region negative. The holes, left behind, make the P-region positive. If an external path is provided, current will flow through it from N-region to P-region.

Thus light energy is transformed directly to electrical energy. The more intense the light, the greater will be the current.

These cells may be constructed of large or small crystals. For example, a tiny crystal, a lens to focus the light upon it, and a transistor amplifier may be enclosed in a case less than an inch long. This unit may be used to operate some device, such as a relay. See Figure 30-20.

Weston Instruments Div., Daystrom, Inc.

Fig. 30–19.

Front and rear views of exposure meter.

Batteries of these photovoltaic cells employing large crystals are used to generate the electric power required to operate the various instruments contained in some of the earth-circling satellites that have been sent aloft. These cells are encased in transparent plastic for protection and receive their light directly from the sun.

Similar solar batteries have been used experimentally to supply electricity to telephone systems. In one such test, a battery of 432

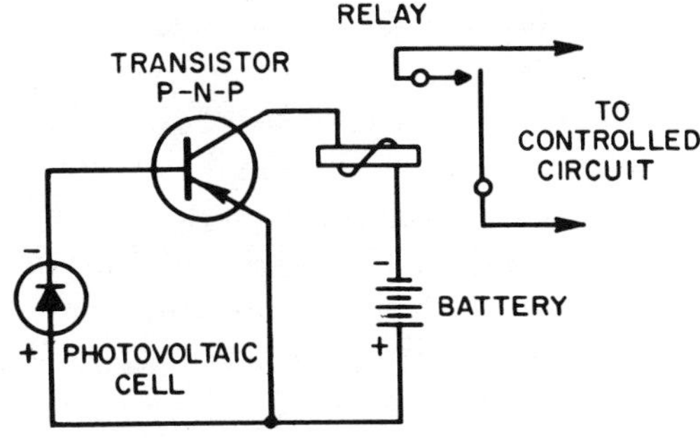

Fig. 30–20.

Relay circuit using a photovoltaic cell and a transistor amplifier.

individual cells was mounted on a pole and exposed to sunlight. On a bright day it generated 10 watts. Part of the electricity went to the telephone system and the rest was used to charge up a storage battery that was a reserve to be used at night or on a cloudy day when the sun was obscured.

Hoffman Electronics Corp.

Fig. 30–21.

Explorer VII satellite launched October 13, 1959. Power for its instruments and transmitters is furnished by the banks of solar cells mounted on its sides.

3. PHOTOTRANSISTORS

In the course of studying semiconductors and transistors, in particular, a number of photosensitive devices employing P-N junctions have been developed. Such devices are known as *phototransistors.* As you know, if a P-N junction is biased in the reverse direction (N-region positive and P-region negative), a small reverse current will flow across the junction due to the presence of minority carriers. If light rays strike the area around the junction, the light energy will cause the breaking of covalent bonds, producing a free electron and hole for each bond broken. Because of the increase in minority carriers, a greater reverse current will flow and the increase in current flow is proportional to the intensity of the light rays.

One such device is the *point-contact phototransistor* illustrated in Figure 30-22A. A small slab of an N-type semiconductor (the *base*) has a depression etched in one of its sides and is mounted in a metal case with which it makes electrical contact. A thin wire catwhisker (the *collector*) presses against the center of the depression. Connections are made to the metal case and the catwhisker.

Note the resemblance to the point-contact diode previously discussed. (See Chapter 27, subdivision A, 1.) As in that case, a sort of P-region is established where the catwhisker touches the base. Thus we have a P-N junction. The phototransistor is biased in the reverse direction and, as a result, a very small reverse current flows through the circuit.

Light enters through a small lens mounted in the front of the case and is focused at the center of the slab directly opposite the point of the catwhisker. This is important because light striking even a small distance from the junction area produces electron-hole pairs that will recombine (and so neutralize each other) before they can cross the junction and add their charges to the current.

The resulting increase in reverse current may be amplified and used to operate some device such as a relay. The entire phototransistor (case, lens, and all) may be made very small and frequently is employed with punched-card machines where the light shining through a hole in a card can actuate a counter or some other device.

Fig. 30—22. Point-contact phototransistor.

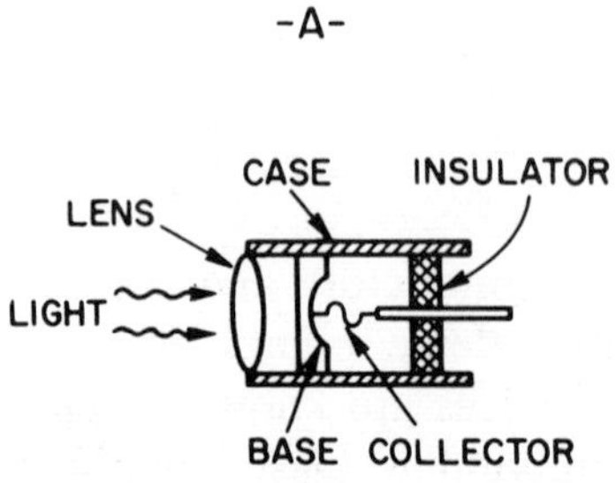

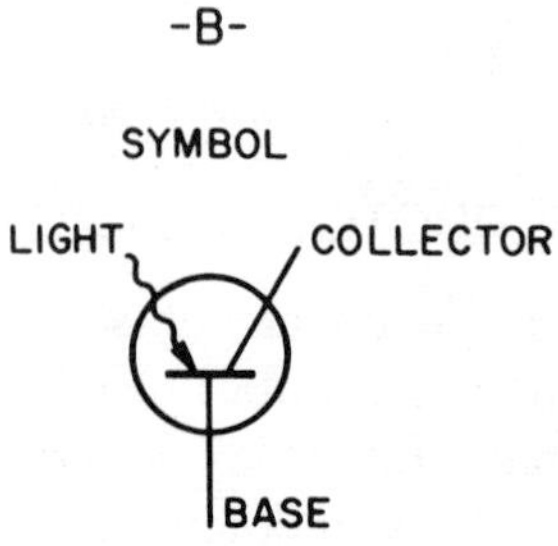

The symbol for the point-contact phototransistor appears in Figure 30-22B. The wiggly arrow indicates the light rays.

The *P-N junction phototransistor* operates as the point-contact type. It consists of an N-region and P-region, separated by a potential barrier at the junction and enclosed in an evacuated, hermatically-sealed plastic case. Biased in the reverse direction (positive to

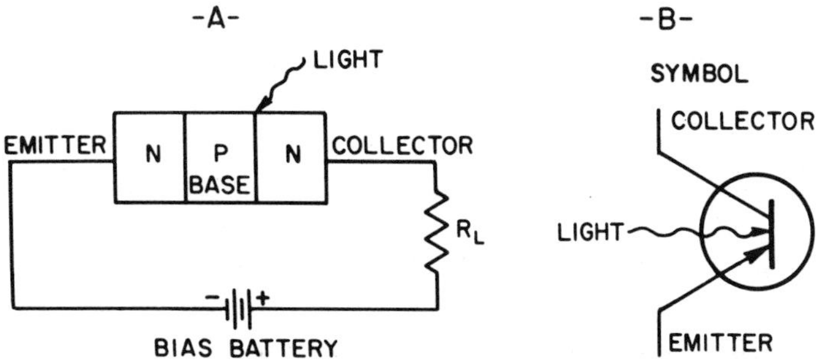

Fig. 30—23. Two-junction N-P-N phototransistor.

the N-region and negative to the P-region), a very small reverse current flows. When light rays are focused at the junction, the breaking of covalent bonds causes an increase in this reverse current.

Note that both the point-contact and P-N junction phototransistors are, in reality, diodes (base and collector). However, the light rays act somewhat as does the emitter of an ordinary transistor by increasing the number of carriers and reducing the resistance of the device, hence the name phototransistor.

The circuit of a *two-junction N-P-N phototransistor* is illustrated in Figure 30-23A. Note that the emitter-base junction is biased in the forward direction whereas the collector-base junction is biased in the reverse direction. Thus free electrons in the emitter region may pass to the base region, but electrons in the collector region cannot pass to the base region.

Even in the absence of light, however, a very small current will flow. Some of the electrons of the emitter region will diffuse through the base of the collector region. At the same time, some of the holes of the base region will move into the emitter region.

When the collector-base junction is struck by light rays, the flow of current through the phototransistor is greatly increased. The light energy ruptures some of the covalent bonds, thus producing electron-hole pairs in the base region. The holes move into the emitter region and the electrons flow into the collector region. Since the movement of electrons is between 100 and 1,000 times as rapid as the movement of the holes, the current flow is, essentially, the movement of electrons. The gain of the phototransistor is a measure of the ratio between the speed of travel of the electrons and the speed of the holes.

The phototransistor, then, is a light-operated device that will permit a greater flow of current when it is illuminated. Thus it acts somewhat as the phototube and the photoconductive cell. It has, however, a number of advantages.

Compared to the photoconductive cell, the phototransistor permits the flow of a smaller reverse current, the *dark current,* in the absence of illumination. Compared to the phototube, it is smaller in size, consumes less power, and has a greater sensitivity. Also, it is more rugged and has a longer life.

Among the phototransistors, the point-contact type is more sensitive than the P-N junction type. And the two-junction type is a great deal more sensitive than either of the others.

G. *Miscellaneous devices*

As we have learned, the junction diode consists of an N-region and P-region. The excess electrons of the N-region and the excess holes of the P-region attract each other. Thus some of the excess electrons of the N-region travel to the P-region, leaving the N-region charged positively. Similarly, some of the excess holes of the P-region travel to the N-region, leaving the P-region charged negatively. These internal charges create a potential barrier at the junction between the two regions that prevents any further intermixing of electrons and holes.

When the diode is biased in the reverse direction (positive to the N-region and negative to the P-region), the electrons and holes are attracted away from the junction, leaving a zone of depletion there. Thus we have a sort of capacitor with two charged plates (the N- and P-regions) separated by a dielectric or nonconductor (the zone of depletion).

The capacitance of a capacitor depends (in part) upon the thickness of the dielectric. The thicker the dielectric, the smaller will be the capacitance; the thinner the dielectric, the larger will be the capacitance. In the diode, the width of the depletion zone depends upon the reverse-bias voltage. The larger this voltage, the wider will be the depletion zone, and the smaller will be the capacitance of the diode. Thus, in effect, we have a variable capacitor whose capacitance depends upon the magnitude of the reverse-bias voltage applied to it.

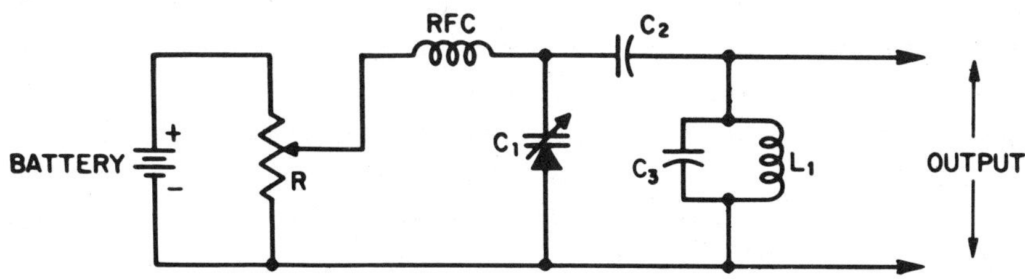

Fig. 30–24. Circuit showing how a silicon capacitor may be used for resistance tuning.

Such diode capacitors have been made of silicon. In size and shape they resemble ordinary junction diodes. The capacitance is low; in the order of micromicrofarads. However, capacitance variations of over 10 to 1 have been obtained. Although there does not appear at this time to be full agreement, the symbol for the silicon capacitor usually is shown as or

There are a number of applications to which the silicon capacitor may be put, but one of the most obvious is resistance tuning. How this may be accomplished is illustrated in Figure 30-24. The tuned circuit consists of L_1 and C_3, which is shunted by the silicon capacitor C_1. C_2 is a blocking capacitor employed to keep the direct current of the battery out of the tuned circuit. The r-f signal of the tank circuit is kept from the battery by the r-f choke coil RFC.

Potentiometer R varies the reverse-bias voltage placed upon C_1, thus varying its capacitance. This, in turn, varies the resonant frequency of the tank circuit. One advantage of resistance tuning is the possibility of remote control since the voltage may be applied over appreciable distances.

It was found that a crystal of a semiconductor, such as silicon, exhibits a property called *piezoresistance* wherein the electrical resistance of the crystal varies with changes in pressure, strain, or stress upon it. This suggests its use as a strain gage in applications similar to those discussed in Chapter 19.

The semiconductor strain gage was found to be from about 10 to 100 times as sensitive as the wire strain gage. Also, it is smaller, more rugged, and is not affected as much by changes in temperature.

In our discussion of the motor (see Chapter 21) we learned how a conductor carrying a current (and, hence, surrounded by a magnetic field) is made to move when placed within a second magnetic field. Actually, it is not the conductor that is pushed. The force acts upon the moving electrons within the conductor. However, since these electrons are confined within the conductor, it is the conductor that is made to move.

Within the conductor, the external magnetic field pushes the electrons to one side, producing an excess of electrons (negative charge) at that side and a deficiency (positive charge) at the other side. Thus a potential difference exists between the sides of the conductor. This phenomenon is called the *Hall effect.* In good conductors the voltage is small and exceedingly difficult to detect.

In semiconductors, on the other hand, the Hall effect is appreciable and the voltage can be readily measured. The relationships between the applied magnetic field, the current flowing through the semiconductor, and the voltage produced by the Hall effect is illustrated in Figure 30-25. The magnitude of the voltage is proportional to the product of the magnetic field and the current. If

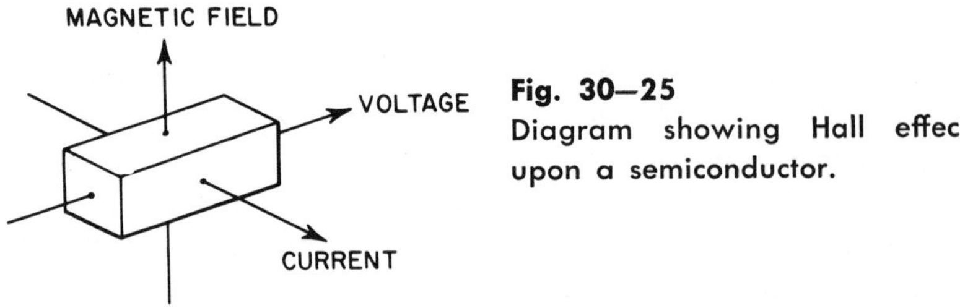

Fig. 30–25
Diagram showing Hall effect upon a semiconductor.

the current is kept constant, the voltage will vary directly with the strength of the magnetic field that is perpendicular to the semiconductor.

As the charge carriers are pushed aside by the magnetic field, the resistance of the semiconductor is increased. If the current is kept constant, the resistance will vary directly with the strength of the magnetic field that is perpendicular to the semiconductor. This phenomenon is called *magnetoresistance.*

It becomes obvious that the Hall effect and magnetoresistance of semiconductors may be used in applications where the strength of a magnetic field is to be measured.

Among the fission products produced by a nuclear reactor are neutrons, particles bearing no electrical charge. Because of their uncharged condition, they are hard to detect. However, neutron detectors are made by coating the top surface of a P-N junction diode with a very thin layer of uranium. The whole may be no larger than the head of a pin.

In the main, the uranium atoms are of the nonfissionable U_{238} type. A small percentage of the atoms, however, are of the fissionable U_{235} type. As slow-speed neutrons strike the U_{235} atoms of the uranium layer, fission occurs, releasing a number of high-energy particles. These particles cause some of the covalent bonds of the semiconductor to be broken, releasing a number of electrons and holes, and producing pulses of electricity which can be detected and counted. Where high-speed neutrons are to be detected, they may be converted to the low-speed type by making them first pass through a thin layer of paraffin over the uranium layer.

As indicated, these are but a few of the many applications to which semiconductor devices may be put. We may be sure we shall learn more about these and many others in the near future.

QUESTIONS

1. a) Describe and explain the theory of the *double-base junction diode.*
b) Explain its action as a high-frequency switch.

2. a) Describe and explain the theory of the *hook transistor.*
b) What is its advantage over an ordinary transistor?

3. Describe and explain the theory of a P-N-P-N device operating as a *controlled rectifier.*

4. Describe and explain the theory of the *thermistor.*

5. Explain how the thermistor may be employed as
a) a thermometer;
b) a time-delay device;
c) a temperature-compensating device.

6. a) By means of a graph explain the action of the *tunnel diode.*
   b) Explain its theory of operation.
7. Draw and explain the circuit of the tunnel diode used as
   a) a sine-wave oscillator;
   b) a sawtooth oscillator.
8. Describe and explain the theory of the *selenium photoconductive cell.*
9. a) Describe and explain the theory of the *cadmium-sulfide cell.*
   b) Explain the advantages and disadvantages of the cadmium-sulfide cell compared to the phototube.
10. Draw and explain the circuit of a light-sensitive relay using a cadmium-sulfide cell.
11. Describe and explain the theory of the *selenium photronic cell.*
12. Describe and explain the theory of the *solar cell* used in satellites for converting sunlight to electricity.
13. Describe and explain the theory of the *point-contact phototransistor.*
14. Describe and explain the theory of the *P-N junction phototransistor.*
15. Describe and explain the theory of the *two-junction N-P-N phototransistor.*
16. What are the advantages of the phototransistor over the phototube?
17. a) Describe and explain the theory of the *semiconductor variable capacitor.*
    b) What are its advantages over the ordinary variable capacitor?
18. a) Describe and explain the theory of the *semiconductor strain gage.*
    b) What are its advantages over the wire strain gage?
19. a) Explain the *Hall effect.*
    b) Explain how it may be employed by a semiconductor device to measure the strength of a magnetic field?
20. Explain how a P-N junction diode may be used as a neutron detector.

Appendix A

EIA color code

To identify the various values of standard radio components, the Electronics Industries Association (EIA) has adopted a color code. Numbers are represented by the following colors:

0	black	5	green
1	brown	6	blue
2	red	7	violet
3	orange	8	gray
4	yellow	9	white

EIA COLOR CODE FOR RESISTORS

The value in ohms of a resistor may be learned from three bands of color. The first color band (the one nearest to one end) represents the first figure of the resistance value. The next color represents the second figure. The third color represents the decimal multiplier, or number of zeroes, following the first two figures. Sometimes a fourth band of color is shown to indicate the degree of accuracy, or tolerance, of the resistor and occasionally a band at the other end stands for the reliability.

The complete color code for resistors is given in the following table.

Color	*1st Figure*	*2nd Figure*	*Multiplier*	*Tolerance*
Black	—	0	1	—
Brown	1	1	10	1%
Red	2	2	100	2%
Orange	3	3	1,000	3%
Yellow	4	4	10,000	4%
Green	5	5	100,000	—
Blue	6	6	1,000,000	—
Purple	7	7	10,000,000	—
Gray	8	8	100,000,000	—
White	9	9	1,000,000,000	—
Silver	—	—	———	10%
Gold	—	—	———	5%
No color	—	—	———	20%

Let us consider an example. Assume that a resistor shows four bands of color which, starting at one end, appear as yellow, purple, orange, and gold.

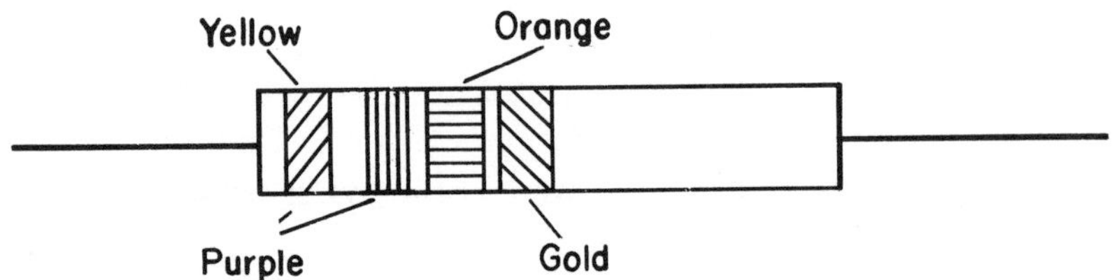

These colors, when translated into the code, stand for

4 - 7 - 000 or 47,000 ohms.

The tolerance (gold) is 5 per cent.

EIA COLOR CODE FOR MICA CAPACITORS

The EIA has introduced two systems of color-coding fixed mica capacitors. One is a three-dot code to be applied to mica capacitors rated at 500 volts, d-c working voltage, and ±20% tolerance only. The first color dot indicates the first figure, the second dot the second figure, and the third dot indicates the multiplier. The value is read in micromicrofarads ($\mu\mu f$), and the sequence of the dots is indicated by an arrow.

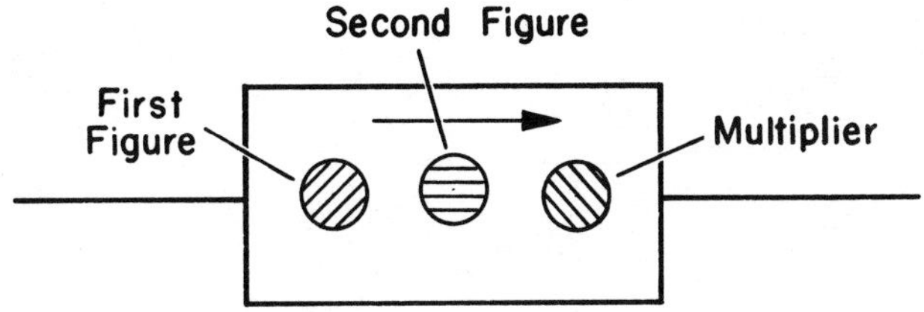

The other system is a six-dot color code with three dots along the top and three dots at the bottom. Again, the value is read in micromicrofarads, and the sequence is shown by means of an arrow. As you can see, we may determine from this code the first three figures, the multiplier, the tolerance, and the direct-current working voltage.

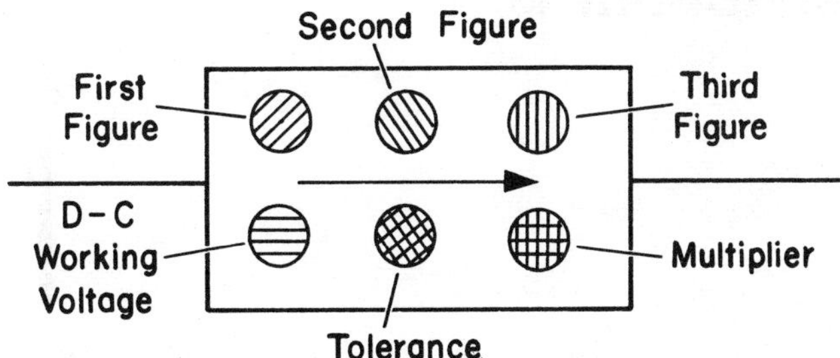

The complete color code for mica capacitors is as follows:

Color	*Figure*	*Multiplier*	*Voltage D-C Working*	*Tolerance*
Black	0	1	—	—
Brown	1	10	100	1%
Red	2	100	200	2%
Orange	3	1,000	300	3%
Yellow	4	10,000	400	4%
Green	5	100,000	500	—
Blue	6	1,000,000	600	6%
Violet	7	10,000,000	700	7%
Gray	8	100,000,000	800	8%
White	9	1,000,000,000	900	9%
Gold	—	0.1	1,000	5%
Silver	—	0.01	2,000	10%
No color	—	———	500	20%

Appendix B

How to identify the base pins of electron tubes

With the pins held facing you, the pin numbers appear as follows:

Octal Base

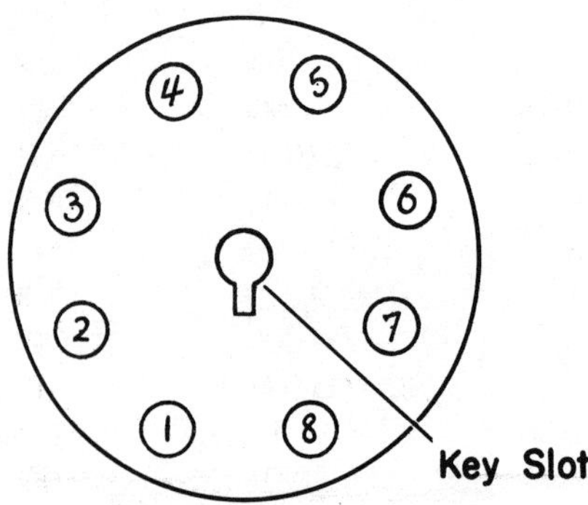

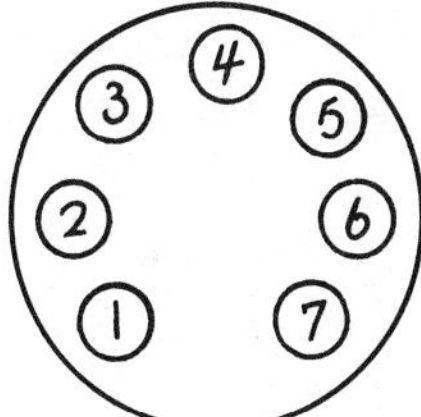

7-pin Miniature Base

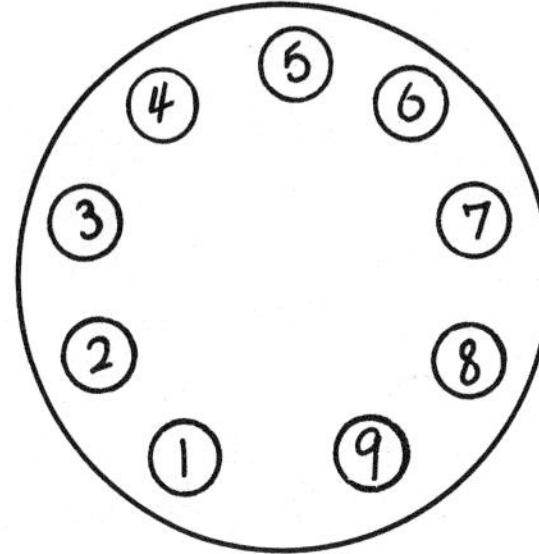

9-pin Miniature Base

Once the pin numbers are known, the electrodes corresponding to these numbers may be found by consulting the tube manuals of the various manufacturers. Listed below are the pin connections for several typical types of tubes. (But note that not all tubes have the same pin connections.)

OCTAL BASE

Type 6J5 (Triode)	Type 6K6 (Pentode)
Pin 1- Shell	Pin 1- No connection
2- Heater	2- Heater
3- Plate	3- Plate
4- (No pin)	4- Grid No. 2
5- Grid	5- Grid No. 1
6- (No pin)	6- (No pin)
7- Heater	7- Heater
8- Cathode	8- Cathode, Grid No. 3

7-PIN MINIATURE BASE

Type 6C4 (Triode)
Pin 1- Plate
2- Do not use
3- Heater
4- Heater
5- Plate
6- Grid
7- Cathode

Type 6AU6 (Pentode)
Pin 1- Grid No. 1
2- Grid No. 3
3- Heater
4- Heater
5- Plate
6- Grid No. 2
7- Cathode

9-PIN MINIATURE BASE

Type 6S4A (Triode)
Pin 1- Do not use
2- Cathode
3- Grid
4- Heater
5- Heater
6- Grid
7- Do not use
8- Do not use
9- Plate

Type 6BK5 (Beam Power)
Pin 1- Plate
2- No connection
3- Grid No. 1
4- Heater
5- Heater
6- Cathode, Grid No. 3
7- Grid No. 1
8- Grid No. 2
9- No connection

Appendix C

Transistor shapes and lead identification

Because of the relative newness of the transistor, it has appeared in a large variety of shapes and sizes (though all are small). Nor has there been any uniformity of lead identification. Although efforts are being made to establish standards for the industry, the large number of different shapes and lead connections among present-day transistors makes for considerable confusion. The best source of information is the manufacturer's notes.

To aid the operators of their transistor testers, the Precision Apparatus Company, Inc., has drawn up a list of some of the most-common types of transistors and pictured their shapes and lead connections. The following is an extract from this compilation.

Lead Identification for Various Types of Small-signal Transistors

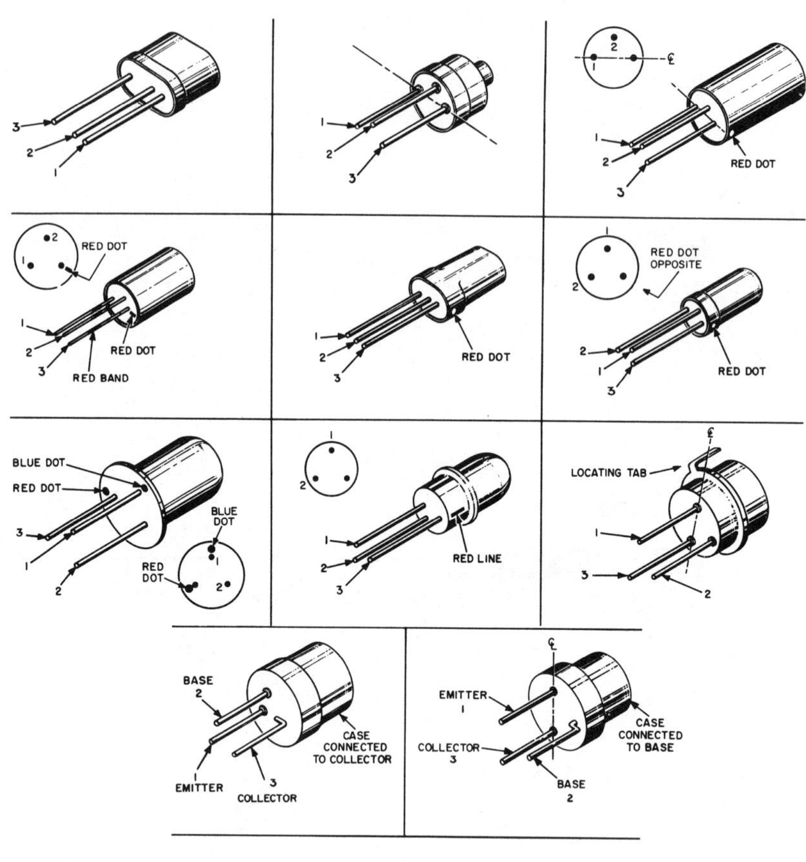

1 = EMITTER

2 = BASE

3 = COLLECTOR

Lead Identification for Various Types of Power Transistors

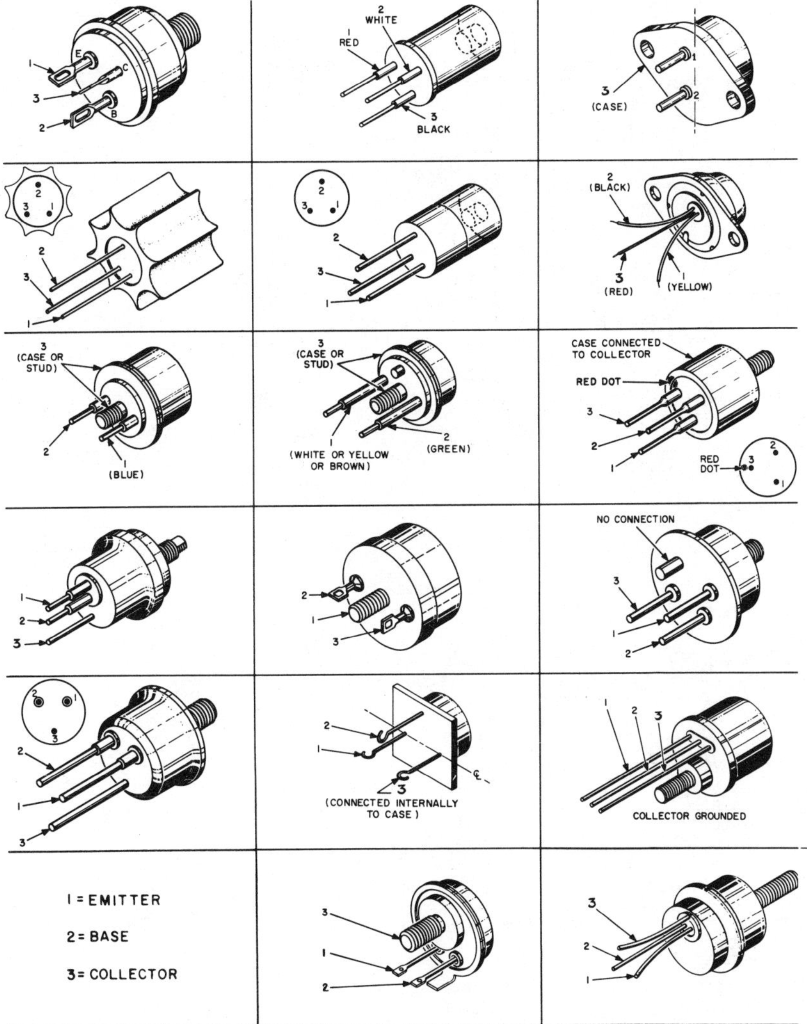

As you can see, there are many different shapes. The electronic industry has compiled a list of standard transistor outlines, a number of which are shown below.

Several Standard Transistor Outlines

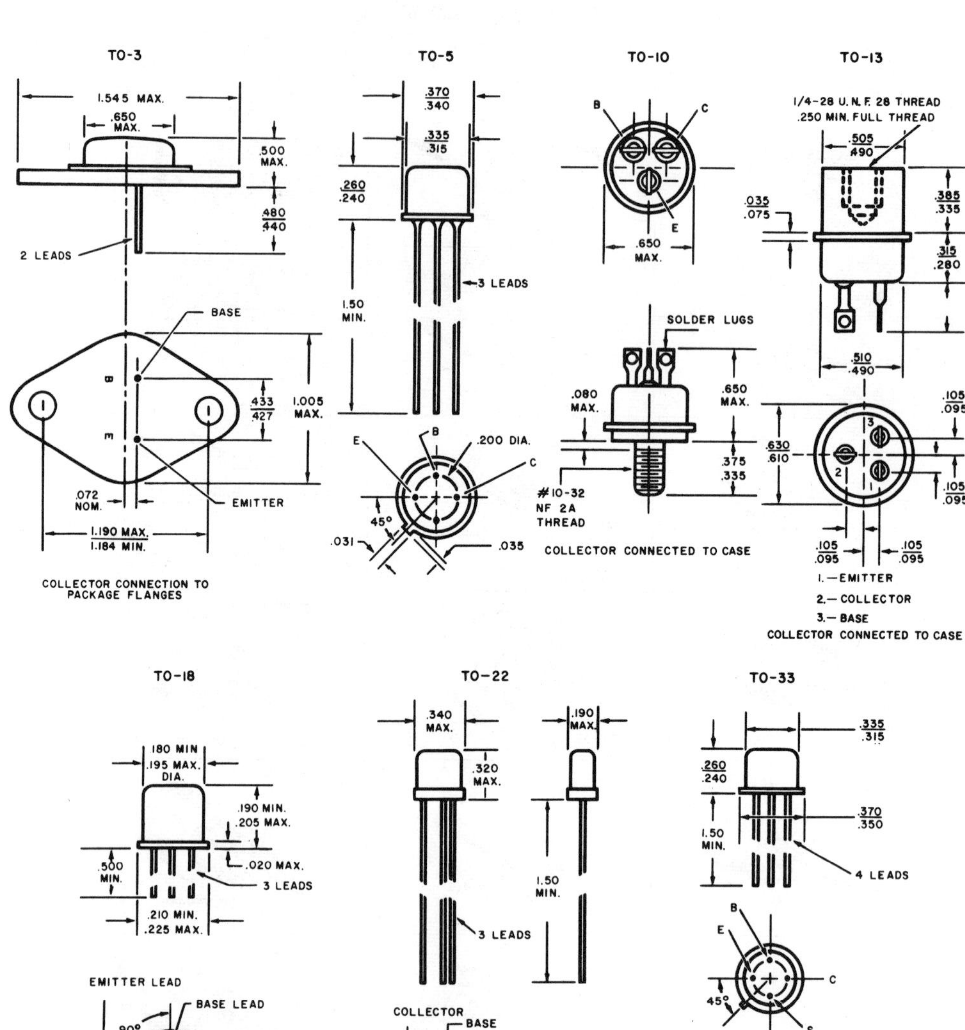

Appendix D

Graphic symbols for electronic diagrams

Antenna

Ground

Capacitor

Capacitor, fixed

Capacitor, variable 

Capacitor, trimmer 

Capacitor, split-stator 

Capacitors, ganged variable 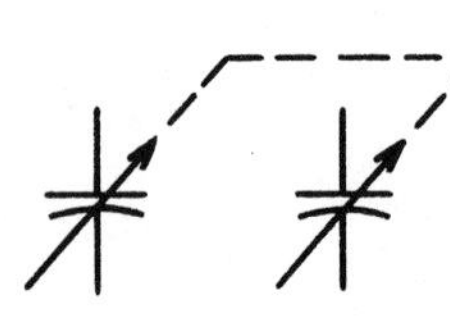

Cell 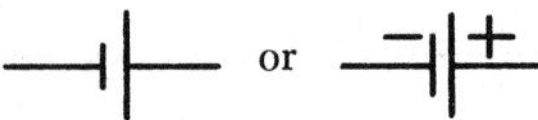

Battery 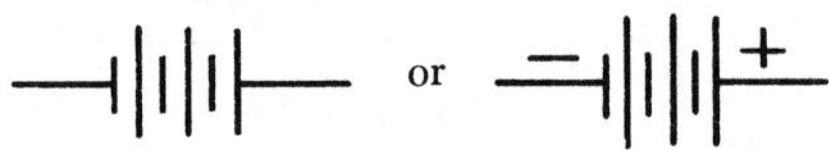

Crystal, piezoelectric 

Fuse 

Generator, a-c 

Inductors, general or

Inductor, air-core 

Inductor, iron-core 

Inductor, powdered iron-core

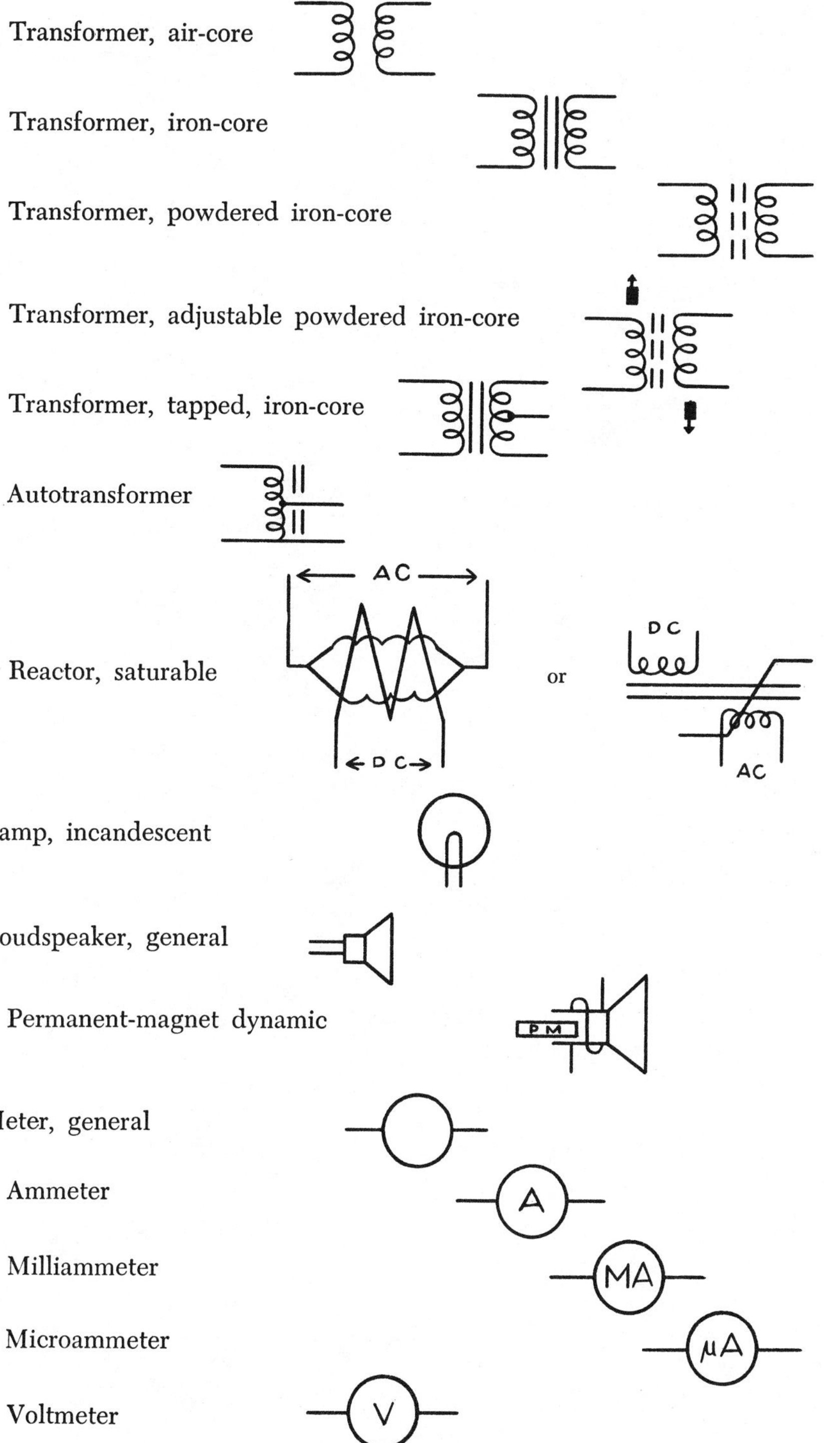
Transformer, air-core
Transformer, iron-core
Transformer, powdered iron-core
Transformer, adjustable powdered iron-core
Transformer, tapped, iron-core
Autotransformer
AC
Reactor, saturable
DC
or
DC
AC
Lamp, incandescent
Loudspeaker, general
Permanent-magnet dynamic
PM
Meter, general
Ammeter
A
Milliammeter
MA
Microammeter
μA
Voltmeter
V

Microphone

Motor

Field winding

D-C armature, commutator, and brushes

Relay

Normally open (NO)

Normally closed (NC)

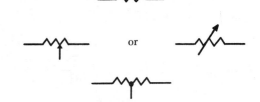

Resistors

Resistor, fixed

Resistor, variable or

Resistor, tapped

Semiconductor Devices

Semiconductor diode Anode Cathode

Tunnel diode

or

Zener diode or

B

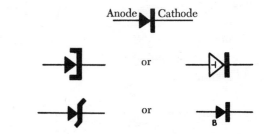

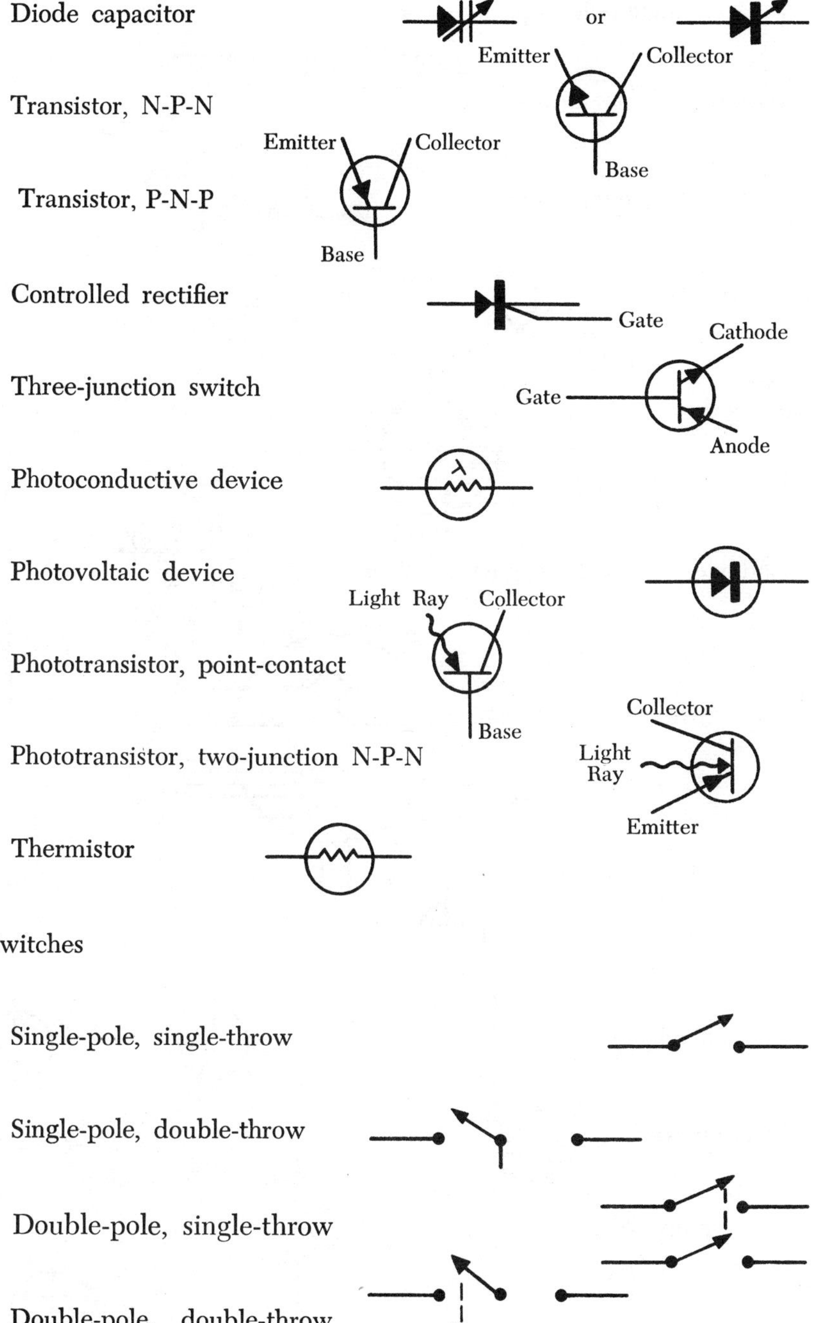
Diode capacitor
or
Transistor, N-P-N
Emitter
Collector
Base
Transistor, P-N-P
Emitter
Collector
Base
Controlled rectifier
Gate
Three-junction switch
Gate
Cathode
Anode
Photoconductive device
λ
Photovoltaic device
Phototransistor, point-contact
Light Ray
Collector
Base
Phototransistor, two-junction N-P-N
Collector
Light
Ray
Emitter
Thermistor
Switches
Single-pole, single-throw
Single-pole, double-throw
Double-pole, single-throw
Double-pole, double-throw

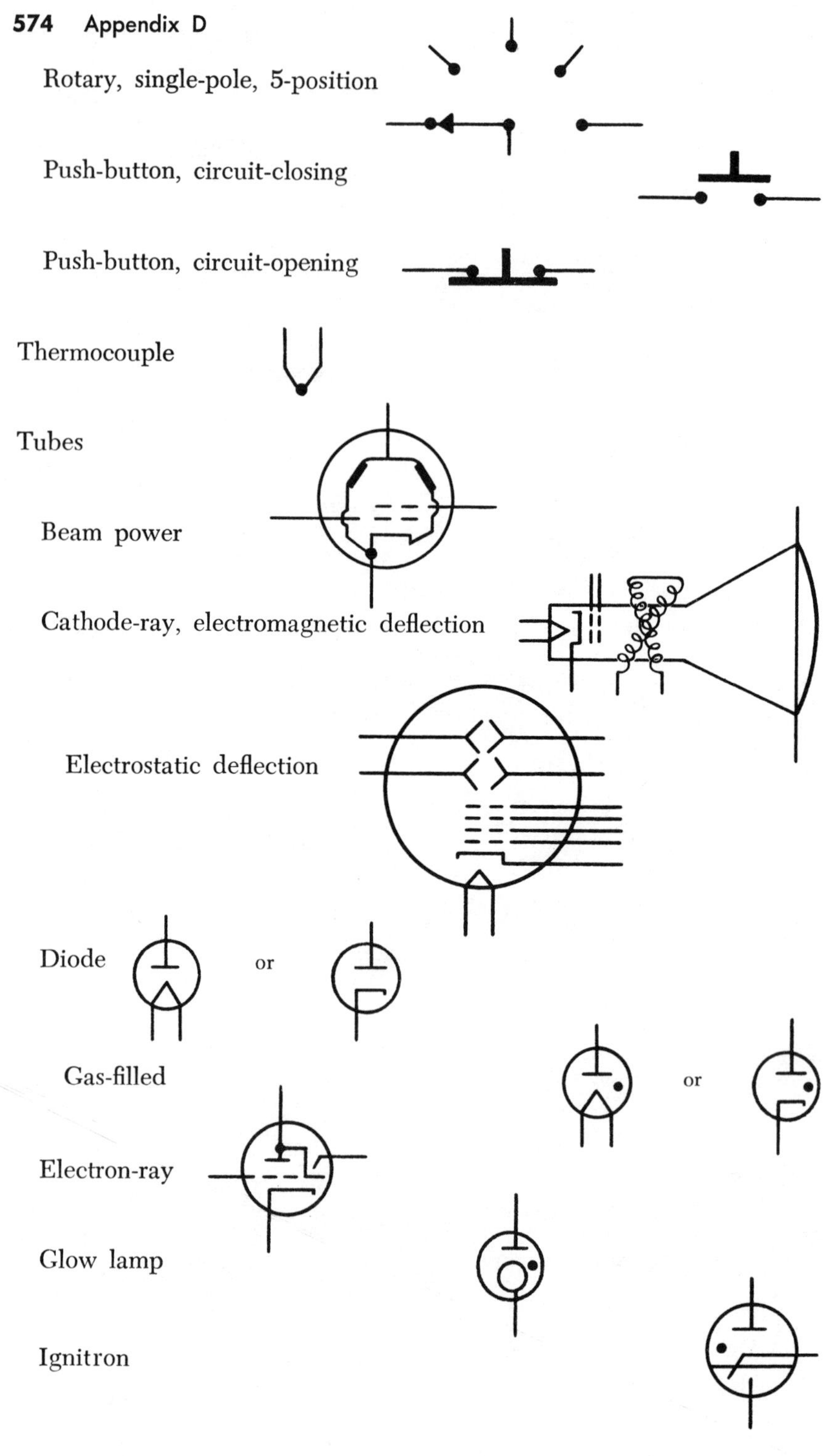
Rotary, single-pole, 5-position
Push-button, circuit-closing
Push-button, circuit-opening
Thermocouple
Tubes
Beam power
Cathode-ray, electromagnetic deflection
Electrostatic deflection
Diode
or
Gas-filled
or
Electron-ray
Glow lamp
Ignitron

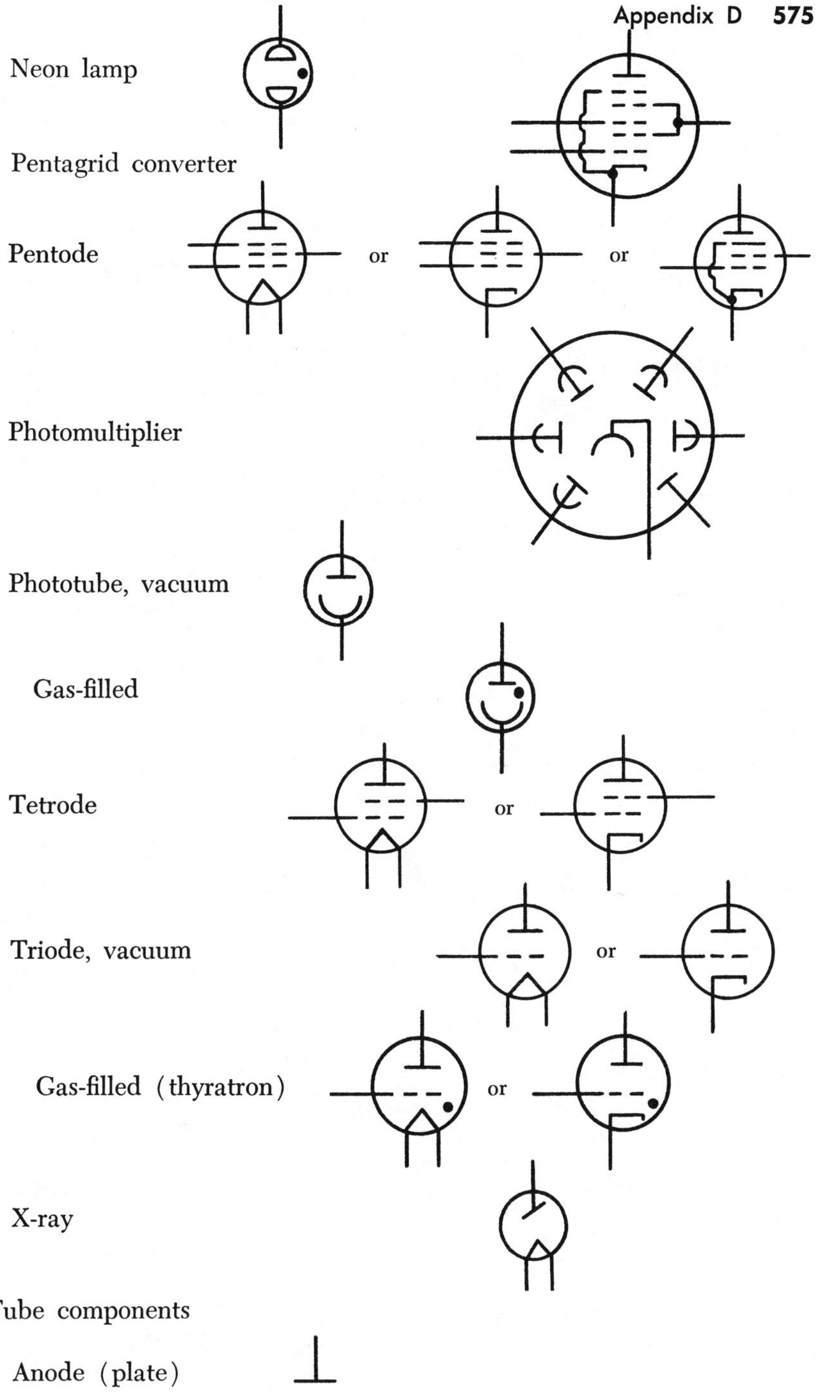
Neon lamp
Pentagrid converter
Pentode
or
or
Photomultiplier
Phototube, vacuum
Gas-filled
Tetrode
or
Triode, vacuum
or
Gas-filled (thyratron)
or
X-ray
Tube components
Anode (plate)

X-ray target

Cathode, cold

Filament

Indirectly-heated

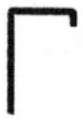

Mercury-pool

Photocathode

Dynode

Envelope, vacuum

Gas-filled

Grid

Ignitor

Wiring

Wires connected

Wires not connected

Index

F

S

T

U

V

W

X

Y

Z